Pulsed Laser Ablation

Pulsed Laser Ablation

Advances and Applications in Nanoparticles
and Nanostructuring Thin Films

edited by
Ion N. Mihailescu
Anna Paola Caricato

Published by

Pan Stanford Publishing Pte. Ltd.
Penthouse Level, Suntec Tower 3
8 Temasek Boulevard
Singapore 038988

Email: editorial@panstanford.com
Web: www.panstanford.com

British Library Cataloguing-in-Publication Data
A catalogue record for this book is available from the British Library.

Pulsed Laser Ablation: Advances and Applications in Nanoparticles and Nanostructuring Thin Films

Copyright © 2018 by Pan Stanford Publishing Pte. Ltd.

All rights reserved. This book, or parts thereof, may not be reproduced in any form or by any means, electronic or mechanical, including photocopying, recording or any information storage and retrieval system now known or to be invented, without written permission from the publisher.

For photocopying of material in this volume, please pay a copying fee through the Copyright Clearance Center, Inc., 222 Rosewood Drive, Danvers, MA 01923, USA. In this case permission to photocopy is not required from the publisher.

ISBN 978-981-4774-23-9 (Hardcover)
ISBN 978-1-315-18523-1 (eBook)

Contents

Preface   xiii

1. Surface Energy and Nucleation Modes   1
Maura Cesaria

    1.1   Introduction   2
    1.2   Thermodynamic Background Concepts   5
          1.2.1   Thermodynamic Potentials and Surface Free Energy   5
          1.2.2   Phase Transformations of a Thermodynamic System and Supersaturation   8
          1.2.3   Strain and Epitaxial Growth   9
    1.3   Thermodynamic Nucleation Theory and Growth Modes   13
          1.3.1   Principles of Nucleation Theory   14
          1.3.2   Growth Modes at Thermodynamic Equilibrium   20
    1.4   Elementary Kinetic Processes on Surfaces and the Energy Landscape   25
          1.4.1   Adsorption, Real Substrates, and Surface Elementary Processes   26
          1.4.2   Characteristic Kinetic Coefficients, Energy Barriers, and Timescales   30
    1.5   Condensation Processes and Kinetic-Driven Growth Modes   34
          1.5.1   Kinetic Control of Nucleation   35
          1.5.2   Intralayer and Interlayer Diffusion, Island Coalescence, and Growth Modes   37
          1.5.3   Microstructure Evolution   42
    1.6   Deposition Techniques in Nanoscience and Advantages of the Pulsed Laser Approaches   46
    1.7   Growth Opportunities by the PLD Technique   50
          1.7.1   Distinctive Characteristics of the PLD Approach   51

	1.7.2	Growth Manipulations by PLD	59
1.8		Conclusions	68

2. Nanosecond Laser Ablation and Processing of Solid Targets in Vacuum or in a Low-Gas Atmosphere 85
Vincenzo Resta, Ramón J. Peláez, and Anna Paola Caricato

2.1	Introduction		85
2.2	Plasma Dynamics and Expansion		86
	2.2.1	Plasma Parameters: Temperature and Density	91
	2.2.2	Plasma Composition: Atom and Ion Distribution/Yields	93
2.3	Production of Metal Nanoparticles by Pulsed Laser Deposition		100
	2.3.1	Dependence with the Number of Laser Pulses in the Metal Target	101
	2.3.2	Dependence with the Laser Fluence	106
	2.3.3	Peculiarities of Pulsed Laser Deposition in Nanoparticle Formation and Dependence with the Substrate	110
2.4	Thermal Process		113
	2.4.1	Substrate Temperatures	113
	2.4.2	Postheating by Laser Irradiation	115

3. Nanosecond Laser Ablation of Solid Targets in a High-Pressure Atmosphere 131
Sebastiano Trusso, Fortunato Neri, and Paolo Maria Ossi

3.1	Introduction	132
3.2	Comparison between Some Basics of Laser Ablation in Vacuum and in a Gas at High Pressure	133
3.3	Nanoparticle Synthesis and Assembling upon Ablation in a High-Pressure Gas: Selected Examples	136
3.4	Deposition of Noble Metal Nanoparticle Arrays for Application in Biomedical Sensing	142
3.5	Conclusions	150

4. Femtosecond Laser Ablation of Solid Targets in Vacuum and Low-Pressure Gas Atmosphere   155

Salvatore Amoruso

4.1	Introduction	156
4.2	Experimental and Theoretical Analyses of the Early Stage of Femtosecond Laser Ablation	161
4.3	Experimental Analysis of Late Stages of Femtosecond Laser Ablation and Plume Propagation	167
	4.3.1 High-Vacuum Expansion	168
	4.3.2 Propagation in a Low-Pressure Background Gas	174
4.4	Femtosecond Laser Ablation of Thin Films	179
4.5	Nanoparticles and Nanoparticle-Assembled Films	180
4.6	Conclusions	182

5. Short-Pulse Laser Near-Field Ablation of Solid Targets under Liquids   193

M. Ulmeanu, P. Petkov, F. Jipa, E. Brousseau, and M. N. R. Ashfold

5.1	Introduction	193
5.2	Working Principle of the LILAC Lithography Technique	195
	5.2.1 Preparing the Si Substrates	196
	5.2.2 Preparing the Colloidal Mask	197
	5.2.3 Laser Processing Parameters	198
	5.2.4 Focusing the Laser Beam through the Liquids	198
	5.2.5 Finite-Difference Time Domain Simulations	200
5.3	Experimental Demonstrations	202
5.4	Conclusions	204

6. MAPLE Deposition of Nanomaterials   207

Enikö György and Anna Paola Caricato

6.1	Introduction	208
6.2	Ultraviolet Matrix-Assisted Pulsed Laser Evaporation	215

	6.3	Infrared Matrix-Assisted Pulsed Laser Evaporation		230
	6.4	Inverse Matrix-Assisted Pulsed Laser Evaporation		234
	6.5	Conclusions		236

7. Thin Films and Nanoparticles by Pulsed Laser Deposition: Wetting, Adherence, and Nanostructuring — **245**

Carmen Ristoscu and Ion N. Mihailescu

	7.1	Introduction		246
	7.2	Wetting		248
		7.2.1	Definitions	248
		7.2.2	Case Examples	251
	7.3	Adherence		254
		7.3.1	Basic Mechanisms	254
		7.3.2	Investigations and Examples	255
	7.4	Nanostructuring		258
		7.4.1	Definitions	258
		7.4.2	Imaging of Nanostructures	259
			7.4.2.1 Conventional imaging	259
			7.4.2.2 Differential evanescent light intensity imaging	260
		7.4.3	Nanostructuring with Advanced PLD Techniques	262
		7.4.4	Applications	265
			7.4.4.1 Metal oxides for gas sensing	265
			7.4.4.2 Fuel cell elements	266
			7.4.4.3 Nanoparticles for SERS	267
	7.5	Conclusions		268

8. Core-Shell Nanoparticles for Energy Storage Applications — **277**

Manish Kothakonda, Briley Bourgeois, Brian C. Riggs, Venkata Sreenivas Puli, Ravinder Elupula, Muhammad Ejaz, Shiva Adireddy, Scott M. Grayson, and Douglas B. Chrisey

	8.1	Introduction		278
		8.1.1	Nanoparticle Property Selection	280
		8.1.2	Nanoparticle Synthesis	281
			8.1.2.1 Core-shell nanoparticles prepared by the grafting-from route	281

		8.1.2.2	Core-shell nanoparticles prepared by the grafting-to route	283
8.2	Experimental Section			286
	8.2.1	Nanoparticle Synthesis		286
	8.2.2	Nanoparticles Synthesis by Pulsed Laser Ablation		288
	8.2.3	Synthesis of $BaTiO_3$ Nanoparticles by the Solvothermal Method		289
	8.2.4	Polymerization of Nanoparticles		292
		8.2.4.1	Synthesis of PGMA-$BaTiO_3$ core-shell nanostructures by grafting-from	292
		8.2.4.2	Synthesis of PVDF-HFP-GMA-$BaTiO_3$ core-shell nanostructures by grafting-to	294
8.3	Materials Characterization			295
8.4	Experimental Observation			296
	8.4.1	Dielectric Properties and Leakage Current Behavior		299
8.5	Conclusions			303

9. Nanoparticle Generation by Double-Pulse Laser Ablation — 317

Emanuel Axente, Tatiana E. Itina, and Jörg Hermann

9.1	Introduction		318
9.2	Typical Experimental Design for Laser–Matter Interactions with Double Pulses		321
	9.2.1	Collinear Double-Pulse Interaction Geometry	322
	9.2.2	Orthogonal Double-Pulse Interaction Geometry	322
	9.2.3	Experiment for NP Generation with Delayed Short Laser Pulses	323
9.3	Investigation of Nanoparticles Produced by Short Double-Pulse Laser Ablation of Metals		325
	9.3.1	Correlation between Ablation Efficiency and Nanoparticle Generation in the Single-Pulse Regime	326
	9.3.2	Influence of Interpulse Delay on Plume Composition	327

		9.3.3	Influence of Interpulse Delay on Ablation Depth and Crater Morphology	331
		9.3.4	Overview of Other Investigations in the Field of Double-Pulse Laser–Matter Interactions	334
	9.4		Modeling of Double-Pulse Laser Ablation	338
		9.4.1	Fundamentals of Laser–Matter Interactions	338
		9.4.2	Numerical Simulations of Short Double-Pulse Interaction with Materials	341
	9.5		Conclusions and Perspectives	343

10. Ultrafast Laser-Induced Phenomena inside Transparent Materials — 357

Felix Sima, Jian Xu, and Koji Sugioka

	10.1	Introduction		358
	10.2	Characteristics of Glass Material Processing by Ultrafast Laser Pulses		359
		10.2.1	Interaction Mechanism of Ultrafast Laser Pulses with Glasses	359
			10.2.1.1 Nonlinear multiphoton absorption	359
			10.2.1.2 Heat accumulation effects	362
		10.2.2	Spatial Resolution in Ultrafast Laser Processing of Glass	363
	10.3	Undeformative Processing: ULP-Induced Internal Modifications		366
	10.4	Subtractive Processing: Formation of 3D Micro- and Nanofluidic Structure		370
		10.4.1	Ultrafast Laser-Induced Modification Followed by Selective Wet Etching	370
		10.4.2	Liquid-Assisted ULP Processing	374
		10.4.3	Pros and Cons of the ULP 3D Subtractive Process	375
	10.5	Additive Processing: ULP-Induced Photopolymerization of Photoresists		376
		10.5.1	Mechanisms and Limitations	376
		10.5.2	Applications of Two-Photon Polymerization	379
	10.6	Hybrid ULP 3D Processing		381

		10.6.1	Combination of Subtractive and Undeformative Processing	381
		10.6.2	Combination of Subtractive and Additive Processing	383
	10.7	Challenges and Perspectives		385

11. Ultrafast Processes on Semiconductor Surfaces Initiated by Temporally Shaped Femtosecond Laser Pulses — 399

P. A. Loukakos, G. D. Tsibidis, and E. Stratakis

11.1	Introduction	400
11.2	Experimental Details	402
11.3	Theoretical Details	403
11.4	Results and Discussion I: Si	403
11.5	Results and Discussion II: ZnO	416
11.6	Conclusions	420

12. Atomistic Simulations of the Generation of Nanoparticles in Short-Pulse Laser Ablation of Metals: Effect of Background Gas and Liquid Environments — 425

Cheng-Yu Shih, Chengping Wu, Han Wu, Maxim V. Shugaev, and Leonid V. Zhigilei

12.1	Introduction	426
12.2	Computational Setup for the Simulation of Laser Interactions with Metals in a Background Gas or Liquid Environment	429
	12.2.1 Representation of Laser Interaction with Metals	430
	12.2.2 Representation of Background Gas and Liquid Environments	433
12.3	Large-Scale MD Simulations of Laser Ablation in Vacuum	435
12.4	Ablation in a Background Gas	442
12.5	Ablation in Liquids	445
12.6	Concluding Remarks	454

13. Laser Nanostructuring of Polymers — 471

Esther Rebollar, Tiberio A. Ezquerra, and Marta Castillejo

| 13.1 | Introduction | 472 |

13.2	LIPSS Formation Using Nanosecond Pulses		475
13.3	LIPSS Formation Using Femtosecond Pulses		478
13.4	Formation of LIPSS on Nonabsorbing Polymers		481
13.5	Formation of Alternative Periodic Structures		483
13.6	Applications of Polymer LIPSS		484
	13.6.1 Polymer LIPSS for Cell Culture		484
	13.6.2 Polymer LIPSS for SERS Substrates		485
	13.6.3 Polymer LIPSS for Nonvolatile Organic Memory Devices		486
13.7	Conclusions		488

14. Laser Materials Processing for Energy Storage Applications 499

Heungsoo Kim, Peter Smyrek, Yijing Zheng, Wilhelm Pfleging, and Alberto Piqué

14.1	Introduction	500
14.2	Background and Overview of Materials for Energy Storage	501
14.3	Growth of Energy Storage Materials by Pulsed Laser Deposition	503
	14.3.1 PLD of Cathodes	505
	14.3.2 PLD of Anodes	508
	14.3.3 PLD of Solid-State Electrolytes	510
14.4	Printing of Energy Storage Materials by LIFT	512
	14.4.1 LIFT of Ultracapacitors	514
	14.4.2 LIFT of Li Ion Microbatteries	516
	14.4.3 LIFT of Solid-State Electrolytes	520
14.5	3D Processing of Energy Storage Materials by LS and LA	521
	14.5.1 LA and LS of Thin-Film Electrodes	522
	14.5.1.1 Laser annealing	523
	14.5.1.2 LS of LCO thin films	524
	14.5.1.3 LS of SnO_2 thin films	526
	14.5.1.4 LS of LMO thin films	527
	14.5.2 LS of Thick-Film Electrodes	528
	14.5.3 LS Turns Electrodes into Superwicking	530
14.6	Challenges and Future Directions	533
14.7	Summary	534

Index 545

Preface

Nanotechnology and nanomaterials are at the origin of major progress, breakthroughs, and solutions to a vast number of engineering, biology, and medicine challenges. In fact, at the nanoscale, and at a macroscopic scale as well, the onset of size-dependent phenomena occurs and matter begins to exhibit entirely new properties. The advantage connected to the small feature size of nanostructured materials is exploited in many different applications, both in ordinary life and in high-technology fields.

Obviously, the possibility to valorize the new properties of nanomaterials is strictly connected to the availability of manufacturing processes and characterization techniques.

Among manufacturing methods, pulsed laser–based techniques offer several advantages compared to "traditional" ones for the fabrication of nanomaterials and surface nanostructures, due to the possibility to tune many independent processing parameters. Laser and laser ablation proved important and prospective in various fields, spanning from a better understanding of fundamental physical mechanisms and light–matter interactions to a large range of applications in physics, chemistry, biology, medicine, materials science, manufacturing technology, and even arts and conservation.

This book consists of 14 chapters covering a broad range of topics, written by internationally recognized experts in the field, on the recent advances in the application of laser ablation for the generation of nanoparticles and surface nano- and microstructures and their applications.

It includes a comprehensive overview of the classical theory of growth, with a focus on the importance of kinetic factors and processes in far-from-equilibrium deposition techniques such as pulsed laser deposition (Chapter 1). The wetting, adherence, and nanostructuring properties of the synthesized coatings, by pulsed laser processes, are described and discussed in Chapter 7. A detailed description of the mechanisms and significance of deposition parameters on nanoparticle immobilization and production in different environments (e.g., vacuum, Chapter 2; a high-gas-pressure atmosphere, Chapter 3; liquid, Chapter 5; and matrices, Chapter 6), pulse duration (i.e., the ns regime, Chapters 2, 3, and 5; or the fs regime, Chapters 4 and 5), and use of two time-delayed laser pulses (Chapter 9) is provided. An overview of the results obtained in recent molecular dynamic simulations of laser ablation of metal targets in vacuum, a background gas, and a liquid environment (Chapter 12) is also given.

The processing of materials by fs laser pulses has attracted a great deal of attention because fs pulse energy can be precisely and rapidly delivered to the material without detectable heat perturbation of the neighboring zones. In some cases, periodic micro- and nanostructures can be generated directly (without the use of masks or chemical photoresists to relieve the environmental concerns) in almost any samples from semiconductor to dielectric materials and polymers, supplying relevant results to be used in different applications like nanofluidics, nanophotonics, and biomedical devices.

The effects of some key parameters, including multiple pulses, variable pulse shaping, and fluence, which could be useful in the laser nanostructuring of surfaces and micromachining of different materials (e.g., dielectric, Chapter 10; semiconductor, Chapter 11; and polymer, Chapter 13), are reviewed and discussed from fundamental and/or applicative points of view. The use of laser-based material processing techniques, such as pulsed laser deposition (PLD), laser-induced forward transfer (LIFT), material processing via 3D laser structuring (LS), and laser annealing (LA) techniques for energy storage applications is analyzed in Chapter 14.

Next, Chapter 8 is devoted to the importance of the use of core-shell nanoparticles when different material properties must be merged at the nanoscale to meet the requirements for smart applications.

In the opinion of the authors, the book offers a comprehensive review of the latest advances in top research and development in the laser material processing field for nanoparticles and nanomicrostructure generation and exploitation of different kinds of materials. Theoretical models are discussed by correlation with advanced experimental protocols, to explain the fundamentals and underline physical mechanisms of laser–matter interaction.

The book was conceived as a starting point and guide for students and young researchers who are beginning to initiate in the field of nanostructures and nanoparticles.

Last but not least, the two editors would like to thank all of the chapter contributors for their great efforts and kind cooperation in preparing this book.

Ion N. Mihailescu and Anna Paola Caricato
2018

Chapter 1

Surface Energy and Nucleation Modes

Maura Cesaria
Department of Mathematics and Physics "Ennio De Giorgi," University of Salento, Via Arnesano, I-73100 Lecce, Italy
maura.cesaria@le.infn.it

Deposition is a manufacturing process in which a precursor material is delivered to a surface on which it reorganizes by interplay between thermodynamics and kinetics. The evolution of the microstructure of the deposit (layers with thicknesses between a few atomic layers and several micrometers [films] or nanostructure distributions or nanostructured films) is highly dependent on the physical conditions (i.e., the deposition technique and parameters) under which deposition takes place, apart from the choice of materials. Understanding the mechanisms underlying the structural and morphological evolution of a growing deposit is essential to being able to tailor the properties and performances of functional materials at the atomic scale depending on the deposition approach. In this chapter a comprehensive overview to the classical theory of growth will be given, with a focus on the importance of kinetic factors and processes as well as the growth manipulations enabled

---

Pulsed Laser Ablation: Advances and Applications in Nanoparticles and Nanostructuring Thin Films
Edited by Ion N. Mihailescu and Anna Paola Caricato
Copyright © 2018 Pan Stanford Publishing Pte. Ltd.
ISBN 978-981-4774-23-9 (Hardcover), 978-1-315-18523-1 (eBook)
www.panstanford.com

by a prototypal far-from-equilibrium deposition technique, such as pulsed laser deposition.

1.1 Introduction

The impossible-to-review large number of publications that deal with growth arguments demonstrates a widespread interest at both fundamental and technological levels in understanding the parameters, processes, and kinetic mechanisms that determine and tailor several growth regimes. This purpose has driven a great amount of effort to either model growth theoretically for improving the growth techniques or develop and optimize processing approaches to drive growth far from thermodynamic equilibrium.

In regard to tailoring material functionalities and morphology, a great deal of attention is being focused on nanostructures (systems scaled down to dimensions below 100 nm along at least one spatial direction) that lead to entirely new and challenging applicative perspectives due to quantum confinement effects and a large surface area–volume ratio. Since the fundamental properties of nanoscaled systems are critically affected by size, shape, size and shape dispersion, stoichiometry, and the surrounding environment, the deposition approaches and process parameters play a key role in driving growth and obtaining metastable phases as well as state-of-the-art and tailorable morphology. To fabricate mesoscopic systems two main types of processing approaches are applied, top-down approaches (i.e., patterning performed in bulk material building blocks by combining [photo, e-beam, or nanoimprint] lithography, etching, and direct deposition on substrates) and bottom-up approaches (i.e., structure assembling by self-assembly or self-organization and selective growth from previously chemically or physically synthesized nanoscale building blocks) [1–6].

To explain and predict the observed morphologies and their evolution in growth experiments, equilibrium (thermodynamic) models, which have been the primary stage of investigation historically, may provide the starting information to be combined with the out-of-equilibrium (kinetic) picture [7, 8]. Nucleation, phase transformations, crystallization, and growth are examples of processes requiring thermodynamic as well as kinetic treatment. The kinetic framework is one of the most successful in describing the nonequilibrium morphology of crystals in both bulk and mesoscopic phases. Thermodynamics establishes whether a certain transformation (chemical reaction or phase change) is allowed or not on the basis of a decrease (leading to a global minimum) in the free energy and gives an account of the forces driving such a transformation [9–13]. From a thermodynamic standpoint, growth is determined by the competition between surface and interface free energy, as well as strain energy and strain relief. Such an approach works if growth is driven close to equilibrium, that is, under conditions not dominated by kinetic effects (how fast or slow a process can occur [rate], adsorption, desorption, and diffusion).

Crystallization (i.e., the formation of crystals from the melt, solution, or vapor phase) is a process very much concerned with kinetics because the morphology evolution and shape of crystals are highly influenced by kinetic factors and a complex energy landscape. Indeed, in the thermodynamic picture, the species locally added on the surface of any growing nanocrystal should migrate toward sites with the lowest free energy. However, whenever surface diffusion is not effective enough, a thermodynamically less favorable shape (leading to a local rather than a global minimum of the free energy) may result. All of this indicates that the final (nano)crystal morphology is determined by the competition between kinetic factors, such as the rates corresponding to deposition and surface diffusion. In general, the deposition rate is determined by the rate at which the precursor species are supplied and the diffusion rate is either a thermally or kinetically activated process involving the motion of the deposited species on a solid surface through hopping from site to site. Hence, unbalance between the rates of deposition and surface diffusion by well-controlled manipulation of the experimental parameters can drive the growth toward equilibrium, close-to-equilibrium, or far-from-equilibrium conditions.

Since in practice growth is a nonequilibrium process, competition between thermodynamics and surface kinetics determines the final morphology instead of a purely thermodynamics picture [14, 15].

Basic steps in the growth of materials by deposition are emission of the species to be deposited from a source (heater for thermoionic emission, high voltage for generating charged particle beams, pulsed laser–irradiated target for inducing ablation), their transport from the source to the system supporting the growing deposit (substrate) nucleation, and growth by condensation processes of the deposited species. The depositing species can diffuse under the drive of both their self-energy and substrate thermal energy and then move to a stable position on the substrate, where surrounding atoms may be captured and incorporated, leading to the primary stage of growth, that is, nucleation. Condensation develops through stages: formation of clusters (i.e., groups of deposited species linked by a neighborhood relation), nucleation (formation of stable clusters, termed "nuclei"), dynamics of unstable and stable clusters, formation of an "island" (growing nuclei) and its growth, island coalescence, percolation, and structure development (amorphous, polycrystalline, single crystalline, defects, roughness, and grain growth).

In this chapter these arguments will be discussed in detail in the attempt to provide the basic information and knowledge to a reader new to the growth subject and intending to gain physical insight into mechanisms and processes involved in any deposition experiment. It being impossible to review all the results reported in the literature, discussion will aim at an overview of the field, draw guidelines to establish a foundation for further specific readings, and treat the underlying connections between deposition conditions and the resulting structure and morphology with a focus on growth kinetics driven by out-of-equilibrium approaches.

To start, a thermodynamic background will be given by dealing with free energy, surface free energy, equilibrium criterion for a crystal formation, strain, epitaxy, nucleation, growth modes, and microstructure evolution under different conditions of temperature and supersaturation. Then, deposition, condensation, and morphological evolution will be discussed as related to the realistic physical landscape, kinetic processes, and growth conditions. In this respect, a brief mention of the deposition approaches used for fabricating mesoscopic systems will be made to point out both technological applications of the growth and possibilities for exploiting the surface kinetics offered from growth driven far from the thermodynamic equilibrium. Indeed, the energetic particle beams used in nonthermal growth techniques, such as sputtering and pulsed laser deposition (PLD), enable interaction mechanisms between the substrate and the depositing beams, as well as growth kinetics that are not enabled by near-equilibrium deposition approaches, such as deposition from the gas/vapor phase [16–18].

In particular, PLD will be discussed as a prototypical example of the out-of-equilibrium deposition technique that allows growth manipulations forbidden in the thermodynamic framework.

1.2 Thermodynamic Background Concepts

In this section basics of the classical thermodynamic approach to nucleation theory and classification of growth modes (i.e., the morphology taken on by a system grown close to thermodynamic equilibrium) are introduced and discussed as related to the concepts of free surface energy and strain.

1.2.1 Thermodynamic Potentials and Surface Free Energy

In thermodynamics, two basic thermodynamic potential functions are usually introduced to describe a system: the Gibbs free energy $G = H - TS$ and the Helmholtz free energy $F = U - TS$, where U is the internal energy, T is the absolute temperature, H is the enthalpy, and S is the entropy of the system [19–21]. For a multicomponent system of N entities, the internal energy is given by $U = TS - PV - \Sigma_i \mu_i$, where P and V are the macroscopic variables pressure and volume, respectively, and μ_i is the chemical potential of the component i ($i = 1, \ldots, N$) of the system [19–21].

The possibility of any transformation (such as phase change, nucleation, condensation, or chemical reaction) depends on the sign of the change ΔG of the Gibbs free energy: a transformation is classified as thermodynamically forbidden if $\Delta G > 0$, allowed if $\Delta G < 0$, and in equilibrium if $\Delta G = 0$. A system in thermodynamic nonequilibrium evolves in such a way so as to reach equilibrium through some driving forces, resulting in mass or energy transport. As a thermodynamic criterion to determine operatively the thermodynamically stable phase of a system, for given temperature and pressure (or volume), the system will evolve spontaneously in order to minimize the Gibbs and Helmholtz free energies under the appropriate boundary conditions or to increase entropy at constant energy, volume, and particle numbers in the system [21, 22].

Investigation of the conditions under which structure formation and phase formation/transition processes may proceed is one main focus in thermodynamics. General physical principles leading to the equilibrium form of a crystal were first systematically formulated by J. W. Gibbs on the basis of the total free energy G expressed as the sum of the free energies of the volume, the surface, and the edges and corners of the crystal [23]. It was pointed out that in the case of a nanometer-size crystal, the influence of edges and corners cannot be neglected. For a given volume, the minimization of the surface energy of the crystal was identified as the criterion determining its equilibrium form.

Basically, the surface free energy is defined as the work required to build a surface of a unit area in contact with vacuum at constant temperature, volume, and chemical potential. Microscopically, it originates from surface bonds that are weaker than the corresponding bulk bonds due to a reduced degree of coordination with their neighbors, resulting in lower cohesive energy of the crystal. The main contribution to the enthalpy term of the free energy stems from the chemical bonding, and the stronger the bonds, the smaller the interfacial free energy. Since the difference in the interatomic energy between bulk atoms and surface atoms is the origin of surface energy, altering surface atomic arrangements (e.g., by deposition, diffusion and growth) or straining a surface can change its surface energy.

The surface free energy determined from microscopic interactions (bonds) at an interface (heterogeneous boundary) is termed "interface energy" or "interfacial free energy." Since at a given volume the minimum surface area leads to the minimum of thermodynamic potential, low-energy faces are preferentially expressed.

The total free energy of a system with a surface of area A (e.g., an interface) is given by $G = G_{bulk} + \gamma A$, where G_{bulk} is the free energy of the bulk and γ is the surface energy (i.e., the excess free energy due to broken bonds close to the surface). An increase in the free energy of the system (ΔG) due to an increase in the surface area ΔA can be expressed as $\Delta G = \gamma \Delta A$ under the assumption that γ does not depend on the area A. On the basis of this formalism, mathematically, the

Gibbs condition (minimization of the surface energy of the crystal) demands that, at a fixed volume, the total surface free energy of the crystal $\Sigma_i \gamma_i A_i$ is a global minimum, where γ_i and A_i are the surface free energy and area, respectively, of the face ith of the crystal and the summation is taken over all the faces of the crystal [10, 12, 24, 25].

According to the Gibbs criterion, a crystal forms by assuming a shape of minimal surface energy for a given volume. In isotropic systems, where the surface free energy does not depend on the direction, a spherical shape minimizes the total surface area at a fixed volume. In anisotropic systems the surface free energies depend on the crystallographic planes identified by the Miller indices klm ($\gamma_i = \gamma_{klm,i}$). Usually the densest planes of crystals grow the fastest because of the higher potential energy decrease resulting from incorporation of surface diffusing species due to the higher density of bonding sites.

Since the surface free energy reflects the cost in energy when bonds are broken, if only interactions between the nearest neighbors are considered and the crystal is surrounded by vacuum (broken bond approximation), then the specific surface free energy of a crystallographic plane can be calculated by $\gamma_{klm} = (1/2)N_B \varepsilon \rho_A$, where N_B is the number of broken bonds per surface unit cell (or per surface atom), ε is the bond strength, and ρ_A is the number of surface atoms per unit area [26]. The prefactor 1/2 accounts for the fact that each bond involves two atoms. More accurate calculations of the surface free energies require the inclusion of interactions with the second- and third-nearest neighbors, surface relaxation, and surface adsorption of chemical species ([27] and references therein). By combining crystal anisotropy and free-energy minimization, one may come up with faceted crystal shapes. In fact, while isotropic interface free energy induces a sphere as the equilibrium shape, for anisotropic (i.e., depending on orientations) interface free energy the Wulff construction can be used [28–30] for determining the equilibrium shape of a crystal in vacuum geometrically by a polar plot of surface energies. According to the Gibbs–Wulff theorem, at equilibrium the distances of the borders from the crystal center are proportional to their surface free energy, that is, $\gamma_{klm,i}/h_i$ = constant, where $\gamma_{klm,i}$ and h_i are the specific surface free energy as a function of the orientation and the distance to the center for the ith facet, respectively. It results in an average interface free energy corresponding to a hypothetical spherical cluster having the same volume and the same interface energy as the real one with a faceted shape (closed polyhedron) rather than a spherical equilibrium shape. Therefore, when the growth of a single crystal occurs in vacuum under thermodynamic control, a closed enlarging polyhedron is expected according to the Wulff theorem.

However, since the equilibrium shape of a crystal is determined by the interplay among surface free energies associated with different crystallographic facets, modified ratios between such energies may involve shapes other than the ones predicted by the Wulff theorem. This evidence points to the role played by kinetic versus thermodynamic control. In addition, the surface free energy can act to destabilize the crystal for increasing the surface-to-volume ratio (i.e., decreasing crystal size), which makes surface energy a key quantity in driving morphology and mechanical and thermal stability of nanosystems.

1.2.2 Phase Transformations of a Thermodynamic System and Supersaturation

Another important argument in thermodynamics deals with phase transformations of a thermodynamic system. The occurrence of any transformation from the vapor phase to a condensed phase (liquid or solid) demands a partial pressure (P_V) exceeding the equilibrium vapor pressure with respect to its condensed phase ($P_{V,e}$) at the local temperature T. Notably, the vapor pressure over a curved surface is greater than that over the corresponding flat surface, as described by the so-called Kelvin–Gibbs relation.

The thermodynamically stable phase corresponds to the minimum of G, and transitions of the system correspond to a change in the Gibbs free energy, given as difference between the values of G in the starting and final states, that represents the driving force of the transformation. Any vapor-to-solid transition of a thermodynamic system (e.g., nucleation and condensation) can be described by a change in the Gibbs free energy $\Delta G = \Delta G_V + \Delta G_S$, where ΔG_V and

ΔG_S refer to the volume and surface contribution to ΔG, respectively. Usually, the energy release accompanying the transition is expressed not in terms of total free-energy change but rather as the change in chemical potential $\Delta \mu$ given as the difference of the chemical potentials associated with the initial and final phases (i.e., $\Delta \mu = \mu_{initial} - \mu_{final}$). The larger the $\Delta \mu$, the greater is the driving force for crystallization [31]. For the description of phase formation in solids, $\Delta \mu$ is also a convenient measure of the so-called supersaturation ratio SS expressed by $(1 + SS) = P/P_e$, where P_e is the equilibrium vapor pressure of the infinitely extended condensed phase and P is the actual value of the partial pressure. Therefore, the relationship between $\Delta \mu$ and SS is given by the formula $\Delta \mu = k_B T \ln(1 + SS)$.

The supersaturation is a critical parameter in condensation processes [32]. The situations associated with $P = P_e$, $P < P_e$, and $P > P_e$ are termed "saturated," "undersaturated," and "supersaturated," respectively. Supersaturated conditions, corresponding to $\Delta \mu > 0$ (namely, $\Delta G_V < 0$), are a prerequisite to initiate condensation processes as well as to make possible nucleation and subsequent growth.

1.2.3 Strain and Epitaxial Growth

Crystals may form from liquid (precipitation from an aqueous solution), melt (solidification), gas/vapor (deposition), and another crystal (transformation, replacement, or metamorphism). In any case, the minimization of the free energy is not an absolute criterion for the formation of a crystal because minimization of the strain energy also plays a role [33].

Strain or stress is commonly introduced by the growth of heterostructures, namely systems consisting of different materials, as a result of deposition or growth (growth stress), defects, constraints imposed by mismatch in the thermal expansion coefficients (playing during high-temperature growth or postdeposition annealing), and/or structural misfit. Hence, three main classes of stress can be identified: intrinsic or growth stress (developing due to either defects in the deposit or the choice of growth approach and conditions), thermal stress (caused by differential thermal expansion), and misfit strain (arising from lattice misfit). Internal stress arises due to nonequilibrium growth conditions. It differs from external stresses (thermal and misfit strain) that can affect the growth mode by interplay of surface energy minimization and strain energy minimization. Stress buildup can be useful or harmful depending on the application in today's solid-state electronic and optical state-of-the-art devices. It may affect growth mechanisms and evolution, as well as stability conditions. Strain is a key concept in the case of epitaxial growth.

In thermodynamics, the term "epitaxy" refers to the growth of a crystal (usually termed "film," "overlayer," or "epilayer") deposited over a (foreign) material (termed "substrate"). The choice of the crystalline order, structure, and temperature of the substrate may be exploited to affect, drive, and manipulate the film's final structure [34, 35]. An ideal interface can be depicted as atomically abrupt (i.e., of a submonolayer width), with each of the materials on either side existing in a state unperturbed by the presence of the other. This is generally not the case for most practical interfaces even if state-of-the-art epitaxial interfaces can be obtained. More generally, the principle of epitaxial growth is applied to growth distributions of nanostructures and sequences of crystalline layers superposed on each other with sharp interfaces and control on composition, crystalline quality, and thickness down to the monolayer (below one atomic layer). Epitaxial growth also develops as a function of the growth method and temperature, interfacial energy, and the lattice misfit between the substrate and epilayer. Basic requirements for the substrate are a good crystallographic lattice matching with respect to the film material with eventually selectively chosen chemical termination [36, 37], substrate material chemically compatible to the film, film and substrate having comparable thermal expansion coefficients, and a thermodynamically as well as chemically stable surface.

Epitaxy distinguishes between homoepitaxy and heteroepitaxy to refer to deposit–substrate systems consisting of the same material and of different materials, respectively.

Homoepitaxy is applied to the growth of a high-purity, structurally defect-free, and homogeneous film starting from a non-high-purity, structurally imperfect, and nonhomogeneous substrate acting as a seed-supporting layer. Since the deposited material is identical to the substrate material, the crystalline structure of the substrate is extended into the growing epilayer and both lattice and thermal misfit do not play a role. In contrast, heteroepitaxial films can be in a state of strain due to crystal misfit or, more generally, to constraints imposed by the film-substrate interface and differential thermal expansion. Since lattice-matched crystals are required in principle, heteroepitaxial structures can be deposited by a relatively limited number of combinations of materials. Extension to combinations of lattice-mismatched crystals enables more flexibility in the achievable surface morphologies at the cost of incorporating strain into the growing system and depending on the amount of strain. Heteroepitaxial growth can result in coherent incorporation of the strain into the epilayer if this one and the substrate are lattice matched, with perfectly coherent interfaces and crystal lattice lining up (Fig. 1.1a). This is the case of cubic-on-cubic crystals, such as AlAs and GaAs. A coherently strained epilayer is termed "pseudomorphic." Figure 1.1 depicts the concept of pseudomorphic strain produced by a lattice-matched epilayer-substrate system when the lattice constant of the film material (a_{film}) is larger (Fig. 1.1c) and smaller (Fig. 1.1d) than the substrate one (a_{sub}). Continuous transition of the lattice constant from the substrate value to the film value may occur by shrinking (compressive strain) (Fig. 1.1c) or expanding (tensile strain) (Fig. 1.1d) of the film lattice constant in the interfacial plane. The lattice constant also changes in the direction perpendicular to the film–substrate interface: it shrinks and expands if the in-plane lattice constant expands and shrinks, respectively.

A pseudomorphic epilayer is biaxially strained in the interface plane by a degree that can be quantified through the so-called misfit strain parameter $\varepsilon_{\parallel} = (a_{film} - a_{sub})/a_{sub}$, that is, the relative difference of the lattice parameters. It is worth noticing that a surface-strained finite crystal (supposed isotropic) has a lattice parameter given by $a_0(1 + \varepsilon)$, where a_0 is the lattice parameter of the corresponding infinite crystal and ε is the size-dependent strain of the finite crystal. Because of this, the misfit strain developing in the case of a pseudomorphic epilayer can be expressed as $\varepsilon_{\parallel} = (a_{film} - a_{sub})/a_{sub} = \varepsilon_{\parallel,0} - \varepsilon$ by substituting $a_{film} = a_0(1 + \varepsilon)$ and introducing the misfit $\varepsilon_{\parallel,0} = (a_0 - a_{sub})/a_{sub}$ associated with the infinite phases [13]. The above expression of the misfit strain is valid only when the epilayer is thin enough to be able to assume a lattice constant close to the one of the substrate. Once the epilayer thickens, its lattice constant is expected to approach the value of the corresponding bulk material because the overlayer tends to adopt more and more its bulk lattice constant. However, for the thickening epilayer, the formation of misfit dislocations at the interface between film and substrate becomes energetically favorable to relieve the strain built up in the film (Fig. 1.1b). Therefore, lattice misfit can be accommodated by distortions near the interface. If the lattice mismatch is not too high, the lattice of the epilayer can deform to become pseudomorphic to the substrate (a so-called strained film results). Hence, as long as the epilayer grows below a certain critical thickness, the built-up strain can be accommodated coherently and misfit dislocations do not form [13]. Instead, above the critical thickness, the lattice mismatch can be so significant as to generate edge dislocation defects at the epilayer–substrate interface [38, 39] and surface roughening [40, 41] as a way to relieve the strain. In other words, while for a thin-enough film surface energy dominates, for a thick-enough film strain energy dominates. The maximum thickness of a thermodynamically stable film corresponds to the minimum of the free energy ($F_N - N\mu_{bulk}$), where F_N is the change in the free energy due to adding N atoms to the growing film and μ_{bulk} is the chemical potential of the bulk material of the film [42]. Although a detailed description depends on the nature of the interactions between the film atoms and of the film atoms with the substrate (weak or strong film–substrate interaction), $F_N - N\mu_{bulk}$ as a function of epilayer thickness reaches a minimum after depositing at least one monolayer. The formation of a uniform film not exceeding a critical thickness of a few monolayers may minimize the free energy only in the presence of a strong film–substrate interaction. In the case of a weak film–substrate interaction, the growth of a uniform epilayer is never thermodynamically favored and, instead, the lattice misfit is a driving force for the generation of bulk 3D structures. Therefore, the thermodynamic driving force (interface free energies) for different morphologies and the structural misfit between substrate and epilayer are key parameters for determining the epitaxial thermodynamic growth modes.

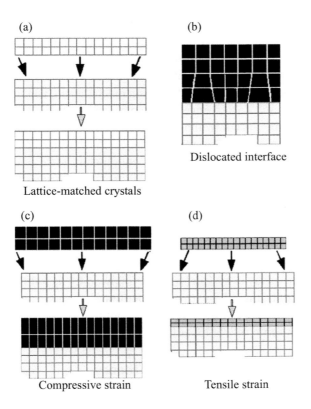

Figure 1.1 Schematics of the various situations occurring in heteroepitaxy depending on the relationship between the lattice constants of an epilayer and the underlying substrate.

1.3 Thermodynamic Nucleation Theory and Growth Modes

Formation of a crystal occurs through two processes mainly, nucleation and growth by condensation processes. Aggregation of the deposited species can form the so-called clusters, which can be either unstable or stable. Nucleation, which is the primary stage of the condensation, process from which growth proceeds, consists of random generation of solid-phase nanometer-size clusters that can irreversibly grow to macroscopically large sizes. In thermodynamics, nucleation originates from local fluctuations from equilibrium of a supersaturated starting phase that give rise to a phase transition (from the vapor or liquid to the solid phase). A supersaturated starting phase is a prerequisite for the occurrence of nucleation events [10]. Thermodynamically stable clusters, termed "nuclei," form the building blocks of further condensation mechanisms (island growth, aggregation, and coalescence). From a thermodynamic standpoint, the key parameters controlling nucleation are the nucleation driving force, the interface free energy, and the condensation rate.

To describe the thermodynamic equilibrium state and growth morphology of the supported thin film, the wetting conditions, the supersaturation, and the degree of misfit between the film and the substrate are the main parameters to be considered. The final morphology is the result of the growth mode adopted from the deposit. The main thermodynamic growth modes are the layer-by-layer growth (2D growth), island growth (3D growth), and the mode starting out as layer-by-layer growth followed by island formation. They are determined by the competition between surface and interface free energy, as well as the competition between strain energy and strain relief [43, 44].

1.3.1 Principles of Nucleation Theory

The main concepts concerning nucleation in phase transformations of a fluid phase within another fluid were introduced by Gibbs, which considered ensuring minimal free energy with respect to the radius as the thermodynamic criterion to form nuclei [45]. The radius satisfying such a condition was termed "critical nucleus radius." Although oversimplified, Gibbs's picture introduced some key assumptions that were later transferred to model the nucleation of crystals from dilute or condensed fluids or from other solid states [46–49]. The basic formulation of classical nucleation theory dates to 1927 by Volmer, Weber, and Farkas [50, 51] and to 1935 by Becker and Döring [52]. Thermodynamically, a nucleation process was modeled as the formation of small embryos of the new homogeneous phase inside a large volume of the old phase (parent phase) due to heterophase fluctuations [53, 54]. The temperature was the controlling parameter driving the transition of an initially stable (at thermal equilibrium) homogeneous phase to a metastable state, that is, stable with respect to small and unstable with respect to sufficiently large thermal fluctuations. In other words, the transformation required overcoming a free-energy barrier ΔG^* termed "nucleation barrier" [11, 55, 56]. In the framework of the so-called homogeneous capillary approximation [50–52], the main assumption was that the cluster free energy to be minimized with respect to the size can be partitioned as the sum of surface and volume contributions. According to such a classical thermodynamic picture, the stability of clusters depends on the balance between the surface and volume free energies that contribute to the formation of free energy. This is easily understood by considering the free-energy changes associated with the formation of a spherical cluster with a radius r from an old phase whose free energy has become higher than that of the emerging bulk new phase. The condensation reaction is driven by the chemical free-energy change per unit volume.

A cluster containing units (termed "monomers"), each occupying a volume V, is composed of $(4/3)\pi r^3/V$ units. Hence, thermodynamically, forming a spherical cluster with a radius r involves a change of Gibbs free energy ΔG_V or a volume energy $\Delta \mu_V$ given by $\Delta \mu_V = (4/3)\pi r^3 \Delta G_V$. In the case of condensation from a supersaturated initial phase

$$\Omega \Delta G_V = -[(k_B T) \ln(1 + SS)], \qquad (1.1)$$

where Ω is the atomic volume and SS is the already defined supersaturation ratio. The energy reduction $\Delta \mu_V$ is counterbalanced by an increase in the surface energy $\Delta \mu_S = 4\pi \gamma r^2$, where γ is the surface energy per unit area. Then, the total change of the chemical potential (i.e., the difference of chemical potentials in the parent and in the nucleating equilibrium phases) [31, 55] resulting from the formation of a cluster of radius r defines the free energy ΔG associated with the formation of a solid spherical cluster in an otherwise homogeneous fluid:

$$\Delta G = \Delta \mu_V + \Delta \mu_S = (4/3)\pi r^3 \Delta G + 4\pi \gamma r^2 \qquad (1.2)$$

In thermodynamics the change in free energy associated with the formation of a cluster is also called work of formation.

The general profile of the change of volume free energy $\Delta \mu_V$, surface free energy $\Delta \mu_S$, and the formation energy ΔG is shown in Fig. 1.2 as a function of cluster radius. The surface contribution to the free energy is always positive and acts to destabilize the cluster—the more unstable the cluster is the larger its surface-to-volume ratio is. Once the size r gets large enough, the drop in the free energy associated with the formation of the bulk phase dominates the surface free energy and every further increase of size lowers the free energy of the system. From there on, the gain in volume drives the growth of a cluster because growth becomes energetically favorable.

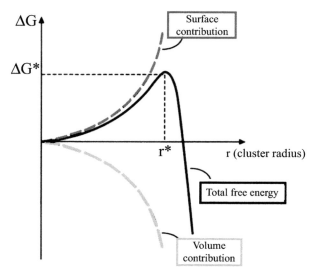

Figure 1.2 General profile of the change of the volume free energy, surface free energy, and formation energy in a nucleation process in the thermodynamic framework.

Because of the competition between the volume and the surface contributions, the formation energy ΔG shows a maximum ΔG^* at a cluster radius r^*, usually termed "nucleus critical size." The formation of a cluster with a critical size (the one for which the free energy ΔG is maximum) is referred to as nucleation. The expressions of ΔG^* and r^* can be easily estimated by setting equal to zero $(d\Delta G/dr)|_{r=r^*}$, solving which gives the following formulas [57]:

$$\Delta G^* = (16\pi r^3)/3(\Delta G_V)^2 \qquad (1.3)$$

$$r^* = -2\gamma/\Delta G_V \qquad (1.4)$$

As the maximum value of the formation energy (ΔG^*) represents an energy barrier that a nucleation process must overcome to form an irreversibly growing nucleus, it is called activation energy for nucleation or nucleation barrier. The dependence of both ΔG^* and r^* on the surface free energy has the important implication that any change of the surface free energy affects the cluster formation. Therefore, below the critical size r^* the surface term $\Delta\mu_S$ dominates the volume contribution $\Delta\mu_V$ and drives the increase of the energy of the growing clusters (Fig. 1.2). As a result, a cluster of radius $r < r^*$ (termed subcritical cluster) is unstable, that is, once formed, it will tend to disintegrate in monomers and/or smaller clusters. Instead, since above the critical size r^*, the drop in free energy is dominated by the volume term $\Delta\mu_V$, a cluster larger than r^* will continue to grow. A cluster with a size at least equal to the critical size is usually termed "nucleus" or "supercritical cluster."

The nomenclature "homogeneous nucleation" is adopted to refer to a cluster formed within a homogeneous phase. In the presence of an additional foreign material acting as a catalyst for the cluster formation (heterogeneous nucleation), the formation energy ΔG is expressed in terms of the interface energy related to the cluster surface ($\gamma_{cluster}$), the substrate surface (γ_{sub}), and the cluster–substrate interface (γ_{cl-sub}). Under the assumption that a hemispherical (spherical cap-shaped) cluster of radius r is formed over a substrate, it results in:

$$\Delta G = \Delta\mu_V + \Delta\mu_S = a_1 r^2 \gamma_{cluster} + a_2 r^2 \gamma_{cl-sub} - a_2 r^2 \gamma_{sub} + a_3 r^3 \Delta G_V, \quad (1.5)$$

where the coefficients are geometric factors, with $a_1 r^2$ the surface area, $a_2 r^2$ the projected surface area, and $a_3 r^3$ the volume of the spherical cap-shaped cluster. The same procedure applied in the case of the homogeneous nucleation lets us obtain the following expressions of the critical nucleus size r^* and the maximum formation energy ΔG^*:

$$r^* = -2(a_1 \gamma_{cluster} + a_2 \gamma_{cl-sub} - a_2 \gamma_{sub})/3 a_3 \Delta G_V \quad (1.6)$$

$$\Delta G^* = 4(a_1 \gamma_{cluster} + a_2 \gamma_{cl-sub} - a_2 \gamma_{sub})^3 / 27(a_3 \Delta G_V)^2 \quad (1.7)$$

The above thermodynamic formalism provides a generalized framework that lets us include in a straightforward way other energy contributions such as strain. In the presence of a lattice mismatch between depositing material and substrate, a term proportional to $r^3 \Delta G_{strain}$, where ΔG_{strain} is the strain free-energy change per unit volume, must be inserted in the formation energy ΔG leading to an increase of the overall energy barrier to nucleation ΔG^*. Stress relief occurring during nucleation favors a reduction of ΔG^*.

To summarize the classic thermodynamic picture, clusters smaller than a critical size appear and disappear spontaneously through thermal fluctuations. A transient regime, lasting a time termed "incubation time," exists before the nucleation rate reaches its stationary value [55, 58–60]. In the case of a nucleation barrier comparable to or higher than the thermal energy $k_B T$, metastable clusters may overcome the critical size and from there on continue to grow and become more and more stable. Classical nucleation theory assumes that the steady-state distribution of a nucleating system slightly deviates from the equilibrium distribution around the critical size. Cluster random size fluctuations around the critical size may cause disintegration of a stable nucleus [52] and only the critical clusters reaching a size large enough to enter the steady-state regime fall in the stable region and can continuously grow (supercritical nuclei). Definitively, the steady state can be reached once the cluster size increases far enough away from the critical size.

An important property of the activation energy ΔG^* is its strong influence on the density of stable nuclei. As in the case of any kinetically limited process, classical nucleation theory assumes that the nucleation probability or rate J (i.e., the number of nuclei formed per unit time per unit volume) obeys an Arrhenius law. That is, supercritical nuclei generate due to thermodynamic fluctuations in the subcritical region by overcoming the nucleation barrier ΔG^* at a rate $J = J_o \exp(-\Delta G^*/k_B T)$ [31, 61]. The prefactor J_o also depends on the supersaturation ratio SS, material constants, and temperature, as well as fluctuations around the critical size by the nonequilibrium Zeldovich factor [52]. Therefore, a high nucleation barrier would involve a small concentration of critical nuclei. Notably, the Zeldovich expression for the nucleation rate implies that J is relatively very small until a critical value of SS is achieved, after which J increases exponentially. Furthermore, the values of both r^* and ΔG^* decrease for increasing SS [62]. If SS is high enough, the nucleation barrier would virtually vanish and the rate of formation and growth becomes limited by the rate of transport of mass or energy. Classical nucleation theory assumes a not too high SS, that implies a large enough critical size. This assumption also has procedure implications because it allows one to treat the cluster size as a continuous variable and introduce derivatives to minimize ΔG as well as make a finite expansion of key quantities around the critical size.

The dependence of the nucleation rate J on the nucleus critical size implies that the nucleation process can be operatively affected by modulating the critical size, for example, through the interfacial energy. In the case of heterogeneous nucleation, the decrease in the value of the surface free energy favors decrease of r^*, ΔG^*, and critical supersaturation [62]. Therefore, under conditions of low SS heterogeneous nucleation is energetically more favored than homogeneous nucleation. On the other hand, it may be energetically favored for the clusters to form heterogeneously on preferred nucleation sites (such as steps, dopant sites, existing impurities, or some lattice defects) than homogeneously because of their acting as catalyzers in lowering the cluster free energy by a gain in the interface free energy. In general, heterogeneous and homogeneous nucleation can compete with each other depending on the number of heterogeneous sites with respect to the total number of sites for homogeneous nucleation. As a result, nucleation can be manipulated by tuning either the supersaturation or the nucleation environment.

As a concluding remark, the main conclusions of classical nucleation theory can be derived as approximated solutions of a kinetic approach that solves for a master equation to describe the nucleation process and model the dynamics of the cluster size distribution (population) [52, 60, 63]. Moreover, to describe the nucleation evolution (nucleation and its whole kinetics) an approach was developed that couples classical descriptions and kinetic approach [64]. In the framework of the cluster dynamic model, a cluster is defined by a single parameter, that is, its size or the number of units it contains. The cluster dynamic formulation removed two limiting assumptions of the classical nucleation theories: first, the assumption that only reactions involving monomers can occur and, second, the assumption that the cluster composition is given (i.e., the nucleating phase at equilibrium with the parent phase). Whenever this information is not known a priori, other modeling techniques can be applied, such as molecular dynamics [65, 66] and kinetic Monte Carlo simulations [67]. Over the years, the "kinetic theory" of homogeneous nucleation has been revised and/or extended [68–70].

Discussions of the conceptual differences between cluster dynamics and classical nucleation theory have evidenced consistency between the results of the two approaches in the dilute limit and coherence concerning the critical size in the limit of large cluster sizes [71–73]. These findings underline some limitations of the classical approach with increasing supersaturation [66]. Indeed, the classical nucleation approach fails in the limit of a very small critical nucleus because of the increased weight of the surface free-energy term and the influence of the sharp curvature on the interfacial energy [74].

1.3.2 Growth Modes at Thermodynamic Equilibrium

Thermodynamic arguments let us introduce a recipe to describe the equilibrium form of a crystal A condensed on a crystal B (acting as a substrate) in terms of the interfacial free energy γ_{AB} between A and B in addition to the surface free energies γ_A and γ_B of the crystals A and B, respectively. In the following, on the basis of the nomenclature usually adopted experimentally, γ_A and γ_B will be called γ_{film} and γ_{sub}, respectively, where the subscripts "film" and "sub" stand for film and substrate, respectively. Also, γ_{AB} will be referred to as $\gamma_{interface}$.

Historically, the attempts to understand and predict the epitaxial growth based on thermodynamics depicted three main scenarios named after their original investigators: the Frank–van der Merwe (FM) growth mode (i.e., 2D morphology, layer-by-layer growth, or step-flow growth) [75], the Volmer–Weber (VW) growth mode (i.e., 3D morphology, island growth) [50], and the Stranski–Krastanow (SK) growth mode (i.e., initially 2D morphology evolving toward a 3D morphology after a critical thickness, layer-plus-island growth) [76].

Unification and more rigorous treatment of the classification of the thermodynamic growth modes in heteroepitaxy was achieved by introducing the so-called wetting factor [9]

$$\Phi = \gamma_{film} + \gamma_{interface} - \gamma_{sub}. \qquad (1.8)$$

According to the competition between surface and interface energies (i) if $\gamma_{film} + \gamma_{interface} \leq \gamma_{sub}$, then a pure 2D layer-by-layer growth results (FM growth mode) (Fig. 1.3a); (ii) if $\gamma_{sub} < \gamma_{film} + \gamma_{interface}$, then a 3D morphology is energetically favored (VW growth mode) (Fig. 1.3b); and (iii) in the presence of strain, the relationship $\gamma_{film} + \gamma_{interface} \leq \gamma_{sub}$ implies a crossover from layer-by-layer 2D growth to 3D island growth ($\gamma_{film} + \gamma_{interface} > \gamma_{sub}$) at a critical thickness of the deposit (SK growth mode) (Fig. 1.3c).

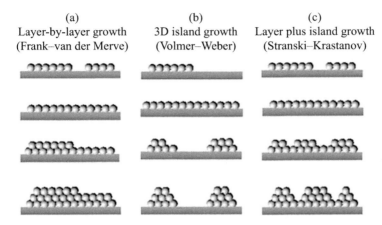

Figure 1.3 Morphology evolution (from top to down) leading to the three fundamental thermodynamic growth modes in heteroepitaxy (from the left, the Frank–van der Merwe growth mode, the Volmer–Weber growth mode, and the Stranski–Krastanow growth mode).

The above classification of the thermodynamic growth modes based on the wetting factor $\Phi = \gamma_{film} + \gamma_{interface} - \gamma_{sub}$ (2D growth for $\Phi < 0$ and 3D growth for $\Phi > 0$) doesn't account for the surface stress effects. Thermodynamics of strained solids and surface thermodynamics have been developed for describing thermodynamic transitions and crystal growth in the presence of epitaxial strain [23, 33, 77] and to connect the surface stress to the surface energy [78, 79]. In general, the epitaxial stress acts against wetting, hence leading to a thickening and morphology evolution toward a 3D equilibrium shape [80, 81]. If an inhomogeneous strain is disregarded that can create preferential sites for 2D or 3D nucleation, the interplay between thermodynamics and elastic effects allows the discussion of the thermodynamic classification of the main growth mechanisms, as follows.

Whenever the sum of the surface free energy of the film and the interface free energy is lower than the free energy of the substrate surface (i.e., $\Phi < 0$), the gain of energy results from the complete coverage of the substrate by the film that thickens uniformly in a pure 2D layer-by-layer manner (FM growth mode) (Fig. 1.3a). The equality in equation $\gamma_{film} + \gamma_{interface} = \gamma_{sub}$ holds for the trivial case of homoepitaxy, where the interface between film and substrate essentially vanishes ($\gamma_{interface} = 0$). Since the surface energy of a crystal depends mainly on the chemical bond energies, the layer-by-layer growth mode is thermodynamically favored when the species of the overlayer can be more tightly bound to the substrate than to each other and bonding dominates surface diffusion [82]. Under this circumstance, the deposited material nucleates on the substrate surface, forming 2D islands, which coalesce with increasing coverage, and then the first complete monolayer on the substrate surface becomes covered with a somewhat less tightly bound second layer (Fig. 1.3a). This growth evolution provides a layer-by-layer growth mode, meaning that a layer is completed before nucleation of a new layer starts. In practice, however, nucleation on higher layers starts before the previous layer has been completed, thus forming steps responsible for a transient roughening. In fact, the surface morphology of the growing crystal changes from smooth over rough by nucleation of 2D islands to smooth by coalescence of 2D growing islands that form a layer without steps.

If the film and substrate materials are lattice mismatched, the 2D growth can continue until the epilayer is able to accommodate elastically the building up (compressive or tensile) strain. In the case $\Phi = 0$, a film grown onto a rigid substrate is stable, provided its thickness is smaller than some critical value depending on the supersaturation [83]. Wetting interactions with $\Phi < 0$ and relief of surface stress can act to stabilize again the film up to a larger thickness [84]. As the accumulated strain energy increases linearly with the thickness of the epilayer, relief of the strain energy can cause the formation of interfacial misfit dislocations for thick (above a critical thickness) films having a small lattice misfit with respect to the substrate. For higher misfits, the growth mode changes to the layer-plus-island SK mode (Fig. 1.3c). Therefore, a layer-by-layer growth mode is the situation occurring in heteroepitaxy if the free-energy minimum of the growing deposit favors an atomically smooth surface. It is observed in the case of rare gases adsorbed on graphite and on several metals, in some metal-metal systems, and in semiconductor growth on semiconductors. The layer-by-layer growth is technologically important for depositing quantum well heterostructures and superlattices by epitaxial growth.

When the surface free energy of the film plus the substrate surface free energy is lower than or comparable to the interface free energy, a partially covered substrate surface is energetically more favorable than the growth of a uniform epilayer (VW growth mode) (Fig. 1.3b). The role of the strain must be also taken into account. While the deposition of material starts with a complete wetting of the substrate (a so-called wetting layer forms), as the deposition progresses strain energy accumulates with increasing thickness of the wetting layer and drives the system to enter a metastable region (i.e., there is a potential for 3D growth with an activation energy or barrier to be overcome). For lattice-mismatched film-on-substrate systems, as the elastic strain energy increases linearly with the film thickness and quadratically with the strain, the formation of coherent 3D islands is energetically favored beyond a critical thickness to release the strain energy. If this one begins to build up as the first few atomic layers wet the substrate, once the wetting layer reaches a critical thickness it decomposes and the growth proceeds with the formation of self-assembled 3D islands (clusters larger than a critical size) (Fig. 1.3b) that grow steadily determined by surface and interface energies and bond strength. In this respect, the island growth mode is associated with the situation where the species of the epilayer are more strongly bound to each other than to the substrate. Once small 3D clusters are nucleated directly on the substrate surface, they grow into islands of the condensed phase because nearly vanishing second-neighbor bonds (as in the case of the edge atoms of a 3D island) involve a rough surface as the lower-free-energy surface. Therefore, in heteroepitaxial growth on a lattice-mismatched substrate, the equilibrium shape (3D islands) is determined by the minimization of the total energy (including the strain contribution) rather than the minimization of the surface energy. The VW growth mode is usually observed in heteroepitaxial systems for strained layers with a lattice mismatch larger than 10%, for example, in the case of many systems of metals growing on insulators [85–89], alkali halides, graphite, and compounds such as mica.

The SK growth mode is an interesting intermediate hybrid growth mode governed by elastic relaxation [90]: after a wetting layer a few nanometers thick forms, subsequent layer-by-layer growth is unfavorable and an island forms over the wetting layer. The condition $\gamma_{film} + \gamma_{interface} \leq \gamma_{sub}$ implies instability at a critical thickness t_c (depending on strain and chemical potential), where a switch to the relationship $\gamma_{film} + \gamma_{interface} > \gamma_{sub}$ drives a crossover from 2D to 3D growth morphology (Fig. 1.3c). From the standpoint of the chemical bond energies, in the case of strong bonding to the substrate of the film species, the growth of one or more monolayers is favored energetically at the beginning. However, as this interaction is over a short range, subsequent evolution to a cluster morphology occurs energetically, driven by a mismatch in the lattice parameter or symmetry or crystal orientation between the bulk material of the epilayer and the substrate. Morphology transition is favored by the accommodation of elastic strain in a pseudomorphic layer that changes the balance between the surface and interface free energies during growth. Notably, the formation of islands on top of a thin wetting layer may be coherent, that is, dislocation free.

According to thermodynamics, the net free-energy change for the nucleation of a hemispherical island with a radius r assumed to be incoherent (or relaxed) on top of a growing, strained partially relaxing epilayer is given by

$$\Delta G = \Delta \mu_V + \Delta \mu_S = (2/3)\pi r^3 \Delta G + 4\pi \gamma r^2 + \Delta G_{strain}, \quad (1.9)$$

where ΔG is the volume free energy of the island, γ is the interface free energy, and the strain energy interaction between the island and the underlying epilayer ΔG_{strain} represents the difference in epilayer strain energy per unit area after the island nucleation ($1/2Y\varepsilon^2$) relative to that in the epilayer prior to island nucleation [91]. On the basis of the theory of elasticity, the epilayer strain energy G_{strain} before and after the island nucleation can be written as $1/2Yf^2$ and $1/2Y\varepsilon^2$, respectively, where Y is the elastic modulus of a layer and ε/f is its strain and f is the lattice mismatch strain defined as $f = (a_{sub} - a_{film})/a_{film}$ (a_{film} and a_{sub} being the lattice parameter of film and substrate, respectively). Under the condition that ΔG has a minimum with respect to r and in the limit of a vanishing critical radius, the critical thickness of the wetting layer can be evaluated as given by $t_c = 2\gamma/[(\varepsilon^2 - f^2)Y]$. This the thickness of the epilayer for the onset of the rough island morphology (e.g., SK growth) [13].

Therefore, in the case of strained epitaxy, the system undergoes a transition from 2D to 3D island growth mode due to interplay between second-neighbor bond strengths and strain energy that both dictate the deposit equilibrium morphology. In this case the lattice mismatch is commonly a few percent compressively strained. The SK growth mode can be observed in semiconductor/metal systems such as Ge/Si(001), InAs/GaAs, CdSe/ZnSe, GaN/AlN, Bi/GaP, Ag/Si, Au/Ni, and Au/Ag. For example, since the lattice constant of Ge is 4.2% larger than that of Si, while growing a Ge/Si(001) system a Ge wetting layer consisting of about 4–5 monolayers can be grown before 3D Ge islands form with sizes of tens of nanometers on its topmost surface. The SK growth mode is applied for fabricating coherently strained 2D systems, as well as arrays of closely spaced quantum dots with control on size and shape dispersion [92].

Definitively, the growth regimes of a film can be classified in terms of surface energy ratio $W = (\gamma_{sub} - \gamma_{film})/\gamma_{sub}$ as a function of the lattice misfit, leading to a plot of the stability regions of the three main growth modes: island growth dominates in the case of $W < 0$ and layer growth is possible only when $W > 0$ and in the presence of a small amount of misfit (strained-layer epitaxy). In between and in competition with the island growth mode and layer morphology is the layer-plus-island SK growth mode. The energy stored in an interface between epitaxial film and substrate is determined by the relative contributions of elastic strain (deformation of the lattice of the film) and formation of edge dislocations.

1.4 Elementary Kinetic Processes on Surfaces and the Energy Landscape

The thermodynamic approach to growth lets one predict the close-to-equilibrium growth modes. However, the applicability of the thermodynamic classifications to modern growth experiments is limited because most deposition experiments are performed under far-from-ideal equilibrium conditions. Therefore, growth is ruled by kinetic parameters and processes, as well as a complex energy landscape.

This section deals with the elementary processes occurring on a substrate surface, during and following deposition of species, that concur to the growth stages and structure formation. The characteristic kinetic parameters and laws governing such processes and the corresponding nomenclature will be introduced and discussed [93].

The main independent experimental variables effective in tuning the growth regimes are the substrate temperature (T_{sub}), the deposition flux F (the number of impinging species per unit surface area and per unit time), and the kinetic energy of the deposited species that strongly depends on the method used to generate the deposition flux (e.g., thermal evaporation, sputtering, and laser ablation).

During growth, the substrate temperature influences the surface diffusivity of the adsorbed species, as well as residence time and condensation processes. In growth mechanisms, the surface mobility, which plays an important role in determining the growth modes, can also be influenced by the energy of the deposition flux. Typically, the deposition kinetic energies are of the order of 0.5–1 eV for the evaporated species, hundreds of electron-volts for the sputtered species, and hyperthermal energies (10–100 eV) in the case of laser-ablated species [18]. In experiments of deposition from the vapor/gas phase, the actual deposition flux F is related to the supersaturation ratio SS by the relationship SS = $k_B T \ln(F/F_0)$, where k_B is the Boltzmann constant, T is the absolute growth temperature, and F_0 is the equilibrium value of F at the temperature T [10, 18]. Thermodynamic growth conditions (so-called close to equilibrium) are associated with small or moderate SS and/or a high T_{sub}.

1.4.1 Adsorption, Real Substrates, and Surface Elementary Processes

As depicted in Fig. 1.4a, given a flux F of precursor species (atoms or molecules, eventually of different materials) impinging on a substrate surface, such species can bounce off the surface or lose their component of velocity perpendicular to the incidence surface and are adsorbed (for a finite time or permanently) or desorbed depending on the substrate temperature T_{sub} and interactions with other surface features/structures (substrate crystallographic directions and morphological features). The density of the deposited species increases over time until surface coverage reaches the threshold value necessary to initiate nucleation processes. Growth may occur whenever the flux of species adsorbed on the deposition surface exceeds the flux of species desorbing from the surface. In a model in which the incident species can bond to the surface with a rate proportional to the incoming flux F, the diffusion rate along the substrate surface depends on the wetting interaction between deposited species and substrate as well as on the surface temperature. Loosely bounded adsorbed species are referred to as "adatoms" (adsorbed atoms): the adsorbed species forming only a few bonds with underlying sites have so weak a bond connection that they can migrate on the deposition surface. Immediately following deposition, adatoms lack any spatial correlations with each other but surface diffusion over the substrate surface and substrate surface peculiarities establish such correlations that start the growth dynamics. The situation in which adatoms are bonded to the surface via weak van der Waals forces and their binding energy is typically comparable to $k_B T$ is termed "physisorption." Also, the impinging species can lose their kinetic energy to a chemical reaction that strongly bonds (covalent/ionic bond) them to a surface atom (chemisorption). Chemisorption is almost always an exothermic process. Hence, adatoms can diffuse for a finite time (termed residence time) by exchanging energy before being desorbed by evaporation or stick permanently to the substrate surface. In the early stages of growth, while diffusing, adatoms can meet other adatoms to form dimers (a system of two adatoms), trimers (a bonded system of three adatoms), or more generally clusters (bounded groups of adatoms), as well as hop to the nearest-neighbor site, being trapped from substrate defect centers, and attach to or detach from existing clusters. Detachment of adatoms from small stable clusters is unlikely in the presence of high supersaturation [94].

Being adsorbates, adatoms can modify physical and chemical surface properties (i.e., the free energy of the system), as well as other aspects of the dynamics of the growing surface. The surface energy of a clean surface is constant in the absence of formation of defects or the adsorption process. Adatoms forming covalent, metallic, and ionic bonds with the substrate change the free energy of the system, which can be expressed as $\gamma^{(cry+ads)} = \gamma^{(cry)} + \gamma^{(ads)}$, where $\gamma^{(cry+ads)}$ is the Gibbs surface free energy postadsorption per unit area, $\gamma^{(cry)}$ is the surface energy of the clean crystal surface, and $\gamma^{(ads)}$ is the change in the Gibbs free energy per unit area of the system. The rate at which adatoms bond to surface sites and move on the surfaces dictates the nonequilibrium morphology and in turn depends on the energy surface landscape.

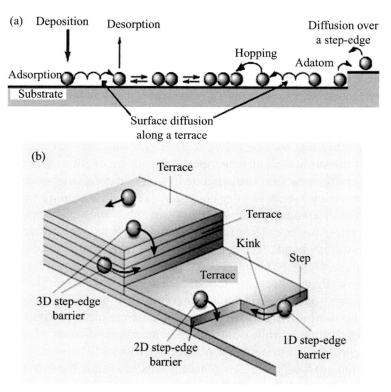

Figure 1.4 Schematics of the elementary surface processes occurring (a) after deposition from the vapor/gas phase on the substrate and (b) on a realistic model substrate exhibiting terraces, edges, steps, and kinks.

The description of the thermodynamics as well as energetics of formation and transformation of crystalline surfaces rely on the so-called terrace step kink (TSK) model (originally proposed by Kossel and Stranski), which considers the adatom energy as determined by its bonding to neighboring atomic sites and growth stages as involved by the interplay and balance between broken and formed bonds. Adatom kinetics and binding ability are strongly influenced first by the substrate's crystallographic directions, chemistry, temperature, and surface morphology and then, while deposition and growth progress, by the topography of the deposition surface.

In the ideal picture, the substrate surface is depicted as an ideally flat (perfect terrace) and chemically inert surface. In practice, real crystalline surfaces have edges, steps, dislocations, point defects, impurities (a site of a foreign material), vacancy sites, and kinks (single-atom-width displacements of the edge), in addition to terraces (Fig. 1.4b) [24, 95]. Therefore, a sketch of the adatom kinetics more realistically has to introduce composite surfaces containing nominally flat surfaces (terraces); defect sites; 1D, 2D, and 3D step-edges; and vicinal surfaces (a number of flat terraces separated by steps of atomic height) [21]. It is worth noticing that fortunately, crystal substrates are far from being perfect and inherently contain breaks in the crystal lattice symmetry as sources of steps [96]. In fact, in practice, inherent defects of crystals and "imperfections" (deviations from the ideally flat surface) of the deposition surface are beneficial in influencing both the ability of adatoms to form bonds and adsorption, diffusion, and formation of small clusters. Indeed, atomically flat surfaces have a lower rate of adatoms sticking than atomically rough surfaces. In the case of atomically flat surfaces, a landing adatom can bond to only one nearest neighbor (Fig. 1.5a) with a relatively weak bond and other adatoms may bond to it but its small binding energy results in a barrier to the growth. Therefore, an adatom diffusing on a terrace moves rather easily because it feels only the attractive forces from the atoms beneath it. In contrast, an adatom that diffuses to the edge of a terrace can form more bonds and the resulting bonding energy may be larger than the one resulting from the substrate sites. In fact, as depicted in Fig. 1.5b, since an adatom reaching the sites A and B of an atomically rough surface can form bonds with two and three nearest neighbors, the resulting binding energy is larger than the one of a single bond. As an adatom moving along an edge feels forces from different directions, leaving the edge for the terrace is more difficult because breaking more bonds is required. Leaving a kink site for an edge or a terrace is even more difficult because adatoms that attach to kinks can form more bonds with neighboring sites than the ones that attach to the terraces or to sites embedded in the terraces or to flat step-edges (Fig. 1.5c). Since a given coordination number may correspond to surface sites with different numbers of second neighbors, bonding energies also depend on the long or short range of interactions [97]. Additionally, anisotropy of bonds (some strong and some weak, depending on directions) may also play a role. For example, for a crystal possessing strong bonds parallel to a step-edge but weak bonds perpendicular to the edge, a low kink density is expected. The depicted mechanisms introduce the concept of preferential sites of adatom attachment played by breaks in the crystal lattice symmetry that generate edges and kinks. Hence, any adatom diffusing over a rough surface and encountering steps and kinks is expected to have a greater probability of sticking and incorporation than in the case of a smooth, flat adsorption surface. As a result, the growth rates can be tuned by either roughening steps or blocking kink sites as well as by the anisotropy of bonds.

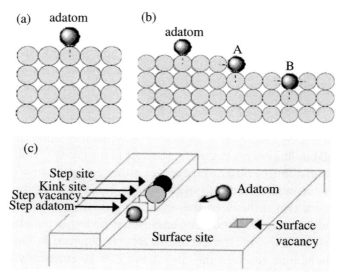

Figure 1.5 Sketch of the number of bonds with neighbors enabled for an adatom (a) on an atomically flat surface, (b) on a rough surface, and (c) due to the presence of preferred attachment sites on a realistic model substrate.

1.4.2 Characteristic Kinetic Coefficients, Energy Barriers, and Timescales

The elementary processes experimented by adatoms during the early stages of growth can be rationalized by introducing a few characteristic kinetic coefficients, timescales, and energy barriers (activation energies). In general, characteristic timescales (τ) are associated with thermally activated processes that can be described by an Arrhenius law containing a characteristic surface vibration frequency (v) related to the thermal oscillation frequency of the lattice, an activation energy barrier E associated to the process under description, and the thermal energy $k_B T$ at the absolute temperature T, that is, $\tau = (1/v)\exp(E/k_B T)$.

The main parameters of the kinetic picture of the growth by deposition are the accommodation coefficient, the sticking coefficient, the rate of physisorption, the mean residence time, the adatom diffusion length and coefficient, the hopping rate, and the step barrier energy.

The accommodation coefficient is indicative of energy exchange linked to the temperature and is defined as $\alpha = (T_{inc} - T_{des})/(T_{inc} - T_{sub})$, where T_{inc} is the temperature corresponding to the kinetic energy of the incident species, T_{des} is the temperature of the species eventually desorbing by evaporation from the substrate, and T_{sub} is the substrate temperature.

The condensation or sticking coefficient is the fraction of incident species which adhere and remain bound to the substrate surface. It is low whenever the energy of the adsorbed species is low (thermal source) and/or T_{sub} is high (relevant probability of loss by evaporation) or the deposited species bound loosely to the substrate and growing coating.

The probability of physisorption per unit time per unit area is χF, where the adsorption coefficient χ gives the percentage of atoms from the impinging deposition flux F that actually become adatoms.

The mean residence time for desorption by evaporation (meaning the mean adatom lifetime on the surface before re-evaporation) is given by $\tau_R = \tau_0 \exp(E_a/k_B T)$, where $\tau_0 = (1/v_0)$ is a characteristic time (lattice vibration time, $10^{11}-10^{13}$ s), E_a is the adsorption binding energy, and T is the absolute surface temperature. The probability of residence of an adsorbed unit is determined by the bond strength to its neighbors, which is also a function of T_{sub}.

The adatom diffusion length l_D is the average distance traveled by the adatom on a bare substrate before it desorbs or sticks. As the activation energy for desorption is usually larger than the activation energy for diffusion, an adatom can travel by diffusion over several lattice sites before its eventual desorption.

The adatom diffusion coefficient D is a kinetic parameter that determines the average distance an adatom can travel on a terrace before being trapped by other adatoms, nuclei, islands, impurity centers, kinks, and steps. It is related to l_D by the relationship $l_D = (D\tau_R)^{1/2}$, where τ_R is the residence time and is given by $D = D_o \exp(-E_{diff}/k_B T)$, where D_o is the diffusion pre-exponential factor (depending on the 2D nature of the surface diffusion and the vibration lattice frequency), E_{diff} is the energy barrier for surface diffusion, and T is the absolute temperature. Hence, surface diffusion is usually a thermally activated process. The activation energy of the adatoms diffusion on a terrace, that is, the energy required to transfer one adatom to an adjacent adsorption site, is just one contribution to E_{diff}. In fact, in general, E_{diff} is determined by a number of factors, including the strength of the bond between the surface atom and the adatom, the crystallographic plane of the surface, and the surface chemistry and morphology, as well as the deposition kinetic energy. Desorption of adatoms is contrasted if the surface binding energy of an adatom is much larger than E_{diff}. Therefore, the energy surface landscape also controls the adatom diffusivity apart from the deposition temperature. Due to the exponential law, small changes in the diffusion barriers can result in large changes in the diffusion rates.

In the presence of strongly localized adsorbates, adatoms oscillate near the points of minima of the surface potential and hop to the neighboring sites with an activation energy determined by the occupancy of their nearest neighbors as well as the physical and energy landscape. An adatom randomly occupies a site on the deposition surface, where it stays temporarily, and then it hops between sites by getting energy thermally and from adatom–adatom collisions until it reaches a lower-energy site. The condensation stage on a surface during growth can be described by the diffusion of the deposited material in terms of single-particle lattice hopping according to the solid-on-solid (SOS) model [98, 99].

The hopping kinetics is described by an Arrhenius process determined by the diffusion barrier that includes two terms, the diffusion barrier of a free adatom and the energy of each bond formed with a nearest neighbor weighted by the nearest-neighbor coordination of each particle along the surface. That is, the rate of an adatom hop to a nearest-neighbor site on a nominally flat terrace is expressed as $1/\tau_H = v_{0H} \exp[-(E_{diff} + nE_n)/k_B T]$, where $v_{0H} = 2k_B T/h$ (~10^{-11} to 10^{-13} s^{-1}) is a characteristic vibration frequency of an adatom site-to-site hopping, E_{diff} is the activation energy for the diffusion of a single adatom on a flat surface, n is the integer number of in-plane nearest neighbors, T is the absolute temperature of the substrate, and E_n is the energy of a bond formed between the hopping adatom and a nearest-neighbor vacant site labeled by the integer n.

If a pulsed flux of species with an average kinetic energy E_k arrives at the substrate surface, the hopping rate of an adatom diffusing along random directions is given by $1/\tau_H = v_{0H} \exp[-(E_{diff} + nE_n - E_k)/k_B T]$. Therefore, the probability of hopping of an adatom depends on both its kinetic energy and the number of neighbors. The activation barrier to overcome for a successful adatom hop typically amounts to 0.1–1 eV for most materials (in comparison $k_B T \approx 26$ meV at room temperature).

The above discussion indicates that kinetic control of growth is mainly influenced by the complexity of the physical and energy landscape. Consideration of the energy barriers related to the processes experienced from adatoms lets us define a potential energy landscape. In this respect, Fig. 1.6 shows a prototypical physical landscape seen by an adatom and the corresponding potential energy diagram with the activation energies of different processes: surface diffusion on a terrace, trapping to sites embedded in a terrace, descending a 1D step-edge, and diffusion to a step-edge. Moving over a terrace from one adsorption site to another requires energy amounting to the diffusion barrier E_{diff}. Sticking and surface incorporation of a diffusing adatom mean that a binding potential well forms, preventing trapped adatoms from further diffusion. Trapping can be irreversible (a trapped adatom cannot leave the trap) or reversible depending on the relationship between the thermal energy $k_B T$ and the trapping energy barrier E_t. The step-edge barrier encountered by an adatom descending a step is given by the energy barrier to descend the step to a lower terrace minus the diffusion energy barrier on a flat terrace. Its origin can be understood by first the stretching and breaking of the surface bonds of adatoms descending a step and then their reconstruction at the bottom of the step (Fig. 1.6). Since forming bonds is energetically less expensive than breaking them, a potential well forms at the bottom of a step-edge that traps adatoms and lowers the rates of interlayer diffusion at low temperatures. Hence, moving down a 1D step-edge introduces an energy barrier for diffusion over the step-edge, known as the Ehrlich–Schwoebel barrier [100–104]. Adatoms at the bottom of the step-edge being strongly bound to the step, this acts as a preferred adsorption site. Since the step-edge energy can also depend on the orientation (2D and 3D steps), multiple adsorption and nucleation pathways can be available. The Ehrlich–Schwoebel barrier is of the order of 0.1 eV in metals [105]. In regard to traps, three regions can be introduced: (i) a high-temperature region where adatoms are not trapped or become detached from them once trapped, (ii) a low-temperature region where the traps are occupied with a density much smaller than the nucleation density, and (iii) an intermediate-temperature region where the density of traps and nucleation are comparable and the system behavior depends on the relationship between E_t and E_{diff}.

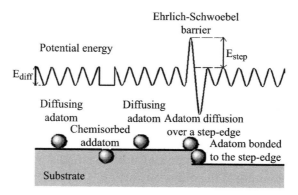

Figure 1.6 Schematic of a typical physical and energy surface landscape experienced by an adatom.

1.5 Condensation Processes and Kinetic-Driven Growth Modes

Two independent processes, that is, nucleation and growth of islands, play an important role during growth by deposition: nucleation leads to the formation of surface steps and subsequent growth causes the lateral movement of these steps and morphology evolution. Under the common experimental situation of growth from the gas/vapor phase, crystal growth is a nonequilibrium process governed by the competition between kinetics and thermodynamics, with the final morphology driven by the sequence of kinetically determined pathways. In this section the interplay between thermodynamic picture and kinetic control on the growth mechanisms is explored in general terms. As already outlined, basic stages in growth by deposition are deposition of species from a source, adatom diffusion processes, nucleation, and further condensation processes involving the transport of species and their incorporation in the growing structures. Conditions of enough high SS and growth temperature with negligible desorption of the deposition flux are necessary to start condensation.

1.5.1 Kinetic Control of Nucleation

The basic condensation process from which growth and crystallization proceed is nucleation, which is mainly controlled by three parameters: the adsorption energy (E_a) between the adatoms and the substrate or growing deposit, the surface diffusion energy (E_{diff}), and the binding energy between adatoms in the deposit material (E_b). Given an ideally flat substrate (terrace), if the desorption flux is negligible as compared to the deposition flux and surface diffusion is active to not allow the arriving species to stay where they arrive, various regimes can be identified. If $E_a > E_b$ then the adatoms condense to form a single layer completely covering the substrate. Instead, if $E_a < E_b$ adatoms can form nuclei under the condition of high-enough supersaturation (critical condensation flux). As a general rule, small supersaturation leads to a distribution of isolated large nuclei and a low surface adatom density implies a growth rate limited by nucleation events. Instead, high supersaturation involves shrinking of the nucleus critical size and increase of the nucleation density. Under this circumstance, a high surface adatom density favors condensation because systems with more nearest neighbors (as it occurs in forming clusters) have an energy binding higher than isolated adatoms.

As already pointed out, the diffusion kinetics is described by an Arrhenius law including a diffusion barrier that accounts for the diffusion barrier of a single adatom, the number of nearest neighbors, and the energy of each bond formed with a nearest neighbor. A change in the total diffusion barrier is expected while deposition continues, due to the formation of different atomic terminations or epitaxial misfit strain [106, 107]. A decrease of the energy barrier acts to increase the rate of surface diffusion, thus causing atoms to reach the edges of the substrate steps. Since diffusion toward these steps of the deposited atoms is favored under high-enough mobility conditions, vicinal substrates can work as a sink for diffusing adatoms and prevent nucleation on the terraces while favoring step-flow growth. Conversely, a decrease of the rate of surface diffusion, due to an enhancement of the energy barrier, enhances the nucleation probability on the terraces. Therefore, it may be energetically favored (due to a gain in the interface free energy) for the nuclei to form heterogeneously on preferred nucleation sites than homogeneously. Homogeneous nucleation occurs at random positions if a high-enough number of adatoms (species deposited from a homogeneous phase) meet through diffusion to form stable clusters homogeneously (i.e., without the presence of defective sites as low-energy nucleation sites). Heterogeneous nucleation takes place at special sites (preferential adsorption sites such as step-edges or kinks, impurities, and defects on the substrate surface) that the adatoms can reach by diffusion. In fact, when the mean free path of the landing adatoms is comparable to or larger than the mean distance between point or line defects, these ones may act as nucleation sites and influence the ability to migrate of the adatoms. Therefore, in practice, patterned substrates can be exploited to introduce a suitable heterogeneous distribution of sites of preferred nucleation for enhancing the nucleation density. In this situation, heterogeneous nucleation would dominate the nucleation process that may effectively occur even at high deposition temperatures or low coating fluxes.

The nucleation density depends on the supersaturation ratio SS (i.e., the deposition flux F): for an increasing density of adatoms the likelihood that two adatoms can meet increases. In addition, whenever adatoms can diffuse over longer distances (i.e., a large enough D), the probability of adatoms attaching to larger clusters instead of forming dimers increases. Quantitatively, the saturation density N_i of stable clusters with a size R_i is expressed by the scaling relationship $N_i = (F/D)\chi$, where $\chi = R_i/(R_i + 2)$ for 2D growth and $\chi = R_i/(R_i + 2.5)$ for 3D growth [8]. Therefore, the nucleation saturation density can be adjusted by increasing the flux or decreasing the diffusion constant (e.g., by lowering the substrate temperature). As will be discussed in detail, the above scaling relationship does not apply when the deposition flux is pulsed [108].

1.5.2 Intralayer and Interlayer Diffusion, Island Coalescence, and Growth Modes

Another important topic is the correlation between morphology evolution and diffusion ability. Notably, surface diffusion occurs at several stages of the growing process: the motion of the adatoms on terraces and along a step-edge in forming clusters and nuclei, the mobility of clusters themselves in forming nuclei, the rearrangement of nuclei and adatoms in forming larger nuclei, and growth of nuclei by adatom capture. Even if the deposition temperature is important for affecting the adatom diffusivity, two kinetic processes, determined by kinetic parameters, have to be considered to understand the possible growth modes on nonideal substrates. Such processes are the diffusion of adatoms on both singular and vicinal surfaces, which leads to the so-called intralayer mass transport (diffusion on a terrace) (Fig. 1.7a) and the interlayer mass transport (diffusion to a lower terrace by descending a step-edge) (Fig. 1.7b). Intralayer diffusion is inhibited when the substrate temperature is low enough to result in deposited species staying more or less where they impinge on the deposition surface. In this circumstance adatoms are trapped on top of nuclei, either due to a low deposition temperature or a high step-edge barrier and nucleation of a new atomic layer is favored while deposition continues. Nucleation of new layers on top of nuclei promotes surface roughness and its increase for increasing deposition duration. Instead, fast intralayer adatom diffusion means that the mobility of adatoms is high enough to enable them to reach step-edges while diffusing on a terrace. In fact, whenever diffusion is fast compared to deposition and the distance between surface steps is smaller than the average distance between nuclei, adatoms preferentially attach to the step-edges because nucleation on the terraces is prevented because of step-edges and kinks acting as traps for the diffusing adatoms. The propagation of steps during growth, that results in lateral growth of the terraces, is technically termed "step flow" growth (Fig. 1.7b) [109–111]. In principle, step-flow growth is also possible when interlayer diffusion is restricted because it can proceed through adatoms sticking to the ascending step.

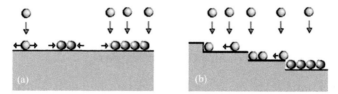

Figure 1.7 Schematics of (a) intralayer and (b) interlayer diffusion.

In contrast, a not-fast-enough intralayer adatom diffusion (or the average distance between nuclei smaller than the terrace width) favors nucleation on the terraces and nuclei can grow in their number and size until reaching a saturation nucleation. After that, it is more probable for adatoms to attach to an existing nucleus than to form a new nucleus. In this case, the interlayer diffusion associated with adatoms deposited on top of a growing nucleus that descend along the nucleus edge to a lower terrace plays a dominant role to favor the layer-by-layer growth mode.

In general, competition and interplay between intralayer and interlayer diffusion act to drive the morphological evolution. The nucleation and growth depend on the relative distance of the terrace length l_T and the surface diffusion length l_D (i.e., $l_D = (D\tau_R)^{1/2}$, where D is the surface diffusion coefficient and τ_R is the residence time): if $l_D > l_T$, nucleation on the terrace is prevented and step-flow-like growth results; if $l_D < l_T$, nucleation on the terrace occurs, resulting in 2D growth. Another important condensation step is the formation of islands, that is, growing nuclei (Fig. 1.8). In the island dynamic model, the islands that initially form may be unstable (i.e., they can disintegrate into smaller islands and/or nuclei) or grow further by attachment of adatoms (diffusive capture or direct impingement) as well as by coalescence of islands and/or nuclei. Exactly as in the case of adatoms, island growth is sensitive to the characteristics of the substrate that, in practice, is far from being perfect (ideally flat, smooth, and without impurities) and can promote nucleation through preferred nucleation sites. The effectiveness of the capture of adatoms by other adatoms and islands can be described by the so-called capture numbers [112]. In particular, a capture area of an island can be introduced, which is defined as the area surrounding an island containing adatoms that can on average diffuse toward the island and attach to it [113]. Because of spatial fluctuations during the formation of islands and the occurrence of nucleation events in regions between islands, islands of an equal size may have different capture areas. The adatom density around small islands is expected to be higher than around large islands because the dependence of the binding energy of surface atoms on the curvature through the island shape implies that the detachment of an adatom from an island is more probable the smaller and more faceted the island is. The presence of a gradient in the adatom density between neighbor islands of different sizes drives the adatom diffusion from the zone with a higher adatom density (around the smaller island) to the one with a lower adatom density (around the larger grain) (Fig. 1.9a). As a result, the gain of adatoms attached to large islands speeds both the growth of larger island at the expense of the smaller ones and the coalescence of smaller islands to form larger islands in a process known as Ostwald ripening (Fig. 1.9b). [114]. Coalescence proceeds through three main mechanisms: Ostwald ripening (adatoms are lost more readily from small islands than from large islands), sintering (a neck forms between two growing nuclei and the neck's curvature allows faster growth and merging by reduced surface energy), and migration of small clusters concurring to island growth. The driving force for coalescence is a reduction in surface energy by curvature-driven diffusion, causing the islands to become taller and more compact.

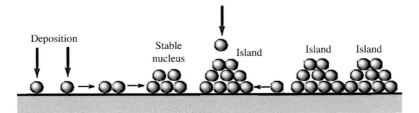

Figure 1.8 Formation of islands and their growth by coalescence.

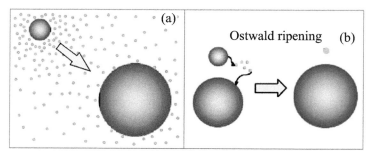

Figure 1.9 (a) Gradients of adatom diffusion around large and small islands. (b) Ostwald ripening.

To summarize, a diffusing adatom has two fates: It either meets another adatom, forming a stable nucleus (nucleation) that grows by adatom attachment leading to an island, or it meets an existing nucleus or island and sticks to it (growth). For subsequent condensation, characteristic length scales become the mean island distance and the mean free path of adatoms before they can form a new nucleus or attach to an existing island. Two regimes can be distinguished: the pure nucleation regime and the pure growth regime. In the former situation, the island mean size does not change for increasing coverage and additional deposition first results mainly in nucleation processes and then in the transition from nucleation to growth. In the case of the pure growth regime, a saturation island density is obtained with increasing coverage and adatoms with enough mobility are able to reach the closely spaced islands instead of concurring to another nucleation event. The competition between nucleation and growth is determined by the adatom diffusion coefficient: large adatom diffusivity means a probability for an adatom to meet another (deposited or diffusing) adatom lower than the probability for an adatom to find an existing island along its walk. Therefore, the larger the adatom diffusion coefficient, the lower the island density.

Island density and size primarily depend on the deposition temperature and deposition rate [115, 116]. On a quantitative level, mean-field nucleation theory can be used to derive the stable island density [100, 117].

The growing islands will eventually meet and once two islands are close enough to interact with each other, they will begin to merge to minimize the surface area and strain energy. Adatom surface diffusion continues to change the shape of the island edges until the new island reaches its equilibrium shape. Size and curvature differences between the coalescing islands will lead to growth of the large island at the expense of the smaller one [118]. Thus, coalescence between islands reduces the island density at small sizes and increases the density at large sizes.

For a given system of two or more coalescing islands, above a certain island size the time required for coalescence can be larger than the average time interval before an additional island impinges with one of the two coalescing islands. Hereafter, coalescing islands remain elongated so that the deposit consists of multiply-connected untouching islands. Such surface morphology is said to be "kinetically frozen" [89, 119] and is described within the framework of the "kinetic freezing" model as the result of a competition between island–island coalescence and island–island impingement when the deposition is in progress. Indeed, during deposition, the elongated clusters of islands join to form a tortuous network of bridged elongated islands separated by regions of uncovered substrate. This coalescence regime is termed "percolation transition" [119]. In the kinetic freezing model, while percolation drives the island growth toward impingement and filling of voids, coalescence tends to drive the system away from percolation. Evidence of the phenomenon of percolation is that the deposit as a whole becomes conductive (the islands are in contact) and exhibits magnetostatic interactions (the islands do not necessarily percolate from a geometrical point of view). At longer deposition times, once bare channels and holes become filled, the percolating deposit transforms into a continuous film.

Once islands start to grow, kinetic effects, such as the interlayer adatom diffusion, play an important role in affecting the growth mode because the diffusing adatoms can stick to the existing islands and the adatoms bounded to an island can also detach from the island edge or diffuse along the island edge. Adatoms deposited on top of islands and the corresponding kinetics have to be considered as well. If the species impinging onto coalescing islands and/or on the top of a growing island are able to descend from the islands, reach the island edges by diffusion, and diffuse to the lower layer at any time during growth (i.e., steady interlayer mass transport occurs), then a 2D layer-by-layer growth mode results (Fig. 1.3). In fact, a layer-by-layer growth results as long as the interlayer diffusion is fast enough to allow for the adatoms deposited on top of the islands to overcome the step-edge barrier before forming nuclei on top of the islands. In this situation large islands will grow and coalesce in such a way to cover the underlying layer before nucleation takes place on top of the islands and the cycle starts again for progressing deposition. Interlayer transport can be reduced by the Ehrlich–Schwoebel barrier, which the adatoms have to overcome while hopping down an island edge [101, 120]. In the case of kinetically limited interlayer mass transport, nucleation on top of islands is active before coalescence, hence leading to a 3D growth mode (Fig. 1.3). The built-up 3D islands can grow larger in all directions, including the direction normal to the surface, and eventually they impinge upon each other and begin to coalesce in such a way to lead to the development of a network of contiguous islands and later a continuous rough film.

As already discussed, while introducing the thermodynamic growth modes, lattice-mismatched epitaxial growth can lead to the SK mode, that is, initial layer-by-layer growth followed by 3D growth. It is worth noticing that different thermodynamically and kinetically defined critical wetting layer thicknesses can be introduced [121]. This finding clearly indicates that under nonequilibrium growth conditions kinetic control plays a dominant role in driving morphology evolution and determining the final growth mode.

1.5.3 Microstructure Evolution

The main growth steps by deposition are deposition of the precursor material, thermal accommodation, binding and surface diffusion, clustering and nucleation, formation and growth of islands, coalescence, percolation, and subsequent growth and structural as well as morphological evolution. Once a continuous coating is formed on a substrate, the growth progresses by evolution of the microstructure (amorphous, crystalline, or polycrystalline).

The term "microcrystalline" indicates a deposit consisting of quasi-crystalline ordered structure extending over regions no larger than a few tens of angstrom (Å), embedded in an amorphous disordered network. A polycrystalline deposit shows crystalline regions (termed "grains") an order of magnitude larger separated from each other by the so-called grain boundaries. The ability of adatoms to move along the grain surfaces and between the grains plays a role in driving the microstructural evolution. While in the case of epitaxial films all islands show the same orientation as the substrate, polycrystalline films consist of islands with different orientation relationships. Interface energy minimization can result in the growth of grains with preferred orientations or textures: the crystallographic orientation of the grains is more or less random in the early stages of growth unless the substrate acts as a crystallographic template. In polycrystalline structures a driving force for the development of a polycrystalline grain structure is the minimization of the energies of the free surfaces, interfaces, and grain boundaries. Still more complications are introduced due to strains (homogeneous versus inhomogeneous). During the growth, the structural evolution (i.e., the formation of a single-crystal, amorphous, or polycrystalline deposit) is determined not only by the interplay between mass transport across different atomic sites and layers and the nucleation of new islands but also by the deposition conditions such as the deposition temperature and the characteristics of the deposition flux.

If the substrate temperature is so low that the impinging species may not have sufficient energy for effective diffusivity or to overcome the energy barriers of the substrate energy landscape, then the deposited species may get stuck where they hit or be trapped while landing due to low diffusivity. An amorphous deposit (small, poorly defined crystalline areas dispersed within an amorphous background) results under this circumstance. At higher substrate temperatures, the enhanced surface mobility and diffusion distance of the adatoms promote more collision processes and more nucleation events may result in grain-structured growth by subsequent island growth and coalescence. In what concerns the deposition flux F, its intensity and nature (pulsed or continuous and energetic or hyperthermal) involve different scaling laws of the nucleation density and, as a consequence, different growth evolution, as will be discussed in detail.

Under conditions of continuous deposition flux, the systematic analysis of trends in the microstructural evolution of films in later stages of growth allow the development of the so-called structure zone models [97, 122–124], which explain and rationalize the morphology evolution in terms of the interplay between shadowing and adatom mobility controlled by the process parameters such as the energy of the depositing species and the normalized temperature T_{sub}/T_m (T_{sub} is the deposition/growth temperature and T_m is the melting temperature of the coating material). According to such a picture, materials deposited at temperatures above or below certain temperature thresholds in comparison to their melting temperatures display similar microstructures. Structure zone models introduce a classification into zones (briefly referred to as I, T, II, and III) characterized by different film morphologies depending on the value of the ratio T_{sub}/T_m. The structure zone diagram including the zones I, II, and III was built by raising the deposition temperature. For depositions with $T_{sub}/T_m < 0.3$ low surface diffusion causes the formation of small-diameter columns with poor crystallinity, a fibrous structure, voided boundaries, and a high dislocation density. For an increasing substrate temperature ($0.3 < T_{sub}/T_m < 0.5$) columns with tight grain boundaries, fewer defects, and faceted tops form due to improved surface diffusion. Depositions at $0.5 < T_{sub}/T_m < 0.8$ result in large columnar grains with dense grain boundaries, smoother surfaces, and few dislocations, leading to smoother films. Close to the melting point ($T_{sub}/T_m > 0.8$), the deposit consists of large equiaxed grains formed due to recrystallization and segregation of impurities to the surface and grain boundaries. A poorly defined structure, a fibrous column structure, voids, porosity, and defects are characteristic of the low-mobility zone I, which can develop in amorphous and crystalline deposits at very low deposition temperatures ($T_{sub}/T_m < 0.2$). Zone II morphology characterizes deposits having a fully dense columnar grain structure with long columns extending from the substrate to the film surface and dislocations preferentially located at the boundary zones. In zone III the adatom surface diffusion dominates the growth evolution. Columnar or equiaxed (in the presence of boundary strain) grains form and recrystallization leads to a bulk-like deposit. An additional zone exists between zones I and II called zone T ("T" stands for Thornton) arising from considering the dependence of the microstructure on the energy of the deposing species, too. Increased surface diffusivity (through higher deposition temperature and/ or deposition energy up to a few hundred electron-volts) can cause grains to grow slightly larger (smaller grains coalesce or are incorporated into larger, energetically more favored ones) and more tightly packed, with a smoother surface and denser boundaries. Zone T growth occurs at temperatures not high enough to induce any preferred crystallographic orientation during the nucleation and enabling surface diffusivity large enough for adatoms to cross the grain boundaries and land on grains of any orientation with grain-surface diffusivity.

As a general rule, growth of single crystals requires a single-crystal substrate being lattice matched with the epilayer material, a clean substrate surface to avoid secondary/heterogeneous nucleation, a high-enough growth temperature to favor surface mobility, and low deposition flux F to ensure enough time for surface diffusion and incorporation into the crystal structure of the deposition species before localized crowding of adatoms due to the continuous deposition flux. In general, growth and/or postgrowth annealing temperature affects the diffusivity of both the deposited adatoms that migrate through several atomic distances before sticking to a stable position, and the adatoms already incorporated that can readjust their incorporation sites within the newly formed deposit by diffusion along its surface features. An amorphous structure of the growing coating is expected under conditions of a low growth temperature (involving low surface mobility) and/or very high deposition F (meaning that the deposited species do not have enough time to find the growth-stable sites before their crowding occurs). A polycrystalline structure is an intermediate case that can be obtained when the growth temperature ensures a moderate surface mobility and the impinging flux is not very high. An amorphous structure and a polycrystalline structure can be observed in covalently bonded materials grown at low and high substrate temperatures, respectively.

Grain growth and texture development in a polycrystalline film is affected by the deposition temperature (high to form large grains), adatom diffusivity (high), annealing temperatures (high), deposition flux (low), impurity content (low), film thickness (high), energy of the deposited species (high), and melting point of the depositing material (low).

The structure zone model gives a representation of the cumulative effects of the kinetic factors that control change in the morphology, texture, and surface topography of deposits obtained by steady-state deposition. Such a background is a starting point to understand the powerful growth manipulations enabled by far-from-equilibrium deposition techniques.

1.6 Deposition Techniques in Nanoscience and Advantages of the Pulsed Laser Approaches

To tailor material functionalities and morphology at the nanoscale, well-controlled processing routes and far-from-equilibrium growth techniques are key strategies. The choice of the growth technique is critical because the final growth mode and functional properties are strictly dependent on the adopted fabrication technique.

In the framework of the bottom-up deposition approaches, two main classes are commonly applied, chemical vapor deposition (CVD) and physical vapor deposition (PVD) processes [13].

In CVD techniques volatile precursors are transported close to a substrate surface, where they react with streams of reactive gases to form species of the material to be deposited that are deposited and get rearranged on the substrate.

Although CVD approaches are suitable for depositing over large areas with a good degree of coverage, some disadvantages are the use of toxic chemical agents, contaminations of the deposit, use of pyrophoric (ignite on contact with air) metalorganics, complex designing of chemical and decomposition reactions, slow growth rate, high cost of precursor compounds with enough purity, and costly equipment.

PVD methods are deposition processes of films and nanostructures through physical mechanisms (such as thermal evaporation, sublimation, or ion bombardment) applied to solid precursors followed by condensation of the desired material and structure onto a substrate. The working principle of PVD is the production of a steady-state or pulsed flux of the precursor species transferred to a substrate on which the (thermally activated or driven by the deposition energy) migration of the arriving species to reach equilibrium sites occurs and structural development of the deposit depends on the balance between arrival rate and surface mobility [125].

PVD processes include high-vacuum (background pressures in the range of 10^{-10} to 10^{-4} Pa to minimize contamination effects) thermal and electron-beam evaporation, sputtering (i.e., the ejection of energetic species from a solid surface [the target] subjected to ion bombardment to form the deposition flux), molecular beam epitaxy (MBE) [126, 127], and pulsed laser deposition (PLD) that exploits laser ablation [18, 128–131].

The working principle of MBE is that the source elemental materials to be deposited are placed within Knudsen (evaporation) cells, heated to be evaporated and allowed to impinge on a heated substrate, where the deposit grows. Two main parameters can be changed during the MBE operation: the deposition rate and the substrate temperature. High substrate temperatures and low deposition rates promote the formation of high-quality single-crystal films. MBE requires sophisticated and expensive ultrahigh vacuum equipment (a necessary condition to avoid contaminations), operates at slow deposition rates (about 1 monolayer/s), and allows control on chemical composition at the monolayer level (a few angstroms) and growth of abrupt epitaxial interfaces.

The evaporation-based PVD methods, such as MBE, suffer from a too-low deposition rate and are problematic in the case of materials with high melting temperatures and when the precursor materials have different evaporation rates. In this respect, a vapor composition depending on the vapor pressures of elements in the target material leads to wrong stoichiometry of the growing deposit. Incongruent transfer is also a characteristic of the energetic flux sputtering approaches.

To deposit multicomponent inorganic materials, including alloys and mixtures, with preserved stoichiometry, PLD is a valid alternative [130, 132]. This laser-based technique has experienced a rapid spread since 1987, when it was revealed very in depositing thin films of high-critical-temperature superconductors [133], and is presently well established as the best-suited method for depositing a very wide range of materials [18, 128], including metals, oxides, ceramics, ferroelectrics, high-quality refractory metals, unstable alloys, high-T_c superconductors, and, importantly, manganite ferromagnetic phase compounds [134, 135], due to its unique ability to "transfer" the stoichiometry of complex multielemental targets on different kinds (flat, patterned, heat sensitive, etc.) of substrates. The success of PLD also relies on its letting one obtain improved structural quality, higher packing densities, and better adhesion as compared to thermal evaporation techniques, as well as growth of highly metastable phases and deposition on temperature-sensitive substrates such as polymers. Presently PLD is one of the most versatile techniques to deposit stoichiometrically high-quality thin films and multilayers, with a fine control on film thickness as well as composition and at much lower temperature than other film growth techniques [18, 136–138]. Implementation of PLD enables the deposition of magnetic ultrathin films and multilayers showing improved magnetic behavior as compared to the MBE-grown counterpart structures [139]. In addition, improved epitaxial growth characteristics and smoother surfaces are reported in PLD as compared to a well-refined thermal deposition technique such as MBE [140–142].

Different thermal expansion coefficients and the presence of a polar/nonpolar interface make heteroepitaxial growth challenging [143, 144]. Such drawbacks can be overcome by PLD, which lets one deposit highly ordered heterostructures at lower temperatures than other PVD approaches [145, 146]. This evidence indicates that the growth kinetics of a PLD plays a role different with respect to either other energetic approaches (e.g., sputtering) or thermal (evaporation) deposition techniques [146–149]. Also, PLD has demonstrated to be a versatile approach for fabricating nanotubes [150], nanopowders [151], as well as high-quality nanostructures and nanoparticles with a narrower distribution of size and reduced porosity [152–156].

A few fundamental peculiarities distinguish PLD and MBE, in addition to the larger space of experimental parameters of PLD and its being an inherently far-from-equilibrium deposition and growth approach. That is, while the deposition flux of MBE is continuous and thermal (kinetic energy less than 1 eV), PLDs occur in short pulses (typically of the order of 10–100 µs), with an instantaneous deposition flux of several orders of magnitude more intense than the steady-state one in MBE (typical growth rate of 1 monolayer/s) [119] and hyperthermal (kinetic energy from nearly 1 eV to a few hundreds of electron-volts) [18]. Growth control in situ by high-pressure reflection high-energy electron diffraction (RHEED) allows for atomic engineering of oxide materials and growth of heterostructures with atomically smooth interfaces and well-calibrated thicknesses.

In regard to the working principle of PLD, a short (typically from a nanosecond to a femtosecond) pulsed high-energy laser beam is focused inside a vacuum chamber on a very small area of a target material that is required to highly absorb the laser wavelength. Pulsed laser ablation, that is, highly localized matter removal due to high energy delivered in short times on a small area, is the basic mechanism exploited by PLD that forms a high-density plasma phase with high directionality along the normal of the target (plasma plume). The ablation thermal regime depends on the laser pulse duration as compared to relevant timescales, such as the electron–electron and electron–phonon thermalization times of the target material [128, 157]. Laser key processing parameters are wavelength, pulse duration, fluence (i.e., the laser energy delivered per unit surface area of the irradiated material), repetition rate (i.e., the time interval elapsing between subsequent pulses), and number of deposition pulses.

Laser-induced ablation can take place either in vacuum or in an inert or reactive background atmosphere (reactive PLD) or in liquid (laser ablation synthesis in liquid: liquid pulsed laser deposition [LPLD]). In general, control of the ambient gas/liquid parameters allows tuning of the stoichiometry of the deposit, the energy distribution of the plasma plume, and the spatial distribution, elongation, and chemical composition of the ablation plume.

In the case of LPLD in a solvent, which is widely applied to synthesize metal nanoparticles, the liquid buffer acts to effectively confine the plasma plume close to the irradiated target, a fact that favors nucleation during the plasma plume cooling followed by nuclei growth and coalescence [158–160]. The plasma plume confinement by liquids could drive the laser-induced plasma into the thermodynamic state that definitely differs from that of laser ablation in gas environments [160].

Another laser-based approach emerging as a very attractive alternative to deposit high-quality nanoparticle films even on thermally sensitive, patterned, or nonplanar substrates is the matrix-assisted pulsed laser evaporation (MAPLE) technique [161–164].

As a concluding remark, it is worth pointing out that common characteristics and distinctive peculiarities of the laser-based deposition approaches are the pulsed and energetic (kinetically hyperthermal species) nature of the deposition flux, as well as growth under nonequilibrium conditions. All of this involves growth regimes and metastable phases, which are not accessible from other PVD and CVD approaches, by tuning of the growth kinetics through laser parameters and deposition conditions.

In the following, growing opportunities and growth manipulations enabled by pulsed-laser-based deposition techniques will be illustrated by considering PLD as a prototypal example of far-from-equilibrium deposition techniques.

1.7 Growth Opportunities by the PLD Technique

In thermodynamics, a film growing on a newly exposed crystalline surface can evolve according to several distinct growth modes highly dependent on the free surface and interface energy, as well as the strain between the film material and the substrate. The morphology and growth evolution of films grown by PLD have been observed to be drastically different as compared to steady processing conditions [17, 147, 165, 166]. By comparison, several studies are available in the literature documenting the PLD growth processes and elucidating the implications on the growth progressing (structural and morphology evolution) of the characteristic parameters and working principle of PLD [17, 142, 166–168]. Understanding why the two growth approaches (steady and pulsed) involve such relevant differences requires a comparison and discussion of their working principles and processing parameters. In this paragraph, the distinctive properties, implications, and opportunities of growth by PLD will be discussed as a function of short interaction times and strong nonequilibrium conditions. In this perspective, the comparison between thermal and nonthermal and near-equilibrium (such as MBE) deposition conditions is an excellent model with which insight can be gained into the role played by the surface kinetics and deposition conditions that characterize PLD.

A PLD consists of four stages: (i) a nonequilibrium interaction between the laser radiation and the target material, (ii) a strongly forward-directed expansion of the ablation products (plasma plume) driving transport of the ablated precursor materials from the target to the substrate, (iii) an interaction of the energetic ablation products with the substrate following deposition, and (iv) nucleation events and growth kinetics on the substrate surface driven by the hyperthermal energy regime and pulsed nature of the deposition flux.

Parameters of interest to describe the dynamics of a PLD are the intensity I of each pulse (i.e., the number of particles deposited per unit area at each pulse) [169], the duration of the laser pulse τ_p, the duration of the deposition pulse τ_d, the average flux density of the depositing species F (i.e., average number of species per unit area per unit time), the incident average kinetic energy of the deposition flux E_k, the time interval elapsing between two subsequent pulses $\Delta t_p =$ 1/RR (where RR is the pulse repetition rate), the adatom diffusion constant D, and the supersaturation ratio SS = $(R_a[T_{sub}] - R_d[T_{sub}])/R_d(T_{sub}) \approx R_a(T_{sub})/R_d(T_{sub})$, where R_a is the actual deposition rate (adsorption rate) and R_d is the desorption rate at the absolute substrate temperature T_{sub}.

1.7.1 Distinctive Characteristics of the PLD Approach

PLD and thermal deposition may behave nearly in the same way (once statistical differences from finite-size effects are disregarded) only in the limit of the lowest intensity, that is, when only one unit (atom or molecule) per pulse is deposited [169]. Because of short deposition pulses and high deposition rates, for a given growth rate, the instantaneous peak mass arrival of PLD is remarkably more intense than the continuous one of the thermal evaporation techniques. Hence, although PLD and MBE can have similar average deposition rates, the pulsed short-time-lasting PLD induces a high instantaneous flux of species impinging on the substrate at each pulse. A threshold intensity exists, where the number of deposited species per pulse by PLD is of the same order of magnitude as the average density of deposited species in thermal deposition, above which the growth morphology involved by PLD begins to be remarkably different from the one of the thermal deposition [169].

A few fundamental peculiarities distinguish PLD and thermal deposition or MBE, in addition to the larger space of experimental parameters of PLD and its being an inherently far-from-equilibrium deposition approach. Operatively, PLD exploits a periodically modulated deposition flux generated by laser ablation: the short-time interaction of a pulsed laser beam with a target material causes the formation of an energetically hyperthermal, highly directional plasma plume in front of the target, which allows the ablated species to be deposited on a substrate where kinetics drive growth. While the deposition flux of MBE is continuous and (quasi-)thermal (kinetic energy less than 1 eV), PLDs are periodically modulated in short pulses (typically of the order of 10–100 μs), with an instantaneous deposition flux of several orders of magnitude more intense than the steady-state one in MBE (typical growth rate of 1 monolayer/s) [119] and hyperthermal (kinetic energy from nearly 1 eV to a few hundreds of electron-volts) [18]. In a PLD experiment, periodic deposition bursts with a high rate (>1000 monolayers/min) are separated by time intervals of no deposition, during which surface relaxation processes rearrange the depositing species. Since the timescale of the processes involved in the diffusion and condensation processes is longer than the deposition pulse duration, the deposition of the ablated material on the collecting surface can be assumed as instantaneous for every deposition pulse. The possibility to operate through intervals of no deposition in a fast, periodic sequence, with periodicity determined and tunable by the laser repetition rate, is a unique characteristic of PLD with important implications, as will be discussed.

PLD has several chief advantages and peculiarities with respect to other PVD techniques, such as stoichiometric (congruent) transfer of composite materials, hyperthermal energy regime of the deposition flux, a high degree of supersaturation due to fast periodic deposition rates (typically ~10 μm/s within one short [nanosecond] pulse at intermediate fluence), and unsteady (pulsed, periodic, and instantaneous) deposition flux with a tunable repetition rate.

The unique ability of PLD to achieve a congruent (stoichiometric) deposition of virtually any complex multielemental material is due to an extremely high heating rate of the irradiated target surface (10^8 K/s) leading to the congruent evaporation of the target irrespective of the evaporating point of the elements composing the ablation target.

In what concerns the energy regime, the ablation products may easily exceed thermal energy by more than 2 orders of magnitude. This fact is undesirable if sputtering, implantation effects, displacement or creation of vacancies or interstitials, and morphological degradation of the substrate and growing deposit can be induced [18]. Tuning the kinetic energy of the depositing flux is a valid strategy to induce enhanced low-temperature epitaxial growth [140] as well as to avoid surface sputtering-like reactions [18].

Fluence, target-to-substrate distance, and inert background atmosphere are key parameters to quench the kinetic energy distribution of the plasma constituents as well as to modulate confinement and elongation of the plasma plume. In fact, background inert atmosphere is routinely used to reduce the excess energy of the plasma plume from the nonthermal regime (above 10^3 eV) to the so-called hyperthermal regime (from nearly 1 to 10^3 eV), which is intermediate between the thermal regime (low energy) resulting from evaporation (deposition kinetic energy of 10^{-3}–1 eV) and the sputtering-regime (deposition kinetic energy larger than 10^3 eV) [18]. The additional energy delivered to the growing deposit by impingement of hyperthermal species influences the surface diffusivity by introducing a nonthermal (also termed "hyperthermal") diffusivity. As a result, the adatom surface mobility is driven by the deposition kinetic energy rather than thermal mobility, a fact that allows an effective superficial diffusivity regime even at low substrate temperatures and favors the overcoming of diffusivity barriers (interlayer and intralayer diffusion) [17, 120, 170]. In terms of beneficial effects on the morphology, the energetic (hyperthermal) character of the PLD flux can result in improved densification, enhanced film–substrate adhesion, growth of metastable phases, increased dopant incorporation and a higher degree of ordering, epitaxial growth at lower temperatures compared to other deposition methods, transition among growth modes, and reduced superficial roughness and smoothening at low substrate temperatures [17, 142, 171, 172]. It is worth noticing that a pulsed thermal deposition flux may induce increased roughness in contrast to an energetic one that promotes smoothening by energetic mechanisms [17, 119, 169]. The evidence of smoother surfaces by PLD as compared to MBE is also supported by theoretical studies [167, 169].

To deeply understand the implications of the PLD-induced superficial diffusivity regime, the role of the growth temperature has to be considered because the influences of substrate temperature and kinetic energy of the PLD flux on changing microstructure are different.

Since depositions performed on room-temperature substrates usually yield an amorphous deposit, postannealing treatments and/or high-temperature growth are needed to improve structural order. The main effect of the substrate temperature is influencing the adatom diffusion rate according to an Arrhenius-like law. However, growth temperature also affects the nucleation density. Indeed, since both supersaturation and diffusivity decrease with an increasing temperature, the nucleation density is not expected to be constant in a wide temperature range. A higher nucleation density at a lower temperature decreases the average travel length of the deposited material and the corresponding relaxation times will decrease. Further, at low a substrate temperature the long-range adatoms are hardly captured by the existing islands and, instead, an increasing density of small islands is favored.

Turning to PLD, if a pulsed flux of species with an average kinetic energy E_k impinges on the substrate surface at the temperature T_{sub}, an adatom diffusing along random directions has a hopping frequency given by the law $v_H = v_{0H} \exp[-(E_{diff} + nE_n)/k_B T_{sub}]$, where v_{0H} is a characteristic vibration frequency of an adatom site-to-site hopping; E_{diff} is the activation energy for diffusion for a single adatom on a flat surface (terrace); $n = 0, 1, 2, 3, 4 \ldots$ is the number of nearest neighbors around the adatom on the same terrace; and E_n is the interaction energy with n neighbors. Since the diffusion probability depends on the kinetic energy E_k, as a result enhanced diffusivity can be obtained by more energetic deposited species. The impinging hyperthermal flux of a PLD transfers kinetic energy to the substrate that is mostly converted into thermal energy (i.e., transferred to the lattice as phonon excitations), causing a localized transient increase of the temperature called a thermal spike [18, 173, 174]. Studies on the cooling time of such thermal spikes report a short thermalization period (i.e., time after which the energy corresponding to the substrate temperature is recovered) lasting from less than a picosecond at low kinetic energies (from 40 to 400 eV) to 5 ps [17, 173]. Therefore, the interaction of the hyperthermal incident species with the surface occurs on the much shorter (picosecond) timescale [175–177], which is several orders of magnitude lower than the characteristic times of thermally activated processes such as diffusion, dissociation, and evaporation. Modeling by molecular dynamics predicts an enhanced surface diffusion length of the order of 10 atomic spacings during the first few picoseconds after impinging of the deposition pulse [174], meaning that the deposition energy acts as an activation energy for long-range diffusion more effective than the growth temperature. As a consequence, even under conditions of high coverage, the diffusion induced by a hyperthermal deposition regime can let the adatoms find stable sites within fractions of seconds in between laser shots [148]. Notably, due to the PLD flux, the enhancement of mobility (i.e., the landing of energetic adatoms with energy excess on the growing surface) is transient rather than continuous. During their surface motion, the hyperthermal adatoms transfer their excess translational energy to the substrate by interactions depending on the potential energy surface of the growing surface. Once thermalization with the substrate is reached, the following stage is controlled by the thermal diffusivity, which continues until a next deposition pulse arrives. This picture points out that in PLDs the kinetic control and morphological evolution are governed not only by energetic characteristics of the deposition flux but also by the kinetics associated with the pulsed and instantaneous deposition flux. Notably, the mean diffusion time (given by an Arrhenius law, depending on the activation energy for diffusion) sets the timescale for the atomistic processes, including collision and nucleation, and must be compared with the pulse duration. As for a wide range of deposition conditions the mean diffusion time is longer than the deposition pulse duration and the deposition can be regarded as instantaneous for every pulse and followed by a relatively long time interval of interrupted deposition set by the pulse repetition rate. This peculiarity of a PLD implies that the two processes of deposition and rearrangement/diffusion in time of the hyperthermal species can be decoupled in time. Basically, the pulsed effect would let, during the time interval of no deposition, the adatoms have enough time to adequately migrate and diffuse on the deposition surface until they stick permanently, before the arrival of the next deposition pulse [140, 178]. The parameter that determines the time given to the landing species to rearrange on the substrate surface before the arrival of the next burst of species is the pulse repetition rate, RR. Its variation can play an important role in influencing the average growth speed, the crystalline quality, the island density, and the surface morphology of a growing deposit [179, 180]. For increasing (decreasing) pulse frequency per unit time, the time interval between subsequent laser shots is reduced (increased). In other words, the time interval between an arrival of a deposition burst on the substrate surface by laser shot and the arrival of another deposition burst by the next laser shot decreases (increases) as RR increases (decreases). Even if the amount of species deposited per pulse is the same, in the case of a high RR most adatom–adatom encounters can occur that make it easier to form islands of a smaller size and larger density as compared to the circumstance of a lower RR. High-density islands tend to grow quickly by capture of adatoms, and smaller islands favor the diffusion of the adatoms from their top to the substrate level (interlayer diffusion). Instead, under conditions of a low RR, reduced island density and a large average island size result as well as more time is available to islands to ripe. More compact/regular islands leading to a larger grain size can form because of the allowed detachment of adatoms and active edge diffusion during the interval of no deposition.

Therefore, while the substrate temperature tunes the adatom migration rate continuously during the growth, the hyperthermal energy regime of the PLD-depositing species improves transiently the adatom's diffusivity. This is a unique property of the PLD approach, having important implications in terms of growth evolution and changes of the surface microstructure by transient collisions, enhanced probability of nucleation, detachment of adatoms from islands, and overcoming of surface diffusion barriers [181, 182].

Another critical issue is the high supersaturation of the deposition flux. During a nanosecond PLD process, deposition occurs microseconds after the laser pulse with an instantaneous deposition rate of 10^{-2} to 10^{-1} nm/pulse [142, 183], which is orders of magnitude larger than in the case of thermal deposition (10^{-2} to 10^{-1} nm/s) [184]. As a consequence of a high density of energetic species deposited in short time periods, a very high supersaturation of the deposition intensity is achievable by PLD. This is a characteristic of the PLD technique that enhances the density of the nucleation sites. In the case of MBE growth at the same average deposition rate as a PLD, the continuous deposition flux causes much lower supersaturation on the timescale of the PLD. This circumstance involves a much reduced probability of nucleation events along short distances because each single adatom has enough time to wander a long path before encountering another adatom. A large distance between individual nucleation centers means a low nucleation density. On the other hand, in the case of the thermal deposition a high nucleation density is often associated with limited surface mobility and/or low substrate temperature, causing adatoms to stick and nucleate nearly where they impinge. In the early growth stages, the instantaneous (short interval) deposition rate, higher in PLD than in MBE, favors a higher total island density as well as promotes the formation of small rather than large islands and more densely spaced 2D islands [167, 185]. Under conditions of high-enough growth rates and temperature, change in the island shape can occur, making islands taller and more vertical than under equilibrium growth [167].

As a PLD continues, the energy transfer from the energetic depositing species that collide with a small island causes the species receiving such an energy to either detach from the island and diffuse on the substrate among islands or diffuse along the island edge without leaving it and concurring to the formation of a larger island. The probability of nucleation on the islands increases for increasing the island average diameter. When the size of large islands increases to a certain degree, these large islands begin to coalesce together and eventually cover the substrate surface. Then, again, the supersaturation induces a high density of nucleation sites on the first layer and small islands form that subsequently coalesce and lead to overall coverage of the underlying layer. If the first layer is not fully covered, the diffusing species can jump from the second layer to the first layer (interlayer jump) due to their hyperthermal diffusivity [120, 140, 170] and promote 2D growth. Therefore, as compared to having a low density of larger islands and limited adatom mobility (thermal deposition at low growth temperature and surface diffusivity), a large density of small islands and enhanced interlayer mass transport in PLDs allows the improvement of the structural order by favoring layer-by-layer growth and surface smoothening for increasing coverage.

Notably, depending on the growth temperature, fluence, and average deposition flux, the energetic character of PLD may induce islands taller and with more vertical edges that coalesce more easily than percolate as well as achieve percolation transition at a shorter deposition time than in a continuous deposition [167].

Quantitatively, a fundamental quantity in the description of condensation processes is the time-dependent nucleation density $n(t)$, that is, the number of nucleation events per unit area in the first layer integrated over time. It is expected to increase monotonically with time and saturate once the formation of the first monolayer is completed. In the case of a thermal deposition, the average island distance l_0 can be expressed in terms of the ratio D/F for intensities below a critical intensity I_c and by a power law $1/I^v$ with an exponent $v = 0.26$ for intensities I above I_c [136, 169]. The critical intensity I_c is introduced by the crossover from the situation dominated by nucleation events to the situation when it becomes more likely that an adatom attaches to an already existing island instead of forming a nucleus by adatom–adatom encounters. Several models have been proposed to calculate the scaling law of the nucleation density in thermal deposition and PLD [108, 186–188]. Distinctive characteristics result from the pulsed nature of the incoming flux. In PLD the island density is strongly dependent on the deposition time per pulse and laser frequency and only when D/F is very low (i.e., the nucleation indistinguishably occurs during and between laser pulses) PLD and MBE have similar nucleation densities [108, 186, 188–191]. Under investigation of the island statistics it was found that the scaling of the island size distribution in the coalescence regime of PLD differs from the MBE case for high deposition energies favoring multiple island coalescence [192]. While the time-dependent nucleation density exhibits standard power law scaling in the case of a thermal deposition, a fundamentally different behavior characterizes PLD given by a logarithmic scaling law of the form [169, 181, 186, 189]

$$\ln M(I, \Theta) = (\ln I)\, g(z)\, (\ln\Theta/\ln I), \qquad (1.10)$$

where I is the intensity of each impulse, $M(I, \Theta) = n(I, \Theta)/n(I, \Theta_{max})$ is the normalized nucleation density, $\Theta = Ft$ is the time-dependent coverage, Θ_{max} is the maximum coverage, and the scaling function g can be well approximated by $g(z) = cz^2$ [186]. More generally, the scaling function can be approximated by a power law $g(z) = C\, z^\beta$, where C is a constant and β ranges from 2.0 to 2.4 depending on the thickness. This scaling behavior was obtained in simulations of PLD assuming D/F tending to infinite (i.e., when all adatoms nucleate or attach to an existing island before the next pulse arrives) [186] and not in the case of finite D/F values [189].

Peaks in the island density versus time can be observed in PLD [181] and explained on the basis of the following model. In the case of a high nucleation density of small islands, with increasing island density the existing islands tend to grow in size by capture of more and more diffusing and incident species. For increasing coverage, the formation of new islands is not favored and coalescence dominates, thus decreasing the island density.

Another difference between PLD and thermal deposition is the influence of the Ehrlich–Schwoebel barriers on the interlayer transport. While an Ehrlich–Schwoebel barrier is responsible for reduced interlayer transport and growth instability in the case of thermal deposition [193–198] because of the smaller size of the islands at the early stages of nucleation, better results were observed in the case of PLD [142]. Indeed, while high deposition intensities are expected to increase the influence of Ehrlich–Schwoebel barriers, in PLD, even if many adatoms are deposited on each island due to supersaturation, the nucleation time of adatoms on top of islands is much smaller than their residence time because small islands with vertical edges favor adatom migration along the island edges and adatoms leaving the island. All of this reduces the influence of Ehrlich–Schwoebel barriers in PLD as compared to thermal deposition [169].

1.7.2 Growth Manipulations by PLD

The success of PLD also relies on the possibility of it manipulating growth by flexible tuning of the deposition conditions. For example, PLD drives growth toward modes that would not be allowed by thermodynamics [142, 185, 199, 200], such as metastable phases and novel microstructures [166], as well as to obtain growth mode change [142, 201, 202] and delay of epitaxial breakdown (no longer sustained epitaxial growth with formation of amorphous phase) [119, 203].

For example, in the homoepitaxial growth of $SrTiO_3$, tuning of the deposition temperature and RR enables the introduction of gradual transition of the growth mode from 2D layer-by-layer to step-flow at low RR (5 Hz) and for increasing growth temperature [201, 202].

On the other hand, while deposition of Fe on Cu(111) single crystals leads to an initial bilayer nucleation and growth by thermal deposition, 2D nucleation and growth mode result from the beginning of the PLD process [142]. Inspection of the progressing growth shows deposits with completely different morphological features. In the case of thermal evaporation, one can observe one monolayer with high decoration of the Cu terrace edges, Fe islands with a height of two or three monolayers on the terraces, and no one-monolayer-high island in the initial stages of growth, as well as, with increasing degree of coverage, continuous decoration of the step-edges and larger islands together with increased roughness. Turning to the PLD-deposited sample, complete absence of decoration and random distribution on the terraces of small (1 nm in diameter) and large (10 nm in diameter) one-monolayer-high islands are observable. Statistics of islands would indicate absence of saturation nucleation in the PLD, where the formation of small islands proceeds for increasing coverage. In contrast, in the thermal deposition, a trend toward growth of the existing islands dominates, resulting in a relatively low island density. Very interestingly, the surface kinetics are completely different in the two cases under examination. In the deposition from thermal evaporation, nucleation and growth of the islands are induced by collisions of the depositing Fe atoms that land on the Cu substrate, being sensitive to the presence of terraces and step-edges. Instead, the improved layer-by-layer growth by PLD can be ascribed to the higher deposition rate as compared to thermal evaporation that favors a much higher probability for nucleation due to closely spaced energetic impinging species as well as enhanced transient surface mobility. In addition, the improved diffusivity also minimizes the probability of nucleation on top of islands because the as-deposited adatoms can migrate to and along the island step-edges (interlayer mass transport) and favor nucleation on layers. Nearly ideal layer-by-layer growth by PLD is also demonstrated at low deposition temperatures (around 180 K) and for other combinations of materials (Co/Cu[111] and Co/Cu[100]).

To favor the formation of a metastable phase over the equilibrium phase (epitaxial stabilization) by shifting the energetics of phase stability, lattice, misfit strain energies, and interfacial energies, which play a role in epitaxial growth, can be used [18, 166].

The far-from-equilibrium conditions of growth by PLD offer the possibility to grow structures with highly metastable phases, such as nanocrystalline highly supersaturated solid solutions or amorphous films over a wide composition range, that are hard to stabilize or impossible to obtain by other deposition techniques. This issue is one of the peculiarities making PLD the technique of choice to grow ferromagnetic metallic (single-element and alloy) systems and layered metastable structures with tunable composition. Importantly, the improved morphology and structure enabled by PLD implies improved ferromagnetic properties (higher magnetic stability and larger magnetic moment) than the MBE-grown counterpart. For example, while Ag_xNi_{1-x} in an amorphous phase or highly disordered was obtained by any other techniques due to demixing, polycrystalline Ag_xNi_{1-x} can be deposited by PLD [166]. Another famous metastable phase is a face-centered cubic (fcc) γ-Fe, which does not exist at temperatures below 914°C in natural bulk form and can be stabilized at room temperature by quenching of the γ-Fe precipitates in a Cu matrix or by epitaxial growth on a lattice-matched substrate, such as Cu [166]. Growth of fcc Fe films by MBE leads to the formation of body-centered cubic (bcc)-like stripes and needles that adversely affect the magnetic properties of Fe films on both Cu(100) and Cu(111) substrates [166]. In contrast, PLD-deposited Fe/Cu(100) films show a much more fcc-like structure at low thicknesses and different morphology, with near-ideal 2D growth in the submonolayer regime [166]. The layer-by-layer growth on an fcc(111) substrate is complicated by the potential formation of structural defects (fcc/hexagonal close-packed [hcp] stacking faults or fcc twin boundaries and the usual step-edge barrier) that, instead, can be effectively removed when depositing Fe films on a Cu(111) substrate by using PLD [166]. Another example of the metastable phase successfully grown by PLD is the Fe-Ag system [204, 205].

Another circumstance where PLD works better than MBE is the growth of Co/Cu(111) films. In bulk, the room-temperature stable hcp Co phase undergoes an hcp → fcc Martensitic phase transition at about 415°C that makes easier the stabilization of the fcc Co phase when an fcc substrate (e.g., Cu) is chosen for growing Co at a low temperature. Growth by MBE of fcc Co on a Cu(100) substrate develops as a layer-by-layer mode up to at least 100 monolayers and in the case of a Cu(111) substrate fcc/hcp stacking faults form that speed the fcc → hcp phase transition. In contrast, thin Co/Cu(111) films grown by PLD are stacking-fault-free, grow layer by layer, and do not exhibit islands either before or after the fcc → hcp transition that occurs at higher thicknesses as compared to MBE [166].

Artificial ordered structures consisting of a binary alloy fabricated by stacking ultrathin films of two different kinds of metallic materials (termed superlattices) are another class of metastable materials that can be successfully grown by PLD. In some cases, such as when monolayer fcc-like Fe is one of the constituents, PLD would be the method of choice to grow high-quality superlattices because fcc-like Fe monolayers cannot be grown by MBE on both Cu(100) and Cu(111) substrates [166].

Another applicative field where PLD is the method of choice is the growth of multilayers involving refractory elements that hardly vaporize. Using PLD under well-designed experimental conditions (low fluence in the nanosecond regime or femtosecond regime to avoid particulates) can be also advantageous for fabricating magnetic multilayers with control on composition and alloying because of the possibility of tuning both deposition rate and thickness by controlling the laser repetition rate and the number of laser pulses, respectively [166].

In PLDs, coalescence processes and morphological transitions can be manipulated by exploiting independent control on the amount of deposited material per pulse and pulsing frequency, which must be compared to other timescales related to deposition and diffusion processes.

Pulsing of the deposition flux and changes of its frequency are a valid strategy to manipulate island nucleation and growth [108, 188]. As already mentioned, the saturation density of stable clusters does not have the same dependence on the flux F and the diffusivity D as in the case of a continuous deposition flux. Instead it depends on the pulsing frequency, width of the deposition pulsed flux, and average adatom lifetime ([108] and references therein).

A pulsed flux can be characterized by the duration of the deposition pulse τ_d, the RR, and the instantaneous deposition rate F_i. For typical PLDs τ_d is shorter than the time interval between subsequent pulses Δt_d = 1/RR. The total amount of material deposited per pulse is then $FP = F_i\, \tau_d$, and the average deposition rate is $F_{aver} = F_P$ RR. The timescales τ_d and RR can be compared to the adatom lifetime τ_{ad}, which is the time for which the adatom density persists on the substrate once the pulse is turned off. In the absence of desorption $\tau_{ad} \approx 1/(DN_{isl})$ where N_{isl} is the island concentration and D is the diffusion constant of the adatoms.

A deposition flux impinging on the substrate causes an increase of the adatom density, which saturates when equilibrium is established between the number of species lost by nucleation events and the newly deposited ones. In a pulsed deposition, such a stable adatom density can be reached and retained during the deposition pulse and then it decays until the arrival of the next deposition pulse. If the pulsing frequency is low enough for the time between pulses to be larger than the adatom lifetime, during the deposition pulse a nucleation density proportional to the steady state adatom density can be reached. In this circumstance, the island saturation concentration can be approximated by the same scaling law that holds for a continuous flux. Quantitatively, if $\tau_{ad} \ll \tau_d$ the adatom density vanishes as soon as the pulse is turned off and nucleation occurs during the pulsed deposition, with an island saturation density proportional to $(F_i/D)\chi$, where $\chi = R_i/(R_i + 2)$ and $\chi = R_i/(R_i + 2.5)$ for 2D and 3D growth, respectively, of a stable island having a size R_i.

If $\tau_d \ll \tau_{ad} < \Delta t_d$, the adatoms contribute to the nucleation either during the pulsing on or during the time of no deposition between pulses, leading to an island saturation density independent of D and instead depending on the deposition flux by the scaling law $F_P(\chi/1 + \chi)$ [108]. In the regime of high RR, the adatom density does not decay to zero between subsequent pulses. Since the adatom equilibration occurs during many deposition cycles, the adatom density builds up and reaches its steady state proportional to the average deposition after several deposition cycles. [108]. Quantitatively, if $\tau_d \ll \Delta t_d \ll \tau_{ad}$ nucleation may essentially occur at any time during deposition and the island nucleation density scales according to the same law as continuous deposition, with deposition flux given by the average deposition flux F_{aver}.

Therefore, chopping of the incident flux is a practical strategy to remarkably influence the formation of an island, whose density can change as a function of the frequency of the periodically modulated flux.

Another aspect to be considered is that a high nucleation density in PLD does not automatically involve improved layer-by-layer growth because an efficient interlayer mass transport during the growth is needed as well. Under conditions minimizing the probability of nucleation on top of a 2D island, the adatoms can migrate to the step-edges of the 2D islands and nucleation only takes place on fully completed unit-cell layers, meaning that efficient interlayer mass transport drives a layer-by-layer growth manner. In the absence of nucleation on top of the 2D island, during the time interval of no deposition after the deposition pulse an exponential decrease in the density of diffusing particles is expected that depends on the coverage. In the case of the diffusing species deposited between the islands, their density decays as a function of diffusivity and nucleation density. Substrate coverage affects the maximum travel distance of the diffusing adatoms that decreases for increasing coverage. Although the presence of a large number of small islands may enhance the interlayer diffusion for an adatom on top of an island, the PLD supersaturation also implies an increased probability of nucleation on top of an island. Indeed, as more species arrive at the surface simultaneously for increasing the intensity of the deposition flux, more of the deposited species can meet and form new nuclei. Inspection of the competition between such phenomena evidenced that a surface rougher in PLD than in MBE and the opposite trend would be expected for D/F tending to infinite [169] and at finite D/F with a very large step-edge barrier (Ehrlich–Schwoebel barrier), respectively. All of this would suggest that the observed layer-by-layer growth in PLD [142] may be involved by the interplay between a high nucleation density and some other PLD parameter. This leads to the other important feature of the PLD growth, that is, the hyperthermal character of the deposition flux and the possibility to change the laser repetition rate. In fact, control on the surface roughness in PLDs can be explained by a balance between roughening associated with pulsed deposition and smoothening associated with energetic deposition [17].

Modulation of the morphological evolution of metal-on-insulator films grown by PLD can be demonstrated experimentally by changing the laser RR: deposition at higher RR causes the film transition to the percolation regime at a lower transition thickness, and a smaller in-plane length scale of the film features for increasing

RR [119, 185]. On the other hand, to control the roughening at a low growth temperature due to increased probability of nucleation on top of 2D islands, the use of periodically interrupted growth favoring smoother surfaces can be applied [206].

A key characteristic of PLD in enabling enhancement of epitaxial growth (smoothening and epitaxial growth to a greater thickness without epitaxial breakdown) with respect to MBE is the hyperthermal regime of the plasma plume impinging on the substrate [119, 203]. A comparative study about the morphology progressing of homoepitaxial Ge(001) films grown under identical conditions by a PLD with a peak kinetic energy of 300 eV, a thermal PLD (i.e., a PLD with a peak kinetic energy of 0.01 eV), and a MBE deposition indicates a similar morphology evolution with, however, more irregularly shaped islands and the presence of large surface features as well as higher roughness and epitaxial breakdown at lower thicknesses in the case of the thermal PLD (27 nm for thermal PLD versus nearly 80 nm of MBE and more than 270 nm for kinetic PLD). The improved roughness and delayed epitaxial breakdown enabled by energetic PLD as compared to MBE demonstrate that the surface energetic mechanisms (such as filling of gaps between the deposit mounds, momentum transfer to surface sites, and transient enhanced diffusivity) involved by the energetic character of the PLD flux are decisive in determining the threshold epitaxial thickness.

Growth manipulation by enhancement of the interlayer mass transport is another important application of the unique features of the PLD approach to get a layer-by-layer growth even in conditions that would favor island formation. Two different temperatures, two different growth rates, and periodical ion bombardment are strategies applied to increase the number of nucleation sites and decrease the average island size by enhanced transport of material from an island to a lower level [207, 208]. However, epitaxy of complex oxides imposes a limited range of growth temperatures and pressures because of the phase stability issue and the difficulty to accomplish ion bombardment periodically. Instead, both instantaneous and average growth rate of the PLD flux can be used to impose layer-by-layer growth based on the so-called pulsed laser interval, which aims at controlling the island sizes. The instantaneous deposition rate depends on the laser energy density delivered at the target, pulse energy, target–substrate distance, and ambient gas properties (i.e., pressure and mass). The average growth rate can be changed by the pulse repetition rate independently from the instantaneous deposition rate.

As already discussed, a consequence of the PLD supersaturation is a high density of small 2D nuclei just after the deposition pulse. Since subsequently larger islands are formed through kinetic processes in between deposition pulses, to decrease the probability of second-layer nucleation on the growing islands coarsening can be avoided by maintaining the high supersaturation for a longer time, that is, by setting high repetition rates. A periodic sequence of a short-interval fast deposition followed by a long interval in which no deposition takes place and the deposit can reorganize by recrystallization makes it possible for a unit-cell layer-by-layer growth under temperature and pressure conditions that would lead to dominance of island formation. More in detail, the periodic pulsing of the deposition flux is tuned in such a way that fast deposition of the amount of material needed to form a monolayer is performed by a periodic sequence of a number of cycles of high RR deposition followed by a no-deposition interval (longer than the characteristic relaxation times), enabling effective film reorganization [140, 209]. The underlying principle is that deposition of every unit-cell layer at a very high deposition rate followed by a relaxation interval enables one to exploit the high supersaturation character of the PLD flux to keep the average island size small enough to control the probability of nucleation on the islands, retain enhanced interlayer mass transport, and obtain nucleation on the next level when a cycle of deposition starts. Deposition delivered in time intervals of the order of the characteristic relaxation times (tunable by coverage and growth temperature) allows nucleation to occur after deposition and prevents a multilevel growth by enhanced nucleation of small islands. Therefore, under well-calibrated conditions of interval deposition that optimize mobility and material rearrangement (island growth through recrystallization when the deposition plume is off), it is possible to impose a single-level 2D growth or layer-by-layer growth mode instead of the multilevel growth expected in the case of thermal deposition. All of this is particularly useful in the case of complex oxides with phase stability strongly depending on the oxygen content and temperature (such as manganites). In fact, for these materials, once oxygen and growth temperature are fixed, manipulation of the growth acting to reduce the island size is the only way to drive layer-by-layer growth.

A PLD-based mechanism for promoting 2D growth was demonstrated in the case of heteroepitaxy of a complex metal-oxide film such as $La_{1-x}Sr_xMnO_3$ (LSMO) grown on $SrTiO_3$ (STO) [148]. A so-called interrupted PLD is adopted, in which short bursts of ablation consisting of the minimum number of laser shots needed to retain the composition x constant are alternated with intervals of no deposition lasting a few tens of seconds (i.e., the timescale of thermal relaxation processes). In this circumstance, the thermal relaxation occurs for times much longer than the ones of conventional PLD (fixed RR) and of the time needed to grow a single monolayer by conventional PLD. Film smoothening is explained by energetic mechanisms affecting long-term 2D growth by PLD. Below 50% coverage, the PLD supersaturation favors nucleation of small and densely spaced 2D islands that can split into daughter islands due to the hyperthermal impinging species that land on their top and promptly get inserted into them. This mechanism of breaking up of small islands induces smoothening of the growing deposit, inhibits the formation of larger-size islands as well as future evolution to 3D growth, and increases the coverage. Once the separation between small islands becomes comparable to their average size, island coalescence onsets. Then, while deposition continues, the impinging species must diffuse to descending edge steps and these processes are speeded up by the coupling between the energy of the impinging species and the diffusing adatoms. A consequence is a transiently enhanced surface temperature and surface diffusion, where surface diffusion is improved not only because of the hyperthermal character of the PLD pulse but also because of the kinetic energy transferred to the adatoms still diffusing to find a stable site when the deposition pulse arrives. Therefore, smoothening is linked to the improved degree of ordering that results from enhanced diffusivity and increased probability that the wandering adatoms form an ordered closed packed layer of minimum surface energy.

The PLD flux can affect the nucleation process not only by its energetic regime and high supersaturation but also by tuning of the laser pulse width [157, 210, 211], which can be an efficient strategy to produce size-controlled nanoparticles and nanoparticle films. Depending on the laser pulse width (from nanosecond to femtosecond timescales), nanoparticle formation by condensation and growth can occur essentially during the quasi-adiabatic plume expansion into a vacuum or liquid or low-pressure gas (inert or reactive) atmosphere under a nanosecond ablation regime or directly within the ablating target due to rapid expansion and cooling via different mechanisms of solid density matter irradiated by subpicosecond laser pulses [212–216]. Implementation of PLD exploiting femtosecond laser pulses enables better morphology and reduced extension of heat-affected zones as compared to nanosecond pulses. [211, 216–219].

1.8 Conclusions

Most PVD approaches exploit a continuous (steady) low-intensity and thermalized deposition stream generated by thermal evaporation of precursor material(s) or an energetic ion deposition beam obtained by sputtering. Instead, laser-based techniques are an increasingly used growth method that exploit laser ablation to deliver periodically modulated and hyperthermal deposition flux. As thoroughly discussed, kinetic control of growth is very different in this case as compared to the conventional continuous deposition approaches and drives growth evolution far from equilibrium conditions.

Since its discovery, a huge amount of literature and studies have been conducted about PLD and its variants for soft materials (MAPLE). However, pulsed laser approaches still offer fascinating experimental and theoretical perspectives as well as challenging opportunities in the field of nanofabrication and nanostructuring.

References

1. Burda, C., Chen, X., Narayanan, R., and El-Sayed, M. A. (2005). Chemistry and properties of nanocrystals of different shapes, *Chem. Rev.*, **105**, pp. 1025–1102.

2. Pichumani, M., Bagheri, P., Poduska, K. M., Gonzales-Vinas, W., and Yethiraj, A. (2013). Dynamics, crystallization and structures in colloid spin coating, *Soft Matter*, **9**, pp. 3220–3229.

3. Adachi, M., and Lockwood D. J. (2006). *Self-Organized Nanoscale Materials*, ed. Lockwood, D. J. (Springer, New York, USA).

4. Sakamoto, M., Fujistuka, M., and Majima, T. (2009). Light as a construction tool of metal nanoparticles: synthesis and mechanism, *J. Photochem. Photobiol. C*, **10**, pp. 33–56.

5. Biswasa, A., Bayerb, I. S., Birisc, A. S., Wanga, T., Dervishic, E., and Faupeld, F. (2012). Advances in top–down and bottom–up surface nanofabrication: techniques, applications & future prospects, *Adv. Colloid Interface Sci.*, **170**, pp. 2–27.

6. Brune, H., Giovannini, M., Bromann K., and Kern, K. (1998). Self-organized growth of nanostructure arrays on strain-relief patterns, *Nature*, **394**, pp. 451–453.

7. Hurle, D. T. J. (1993). *Fundamentals: Thermodynamics and Kinetics*, ed. Hurle, D. T. J. (North-Holland, Amsterdam).

8. Michely T., and Krug J. (2003). *Islands, Mounds, and Atoms: Patterns and Processes in Crystal Growth Far from Equilibrium* (Springer, USA).

9. Bauer, E. (1958). Phaenomenologische theorie der kristallabscheidung an oberflaechen I, *Z. Kristallogr.*, **110**, pp. 372–394.

10. Markov, I. V. (1995). *Crystal Growth for Beginners: Fundamentals of Nucleation, Crystal Growth and Epitaxy*, 1st ed. (World Scientific, Singapore).

11. Tromp, R. M., and Hannon, J. B. (2002). Thermodynamics of nucleation and growth, *Surf. Rev. Lett.*, **9**, pp. 1565–1593.

12. Burton, W. K., Cabrera, N., and Frank, F. C. (1951). The growth of crystals and the equilibrium structure of their surfaces, *Philos. Trans. R. Soc. London, Ser. A*, **243**, pp. 299–358.

13. Ohring, M. (2002). *Materials Science of Thin Films*, 2nd ed. (Academic Press, San Diego, CA).

14. Smith, D. L. (1995). *Thin-Film Deposition Principles & Practice* (McGraw Hill, Boston).

15. Caflisch, R. E., Weinan, E., Gyure, M., Merriman, B., and Ratsch, C. (1999). Kinetic model for a step edge in epitaxial growth, *Phys. Rev. E*, **59**, pp. 6879–6887.

16. Hubler, G. K., and Sprague, J. A. (1996). Energetic particles in PVD technology: particle-surface interaction processes and energy-particle relationships in thin film deposition, *Surf. Coat. Technol.*, **81**, pp. 29–35.

17. Taylor, M. E., and Atwater, H. A. (1998). Monte Carlo simulations of epitaxial growth: comparison of pulsed laser deposition and molecular beam epitaxy, *Appl. Surf. Sci.*, **127–129**, pp. 159–163.

18. Eason, R. (2007). *Pulsed Laser Deposition of Thin Films Applications-Led Growth of Functional Mate*rials (Wiley-Interscience-John Wiley & Sons, Hoboken).
19. Perrot, P. (1998). *A to Z of Thermodynamics* (Oxford University Press, Oxford).
20. Gibbs, J. W. (1873). A method of geometrical representation of the thermodynamic properties of substances by means of surfaces, *Trans. Conn. Acad. Sci.*, **2**, pp. 382–404.
21. Reiss, H. (1996). *Methods of Thermodynamics* (Dover Publications, New York).
22. Kondepudi, D., and Prigogine, I. (1998). *Modern Thermodynamics: From Heat Engines to Dissipative Structures* (John Wiley & Sons, Chichester, West Sussex).
23. Gibbs, J. W. (1928). The equilibrium of heterogeneous substances, *The Collected Works of Gibbs, J. W.* (Longmans, Grenn, New York; reprinted by Dover, New York, 1961).
24. Chernov, A. A. (1961). The spiral growth of crystals, *Sov. Phys. Uspekhi*, **4**, pp. 116–148.
25. Roosen, A. R., McCormack, R. P., and Carter, W. C. (1998). Wulffman: a tool for the calculation and display of crystal shapes, *Comp. Mater. Sci.*, **11**, pp. 16–26.
26. Mackenzie, J. K., Moore, A. J. W., and Nicholas, J. F. (1962). Bonds broken at atomically flat crystal surfaces-I face-centered and body-centered cubic crystals, *J. Phys. Chem. Solids*, **23**, pp. 185–196.
27. Xia, Y., Xia, X., and Peng, H.-C. (2015). Shape-controlled synthesis of colloidal metal nanocrystals: thermodynamic versus kinetic products, *J. Am. Chem. Soc.*, **137**, pp. 7947–7966.
28. Wulff, G. (1901). Zur Frage der Geschwindigkeit des Wachstums und der Auflösung der Kristallflächen, *Z. Kristallogr.*, **34**, pp. 449–530.
29. Herring, C. (1951). Some theorems on the free energies of crystal surfaces, *Phys. Rev.*, **82**, pp. 87–93.
30. Henry, C. R. (2005). Morphology of supported nanoparticles, *Prog. Surf. Sci.*, **80**, pp. 92–116.
31. Mullin, J. W. (2001). *Crystallization*, 4th ed. (Butterworth-Heinemann, Oxford).
32. Gutzow, I., and Schmelzer, J. (1995). *The Vitreous State: Thermodynamics, Structure, Rheology, and Crystallization* (Springer-Verlag, Berlin).

33. Larché, F., and Cahn, J. W. (1973). A linear theory of thermochemical equilibrium of solids under stress, *Acta Metall.*, **21**, pp. 1051–1063; (1978). Thermochemical equilibrium of multiphase solids under stress, *Acta Metall.*, **26**, pp. 1579–1589; (1985). The interactions of composition and stress in crystalline solids, *Acta Metall.*, **33**, pp. 331–357.
34. Thomassen, J., Feldmann, B., and Wuttig, M. (1992). Growth, structure and morphology of ultrathin iron films on Cu(100), *Surf. Sci.*, **264**, pp. 406–418.
35. Herman, M. A., Richter, W., and Sitter, H. (2004). *Epitaxy: Physical Principles and Technical Implementation* (Springer-Verlag, Berlin).
36. Kawasaki, M., Takahashi, K., Maeda, T., Tsuchiya, R., Shinohara, M., Ishiyama, O., Yonezawa, T., Yoshimoto, M., and Koinuma, H. (1994). Atomic control of the $SrTiO_3$ crystal surface, *Science*, **226**, pp. 1540–1542.
37. Koster, G., Kropman, B. L., Rijnders, G., Blank, D. H. A., and Rogalla H. (1998). Quasi-ideal strontium titanate crystal surfaces through formation of strontium hydroxide, *Appl. Phys. Lett.*, **73**, pp. 2920–2922.
38. Frank, F. C., and van de Merwe, J. H. (1949). One-dimensional dislocations. I. Static theory, *Proc. R. Soc. London, Ser. A*, **198**, pp. 205–216.
39. Matthews, J. W., and Blakeslee, A. E. (1974). Defects in epitaxial multilayers: I. Misfit dislocations, *J. Cryst. Growth*, **27**, pp. 118–125.
40. Asaro, R. J., and Tiller, W. A. (1972). Surface morphology development during stress corrosion cracking: part I; via surface diffusion, *Metall. Trans.*, **3**, pp. 1789–1796.
41. Grinfeld, M. A. (1986). Instability of the separation boundary between a non-hydrostatically stressed elastic body and a melt, *Sov. Phys. Dokl.*, **31**, pp. 831–834.
42. Gilmer, G. H., and Grabow, M. H. (1987). Models of thin film growth modes, *J. Metals*, **39**, pp. 19–23.
43. Nieminen, J. A., and Kaski, K. (1987). Criteria for different growth modes of thin films, *Surf. Sci.*, **185**, pp. L489–L496.
44. Cammarata, R. C., and Sieradzki, K. (1994). Surface and interface stresses, *Ann. Rev. Mater. Sci.*, **24**, pp. 215–234.
45. Gibbs, J. W. (1876). On the equilibrium of heterogeneous substances, *Trans. Conn. Acad. Sci.*, **3**, pp. 108–248; (1878). On the equilibrium of heterogeneous substances, *Trans. Conn. Acad. Sci.*, **16**, pp. 343–524.

46. Katz, J. L., and Ostermier, B. J. (1967). Diffusion cloud chamber investigation of homogeneous nucleation, *J. Chem. Phys.*, **47**, pp. 478–487.
47. Neilsen, A. E. (1967). *Nucleation in Aqueous Solutions*, ed. Peiser, S. (Pergamon, Oxford), pp. 419–426.
48. Kahlweit, M. (1969). Nucleation in liquid solutions, in *Physical Chemistry*, Vol. VII, ed. Eyring, H. (Academic Press, New York).
49. Walton, A. G. (1969). Nucleation in liquids and solutions, in *Nucleation*, ed. Zettlemoyer, A. C. (Dekker, M. Inc., New York), pp. 225–307.
50. Volmer, M., and Weber, A. Z. (1926). Nucleation of supersaturated structures, *Z. Phys. Chem. (Leipzig)*, **119**, pp. 277–301.
51. Farkas, L. Z. (1927). Keimbildungsgeschwindigkeit in Übersättigten Dämpfen, *Z. Phys. Chem. (Leipzig)*, **125**, pp. 236–242.
52. Becker, R., and Döring, W. (1935). Kinetische Behandlung der Keimbildung in Übersättigten Dämpfen, *Ann. Phys. (Leipzig)*, **24**, pp. 719–752.
53. Hohenberg, P. C., and Halperin, B. I. (1977). Theory of dynamic critical phenomena, *Rev. Mod. Phys.*, **49**, pp. 435–479.
54. Chaikin, P. M., and Lubensky, T. C. (1995). *Principles of Condensed Matter Physics* (Cambridge University Press, Cambridge).
55. Kashchiev, D. (1999). *Nucleation: Basic Theory with Applications* (Butterworths, Heinemann, Oxford).
56. Kashchiev, D. (2003). Thermodynamically consistent description of the work to form a nucleus of any size, *J. Chem. Phys.*, **118**, pp. 1837–1851.
57. Xu, D., and Johnson, W. L. (2005). Geometric model for the critical-value problem of nucleation phenomena containing the size effect of nucleating agent, *Phys. Rev. B*, **72**, pp. 052101 (1–4).
58. Wu, D. T. (1992). The time lag in nucleation theory, *J. Chem. Phys.*, **97**, pp. 2644–2650.
59. Shneidman, V. A., and Weinberg, M. C. (1992). Transient nucleation induction time from the birth-death equations, *J. Chem. Phys.*, **97**, pp. 3629–3638.
60. Kashchiev, D. (1969). Solution of the non-steady state problem in nucleation kinetics, *Surf. Sci.*, **14**, pp. 209–220.
61. Abraham, F. F. (1974). *Homogeneous Nucleation Theory* (Academic Press, New York).

62. Kashchiev, D., and van Rosmalen, G. M. (2003). Review: nucleation in solutions revisited, *Cryst. Res. Technol.*, **38**, pp. 555–574.
63. Soisson, F., and Fu, C.-C. (2007). Cu-precipitation kinetics in α-fe from atomistic simulations: vacancy-trapping effects and Cu-cluster mobility, *Phys. Rev. B*, **76**, pp. 214102–214112.
64. Perez, M., Dumont, M., and Acevedo-Reyes, D. (2008). Implementation of classical nucleation and growth theories for precipitation, *Acta Mater.*, **56**, pp. 2119–2132.
65. Frenkel, D., and Smit, B. (2001). *Understanding Molecular Simulation: From Algorithms to Applications* (Academic Press, San Diego).
66. Kožíšek, Z., Demo, P., and Sveshnikov, A. (2015). Limits of the applicability of the classical nucleation theory, *Adv. Sci., Eng. Med.*, **7**, pp. 316–320.
67. Martin, G., and Soisson, F. (2005). Kinetic Monte Carlo method to model diffusion controlled phase transformations in the solid state, in *Handbook of Materials Modeling*, ed. Yip, S. (Springer, The Netherlands), pp. 2223–2248.
68. Girshick, S. L., and Chiu, C.-P. (1990). Kinetic nucleation theory: a new expression for the rate of homogeneous nucleation from an ideal supersaturated vapor, *J. Chem. Phys.*, **93**, pp. 1273–1277.
69. Durán-Olivencia, M. A., and Lutsko, J. F. (2015). Unification of classical nucleation theories via a unified Itô-Stratonovich stochastic equation, *Phys. Rev. E*, **92**, p. 032407.
70. Lutsko, J. F., and Durán-Olivencia, M. A. (2013). Classical nucleation theory from a dynamical approach to nucleation, *J. Chem. Phys.*, **138**, pp. 244908 (1–14).
71. Martin, G. (2006). Reconciling the classical nucleation theory and atomic scale observations and modeling, *Adv. Eng. Mater.*, **8**, pp. 1231–1236.
72. Lépinoux, J. (2006). Contribution of matrix frustration to the free energy of cluster distributions in binary alloys, *Philos. Mag.*, **86**, pp. 5053–5082.
73. Maibaum, L. (2008). Phase transformation near the classical limit of stability, *Phys. Rev. Lett.*, **101**, pp. 256102 (1–4).
74. Radhakrishnan, R., and Trout, B. L. (2002). A new approach for studying nucleation phenomena using molecular simulations: application to CO_2 hydrate clathrates, *J. Chem. Phys.*, **117**(4), pp. 1786–1796.

75. Frank, F. C., and van der Merwe, J. H. (1949). One-dimensional dislocations. II. Misfitting monolayers and oriented overgrowth, *Proc. R. Soc. London, Ser. A*, **198**, pp. 216–225.
76. Stranski, I. N., and Krastanov, L. (1938). Zur Theorie der orientierten Ausscheidung von Ionenkristallen aufeinander, *Sitz. Akad. Wiss. Wien, Math.-Nat. Kl. IIb*, **146**, pp. 797–810.
77. Rusanov, A. (2005). Surface thermodynamics revisited, *Surf. Sci. Rep.*, **58**, pp. 111–239.
78. Müller, P., and Saul, A. (2004). Elastic effects on surface physics, *Surf. Sci. Rep.*, **54**, pp. 157–258.
79. Nozières, P., and Wolf, D. (1988). Interfacial properties of elastically strained materials, *Z. Phys. B*, **70**, pp. 507–513.
80. Tersoff, J., and Legoues, F. K. (1994). Competing relaxation mechanisms in strained layers, *Phys. Rev. Lett.*, **72**, pp. 3570–3573.
81. Müller, P., and Kern, R. (2000). Equilibrium nano-shape changes induced by epitaxial stress (generalised Wulf–Kaishew theorem), *Surf. Sci.*, **457**, pp. 229–253.
82. Sullivan, D. E. (1979). Van der Waals model of adsorption, *Phys. Rev. B*, **20**, pp. 3991–4000.
83. Spencer, B., Voorhes, P., and Davis, S. (1993). Morphological instability in epitaxially strained dislocation-free solid films: linear stability theory, *J. Appl. Phys.*, **73**, pp. 4955–4970.
84. Savina, T., Voorhees, P., and Davis, S. (2004). The effect of surface stress and wetting layers on morphological instability in epitaxially strained films, *J. Appl. Phys.*, **96**, pp. 3127—3133.
85. Yu, X., Duxbury, P. M., Jeffers, G., and Dubson, M. A. (1991). Coalescence and percolation in thin metal films, *Phys. Rev. B*, **44**, pp. 13163–13166.
86. Brault, P., Thomann, A.-L., and Andreazza-Vignolle, C. (1998). Percolative growth of palladium ultrathin films deposited by plasma sputtering, *Surf. Sci.*, **406**, pp. L597–L602.
87. Pashley, D. W., Jacobs, M. H., Stowell, M. J., and Law, T. J. (1964). The growth and structure of gold and silver deposits formed by evaporation inside an electron microscope, *Phil. Mag.*, **10**, pp. 127–158.
88. Baski, A. A., and Fuchs, H. (1994). Epitaxial growth of silver on mica as studied by AFM and STM, *Surf. Sci.*, **313**, pp. 275–288.
89. Jeffers, G., Dubson, M. A., and Duxbury, P. M. (1994). Island-to-percolation transition during growth of metal films, *J. Appl. Phys.*, **75**, pp. 5016–5020.

90. Müller, P., and Kern, R. (1996). The physical origin of the two-dimensional towards three-dimensional coherent epitaxial Stranski-Krastanov transition, *Appl. Surf. Sci.*, **102**, pp. 6–11.
91. Wessels, B. W. (1997). Morphological stability of strained-layer semiconductors, *J. Vac. Sci. Technol. B*, **15**, pp. 1056–1058.
92. Li, X., Cao, Y., and Yang, G. (2010). Thermodynamic theory of two-dimensional to three-dimensional growth transition in quantum dots self-assembly, *Phys. Chem. Chem. Phys.*, **12**, pp. 4768–4772.
93. Zhang, Z., and Lagally, M. G. (1997). Atomistic processes in the early stages of thin-film growth, *Science*, **276**, pp. 377–383.
94. Meyers, M., Mishra, A., and Benson, D. (2006). Mechanical properties of nanocrystalline materials, *Prog. Mater. Sci.*, **51**, pp. 427–556.
95. Chernov, A. A. (1998). Theoretical and technological aspects of crystal growth, in *Materials Science Forum*, ed. Fornari, R., and Paorichi, C. (TransTech Publications, USA), pp. 71–78.
96. Frank, F. C. (1949). The influence of dislocations on crystal growth, *Discuss. Faraday Soc.*, **5**, pp. 48–54.
97. Gilmer, G. H., Huang, H., de la Rubia, T. D., Dalla Torre, J., and Baumann, F. (2000). Lattice Monte Carlo models of thin film deposition, *Thin Solid Films*, **365**, pp. 189–200.
98. Vvedensky, D. D., Clarke, S., Hugill, K. J., Myers-Beaghton, A. K., and Wilby, M. R. (1990). *Kinetics of Ordering and Growth at Surface*, ed. Lagally, M. G. (Plenum Press, London), pp. 297–311.
99. Weeks, J. D., and Gilmer, G. H. (1979). Dynamics of crystal growth, *Adv. Chem. Phys.*, **40**, pp. 157–227.
100. Brune, H. (1998). Microscopic view of epitaxial metal growth: nucleation and aggregation, *Surf. Sci. Rep.*, **31**, pp. 121–229.
101. Ehrlich, G., and Hudda, F. G. (1966). Atomic view of surface self-diffusion: tungsten on tungsten, *J. Chem. Phys.*, **44**, pp. 1039–1049.
102. Schwoebel, R. L., and Shipsey, E. J. (1966). Step motion on crystal surfaces, *J. Appl. Phys.*, **37**, pp. 3682–3686.
103. Einax, M., Dieterich, W., and Maass, P. (2013). Colloquium: cluster growth on surfaces: densities, size distributions, and morphologies, *Rev. Mod. Phys.*, **85**, pp. 921–939.
104. Leal, F. F., Ferreira, S. C., and Ferreira, S. O. (2011). Modelling of epitaxial film growth with an Ehrlich-Schwoebel barrier dependent on the step height, *J. Phys. Condens. Matter*, **23**, p. 292201 (10 pp).

105. Smilauer, P., and Harris, S. (1995). Determination of step-edge barriers to interlayer transport from surface morphology during the initial stages of homoepitaxial growth, *Phys. Rev. B*, **51**, pp. 14798–14801.

106. Ratch, C., and Zangwill, A. (1993). Step-flow growth on strained surfaces, *Appl. Phys. Lett.*, **63**, pp. 2348–2250.

107. Ratch, C., Nelson, M. D., and Zangwill, A. (1994). Theory of strained-layer epitaxial growth near step flow, *Phys. Rev. B*, **50**, pp. 14489–14497.

108. Jensen, P., and Niemeyer, B. (1997). The effect of a modulated flux on the growth of thin films, *Surf. Sci.*, **384**, pp. L823–L827.

109. Stäuble-Pümpin, B., Matijasevic, V. C., Ilge, B., Mooij, J. E., Peterse, W. J. A. M., Scholte, P. M. L. O., Tuinstra, F., Venvik, H. J., Wai, D. S., Træholt, C., Wen, J. G., and Zandbergen, H. W. (1995). Growth mechanisms of coevaporated SmBa$_2$Cu$_3$O$_y$ thin films, *Phys. Rev. B*, **52**, pp. 7604–7628.

110. Lu, J., Liu, J.-G., and Margetis, D. (2015). Emergence of step flow from an atomistic scheme of epitaxial growth in 1+1 dimensions, *Phys. Rev. E*, **91**, p. 032403 (1–4).

111. Kim, J., Chrisey, D. B., Horwitz, J. S., Miller, M. M., and Gilmore, C. M. (2000). Growth mechanism of YBa$_2$Cu$_3$O$_7$-delta thin films and precipitates on planar and vicinal SrTiO$_3$ substrates, *J. Mater. Res.*, **15**, pp. 596–604.

112. Körner, M., Einax, M., and Maass, P. (2012). Capture numbers and islands size distributions in models of submonolayer surface growth, *Phys. Rev. B*, **86**, p. 085403 (1–8).

113. Bartelt, M. C., and Evans, J. W. (1996). Exact island-size distributions for submonolayer deposition: influence of correlations between island size and separation, *Phys. Rev. B*, **54**, pp. R17359–R17362.

114. Vetter, T., Iggland, M., Ochsenbein, D. R., Hänseler, F. S., and Mazzotti, M. (2013). Modeling nucleation, growth, and Ostwald ripening in crystallization processes: a comparison between population balance and kinetic rate equation, *Cryst. Growth Des.*, **13**, pp. 4890–4905.

115. Kamins, T. I., Medeiros-Ribeiro, G., Ohlberg, D. A. A., and Stanley Williams, R. (1999). Evolution of Ge islands on Si(001) during annealing, *Appl. Phys.*, **85**, pp. 1159–1171.

116. Meyer zu Heringdorf, F.-J., Reuter, M. C., and Tromp, R. M. (2001). Growth dynamics of pentacene thin films, *Nature*, **412**, pp. 517–520.

117. Venables, J. A. (1973). Rate equation approaches to thin film nucleation kinetics, *Philos. Mag.*, **17**, pp. 697–738.

118. Iijima, S., and Ajayan, P. M. (1991). Substrate and size effects on the coalescence of small particles, *J. Appl. Phys.*, **70**, pp. 5138–5140.
119. Aziz, M. (2008). Film growth mechanisms in pulsed laser deposition, *Appl. Phys. A*, **93**, pp. 579–587.
120. Schwoebel, R. L. (1969). Step motion on crystal surfaces II, *J. Appl. Phys.*, **40**, pp. 614–618.
121. Seifert, W., Carlsson, N., Miller, M., Pistol, M.-E., Samuelson, L., and Wallenberg, L. R. (1996). In-situ growth of quantum dot structures by the Stranski-Krastanow growth mode, *Prog. Crystal Growth Charact.*, **33**, pp. 423–472.
122. Mukherjee, S., and Gall, D. (2013). Structure zone model for extreme shadowing conditions, *Thin Solid Films*, **527**, pp. 158–163.
123. Anders, A. (2010). A structure zone diagram including plasma-based deposition and ion etching, *Thin Solid Films*, **518**, pp. 4087–4090 and references therein.
124. Barna, P. B., and Adamik, M. (1998). Fundamental structure forming phenomena of polycrystalline films and the structure zone models, *Thin Solid Films*, **317**, pp. 27–33.
125. Barber, Z. H. (2006). The control of thin film deposition and recent developments in oxide film growth, *J. Mater. Chem.*, **16**, pp. 334–344.
126. McGray, W.-P. (2007). MBE deserves a place in the history books, *Nat. Nanotech.*, **2**(5), pp. 259–261.
127. Warusawithana, M., Zuo, J., Chen, H., Eckstein, J. N., Ohtomo, A., and Hwang, H. Y. (2002). Superlattices grown by MBE, *Nature*, **419**, pp. 378–380.
128. Marcu, A., Stafe, M., and Puscas, N. N. (2014). *Pulsed Laser Ablation of Solids: Basics, Theory and Applications*, Springer Series in Surface Sciences (Springer-Verlag, Berlin, Heidelberg).
129. Singh, R., and Narayan, J. (1990). Pulsed-laser evaporation technique for deposition of thin films: physics and theoretical model, *Phys. Rev. B*, **41**, pp. 8843–8859.
130. Schou, J. (2009). Physical aspects of the pulsed laser deposition technique: the stoichiometric transfer of material from target to film, *Appl. Surf. Sci.*, **255**, pp. 5191–5198.
131. Venkatesan, T. (2014). Pulsed laser deposition—invention or discovery?, *J. Phys. D: Appl. Phys.*, **46**, p. 034001.
132. Arnold, C. B., and Aziz, M. J. (1999). Stoichiometry issues in pulsed laser deposition of alloys grown from multicomponent targets, *Appl. Phys. A*, **69**, pp. 23–27.

133. Dijkkamp, D., Venkatesan, T., Wu, X. D., Shaheen, S. A., Jisrawi, N., Min-Lee, Y. H., McLean, W. L., and Croft, M. (1987). Preparation of YBaCu oxide superconductor thin films using pulsed laser evaporation from high Tc bulk material, *Appl. Phys. Lett.*, **51**, pp. 619–621.

134. Cesaria, M., Caricato, A. P., Maruccio, G., and Martino, M. (2011). LSMO – growing opportunities by PLD and applications in spintronics, *Int. Conf. on Trends in Spintronics and Nanomagnetism (TSN 2010), J. Phys.: Conf. Ser.*, **292**, p. 012003 (15 pp) and references therein.

135. Majumdar, S., and van Dijken, S. (2014). Pulsed laser deposition of $La_{1-x}Sr_xMnO_3$: thin-film properties and spintronic applications, *J. Phys. D: Appl. Phys.*, **47**, p. 034010 (15 pp).

136. Lowndes, D. H., Geohegan, D. B., Puretzky, A. A., Norton, D. P., and Rouleau, C. M. (1996). Synthesis of novel thin film materials by pulsed laser deposition, *Science*, **273**, pp. 898–903.

137. Xu, C., Wicklein, S., Sambri, A., Amoruso, S., Moors, M., and Dittmann, R. (2014). Impact of the interplay between nonstoichiometry and kinetic energy of the plume species on the growth mode of $SrTiO_3$ thin films, *J. Phys. D: Appl. Phys.*, **47**, p. 034009 (11 pp).

138. Cesaria, M., Caricato, A. P., Leggieri, G., Luches, A., Martino, M., Maruccio, G., Catalano, M., Manera, M. G., Rella, R., and Taurino, A. (2011). Structural characterization of ultrathin Cr-doped ITO layers deposited by double-target pulsed laser ablation, *J. Phys. D: Appl. Phys.*, **44**, p. 365403 (8 pp).

139. Shen, J., Ohresser, P., Mohan, Ch. V., Klaua, M., Barthel, J., and Kirschner, J. (1998). Magnetic moment of fcc Fe(111) ultrathin films by ultrafast deposition on Cu(111), *Phys. Rev. Lett.*, **80**, pp. 1980–1984.

140. Koster, G., Rijnders, G. J. H. M., Blank, D. H. A., and Rogalla, H. (1999). Imposed layer-by-layer growth by pulsed laser interval deposition, *Appl. Phys. Lett.*, **74**, pp. 3729–3731.

141. Taylor, M. E., Atwater, H. A., and Murty, M. V. R. (1998). Role of energetic flux in low temperature Si epitaxy on dihydride-terminated Si(001), *Thin Solid Films*, **324**, pp. 85–88.

142. Jenniches, H., Klaua, M., Hoche, H., and Kirschner, J. (1996). Comparison of pulsed laser deposition and thermal deposition: improved layer-by-layer growth of Fe/Cu(111), *Appl. Phys. Lett.*, **69**, pp. 3339–3341.

143. Hernández-Calderón, I. (2013). Epitaxial growth of thin films and quantum structures of II-VI visible-bandgap semiconductors, Chapter 14, in *Molecular Beam Epitaxy*, ed. Henini, M. (Elsevier, Oxford).

144. Romano, L. T., Bringans, R. D., Zhou, X., and Kirk, W. P. (1995). Interface structure of ZnS/Si(001) and comparison with ZnSe/Si(001) and GaAs/Si(001), *Phys. Rev. B*, **52**, pp. 11201–11205.
145. Yoo, Y. Z., Osaka, Y., Fukumura, T., Jin, Z., Kawasaki, M., Koinuma, H., Chikyow, T., Ahmet, P., Setoguchi, A., and Chichibu, S. F. (2001). High temperature growth of ZnS films on bare Si and transformation of ZnS to ZnO by thermal oxidation, *Appl. Phys. Lett.*, **78**(5), pp. 616–618.
146. Vasco, E., and Zaldo, C. (2004). Growth kinetics of epitaxial Y-stabilized ZrO_2 films deposited on InP, *J. Phys.: Condens. Matter*, **16**, pp. 8201–8211.
147. Mayr, S. G., Moske, M., Samwer, K., Taylor, M. E., and Atwater, H. A. (1999). The role of particle energy and pulsed particle flux in physical vapor deposition and pulsed–laser deposition, *Appl. Phys. Lett.*, **75**, pp. 4091–4093.
148. Willmott, P. R., Herger, R., Schlepütz, C. M., Martoccia, D., and Patterson, B. D. (2006). Energetic surface smoothing of complex metal-oxide thin films, *Phys. Rev. Lett.*, **96**, pp. 176102–176105.
149. Tischler, J. Z., Eres, G., Larson, B. C., Rouleau, C. M., Zschack, P., and Lowndes, D. H. (2006). Nonequilibrium interlayer transport in pulsed laser deposition, *Phys. Rev. Lett.*, **96**, p. 226104.
150. Zhang, Y., Gu, H., and Iijima, S. (1998). Single-wall carbon nanotubes synthesized by laser ablation in a nitrogen atmosphere, *Appl. Phys. Lett.*, **73**, pp. 3827–3829.
151. Geohegan, D. B., Puretzky, A. A., and Rader, D. L. (1999). Gas-phase nanoparticle formation and transport during pulsed laser deposition of $Y_1Ba_2Cu_3O_{7-d}$, *Appl. Phys. Lett.*, **74**, pp. 3788–3790.
152. Kabashin, A. V., Delaporte, Ph., Pereira, A., Grojo, D., Torres, R., Sarnet, Th., and Sentis, M. (2010). Nanofabrication with pulsed lasers, *Nanoscale Res. Lett.*, **5**(3), pp. 454–463.
153. Kukreja, L. M., Verma, S., Pathrose, D. A., and Rao, B. T. (2014). Pulsed laser deposition of plasmonic-metal nanostructures, *J. Phys. D: Appl. Phys.*, **47**, p. 034015 (14 pp).
154. Geprägs, S., Opel, M., Goennenwein, S. T. B., and Gross, R. (2007). Multiferroic materials based on artificial thin film heterostructures, *Phil. Mag. Lett.*, **87**, pp. 141–154.
155. Nakamura, D., Shimogaki, T., Nakao, S., Harada, K., Muraoka, Y., Ikenoue, H., and Okada, T. (2014). Patterned growth of ZnO nanowalls by nanoparticle-assisted pulsed laser deposition, *J. Phys. D: Appl. Phys.*, **47**, p. 034014 (8 pp).

156. Tabata, H., Tanaka, H., and Kawai, T. (1994). Formation of artificial BaTiO$_3$/SrTiO$_3$ superlattices using pulsed laser deposition and their dielectric properties, *Appl. Phys. Lett.*, **65**, pp. 1970–1972.

157. Rethfeld, B., Sokolowski-Tinten, K., von der Linde, D., and Anisimov, S. I. (2004). Timescales in the response of materials to femtosecond laser excitation, *Appl. Phys. A*, **79**, pp. 767–769.

158. Soliman, W., Takada, N., and Sasaki, K. (2010). Growth processes of nanoparticles in liquid-phase laser ablation studied by laser-light scattering, *Appl. Phys. Express*, **3**, p. 035201 (1–3).

159. Amendola, V., and Meneghetti, M. (2009). Laser ablation synthesis in solution and size manipulation of noble metal nanoparticles, *Phys. Chem. Chem. Phys.*, **11**, pp. 3805–382 and references therein.

160. Yang, G. W. (2007). Laser ablation in liquids: applications in the synthesis of nanocrystals, *Prog. Mater. Sci.*, **52**, pp. 648–6981 and references therein.

161. Caricato, A. P., Arima, V., Catalano, M., Cesaria, M., Cozzoli, P. D., Martino, M., Taurino, A., Rella, R., Scarfiello, R., and Taurino, A. (2013). Maple deposition of nanomaterials, *Appl. Surf. Sci.*, **302**, pp. 92–98 and references therein.

162. McGill, R. A., and Chrisey, D. B. (2000). Method of producing a film coating by matrix assisted pulsed laser deposition, Patent No. 6,025,036.

163. Torres, R. D., Johnson, S. L., Haglund, R. F., Hwang, J., Burn, P. L., and Holloway, P. H. (2011). Mechanisms of resonant infrared matrix-assisted pulsed laser evaporation, *Crit. Rev. Solid State Mater. Sci.*, **36**, pp. 16–45.

164. Cesaria, M., Caricato, A. P., Taurino, A., Resta,V., Cozzoli, P. D., Belviso, M. R., and Martino, M. (2015). Role of solvents in MAPLE-deposition of Pd nanoparticles, *Sci. Adv. Mater.*, **7**, pp. 1–13 and references therein.

165. Dam, B., and Staucle-Pumpin, B. (1998). Growth mode issues in epitaxy of complex oxide thin films, *J. Mater. Sci. - Mater. Electron.*, **9**, pp. 217–226.

166. Shen, J., Gaib, Z., and Kirschner, J. (2004). Growth and magnetism of metallic thin films and multilayers by pulsed-laser deposition, *Surf. Sci. Rep.*, **52**, pp. 163–218 and references therein.

167. Warrender, J. M., and Aziz, M. J. (2007). Kinetic energy effects on morphology evolution during pulsed laser deposition of metal-on insulator films, *Phys. Rev. B*, **75**, p. 085433 (1–11).

168. Pun, A. F., Wang, X., Meeks, J. B., Zheng, J. P., and Durbin, S. M. (2004). Initial growth dynamics of homo epitaxial (100) GaAs using pulsed laser deposition, *J. Appl. Phys.*, **96**, pp. 6357–6361.
169. Hinnemann, B., Hinrichsen, H., and Wolf, D. E. (2003). Epitaxial growth with pulsed deposition: submonolayer scaling and Villain instability, *Phys. Rev. E*, **67**, p. 011602 (1–9).
170. Levanov, N. A., Stepanyuk, V. S., Hergert, W., Bazhanov, D. I., Dederichs, P. H., Katsnelson, A. and Massobrio, C. (2000). Energetics of Co adatoms on the Cu(001) surface, *Phys. Rev. B*, **61**, pp. 2230–2234.
171. Huang, H., Gilmer, G. H., and Diaz de la Rubia, T. (1998). An atomistic simulator for thin film deposition in three dimensions, *J. Appl. Phys.*, **84**, pp. 3636–3649.
172. Son, J. Y., Kim, B. G., and Cho, J. H. (2005). Kinetically controlled thin-film growth of layered β- and γ-Na$_x$CoO$_2$$\gamma$-Na$_xCoO_2$ cobaltate, *Appl. Phys. Lett.*, **86**, pp. 221918–221920.
173. Marks, N. A. (1997). Evidence for subpicosecond thermal spikes in the formation of tetrahedral amorphous carbon, *Phys. Rev. B*, **56**, pp. 2441–2446.
174. Diaz de la Rubia, T., Averback, R. S., Benedek, R., and King, W. E. (1987). Role of thermal spikes in energetic displacement cascades, *Phys. Rev. Lett.*, **59**, pp. 1930–1933.
175. Jacobsen, J., Cooper, B. H., and Sethna, J. P. (1998). Simulations of energetic beam deposition: from picoseconds to seconds, *Phys. Rev. B*, **58**, pp. 15847–15865.
176. Pomeroy, J. M., Jacobsen, J., Hill, C. C., Cooper, B. H., and Sethna, J. P. (2002). Kinetic Monte Carlo–molecular dynamics investigations of hyperthermal copper deposition on Cu(111), *Phys. Rev. B*, **66**, p. 235412 (1–8).
177. Barth, J. V. (2000). Transport of adsorbates at metal surfaces: from thermal migration to hot precursors, *Surf. Sci. Rep.*, **40**, pp. 75–149 and references therein.
178. Blank, D. H. A., Koster, G., Rijnders, G. A., Setten, E., Slycke, P., and Rogalla, H. (2000). Epitaxial growth of oxides with pulsed laser interval deposition, *J. Cryst. Growth*, **211**, pp. 98–105.
179. Guan, L., Zhang, D. M., Li, X., and Li, Z. H. (2008). Role of pulse repetition rate in film growth of pulsed laser deposition, *Nucl. Instrum. Methods Phys. Res. B*, **266**, pp. 57–62.

180. Kim, J.-W., Kang, H.-S., and Lee, S.-Y. (2006). Effect of deposition rate on the property of ZnO thin films deposited by pulsed laser deposition, *J. Electr. Eng. Technol.*, **1**, pp. 98–100.

181. Zhang, D., Guan, L., Li, Z., Pan, G., Tan, X., and Li, L. (2006). Simulation of island aggregation influenced by substrate temperature, incidence kinetic energy and intensity in pulsed laser deposition, *Appl. Surf. Sci.*, **253**, pp. 874–880.

182. Vasco, E., and Sacedon, J. L. (2007). Role of cluster transient mobility in pulsed laser deposition-type growth kinetics, *Phys. Rev. Lett.*, **98**, p. 036104.

183. Cheung, J. T., and Sankur, H. (1988). Growth of thin films by laser-induced evaporation, *Crit. Rev. Solid State Mater. Sci.*, **15**, pp. 63–109.

184. Krebs, H. U., Bremert, O., Stormer, M., and Luo, Y. (1995). Comparison of the structure of laser deposited and sputtered metallic alloys, *Appl. Surf. Sci.*, **86**, pp. 90–94.

185. Warrender, J. M., and Aziz, M. J. (2004). Evolution of Ag nanocrystal films grown by pulsed laser deposition, *Appl. Phys. A*, **79**, pp. 713–716.

186. Hinnemann, B., Hinrichsen, H., and Wolf, D. E. (2001). Unusual scaling for pulsed laser deposition, *Phys. Rev. Lett.*, **87**, p. 135701 (1–4).

187. Hinnemann, B., Westerhoff, F., and Wolf, D. E. (2002). Layer-by-layer growth for pulsed laser deposition, *Phase Transitions*, **75**, pp. 151–157.

188. Combe, N., and Jensen, P. (1998). Changing thin-film growth by modulating the incident flux, *Phys. Rev. B*, **57**, p. 15553.

189. Lam, P.-M., Liu, S. J., and Woo, C. H. (2002). Monte Carlo simulation of pulsed laser deposition, *Phys. Rev. B*, **66**, p. 045408.

190. Lee, S. B. (2003). Scaling of the nucleation density for pulsed layer deposition, *Phys. Rev. E*, **67**, p. 012601.

191. Jubert, P.-O., Fruchart, O., and Meyer, C. (2003). Nucleation and surface diffusion in pulsed laser deposition of Fe on Mo(1 1 0), *Surf. Sci.*, **522**, pp. 8–16.

192. Narhe, R. D., Khandkar, M. D., Adhi, K. P., Limaye, A. V., Sainkar, S. R., and Ogale, S. B. (2001). Difference in the dynamic scaling behavior of droplet size distribution for coalescence under pulsed and continuous vapor delivery, *Phys. Rev. Lett.*, **86**, pp. 1570–1573.

193. Villain, J. (1991). Continuum models of crystal growth from atomic beams with and without desorption, *J. Phys. I*, **1**, pp. 19–42.

194. Siegert, M., and Plischke, M. (1992). Instability in surface growth with diffusion, *Phys. Rev. Lett.*, **68**, pp. 2035–2038.

195. Siegert, M., and Plischke, M. (1994). Slope selection and coarsening in molecular beam epitaxy, *Phys. Rev. Lett.*, **73**, pp. 1517–1520.
196. van Nostrand, J. E., Chey, S. J., Hasan, M.-A., Cahill, D. G., and Greene, J. E. (1995). Surface morphology during multilayer epitaxial growth of Ge(001), *Phys. Rev. Lett.*, **74**, pp. 1127–1130.
197. Ernst, H.-J., Fabre, F., Folkerts, R., and Lapujoulade, J. (1994). Observation of a growth instability during low temperature molecular beam epitaxy, *Phys. Rev. Lett.*, **72**, pp. 112–115.
198. Thuermer, K., Koch, R., Weber, M., and Rieder, K. H. (1995). Dynamic evolution of pyramid structures during growth of epitaxial Fe (001) films, *Phys. Rev. Lett.*, **75**, pp. 1767–1770.
199. Rijnders, G., Koster, G., Leca, V., Blank, D. H. A., and Rogalla, H. (2000). Imposed layer-by-layer growth with pulsed laser interval deposition, *Appl. Surf. Sci.*, **168**, pp. 223–226.
200. Shin, B., Leonard, J. P., McCamy, J. W., and Aziz, M. J. (2005). Comparison of morphology evolution of Ge(001) homoepitaxial films grown by pulsed laser deposition and molecular-beam epitaxy, *Appl. Phys Lett.*, **87**, pp. 181916–181918.
201. Lippmaa, M., Nakagawa, N., Kawasaki, M., Ohashi, S., and Koinuma, H. (2000). Growth mode mapping of $SrTiO_3SrTiO_3$ epitaxy, *Appl. Phys. Lett.*, **76**, pp. 2439–2441.
202. Lippmaa, M., Nakagawa, N., Kawasaki, M., Ohashi, S., Inaguma, Y., Itoh, M., and Koinuma, H. (1999). Step-flow growth of $SrTiO_3SrTiO_3$ thin films with a dielectric constant exceeding 104, *Appl. Phys. Lett.*, **74**, pp. 3543–3545.
203. Shin, B., and Aziz, M. J. (2007). Kinetic-energy induced smoothening and delay of epitaxial breakdown in pulsed-laser deposition, *Phys. Rev. B*, **76**, p. 085431.
204. Stormer, M., and Krebs, H. U. (1995). Structure of laser deposited metallic alloys, *J. Appl. Phys.*, **78**, pp. 7080–7087.
205. Kahl, S., and Krebs, H. U. (2001). Supersaturation of single-phase crystalline Fe(Ag) alloys to 40 at. % Ag by pulsed laser deposition, *Phys. Rev. B*, **63**, pp. 172103–172106.
206. Rijnders, A. J. H. M., Koster, G., Blank, D. H. A., and Rogalla, H. (1997). In situ monitoring during pulsed laser deposition of complex oxides using reflection high energy electron diffraction under high oxygen pressure, *Appl. Phys. Lett.*, **70**, pp. 1888–1890.
207. Rosenfeld, G., Poelsema, B., and Comsa, G. (1995). The concept of two mobilities in homoepitaxial growth, *J. Cryst. Growth*, **151**, pp. 230–233.

208. Markov, V. A., Pchelyakov, O. P., Sokolov, L. V., Stenin, S. I., and Stoyanov, S. (1991). Molecular beam epitaxy with synchronization of nucleation, *Surf. Sci.*, **250**, pp. 229–234.

209. Blank, D. H. A., Koster, G., Rijnders, G. H. M., van Setten, E., Slycke, P., and Rogalla, H. (1999). Imposed layer-by-layer growth by pulsed laser interval deposition, *Appl. Phys. A*, **69**, pp. S17–S22.

210. Wellershoff, S.-S., Hohlfeld, J., Güdde, J., and Matthias, E. (1999). The role of electron–phonon coupling in femtosecond laser damage of metals, *Appl. Phys. A*, **69**, pp. S99–S107.

211. Pronko, P. P., Dutta, S. K., Du, D., and Singh, R. K. (1995). Thermophysical effects in laser processing of materials with picosecond and femtosecond pulses, *J. Appl. Phys.*, **78**, pp. 6233–6240.

212. Hermann, J., Noël, S., Itina, T. E., Axente, E., and Povarnitsyn, M. E. (2008). Correlation between ablation efficiency and nanoparticle generation during the short-pulse laser ablation of metals, *Laser Phys.*, **18**, pp. 374–379.

213. Amoruso, S., Ausanio, G., Bruzzese, R., Vitiello, M., and Wang, X. (2005). Femtosecond laser pulse irradiation of solid targets as a general route to nanoparticle formation in a vacuum, *Phys. Rev. B*, **71**, p. 033406 (1–4).

214. Amoruso, S., Nedyalkov, N. N., Wang, X., Ausanio, G., Bruzzese, R., and Atanasov, P. A. (2014). Ultrashort-pulse laser ablation of gold thin film targets: theory and experiment, *Thin Solid Films*, **550**, pp. 190–198.

215. Lorazo, P., Lewis, L. J., and Meunier, M. (2006). Thermodynamic pathways to melting, ablation, and solidification in absorbing solids under pulsed laser irradiation, *Phys. Rev. B*, **73**, p. 134108 (1–22).

216. Zhigilei, L. V., Lin, Z., and Ivanov, D. S. (2009). Atomistic modeling of short pulse laser ablation of metals: connections between melting, spallation, and phase explosion, *J. Phys. Chem. C*, **113**, pp. 11892–11906.

217. Chichkov, B., Momma, N. C., Nolte, S., von Alvensleben, F., and Tunnermann, A. (1996). Femtosecond, picosecond and nanosecond laser ablation of solids *Appl. Phys. A*, **63**, pp. 109–115.

218. Hirayama, Y., and Obara, M. (2005). Heat-affected zone and ablation rate of copper ablated with femtosecond laser, *J. Appl. Phys.*, **97**, p. 064903 (1–6).

219. Harzic, R., Huot, N., Audouard, E., Jonin, C., Laporte, P., Valette, S., Fraczkiewicz, A., and Fortunier, R. (2002). Comparison of heat-affected zones due to nanosecond and femtosecond laser pulses using transmission electronic microscopy, *Appl. Phys. Lett.*, **80**, pp. 3886–3888.

Chapter 2

Nanosecond Laser Ablation and Processing of Solid Targets in Vacuum or in a Low-Gas Atmosphere

Vincenzo Resta,[a,b,c] Ramón J. Peláez,[c] and Anna Paola Caricato[a]

[a]*Dipartimento di Matematica e Fisica "E. De Giorgi," Università del Salento, Via Arnesano, I-73100 Lecce, Italy*
[b]*CNR-NANO, Istituto Nanoscienze, Euromediterranean Center for Nanomaterial Modelling and Technology (ECMT), Via Arnesano, I-73100 Lecce, Italy*
[c]*Laser Processing Group, Instituto de Óptica, CSIC, Calle Serrano 121, S-28006 Madrid, Spain*
vincenzo.resta@le.infn.it, vincenzo.resta@unisalento.it

2.1 Introduction

Metal nanoparticles (NPs) supported and/or embedded in a dielectric have been receiving special attention due to their potential for applications in areas such as photonics, catalysis, photovoltaics, and chemical sensing and/or biosensing. In general, their feasibility depends on the size, shape, composition, and spatial distribution, as well as on the surrounding medium. Pulsed laser deposition (PLD) with nanosecond (ns) pulses is a multipurpose technique for

Pulsed Laser Ablation: Advances and Applications in Nanoparticles and Nanostructuring Thin Films
Edited by Ion N. Mihailescu and Anna Paola Caricato
Copyright © 2018 Pan Stanford Publishing Pte. Ltd.
ISBN 978-981-4774-23-9 (Hardcover), 978-1-315-18523-1 (eBook)
www.panstanford.com

synthesis of thin films [1]. Basically, a high-power laser pulse impacts on a solid target, leading to the removal of ionized particles (plasma) that expand in vacuum or in a low-gas atmosphere and finally are deposited on a substrate. In the last years, this technique has been diffusively used for the production of nanostructured layers with a versatility not accessible by other deposition techniques. This is due to the possibility of varying a wide range of parameters, such as flux density, deposition rate, ionization degree, kinetic energy of the incoming species, and substrate temperature. Analysis of these parameters is essential to achieve a suitable control on the morphology of the NPs or to moderate the processes related to energetic ions, such as sputtering or implantation of NPs into the substrate.

In this chapter we will give a brief overview of the production of nanostructured layers by using ns laser pulses. The first section focuses on the generation and expansion of the plasma produced by ns laser ablation. The second section deals with the production of metal NPs supported on and embedded in a dielectric layer by PLD. Here, the relations between deposition parameters and film properties are analyzed. The last section is devoted to the production of nanostructured layers by a thermal-assisted process in the substrate, which can be achieved by conventional heating during the growth process or by postgrowth laser annealing with ns laser pulses.

2.2 Plasma Dynamics and Expansion

The interaction between a laser pulse, with enough energy density, and a material produces a vapor with a high density of particles over the target surface as a consequence of the pressure of photoablation, even before the laser pulse itself is vanished. If the vapor temperature is high enough to cause appreciable atomic excitation and ionization, the vapor, too, starts to absorb the incident laser radiation, thus leading to vapor breakdown and plasma formation. The interaction process strongly depends on both laser plasma coupling and plasma kinetics, and in this section, it will be described in the case of a metal target because this class of materials is the most suitable for obtaining nanostructures by ns laser irradiations [1, 2], with only a few exceptions [3–5].

In metals, the absorption of the laser light is triggered by the free electrons through the inverse *Bremsstrahlung* (IB) mechanism [6]. The electrons with energy less than or equal to the incident photon energy (in an occupied level) are excited above the Fermi level, and the elastic scattering inside the electron gas induces the formation of a different Fermi electron distribution with a higher electron temperature [7]. An external thermalization process is then activated through scattering with the lattice phonons. The overall electron thermalization time (time for all the electrons to scatter with each other until a new Fermi distribution is obtained) ranges between ~500 femtoseconds (fs) and a few picoseconds (ps). Similarly, the electron-to-phonon interactions are enclosed in a timescale of <10 ps, their strength varying as a function of the specific metal, being higher for Al, Au, Ag, Cu, and W with respect to Ni and Pt. On the contrary, the phonon-to-phonon interactions of the hot lattice to the surrounding medium last on a higher timescale, ~100 ps, the phonon defect and phonon boundary scattering relaxation time being even higher (see the scheme in Fig. 2.1, where the different recombination phenomena are summarized). Note that in the present description, the interband transitions have been neglected because they last only for tens of fs.

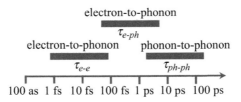

Figure 2.1 Schematic illustration of the timescale related to the scattering phenomena triggered by the laser excitation of a metal.

When the duration of the energy transfer mechanisms is shorter than the laser pulse, the absorbed light is transformed into heat directly and the material response can be treated in a purely thermal way (photothermal mechanism). In this framework, the threshold between quasi-stationary and strongly nonstationary interactions lies around ~100 ns, when common thermal-to-thermomechanical processes are accompanied and overwhelmed by photochemical bond breaking and photomechanical mechanisms [8]. The temporal and spatial evolution of the temperature, T, of metals under ns pulse irradiation is determined by the heat flow equation [9]

$$\rho C_p \frac{\partial T}{\partial t} = \nabla(\kappa \nabla T) + Q_{abs}, \qquad (2.1)$$

where C_p is the specific heat capacity; κ is the thermal conductivity, which is T dependent; and ρ is the density. Q_{abs} is the volumetric heating rate (power density) incorporating the contribution of heat sources and sinks but disregarding the occurrence of phase changes or chemical reactions. Under this assumption, it can be related to the laser absorption as

$$Q_{abs}(x, y, z, t) = I(x, y, t) A \alpha \exp(-\alpha z) \qquad (2.2)$$

through (i) a spatial and temporal shape function of the laser pulse, $I(x, y, t)$, defining the incident laser power density at the surface and (ii) the contribution related to the fraction of photons effectively absorbed by the material (following the Lambert–Beer law). The latter depends on the surface absorbance, $A = (1 - R)$, where R is the thickness-dependent reflectivity of the sample at the laser wavelength, the absorption coefficient $\alpha = 4\pi k_a/\lambda$ (k_a being the imaginary part of the refractive index), and the distance measured from the sample surface is z. Indeed, the optical penetration depth α^{-1} is ~10 nm for both Au and Cu at 248 nm. Figure 2.2a shows the simulated time-dependent temperature profiles at the surface of a Cu target as a function of laser energy surface density, which in the following will be referred to as fluence, F. The temperature rises rapidly during the laser pulse. Once the melting point is reached and the target has absorbed enough heat to overcome the latent heat of fusion, the metal becomes melted in a few nanoseconds for $F \approx 1$ J cm^{-2}. The melted duration period and the peak temperature increase with the fluence up to the saturation level determined by the boiling point. Once the latent heat of vaporization is absorbed, the temperature at the surface of the target reaches the boiling point, which front-propagates inside the more and more as the fluence increases.

For strongly absorbing materials, like metals, the heat penetration depth (depth at which the heat penetrates into the material during the laser pulse), L_{th}, is mainly determined by the heat diffusion length (size of the heated zone) and is defined as [8]

$$L_{th} \approx 2(D\tau_p)^{1/2}, \qquad (2.3)$$

where $D = \kappa/\rho C_p$ is the heat diffusion coefficient of the target and τ_p the laser pulse duration.

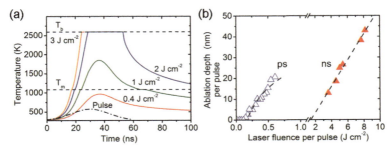

Figure 2.2 (a) Simulated temporal profiles of the temperature evolution induced in a Cu (bulk) target by laser irradiation as a function of the laser fluence (KrF laser, λ = 248 nm, and τ = 20 ns). (b) Ablation depth per pulse, Δz_v in the text, as a function of the laser fluence (Cu target, KrF laser, and λ = 248 nm) for a pulse width of τ = 0.5 ps (open triangles) and τ = 20 ns (solid triangles), respectively (from Ref. [10]. © IOP Publishing. Reproduced with permission. All rights reserved).

Then, for ns laser pulses, a thermal wave propagating into the target is established and the absorbed energy is stored in a layer with a thickness of the order of $L_{th} \gg \alpha^{-1}$ [6]. The threshold value of the laser fluence for the evaporation to occur, F_{th}, can be estimated by calculating the energy needed to melt a surface layer with a thickness of the order of L_{th} [6]

$$F_{th} \approx \rho C_p \Delta T_m L_{th} A^{-1} \propto \tau_p^{1/2}, \qquad (2.4)$$

where ΔT_m is the difference between melting and initial temperatures of the target. In general, the layer thickness ablated per pulse, Δz_v, can be described by either an Arrhenius-type or a linear relationship, depending on whether the fluence is below or above F_{th}, respectively [8]. In the latter, Δz_v can be expressed as a function of latent heat of evaporation, L_v:

$$\Delta z_v \approx \frac{A}{\rho L_v}(F - F_{th}) \qquad (2.5)$$

When the pulse duration is shorter than the electron cooling time, the thermal conduction into the target (electron thermal diffusion and electron-to-lattice coupling during the laser pulse) is negligible, as a first approximation, and the material removal takes place from direct solid-to-vapor transition. In this framework, F_{th} is described by $F_{th} = \rho L_v/(\alpha A)$. Disregarding the heat conduction inside the target, the ablation depth per pulse mostly depends on the optical penetration depth of the material and the fluence [6]:

$$\Delta z_v \approx \alpha^{-1} \ln\left(\frac{F}{F_{th}}\right) \qquad (2.6)$$

As an example of the relationship between the ablation depth, the fluence, and the pulse duration, in Fig. 2.2b the experimental data related to ps and ns laser irradiation of Cu in vacuum are reported together with fits from Eqs. 2.5 and 2.6, respectively.

The hot and dense plasma plume (consisting of neutrals, ions, electrons, and atoms in excited states) gets arranged in a 10–100 μm thick layer that comes into an equilibrium state and isothermally expands away from the target surface during the ns laser pulse. On the contrary, with fs pulses, the plasma formation process (1–10 ps) is still standing after the termination of the laser beam [8]. In the nonstationary regime, the system of the *particles* ejected from the target interacts independently with the incoming radiation, giving rise to flow processes with both reflection and recondensation, even losing memory of the source mechanism. The laser pulse can be absorbed by the plasma through either IB or direct single photoionization (PI) of excited atoms [10]. Both processes allow a preferential transfer of the incident energy into the plume rather than the target material and allow increasing the temperature and the plasma lifetime itself [11].

Driven by the internal pressure gradient the plasma plume increases in size and decreases in density, with basic features, in vacuum, that can be described by an adiabatic expansion, in contrast to the initial isothermal expansion when plasma is formed and heated up [12]. The plasma is finally characterized by a small flow velocity and a forward-peaking distribution that is hemispherical at the leading edge and conical along its longitudinal axis. Indeed, the final angular distribution ranges from $\sim\cos^4\theta$ to $\sim\cos^{50}\theta$ depending on the material removal efficiency (pulse energy), θ being the angle measured from the target normal [13].

2.2.1 Plasma Parameters: Temperature and Density

The kinetic and radiative properties of the expanding plasma greatly depend on the target material and the laser parameters, among which is the fluence that mainly affects the laser target absorption and the laser ablation mechanisms. The plasma production and evolution can be analyzed through the monitoring of the electronic temperature, T_e, and the electron density, N_e. Both parameters can be evaluated from optical emission spectroscopy (OES) measurements with spatial and temporal resolution, where the plasma emission is imaged on the entrance slit of a monochromator and then collected by a photomultiplier or a charge-coupled device (CCD). T_e can be extracted from the intensity ratio of the spectral lines, under the assumption that the plasma is in local thermodynamic equilibrium (LTE), and through the well-known Boltzmann plot method, with the relation [10, 14]

$$\ln\left(\frac{I_{ki}\lambda_{ki}}{g_k A_{ki}}\right) = -\frac{E_k}{k_B T_e} + C, \qquad (2.7)$$

where I_{ki} and λ_{ki} are the relative intensities and wavelengths of the spectral transitions between the k and i energetic levels, g_k are the statistical weights of the upper excited levels with energy E_k, A_{ki} is the transition probability for spontaneous radiative emission, k_B is the Boltzmann constant, and C is a constant.

N_e can be estimated from the Stark width of the emission spectral lines, w_t, which is mainly determined by electron impacts with the radiating atoms and a smaller contribution due to the fields generated by the charged ions. In fact, the emitting species (atoms or ions) in the plasma are under the influence of an electric field by fast-moving electrons and relatively slow-moving ions. This perturbing electric field can act on atoms or ions by shifting their energy levels and, then, broadening the emission lines (Stark broadening). In ionized atom lines, w_t can be related to N_e (in cm^{-3}) through the relation

$$w_t(N_e, T_e) = w_e(T_e) 10^{-17} N_e \qquad (2.8)$$

where $w_e(T_e)$ is the width due to electron impact at $N_e = 10^{17}$ cm^{-3}. Other pressure-broadening mechanisms, such as resonance and van der Waals broadening, are usually negligible and only Gaussian (instrumental and Doppler effects) and Stark broadening must be taken into account in the deconvolution procedure of the spectral lines. The difficulty in the application of these spectroscopic methods arises from the possible self-absorption of the spectral lines (to be tested and corrected) and the spatial inhomogeneity of the plasma. Typical OES measurements have exposure times of tens of ns during a typical lifetime of less than 1 ms and, over the time of interest, N_e and $k_B T_e$ are in the range of 10^{16}–10^{18} cm^{-3} and ~1 eV, respectively. The N_e behavior is reported to be qualitatively pulse width independent, even with the fs laser having a decay density much more rapid than ns pulses [15]. On the contrary, the spectral lines broaden as the laser fluence increases, as a result of the high pressure associated with the plasma, and the plasma itself is assumed to be more energetic and more ionized [16]. The temperature linearly increases as the laser fluence rises, with a slope that increases as the laser pulse width decreases, thus indicating that ns laser ablation is less dependent on laser energy than fs [15].

A typical evolution of T_e over the plasma expansion is shown in Fig. 2.3a, where are reported the values obtained from the irradiation of a LiF single-crystal target with an ArF excimer laser (λ = 193 nm and τ = 20 ns) at a fluence of 1.5 J cm^{-2} [14]. The calculations refer to three spectral emission lines from LiI and demonstrate that T_e decreases sharply along the first millimeter, in the present case from 1.85 eV to 0.66 eV, and then it reaches an almost constant value, around 0.45 eV, both in time and distance. Such behavior is in agreement with similar data obtained with metals, graphite, YBa$_2$Cu$_3$O$_7$, and LiNbO$_3$ [14] and suggests that the expansion of Li atoms mostly takes place without atom-to-atom collisions and can be further favored by the long-lasting delayed release of Li atoms with high velocities [17].

In Fig. 2.3b is reported the variation of the time-integrated N_e, calculated from Eq. 2.8, as a function of the distance to the target, d. The data are obtained for the 610.3 nm triplet observed by LiNbO$_3$ single-crystal ablation through an ArF excimer laser (λ = 193 nm and τ = 20 ns), with a laser fluence at the target surface of 1.2 J cm^{-2} [18]. It is found that, for low values of N_e (~10^{16} cm^{-3}) when the absorption by IB is negligible (typically IB takes place for N_e = 10^{18}–10^{19} cm^{-3}), it decreases by a factor of 2 in the first millimeter from the target surface. Indeed, in such a regime, the contribution of the continuum emission becomes negligible for distances $d > 2$ mm and is characterized by a typical N_e behavior indicating that the initial expansion of the plume is 1D [15, 18]. The time-resolved values of N_e for different distances to the target surface demonstrate that N_e (whose vertical axis refers to N_e in 10^{16} cm^{-3}) first increases up to a maximum and then decreases following a t^{-2} law (becomes nearly constant after ~700 ns) [18].

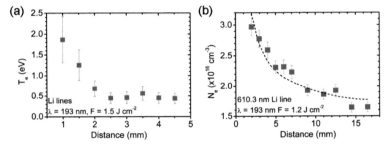

Figure 2.3 (a) Time-integrated electronic temperature, T_e, versus distance from the target, d, for the emission lines of LiI (610.3 nm, 460.2 nm, and 413.2 nm) in a LiF plasma (ArF laser, $\lambda = 193$ nm, $\tau = 20$ ns, and $F = 1.5$ J cm^{-2}). Reprinted from Ref. [14], with the permission of AIP Publishing. (b) Time-integrated electron density, Ne, versus d, related to the 610.3 nm emission in a LiNbO$_3$ plasma (ArF laser, $\lambda = 193$ nm, $\tau = 20$ ns, and $F = 1.2$ J cm^{-2}) calculated with $w_e = 0.261$ Å for Te = 10,000 K. The dashed line is the d^{-1} fit. Reprinted from Ref. [18], with the permission of AIP Publishing.

2.2.2 Plasma Composition: Atom and Ion Distribution/Yields

The ejected particles in the plasma get arranged in an ellipsoidal shape, with an acceleration of the front that is, in any point, inversely proportional to the instantaneous position and goes asymptotically at zero [12, 19]. Then, far enough from the target, the plasma expansion takes place with a constant velocity. In the following, the consideration about plasma dynamics will be restricted to a vacuum-like environment (from a high-vacuum to a low-pressure regime), where the plume expands freely without any external viscous force [20, 21].

The analysis of the traveling species inside the plasma plume can be carried out with different characterization techniques, depending on the species to be analyzed. For electron and/or ion distribution, the plasma diagnostic can be done by means of Langmuir probe (LP) measurements, with a small biased metallic electrode placed inside the plasma. Depending on the applied bias, positive or negative, LP collects electrons (electron saturation region) or positively charged ions (ion saturation region), respectively. The ion current density transient, $f(t)$, is defined as [22]

$$f(t) = \frac{I(t)}{RSe} \qquad (2.9)$$

under the assumption that ions are single-ionized, where $I(t)$ is the current transient, R the circuit resistance, S the LP surface, and e the charge of the electron. The total number of ions per unit area per pulse, that is, the ion yield, is obtained by time integration of $f(t)$. The distribution of the ions can be obtained by assuming the laser plasma as an instantaneous point source of ions and approximating the ion velocity to as $v = d/t$, where t is the time delay with respect to the laser pulse and d the target-to-probe distance.

The corresponding velocity distribution can be obtained as

$$\Phi_i(v) = f(t)|J(v)|, \qquad (2.10)$$

where the Jacobian for the transformation, $J(v)$, can be written as [23]

$$J(v) = \frac{t^2}{d}, \qquad (2.11)$$

consistent with the condition that d is longer than the ion source size (laser spot at the target surface) and the transient lasts much longer than the ion ejection time (i.e., the pulse duration).

On the opposite side, the OES (see Section 2.2.1) is a diagnostic technique reliable for both charged and neutral particles and can be used also by directly imaging the plasma generated by the laser pulse on a time-gated intensified CCD camera. In Fig. 2.4 are shown the 2D time-gated images of the Au plasma generated at 9.0 J cm^{-2}. The spatial profiles of the emission intensity correspond to the emission of excited Au neutrals, Au*, at 479.3 and 481.2 nm, selected after suitable optical filtering and recorded at different delay times. After the laser pulse, at only 1.5 µs, the plasma expands away along the target surface normal and, after its maximum expansion, a severe distortion is observed in the presence of the substrate. The primary front, then, moves backward and, at 2.5 µs, a secondary emission appears close to the substrate, even before the arrival of the Au* distribution. The fronts moving back to the target have been related to self-sputtering of neutrals and backscattering of fast Au⁺ by the substrate and recombination into Au*. These effects have important consequences in the properties of the deposited materials and are analyzed in Section 2.3.

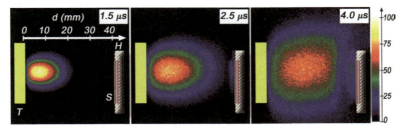

Figure 2.4 2D time-gated images of the Au plasma generated at 9.0 J cm⁻² as from the Au* emission at 479.3 nm and 481.2 nm (ArF laser, λ = 193 nm, τ = 20 ns, and F = 9.0 J cm⁻²). The time delays after the laser pulse; the position of the target, T; the substrate, S; the substrate holder, H; and a scale representing the target-to-substrate distance are shown. Adapted and reprinted (figure) with permission from Ref. [23]. Copyright (2007) by the American Physical Society.

After removing the emission near the target surface (1–2 mm) from Fig. 2.4, which corresponds to the *Bremsstrahlung* continuum, atomic [23] and ionic [24] lines can be analyzed to achieve spatial distribution profiles of the targeted species along the expansion direction. The corresponding velocity distributions obtained from OES can be calculated by dividing the expansion axis units, d, by the corresponding delay time, t, at which each image is recorded, and normalizing to the unit area. Notably, it has been found that the velocity distributions are independent of the delay time for a given laser fluence, as expected for the expansion of species in vacuum.

The kinetic energy distributions from the corresponding velocity distribution, $\phi(v)$, can be expressed as [25]

$$N(E) = \Phi(v)|J(E)|, \qquad (2.12)$$

where $E = 0.5\ mv^2$, in which m is the mass of the selected species and $J(E) = 1/mv$ the Jacobian for the transformation, irrespective of the analysis technique.

The main differences between the velocity distributions related to ions and neutrals inside the same plasma are clearly represented in Fig. 2.5, where the normalized velocity distributions for Au* and Au$^+$ determined by OES and LP measurements, respectively, are reported. These results were obtained ablating an Au target with an ArF excimer laser (λ = 193 nm and τ = 20 ns) at two different fluence values (2.6 J cm^{-2} and 9.0 J cm^{-2}) [23]. The velocity distributions are characterized by a fast rise followed by a slower decay and the maximum of the curve corresponding to the ions is shifted toward higher values with respect to the atoms. The latter indicates that the peak velocity of the ions is considerably higher (twice in the present case) than that of the atoms. Indeed, Au* distributions have a maximum between 3 and 4 km s^{-1} (8–17 eV), while the Au$^+$ ones peak at 9–10 km s^{-1} (80–100 eV). Nevertheless, it is stated that a significant fraction of ions have a velocity considerably higher than the mean value. In the present case, the long-tail high-velocity fraction represents ~17% of the whole set of ions, with velocities >20 km s^{-1} (kinetic energies > 400 eV), whereas the corresponding fraction of neutrals is below 0.5%. As a general agreed picture, the peak positions of both ion and atom distributions vary little with the laser fluence even for wider ranges of values [22, 26], as opposed to the high energetic part of the distributions. In general for metals, neutral atoms always have kinetic energy distributions up to a maximum value of 20 eV, while the ions usually present peaks at higher values, up to 100 eV, and the distributions are typically broader, with a tail that can extend to energies as high as 1500 eV and with a fraction of energetic ions (>200 eV), which increases with the laser fluence. The above-mentioned conditions (at 9.0 J cm^{-2}) determine a density of Au$^+$ ions per pulse of the order of 10^{14} ions cm^{-2}, with a large fraction of them (~2.4 × 10^{13} ions cm^{-2}) having kinetic energies >400 eV. In conclusion, they will bombard the substrate in a time of ~0.8 µs, which, in turn, leads to a transient incident flux of 3.0 × 10^{19} ions cm^{-2} s^{-1} for the most energetic particles of the plasma [23].

The differences between excited neutrals and ions are demonstrated to be irrespective of the target element, the changes between different metals being mostly related to the high energetic part of the distribution and to the different delay time (with respect to the laser pulse). This is shown in Fig. 2.6a, where are reported the time-of-flight measurements under ablation of different metal targets with an ArF excimer laser [22]. Notably, ion yields of the order of 10^{19}–10^{20} ions cm^{-2} s^{-1} are found for different metals, with a distribution that broadens and shifts toward higher delay times as the atomic weight of the element increases. Such a behavior is better depicted in Fig. 2.6b, where the time at which the ion yield distribution reaches the maximum as a function of the element atomic weight is reported [22]. The heavier the element, the longer the time needed to reach the probe.

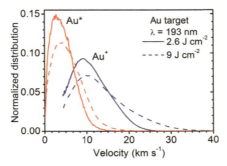

Figure 2.5 Normalized velocity distributions of excited neutrals, Au*, and ions, Au$^+$, obtained for the ablation of a Au target at $F = 2.6$ J cm^{-2} (solid lines) and $F = 9.0$ J cm^{-2} (dashed lines) (ArF laser, $\lambda = 193$ nm, and $\tau = 20$ ns). Reprinted (figure) with permission from Ref. [23]. Copyright (2007) by the American Physical Society.

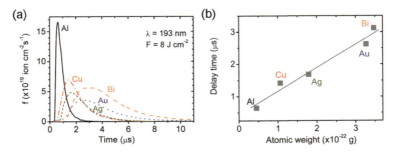

Figure 2.6 (a) Current density transients, $f(t)$, measured from the ablation of different metals (ArF laser, $\lambda = 193$ nm, $\tau = 20$ ns, and $F = 8$ J cm^{-2}) as labeled close to each graph. (b) Time delay, after the laser pulse, at which the maximum of $f(t)$, in (a), occurs as a function of the metal atomic weight. The solid line represents a linear fit of the experimental data. The labels for the different metals are reported. Reprinted from Ref. [22], with the permission of AIP Publishing.

The efficiency of the ablation process depends on the nature of the target in terms of thermal properties. The rate of the ion production increases with the laser fluence, as depicted in Fig. 2.7a, where the values obtained for different metals in a range of fluence values up to 15 J cm^{-2} are reported [22]. Such a behavior is coherent with the relation between ablation depth and fluence (Eq. 2.5), even when, for the ion yield, this dependence shows a change of slope around 4–5 J cm^{-2}. In fact, the increase of the ion yield becomes smoother as the fluence is increased, thus evidencing that the processes inside the plasma become of increased importance as the fluence is increased. Saturation effects by plasma shielding have been found in different laser-induced plasmas and increase as the fluence increases [11]. Moreover, the ion yield also depends on the material to be ablated, decreasing with the melting point, cohesive energy, and volatility of the metal [22]. The yield in Fig. 2.7a decreases approximately 1 order of magnitude from Bi (577 K) to Cu (1357 K), most of the metals having similar values, with the only exception of Bi. In addition, the ablation threshold of the metals increases as the melting temperature increases [22]. This result is in agreement with the relation between the melting point of metals and the fluence described by Eq. 2.4.

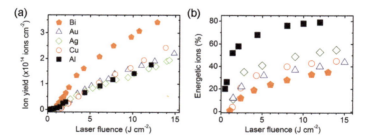

Figure 2.7 (a) Ion yield measured for the ablation of different metals as a function of fluence (ArF laser, λ = 193 nm, and τ = 20 ns). (b) Percentage of ions with kinetic energy >200 eV as a function of laser fluence upon ablation of the metals shown in (a). Reprinted from Ref. [22], with the permission of AIP Publishing.

Finally, the properties of the target also influence the overall ion kinetic energy distribution, not with respect to the maximum position (<100 eV irrespective of the metals), but in terms of a high-energy long tail. Such behavior is summarized in Fig. 2.7b, where the percentage of energetic ions (>200 eV) is reported for different metals as a function of the laser fluence.

The amount of energetic ions, in accordance with the evolution of the ion yield (Fig. 2.7a), increases with the fluence and shows a change of slope around 4–5 J cm^{-2}. Interestingly, when ablating Al (as well as aluminum oxide, Al_2O_3) targets with a laser wavelength of 193 nm, different behaviors have been found. In fact, a kinetic energy distribution peaking at higher values (~130 eV) and broader than other metals was found, both factors promoting a higher fraction of energetic ions (80%) with respect to other metals (20% and 50%). Finally, a high fraction of energetic ions (20%) was measured for fluences for which most metals show no significant number of them (Fig. 2.7b). This is explained because, together with the photoablation process described in Section 2.1, the photon energy associated with the laser wavelength of 193 nm (6.4 eV) overcomes the ionization energy of Al (5.98 eV) [24, 27]. This energy is enough to excite the $3p^{22}S_{1/2}$ level of Al and to produce a high concentration of Al$^+$ through the loss of the outer electron. Notably, the PI process in the present case is also responsible for the reduction of the threshold fluence for ion production.

The shift of the maximum kinetic energy depends on the fact that direct ionization enhances the degree of ionization of the plasma, which consequently shifts the kinetic energy distribution to higher energies due to coulomb acceleration [26]. It indicates that the distribution of ions is forwardly peaked along the direction perpendicular to the target and accelerates the ions with an acceleration that scales with the ion charge. This effect is more important for Al than for the other metals, since triple-ionized species have been detected for the former [28], while single- or double-ionized species have been reported for the latter [29].

The presence of a considerable amount of negative charged ions in the expanded plasma is an additional remarkable property of the ablation of Al_2O_3 with an ArF excimer laser. Mass spectrometry measurements reveal the presence of negative ions at 4 cm from the target for fluences in the range of 1–2 J cm^{-2}. Most of the negative ions (>80%) are O$^-$ and represent one-third of the O$^+$ ions for these experimental conditions. Kinetic energy distribution of O$^-$ is a broadband that extends from 5 eV to 40 eV, that is, more energetic than neutral O. This suggests that O⁻ is produced by the neutralization of O⁺ followed by electron attachment. The presence of negative ions in the ablation process has been overlooked in the last decades, despite their relevance in other depositon processes, like sputtering. Herein, the initial densification in the initial stages of deposition by ion bombardment or damage at the substrate has been linked to the presence of negative ions [30, 31].

It arises from the results shown here that the plasma evolution is the result of a dynamic balance between different processes taking place when high photon energies are involved, such as the PI process from first and second excited states, the production of multiple-charged ions leading to an "apparent" increase of the yield [24, 28], as well as the dissociation of dimers [29]. Indeed, such an effect becomes more and more significant for $F >> F_{th}$ when the thickness of the layer ablated per pulse becomes independent of F.

2.3 Production of Metal Nanoparticles by Pulsed Laser Deposition

The technique of PLD was initially devised to reproduce the composition of complex, mostly oxide, materials in thin films in a quasi-stoichiometric way. However, PLD has turned in the last two decades into an excellent technique for producing metal-dielectric nanocomposites in a thin-film configuration [32]. The ablation plasma produced by the laser pulse directly evolves perpendicular to the target surface and, finally, condensates on the substrate surface, generally separated by a few tens of millimeters. Uniform ablation of the target and film thickness can be improved through the rotation of both the target and the substrate, along with a possible shift of the substrate with respect to the plasma expansion direction.

The laser beam can be focused alternatively onto two or more targets in a single growth process, with the possibility to deposit multilayers and independently control the deposition features for the single material with a feedback over the growth rate and a resolution up to the monolayer (ML) level [33]. One of the limitations, especially for metals, is the ejection of molten droplets of sizes up to a few micrometers into the plumes, which can induce splashing, or the formation of particulates onto the film surface. Their presence is quite significant with Al but negligible in the case of noble metals like Ag or Au [1, 34, 35].

Laser parameters, substrate-to-target distance, and substrate temperature may affect the kinetics of the plasma species and the mobility of the atoms at the substrate. Nucleation and growth of clusters mainly depend on the interaction energies of substrate atoms and atoms/ions reaching the surface. In the case of metal atoms on oxides, the metal atoms should make 3D particles and leave a clean oxide surface between the metal particles if

$$\gamma_{m/ox} > \gamma_{v/ox} - \gamma_{v/m} \leftrightarrow E_{adh} < 2\gamma_{v/m}, \qquad (2.13)$$

where $\gamma_{m/ox}$ is the metal–oxide interfacial free energy, $\gamma_{v/ox}$ and $\gamma_{v/m}$ are the surface free energy of the clean (solid) oxide and the (liquid) metal in vacuum, simultaneously, and $E_{adh} = \gamma_{v/m} + \gamma_{v/ox} - \gamma_{m/ox}$ is the adhesion energy (work needed to separate the metal–oxide interface). These parameters are related to the free enthalpy of the oxide, of the metal, and of the metal–oxide interface. When Eq. 2.13 is satisfied, Volmer–Weber growth mode is established and the adatoms get arranged in 3D islands from the onset of growth [1, 2]. Most of the mid-to-late transition metals on SiO_2 and Al_2O_3 verify Eq. 2.13 and do not wet the oxide surface. Indeed, for Ag, Au, and Cu on a-Al_2O_3 are reported values of 810–930 mJ m^{-2}, 1130–1500 mJ m^{-2}, and 1220–1350 mJ m^{-2}, respectively, as surface free energy of the liquid metals and similar values are found for amorphous SiO_2 as nucleation surface. On the contrary, the corresponding adhesion energies are 323 mJ m^{-2}, 265 mJ cm^{-2}, and 490 mJ m^{-2} for Ag, Au, and Cu, respectively [2]. Analogously, the surface free energy of polycrystalline Al_2O_3 is 650–925 mJ m^{-2} (even lower for a-Al_2O_3), of SiO_2 is 307–605 mJ m^{-2}, of MgO (110) is 800 mJ m^{-2} [36], and of amorphous carbon is 40–170 mJ m^{-2} [37, 38]. Shape, dimensions, and distribution of these NPs can be modified and controlled by changing the deposition parameters.

2.3.1 Dependence with the Number of Laser Pulses in the Metal Target

The amount of material deposited can be controlled by the number of laser pulses on the target. Herein, we analyze firstly the evolution of the properties of metal NPs supported on a dielectric substrate and secondly those of the NPs embedded in a dielectric matrix. Figure 2.8 shows the plan-view transmission electron microscopy (TEM) images of Au NPs deposited, with different numbers of laser pulses, on a 10 nm thick a-Al_2O_3 buffer layer, equal the laser fluence. The dark areas correspond to the Au NPs, and the bright surrounding regions correspond to the substrate. Well-defined Au NPs with a variety of shapes and dimensions can be produced by PLD by varying the number of laser pulses. Small and round Au NPs are produced for a low number of pulses (Fig. 2.8a), while a bimodal distribution, with both small and big NPs, is achieved for the highest number of laser pulses (Fig. 2.8b,c). The small NPs in the bimodal growth produced by the implantation of metal into the substrate are analyzed in the next section. The big NPs have a circular to elongated shape and they are nucleated at the substrate surface (herein defined as "supported"). The corresponding morphological parameters calculated from TEM images are summarized in Fig. 2.9, where the evolution of the in-plane aspect ratio (longer-to-shorter axis ratio, AR), the in-plane mean diameter (mean value between the longer and shorter axes, ϕ), the surface coverage, and the NPs' number density are presented as a function of the number of laser pulses. These results for supported NPs (full symbols of Fig. 2.9) evidence the evolution from small round (AR = 1 and $\phi \leq 3$ nm) to elongated (AR $\geq$ 1.4 and $\phi \geq 6$ nm) Au NPs as the number of laser pulses increases. In addition, these trends are accompanied by a decrease of the number density of NPs and an increase of the metal coverage. The interplay of two different mechanisms contributes to the production of the metal NPs. On one hand, the early-stage deposition is characterized by nucleation and growth of 3D round NPs. On the other hand, as the NPs grow, they impinge upon each other and coalesce into larger, but still compact, quasi-spherical NPs. The time required to complete the coalescence process increases as the NP size increases. There is a size value above which the time required to complete the coalescence exceeds the average time for an additional NP to impinge with the coalescing one [39].

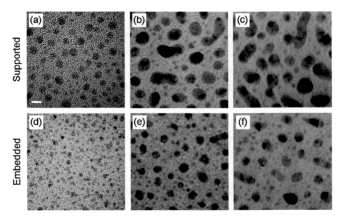

Figure 2.8 Plan-view TEM images of Au NPs supported on (a–c) and embedded in (d–f) a-Al$_2$O$_3$ obtained with a different numbers of laser pulses: 340 (a, d), 640 (b, e), and 720 (c,f). The fluence was 2.7 J cm^{-2} for both Au and Al$_2$O$_3$ targets. The scale bar is 5 nm [45]. (a, d, e) Reprinted from Ref. [45], with the permission of AIP Publishing.

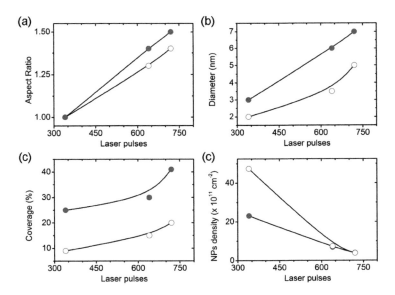

Figure 2.9 Average aspect ratio (a), mean diameter (b), surface coverage (c), and number density (d) of Au NPs supported on (●) and embedded in (○) an a-Al$_2$O$_3$ substrate as a function of the number of laser pulses used to ablate the metal target. The data refer to the NPs in Fig. 2.8. The mean diameters are calculated as the mean value between the length and width of the NP dimensions. The lines are guidelines.

Once this threshold is overcome, the NPs touching each other no longer coalesce into round NPs but form elongated structures, thus explaining also the reduction of NP number density at the expense of the increase of the metal coverage. If further metal atoms are deposited, the elongated NPs are joined and percolated structure are formed. Finally, the channels between the structures are filled in to form a continuous metallic film. The deposition technique affects the equivalent metal thickness at which the film reaches the percolation threshold, and in this sense, the instantaneous and energetic fluxes of species play a crucial role.

Resistance measurements have shown that PLD films reach the percolation with less metal amount than thermally deposited films [39]. The time-averaged flux of ions can be modified not only with the instantaneous flux per pulse but also by the repetition rate of the laser pulses. Kinetic Monte Carlo simulations and depositions of Ag on SiO_2 show that the number density of NPs increases and the percolation thickness decreases as the repetition rate of laser pulses increases [40, 41].

A similar evolution from round to elongated NPs has been identified also for different metal–substrate combinations such as Au on Si [42] or Ag on glass [43], where they found that the NP size increases with the number of laser pulses and it diverges in the percolation threshold in accordance with the formation of elongated nanostructures. Indeed, a power law describes the dependence of the NP diameter on the equivalent metal thickness, as demonstrated for Pt NPs grown on highly oriented pyrolytic graphite (HOPG) [44].

Metal NPs embedded into a dielectric matrix can be successfully produced by PLD. Figure 2.8d–f shows Au NPs embedded in an a-Al_2O_3 host. They are produced in the same conditions as the corresponding supported Au NPs (Fig. 2.8a–c) and then covered by a 10 nm thick layer of a-Al_2O_3. The morphological parameters related to the covered NPs are reported as open symbols in Fig. 2.9, and it is shown that the covering process reduces both the sizes and the size dispersion and, in turn, the metal coverage for the embedded NPs. Two sets of small round (AR = 1 and $\phi \leq 2$ nm) and elongated Au NPs (AR ≥ 1.3 and $\phi \geq 3.5$ nm) can be found in analogy to the case of supported NPs.

In line with the morphological changes, the metal content associated with the NPs, either supported or embedded, increases with the number of laser pulses, the former being always higher than the latter. The reduction of gold content, [Au], from supported to covered NPs is nearly constant (~4 × 10^{15} at. cm^{-2}), as shown in Fig. 2.10a, and is related to the sputtering induced by energetic ions of the host during the covering process. When the energy of the arriving species becomes of the same order of magnitude of the cohesive energy, the rupture of chemical bonds of the lying species is induced (see the next section). Nevertheless, in a range well below the percolation threshold, the overall in-plane mean diameter values follow the same linear dependence with the metal amount, irrespective of whether the NPs are embedded or supported, as shown in Fig. 2.10b. Such behavior reveals that the mean diameter is determined by the effective metal amount deposited, irrespective of the processes the NPs undergo [45]. The out-of-plane dimension, namely the height, of the NPs can be analyzed in the case of embedded NPs by cross-sectional TEM images. The ratios between height and in-plane mean diameter were found to vary from 1 to values anyway not smaller than ~0.7 for Ag NPs embedded in a-Al$_2$O$_3$ [46]. Such an evolution reveals that NPs produced by PLD range between spheres and prolate ellipsoids as the number of laser pulses increases, and states that the coalescence process promotes the formation of the in-plane elongated shapes rather than out-of-plane taller structures.

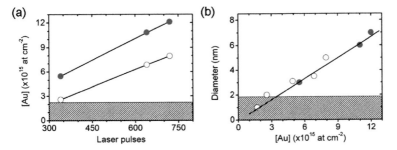

Figure 2.10 (a) Metal content of Au NPs supported on (●) and embedded in (○) a-Al$_2$O$_3$ as a function of the number of laser pulses. The fluence was F = 2.7 J cm^{-2}. The lines are guidelines. (b) Evolution of the mean diameter of NPs as a function of the metal content of supported (●) and embedded (○) NPs in a-Al$_2$O$_3$ from (a) and from Ref. [48] when obtained with the same fluence. The continuous line is a linear fit of the experimental data. Reprinted from Ref. [45], with the permission of AIP Publishing.

2.3.2 Dependence with the Laser Fluence

PLD is characterized by the high kinetic energy of the depositing species that arrive at the substrate in short bursts. Instantaneous flux of ions and amount of energetic species are strongly related to the laser fluence (Section 2.2.2 and Fig. 2.7). The number of ions deposited in the first pulse is not enough to generate stable nucleation centers, and thus nonstable centers are produced. The incoming species from the next laser pulse tend to stabilize these nucleation centers by the addition of new species. After a few pulses, a stationary regime is reached and the arriving species join the stabilized centers that become NPs. According to this description, as the flux of ions per pulse is increased, the number of stable centers and the number density of NPs increases [47].

Depending on the kinetic energy of the incoming species, two regimes can be identified. For low energies, below the displacement threshold (~30 eV), the excess of kinetic energy is used to increase the surface mobility, thus promoting coarsening of NPs [35]. For higher kinetic energies, the nanostructures are modified directly by the arriving species through sputtering of the deposited atoms, subsurface implantation, or mixing at the interface, among others [45, 48]. In general, for fluences above the ablation threshold, the ablation rate of metals increases linearly with the fluence [49, 50], while the percentage of ions with kinetic energy >200 eV is higher than 30% for fluences above ~5 J cm^{-2}, as shown in Fig. 2.7b. In conclusion, together with the laser repetition rate, the most direct way to increase the particle flux is to raise the laser fluence.

Figure 2.11 shows the plan-view and cross-sectional TEM images of Au NPs embedded in a-Al$_2$O$_3$, where a dielectric matrix and metal are produced with the same fluence increasing from the left- to the right-hand side. The size and shape of Au NPs are clearly different, in spite of the fact that they have similar amounts of gold [48]. In all cases, bimodal distributions of small and large NPs are identified. Cross-sectional images evidence that these distributions belong to different layers, the smaller NPs lying in the deeper one. The distance between these layers and the mean diameter of these smaller NPs increase slightly with the fluence, the former varying between 1.2 nm and 2.0 nm and the latter being ~1.9 nm for Au ablation with fluences from 2.7 J cm^{-2} to 9 J cm^{-2} [48]. Furthermore, the films having the lowest amount of metal (not shown here) only show one NP population that is most likely equivalent to the deeper NPs.

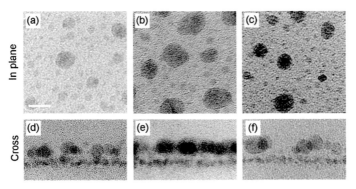

Figure 2.11 Plan-view (a–c) and cross-sectional (d–f) TEM images of Au NPs embedded in a-Al$_2$O$_3$ having similar amounts of Au (~4.5 × 10^{15} at. cm^{-2}) and both metal and dielectric layers produced at increasing fluences: (a,d) 2.7 J cm^{-2}, (b,e) 6.2 J cm^{-2}, and (c,f) 8.9 J cm^{-2}. The scale bar is 5 nm. Reprinted (figure) with permission from Ref. [48]. Copyright (2005) by the American Physical Society.

According to this picture, the production of Au NPs involves different processes. The first leads to the production of homogeneous, round, small NPs implanted into a-Al$_2$O$_3$. Implantation depth scales with the mean kinetic energy of the incoming Au ions, which depends mostly on the atomic weight rather than on the laser fluence [25, 48]. The dashed areas in Fig. 2.10a,b evidence the maximum metal content (2.3 × 10^{15} at. cm^{-2}) and the corresponding diameter (1.9 nm) related to the implanted layer, identified for Au NPs, indicating that the implantation process saturates very fast in PLD.

Similar evidence was found when depositing Bi [51], while they were harder to be observed in the case of Ag [46] and they were not identified for Cu [52], in spite of the presence of energetic ions. The implanted layer is better defined and is located deeper for heavier ions like Bi (208.98 amu) and Au (193.97 amu) rather than Ag (107 amu) or Cu (63.55 amu), in agreement with the estimations of the implantation depth of monoenergetic incident ions with kinetic energy in the range of 10–1000 eV [53].

The second process leads to the nucleation and growth of large NPs and starts once the implantation process becomes saturated. Incoming energetic species produce the self-sputtering of a fraction of the already deposited metal, as shown in Fig. 2.4. Self-sputtering yield increases with the kinetic energy and the degree of ionization of the incoming species. These two parameters are strongly correlated and increase with the laser fluence. The former evolves according to the higher percentage of energetic ions, as shown in Fig. 2.7b. The degrees of ionization calculated under UV ablations of metal are normally larger than 50%: Fe (90% at 4.5 J cm^{-2}), Al (60% at 2 J cm^{-2}), Ag (57% at 2 J cm^{-2}), and Au (60% at 2.7 J cm^{-2} and 75% at 9 J cm^{-2}) [25, 54, 55]. This high flux of energetic ions produces self-sputtering as high as 60%–70% of the arriving species in the case of Au for fluences in the range of 2.7–9 J cm^{-2}, 55% in the case of Ag at 4.5 J cm^{-2}, or 100% in the case of Zn [56].

The third and last process corresponds to the modification of metal NPs when covered with a dielectric. Figure 2.12 shows the plan-view TEM images of Au NPs supported on a-Al$_2$O$_3$ (Fig. 2.12a) and Au NPs embedded into a-Al$_2$O$_3$ produced with increasing fluences to ablate the a-Al$_2$O$_3$ target in the covering process (Fig. 2.12b,c). The difference between Figs. 2.8 and 2.12 is that the fluence used to ablate the metal is constant in the latter case (2.7 J cm^{-2}), and thus the effect of the fluence in the covering layer can be extracted from Fig. 2.12. These images reveal again a bimodal distribution of NPs that are related to implantation (small NPs) and to NPs at the surface (bigger ones). Before the covering process, NPs have a diameter of ~6 nm, whereas after the covering their Au decreases and, then, their diameter is reduced (to ~3 nm). The fraction of sputtering metal is not negligible (~10 ± 5%) still for a laser fluence close to the threshold upon ablation of the Al$_2$O$_3$, and the higher the fluence, the higher the sputtered fraction (up to 70%). This relation between sputtering yield and laser fluence follows the same trend as of the dependence of the population of ions with kinetic energy higher than 200 eV and laser fluence shown in Fig. 2.7b. This suggests that the sputtering process in PLD is directly related to the energetic ions produced during the ablation [24, 57] but also depends on the initial morphology of the NPs. Indeed cohesive energy is effectively size dependent [58], because the lower coordination number of the surface layer atoms makes them less stable than the atoms inside the

NPs. Notably, the cohesive energy reduces, from 3.8 eV for bulk Au to 3.1 eV and 2.2 eV for Au NPs of 7 nm and 3 nm diameter, respectively [45], and thus the smaller the NPs, the higher the fraction of sputtered material [58, 59]. A reliable analysis of the kinetic energy–dependent sputtering yield can be obtained in the framework of the Yamamura and Tawara model [60], all the details about the nuclear-stopping cross section being reported in Ref. [45].

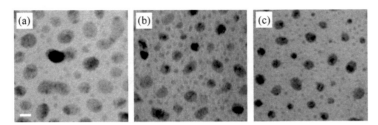

Figure 2.12 TEM images of (a) supported Au NPs (same sample as in Fig. 2.8b) and (b,c) corresponding embedded Au NPs in which the a-Al$_2$O$_3$ covering layer was produced at 2 J cm^{-2} and 9 J cm^{-2}, respectively. The scale bar is 5 nm. Reprinted from Ref. [57], with the permission of AIP Publishing.

In addition, the incoming energetic ions may be implanted into the NPs during the covering process. In the case of Au NPs covered with a-Al$_2$O$_3$, the latter produced at different laser fluences, the implantation process was found to be negligible only for values close to the ablation threshold of Al$_2$O$_3$ (0.8 J cm^{-2}). On the contrary the implantation depth of Al$^+$ into the Au NPs increases up to ~3 nm as the fluence increases in the range of 2–9 J cm^{-2} [57]. For the highest fluence analyzed, the implantation depth is, then, comparable with the whole size of the metal NPs; thus the implantation also modifies the composition of the Au NPs.

Analogously, for Cu NPs covered with a-Al$_2$O$_3$, the latter ablated with a high fluence (4.6 J cm^{-2}), the implantation process leads to the formation of an Al-Cu oxide on the Cu NPs and the degree of mixing was found to depend on the laser fluence used for the ablation of the Al$_2$O$_3$ target and appeared to have a threshold behavior [61]. Also the production of Ag-Cu alloys at the interfaces of Ag/Cu multilayers has been attributed to the implantation of energetic ions produced at a high fluence (7.5 J cm^{-2}) [62]. The relevance of the energetic incident ions in the implantation process has also been analyzed in the PLD of W/Si multilayers, where deposition conditions were selected in order to work with high-energetic Si ions and mainly low-energetic W atoms. Interface mixing of up to 3 nm occurred at the W/Si interface (i.e., Si deposited on W), while the Si/W interface remained sharp. Further, it was shown that using laser fluences close to the corresponding ablation thresholds, multilayers with sharp interfaces can be realized as then the kinetic energy of deposited particles is lower than the energy threshold necessary for implantation [63].

2.3.3 Peculiarities of Pulsed Laser Deposition in Nanoparticle Formation and Dependence with the Substrate

The high transient flux of ions, up to 10^{19}–10^{20} ions cm^{-2} s^{-1}, and the kinetic energies >400 eV have important effects on the morphology of the metal NPs other than the direct action of sputtering and implantation processes described in previous sections. The NP nucleation process itself, in PLD, is strongly affected by the high fluxes of energetic atoms because they generate, on the substrate, high densities of surface point defects, which in turn act as preferential nucleation centers for the clusters. However, it has to be mentioned that with different techniques, even when the flux is many orders of magnitude smaller than in PLD, small amounts of residual charge particles can generate surface defects too [64, 65].

Under the assumption that a significant percentage of nucleation centers occur at point defects, that is, that the nucleation rate is linear with the average flux [23, 66], for an Au average flux of ~0.5 – 1.5 × 10^{16} at. cm^{-2} s^{-1}, a minimum NP nucleation rate of ~5 × 10^{13} cm^{-2} s^{-1} was calculated. As a consequence, number densities of NPs range between 10^{11} and 10^{12} cm^{-2} [33, 48, 67–69], irrespective of the nature of the substrate being either amorphous (carbon, glass, or oxides) or crystalline (MgO [33], CaF$_2$ [69]), or even organics [67]. These values are five times higher than the saturation number density (calculated for Au on MgO [100]) produced by conventional evaporation techniques [70]. Definitely, with PLD it is possible to work in the coalescence regime (threshold of ~0.5 ML) [71], as the typical metal contents involved in Fig. 2.10 are normally higher than 0.5 ML, but still achieve layers of well-separated NPs. Such a peculiarity makes PLD unique with respect to other techniques such as conventional evaporation or molecular beam epitaxy (MBE) that are far from the coalescence regime (~0.1 ML) [72], with typical diffusion rates, D_R, 4 orders of magnitude lower and with less than half the NP densities of PLD [33]. Indeed, the counterbalance between (high) flux of arriving species and (high) adatoms mobility compensate for the differences between PLD and the MBE working regime and, under suitable conditions, allow epitaxial growth to be affordable also for metal structures grown by laser ablation [33, 73]. In fact, a condition generally acknowledged for epitaxy to take place is that the ratio between the diffusion rate and the deposition flux, D_F, that is, the growth speed of the aggregate, has to be higher than 10^5 [72]. Furthermore, such a compensation is clearly stated in the framework of the minimal model approximation, where $D_R/D_F \approx l^6/\ln(l^2)$, in which l is a characteristic length that represents either the mean island separation or the mean free path of diffusing adatoms (before they create a new nucleus or are captured by existing islands) [74]. Indeed, it was verified that for Fe produced on Mo (110), the D_R/D_F values at 300 K for MBE and PLD are similar.

Analogous conclusions were obtained for Au NPs produced by PLD at room temperature on MgO (110) [33] (TEM image shown in Fig. 2.13a). Notably, the nucleation of the NPs was achieved on all the faces of the MgO (110) nanocubes (60 nm average side length) and a number density of 1.7×10^{12} cm^{-2} (Au = 1.2×10^{16} at. cm^{-2}) with a center-to-center separation of $l_{PLD} \approx 4.5$ nm was estimated. A comparison with similar electron beam (EB)-evaporated samples produced at 381°C ($l_{EB} \approx 9.5$ nm) [64] allowed the calculation of the D_R value for PLD as 10^4–10^5 times the corresponding D_R for EB. The differences in D_R values are in line with different studies, demonstrating that decreasing the adatom mobility, the anisotropy of the NPs is enhanced and the deposition rate decreases abruptly [75].

Finally, it has to be mentioned that, irrespective of the deposition technique, the substrate can play an active role in determining the crystalline order (*taxis*) of the thin film grown on it (*epi*) if the substrate surface itself is provided by a crystallographic orientation.

Anyhow, a minimum requirement for the epitaxial growth to be fulfilled is that the lattice mismatch between the material to be deposited and the substrate has to be lower than 9% [76]. Namely, for Au (4.08 Å for bulk) on MgO (110) (4.20 Å) only ~3% of strain of the Au lattice on the MgO one is necessary for the epitaxy to occur, while for Fe on Mo (110) [74], the lattice mismatching is ~7%. Epitaxial growth of Au on MgO (110) was achieved, as stated by the images in Fig. 2.13b, related to the crystalline lattices associated with (002) MgO and (002) Au being parallel, with a 2° tilt [33, 73]. When crystalline substrates are used also the shape of the NPs changes with respect to the amorphous one (carbon as well as Al_2O_3). In the former, faceted NPs appear (Fig. 2.13c), as opposed to the quasi-spheres or nearly cuboctahedral smoother-edged NPs normally observed with the latter (Figs. 2.8, 2.11, and 2.12).

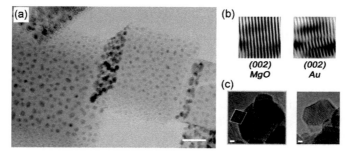

Figure 2.13 (a) TEM image of MgO nanocubes having Au NPs on all their faces with an average metal content of 1.2×10^{16} at. cm and a number density of 1.7×10^{12} cm^{-2}. The scale bar is 20 nm. (b) Reconstructed crystal fringes for (110) MgO and Au by inverse fast Fourier transformation of both an uncovered MgO region and a region with Au NPs. (c) Details of TEM images of Au NPs at the edge of the MgO nanocubes showing small octahedral (left-hand side, area within white diamond) as well as truncated octahedral NPs (right-hand side). The scale bar is 1 nm. Reprinted (figure) with permission from Ref. [33]. Copyright (2009) by the American Physical Society.

In particular, large NPs are truncated octahedral (right-hand side in Fig. 2.13c), with different degrees of truncation, while the small ones have typically an octahedral shape (NP evidenced in the left-hand side of Fig. 2.13c). In fact, the equilibrium shape of large metal NPs with face-centered cubic (fcc) structures on a support is a truncated octahedron. The metal-oxide adhesion energy to surface energy ratio determines the degree of truncation through the expression [77]

$$\frac{E_{adh}}{\gamma_{m/ox}} = \frac{\Delta h}{h_i}, \tag{2.14}$$

where h_i is the central distance to the facet parallel to the interface and Δh is the degree of truncation. From the NP on the right-hand side of Fig. 2.13c, $\Delta h/h_i \approx 0.55$ within 20%, which is in excellent agreement with the prediction of partial wetting as from the Wulff-Kaichew theorem, $\Delta h/h_i \approx 0.62$. Indeed, as the adhesion energy decreases with the size of the NPs [45, 78], the bigger the size of the NPs, the higher the truncation degree.

Besides, the NPs on a crystalline substrate are characterized by a height (out-of-plane)-to-lateral (in-plane)-dimension ratio of <0.7, as opposed to ratios between 0.7 and 1 for amorphous substrates, as discussed in Section 2.3.1. An additional difference is that on a single-crystalline substrate the (faceted) small NPs are attached in many cases to the large ones, as evidenced in Fig. 2.13c (left-hand side). In general, the mean size of the NPs in the crystalline substrate is bigger and the possible sizes are more spread than those on the amorphous substrate. On the contrary, the metal content follows a similar linear dependence on the number of laser pulses as reported in Fig. 2.10b, irrespective of the substrate.

2.4 Thermal Process

2.4.1 Substrate Temperatures

Morphology, crystallinity, and composition of the deposited material depend strongly on the substrate temperature. Metal deposited upon ceramic substrates are arranged in large NPs ($\phi \approx 100$ nm) at high temperatures as opposed to the formation of small and close-packed NPs, as described in the previous section, for room temperature. Such difference depends on the balance between the diffusion rate, D_R, and deposition flux, D_F, introduced in Section 2.3.3. Although D_R of metals on ceramics is not well characterized, it is assumed that it depends strongly on the substrate temperature according to an Arrhenius dependence [79]. The distance traveled by the metal atom on the substrate, L_S, can be estimated as

$$L_S = (D_R \tau_S)^{1/2}, \tag{2.15}$$

where τ_s is the time between laser pulses and $L_s > 1$ μm when $T > 400°C$ in the case of both Ag and Au on ceramic surfaces [80]. Thus, the metal atoms travel long distances and they get incorporated into metal nuclei between consecutive laser pulses and promote the formation of large NPs with $\phi > 100$ nm at high substrate temperatures. In the case of low substrate temperatures, the smaller mobility of the metal atoms promotes the production of smaller and close-packed NPs, which favor transition of island film growth to continuous film and thus lead to early percolation threshold [81–83]. Interestingly, Eq. 2.15 is formally similar to Eq. 2.3, evidencing the relation between the diffusion of the heat inside a solid and the diffusion of atoms on a surface driven by a temperature rise. In the case of bimetallic sequential deposition, the substrate temperature determines the final alloy composition due to the increase of the metal diffusion coefficients with the temperature. Bimetallic Ag-Au NPs with a long range of alloy compositions (ratios between the two metal contents) were produced through a high substrate temperature (300°C). In contrast, room-temperature growth is detrimental for the production of NPs for all the Ag–Au ratios in terms of size distribution and shape regularity [83]. Also, the crystalline orientation of metal films grown on single-crystal ceramic substrates depends on the thermodynamic factors. Indeed, the energetic preference for metals to expose (111) planes deposited on (100)-oriented ceramic substrates becomes lower as the deposition temperature is increased and, thus, a high deposition temperature is desirable for epitaxial growth. In the case of Ag and Au deposited on (100)-oriented ceramic substrates, the amount of (100)-metal increased with increasing temperature [80].

As described previously, the production of large supported NPs by PLD at room temperature is limited by the percolation threshold, the critical film thickness, and the diameter for Ag being ~10 nm [46]. A common method to overcome this limit and to produce surfaces covered with large NPs ($\phi \approx 100$ nm) is the thermal annealing after the deposition of percolated or continuous thin metallic films. This process, known as dewetting of thin films, has attracted attention as an effective low-cost approach to induce self-organized formation of metallic nanostructures, and it is being proposed as a route for producing functional materials for several applications in micro- and nanotechnology [84]. The dewetting process is characterized by the breaking up of a continuous liquid film into holes, followed by hole growth, coalescence, and finally decay into nanostructures on the substrate. Thus, the driving force is the decrease of the system energy by the spontaneous transformation of a continuous layer into discrete NPs. Morphological properties of these NPs depend on metal coverage, surface roughness, film thickness, annealing temperature/time, or substrate properties. Nevertheless, a detailed description of this process is complex due to the interplay of grain growth, capillary diffusion surface/interface energy, texture evolution, and diffusivity anisotropy [85]. However, some trends can be identified. The temperature for the dewetting to occur can be well below the melting temperature of the bulk material, and it decreases with the film thickness, and dimensions and spacing of the NPs deriving from dewetting also decrease with the film thickness [86, 87]. Morphological analysis of Ag NPs formed, by dewetting, on different substrates shows that the average Ag NP size decreases with increasing substrate thermal conductivity and the NP size distribution broadens with increasing surface roughness [88]. In addition, the shape, structure, composition, and functional properties of the NPs are determined by the thermal process. Thus, the annealing temperature and the duration determine the external aspect (from round to faceted) of supported Au NPs [89, 90]. In the case of Ag-Cu NPs produced by PLD on a glass substrate, it was found that the lattice constant of the NPs increases with the annealing temperature and that also their electrical and optical properties are considerably modified, likely due to the precipitation of Cu atoms [91].

2.4.2 Postheating by Laser Irradiation

PLD on hot substrates and thermal annealing of thin films are slow processes, which have no spatial resolution, cause dewetting in the whole sample, and have undesirable surface effects, including metal–substrate chemical interaction and metal diffusion into the substrate. Laser irradiation of thin films with ns pulses can induce an extremely high temperature rise within an extremely short time and could minimize such undesirable surface effects. Moreover, pulsed-laser-induced fast melting/dewetting of the thin metal film allows the NPs to be produced in a well-defined spatial region with resolution down to the micrometer range and minimizes thermal damage of the substrate. In addition, it can act remotely and be focused to different scales in almost any environment as opposed to other surface techniques, such as ion or electron lithography.

Spatial-temporal evolution of the temperature process under a laser pulse can be described by Eq. 2.1. In this framework, the reduced thermal conductivity of the substrate confines the heat in the metal film and the subsequent heat absorbed depends on the relation between the metal layer thickness and the optical penetration depth. Such interplay makes the threshold fluence required for the melting/evaporation of thin films considerably reduced compared to the metal bulk. In metal films with a thickness $h \leq \alpha^{-1}$ and fluences around that one required for melting, the simulated temperature difference in the metal is typically less than 1 K and it is reasonable to assume the film has a uniform temperature in depth (z axis). In this case, the fluence required to melt these ultrathin films decreases with h. On the contrary, for $h > \alpha^{-1}$, the fluence values required for melting are expected to be lower than the previous situation and, for $h \gg \alpha^{-1}$, the temperature difference in the z axis of the metal is not negligible and a partial melting of the metal surface should occur, thus making the differences between thin films and bulk materials disappear. Although the laser pulse duration or wavelength affects the absolute value of the fluence melting threshold, the identification of these regimes stands for all metal films.

In the case of Ag and Au ultrathin films with $h \leq \alpha^{-1}$, the metal melts within a few nanoseconds with fluences of 100–500 mJ cm^{-2}. In the previous decades, many experimental and theoretical studies have focused on the instability mechanisms dominating the breakup of the metal liquid layer generated by laser irradiation. The two generally accepted mechanisms are nucleation and spinodal dewetting. In the former, the instabilities are initialized by the thermal gradients and fluctuations produced by defects in the surface or the substrate, yielding a random spatial distribution of NPs. On the contrary, under a spinodal mechanism the breakup is due to the amplification of surface perturbations, which lead to spatially ordered NPs with a characteristic length scale. Models based on the latter instability process have successfully predicted the power law relation between the NP spacing or the NP diameter and the film thickness [92–95].

The instability mechanisms governing the breakup are basically determined by the initial morphology and the thickness of the thin film and, in turn, affect the dewetting pathway and the time for the formation of the isolated NPs.

The dewetting process is essentially activated during the metal liquid lifetime, which depends strongly on the laser parameters. A single laser pulse with an adequate fluence is typically required to nanostructure the films. However, by adjusting the laser parameters and the layer configuration, the complete dewetting process could end after some hundreds of laser pulses, since individual pulses are essentially additive and contribute to the cumulative lifetime [96]. These conditions are used to analyze the intermediate stages of dewetting. However, laser melting of noble metal films on nonwettable substrates successfully generates metal NPs when h is approximately lower than 25 nm [97–99]. Once the film has broken up into isolated NPs, any subsequent pulse will induce partial or total remelting or fragmentation. If the initial layer consists of irregularly shaped NPs (ϕ = 10–15 nm) on an a-Al$_2$O$_3$ substrate, the laser irradiation with a single ns laser pulse generates a homogeneous distribution of spherical NPs with a mean diameter close to the initial dimension. This process occurs within a fluence interval defined by the size-dependent melting threshold of the NPs and the ablation threshold [68]. In the case of Ni flat structures (circles, squares, and triangles) on SiO$_2$, the ns laser irradiation retracts the edges and corners of the patterned structures when the temperature exceeds the melting threshold of the metal [100]. Flat gold triangles with side lengths of 400–800 nm irradiated with a ns single pulse showed a surprising behavior. The liquid shape contracted toward a sphere, and the center of mass moved upward, which led to detachment of NPs from the substrate. These jumping NPs have radii of ~100 nm and velocities of the order of 10 m s^{-1} [101].

Figure 2.14 shows tilted scanning electron microscopy (SEM) images of the metal NPs generated by a single laser pulse from a starting film of 20 nm for Au, 15 nm for Ni, and 20 nm for Mo grown on SiO$_2$ substrates. The resulting NPs are pseudospheres with diameters in the range of hundreds of nm, which is at least 1 order of magnitude higher than the greatest NPs allowed by the conventional PLD technique. The contact angle between the metal NPs and the substrate, θc, is approximately 120° for Au, 105° for Ni, and 72° for Mo. From a thermodynamic point of view, this angle is roughly related to the enthalpy of formation of the oxides ($-\Delta H_f$) for the metal at issue. In general, θc decreases as $-\Delta H_f$ increases, as verified by these three metals [102]. In Au NPs on native SiO_2 obtained by irradiation with a 532 ns pulsed laser beam, it was found that θc = 140° [103], in agreement with the Young–Drupé equation:

$$\cos\vartheta_c = \frac{\gamma_{v/ox} - \gamma_{m/ox}}{\gamma_{v/m}} \quad (2.16)$$

This equation is verified when the condition described by Eq. 2.13 is satisfied, that is, fulfilled with the high temperatures obtained by laser irradiation, where the atomic diffusion process is activated.

The temperature of the substrate is straightforwardly modified by the laser fluence. In the specific case of Au thin films supported on glass, the metal boiling temperature has been achieved with fluences in the range 250–450 mJ cm^{-2} [104, 105]. Nevertheless, this value depends strongly on the laser wavelength and the initial morphology/thickness of the metal film. Thin films irradiated in this regime of fluences show typically two common features: splitting of the NPs and the generation of craters at the substrate surface. The morphology of NPs generated with fluences high enough for melting the metal but below the boiling threshold is different from NPs produced with fluences above the boiling threshold.

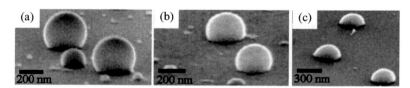

Figure 2.14 Slightly tilted SEM images of metal droplets generated by laser irradiation with a KrF excimer laser (λ = 248 nm and τ = 25 ns) of (a) a 20 nm Au film, (b) a 15 nm Ni film, and (c) a 20 nm Mo film. Reprinted (figure) with permission from Ref. [102]. Copyright (2005) by the American Physical Society.

It is seen that the NPs' mean diameter decreases and the number density of NPs increases strongly as the fluence is increased. This suggests that the larger-diameter droplets fragment into smaller droplets as they boil.

In addition, craters on the substrate surface within planar cross sections of tens of nanometers can be produced with a single laser pulse. The crater formation may be due to the melting of the substrate hit by fragmented Au NPs produced in the explosion induced by the laser heating. The accumulation of pulses generates the enlargement of the explosive splitting of the Au NPs residing in craters.

Finally, laser fluences well above the boiling threshold of the metal are able to produce the implantation of NPs into the substrate. Figure 2.15 shows a TEM cross-sectional image of an Au thin film (h = 22 nm) supported on a SiO_x-coated substrate with SiO_2 and irradiated within a single pulse of 800 mJ cm^{-2}, that is, a fluence well above the threshold of boiling defined previously [106]. It shows a certain surface roughness that can be related to the crater formation. The surface is covered with Au NPs with $\phi \approx$ 10 nm, and underneath there are large NPs (ϕ = 20–60 nm), which are implanted into the substrate at a depth of many tens of nanometers. In addition, laser irradiations with a single pulse (λ = 248 nm and τ = 25 ns) of Au films (h = 4–10 nm) supported on glass have shown that the implantation process occurs for fluences above a threshold value that decreases as the thickness increases, in the range 600–900 mJ cm^{-2} [107]. Laser irradiations in this range of fluences may produce not only the boiling of the metal thin film but also the boiling of a superficial layer of the glass. The condensation of this plasma of glass and Au at the substrate surface and the low solubility of Au in the glass may explain the implantation of these Au NPs in the substrate.

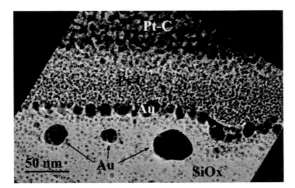

Figure 2.15 TEM cross-sectional image of a 22 nm Au thin film on SiO_x irradiated with a single pulse (ArF laser, λ = 193 nm, τ = 20 ns, F = 800 mJ cm^{-2}). Reprinted from Ref. [106]. Copyright (2016), with permission from Elsevier.

Acknowledgments

The authors VR and RJP wish to thank the present and past members of the Laser Processing Group who have considerably contributed to the study and production of nanostructures with laser techniques and would like to emphasize their special gratitude to C. N. Afonso for her commitment, tenacity, and perseverance.

References

1. Horwitz, J. S., and Sprague, J. A. (1994). Film nucleation and growth in pulsed laser deposition of ceramics, in *Pulsed Laser Deposition of Thin Films*, eds. Chrisey, D. B., and Hubler, G. K. (John Wiley & Sons, New York).
2. Campbell, C. T. (1997). Ultrathin metal films and particles on oxide surfaces: structural, electronic and chemisorptive properties, *Surf. Sci. Rep.*, **27**, pp. 1–111.
3. Núñez-Sánchez, S., Serna, R., García-López, J., Petford-Long, A. K., Tanase, M., and Kabius, B. (2009). Tuning the Er^{3+} sensitization by Si nanoparticles in nanostructured as-grown Al$_2$O$_3$ films, *J. Appl. Phys.*, **105**, p. 013118.
4. Rani, J. R., Mahadevan Pillai, V. P., Ajimsha, R. S., Jayaraj, M. K., and Jayasree, R. S. (2006). Effect of substrate roughness on photoluminescence spectra of silicon nanocrystals grown by off axis pulsed laser deposition, *J. Appl. Phys.*, **100**, pp. 1–6.
5. Liu, W. L., Lee, P. F., Dai, J. Y., Wang, J., Chan, H. L. W., Choy, C. L., Song, Z. T., and Feng, S. L. (2005). Self-organized Ge nanocrystals embedded in HfAlO fabricated by pulsed-laser deposition and application to floating gate memory, *Appl. Phys. Lett.*, **86**, p. 013110.
6. Chichkov, B. N., Momma, C., Nolte, S., von Alvensleben, F., and Tünnermann, A. (1996). Femtosecond, picosecond and nanosecond laser ablation of solids, *Appl. Phys. A*, **63**, pp. 109–115.
7. Hohlfeld, J., Wellershoff, S.-S., Güdde, J., Conrad, U., Jähnke, V., and Matthias, E. (2000). Electron and lattice dynamics following optical excitation of metals, *Chem. Phys.*, **251**, pp. 237–258.
8. Bäuerle, D. W. (2011). *Laser Processing and Chemistry* (Springer, New York).
9. Von Allmen, M. (1987). *Laser-Beam Interactions with Materials*, Springer Series in Materials Science, Vol. 2 (Springer, Berlin).

10. Amoruso, S., Bruzzese, R., Spinelli, N., and Velotta, R. (1999). Characterization of laser-ablation plasmas, *J. Phys. B: At. Mol. Opt. Phys.*, **32**, pp. R131–R172.
11. Aguilera, J. A., Aragón, C., and Peñalba, F. (1998). Plasma shielding effect in laser ablation of metallic samples and its influence on LIBS analysis, *Appl. Surf. Sci.*, **127–129**, pp. 309–314.
12. Singh, R. K., and Narayan, J. (1990). Pulsed-laser evaporation technique for deposition of thin films: physics and theoretical model, *Phys. Rev. B*, **41**, pp. 8843–8859.
13. Kelly, R., and Miotello, A. (1994). Mechanisms of pulsed laser sputtering, in *Pulsed Laser Deposition of Thin Films*, eds. Chrisey, D. B., and Hubler, G. K. (John Wiley & Sons, New York).
14. Gordillo-Vázquez, F. J., Perea, A., McKiernan, A. P., and Afonso, C. N. (2005). Electronic temperature and density of the plasma produced by nanosecond ultraviolet laser ablation of LiF, *Appl. Phys. Lett.*, **86**, p. 181501.
15. Verhoff, B., Harilal, S. S., Freeman, J. R., Diwakar, P. K., and Hassanein, A. (2012). Dynamics of femto- and nanosecond laser ablation plumes investigated using optical emission spectroscopy, *J. Appl. Phys.*, **112**, pp. 1–9.
16. Camacho, J. J., Diaz, L., Cid, J. P., and Poyato, J. M. L. (2013). Time-resolved study of the plasma-plume emission during the nanosecond ablation of lithium fluoride, *Spectrochim. Acta B*, **88**, pp. 203–210.
17. Chaos, J. A., Dreyfus, R. W., Perea, A., Serna, R., Gonzalo, J., and Afonso, C. N. (2000). Delayed release of Li atoms from laser ablated lithium niobate, *Appl. Phys. Lett.*, **76**, pp. 649–651.
18. Gordillo-Vazquez, F. J., Perea, A., Chaos, J. A., Gonzalo, J., and Afonso, C. N. (2001). Temporal and spatial evolution of the electronic density and temperature of the plasma produced by laser ablation of $LiNbO_3$, *Appl. Phys. Lett.*, **78**, pp. 7–9.
19. Anisimov, S. I., Bauerle, D., and Lukyanchuk, B. S. (1993). Gas dynamics and film profiles in pulsed-laser deposition of materials, *Phys. Rev. B*, **48**, pp. 12076–12081.
20. Farid, N., Harilal, S. S., Ding, H., and Hassanein, A. (2014). Emission features and expansion dynamics of nanosecond laser ablation plumes at different ambient pressures, *J. Appl. Phys.*, **115**, pp. 1–9.
21. Harilal, S. S., O'Shay, B., Tao, Y., and Tillack, M. S. (2006). Ambient gas effects on the dynamics of laser-produced tin plume expansion, *J. Appl. Phys.*, **99**, p. 083303.

22. Baraldi, G., Perea, A., and Afonso, C. N. (2011). Dynamics of ions produced by laser ablation of several metals at 193 nm, *J. Appl. Phys.*, **109**, p. 043302.

23. Gonzalo, J., Siegel, J., Perea, A., Puerto, D., Resta, V., Galvan-Sosa, M., and Afonso, C. N. (2007). Imaging self-sputtering and backscattering from the substrate during pulsed laser deposition of gold, *Phys. Rev. B*, **76**, p. 035435.

24. Peláez, R. J., Afonso, C. N., Bator, M., and Lippert, T. (2013). Laser ablation of ceramic Al_2O_3 at 193 nm and 248 nm: the importance of single-photon ionization processes, *J. Appl. Phys.*, **113**, p. 223301.

25. Perea, A., Gonzalo, J., Budtz-Jørgensen, C., Epurescu, G., Siegel, J., Afonso, C. N., and García-López, J. (2008). Quantification of self-sputtering and implantation during pulsed laser deposition of gold, *J. Appl. Phys.*, **104**, p. 084912.

26. Claeyssens, F., Henley, S. J., and Ashfold, M. N. R. (2003). Comparison of the ablation plumes arising from ArF laser ablation of graphite, silicon, copper, and aluminum in vacuum, *J. Appl. Phys.*, **94**, pp. 2203–2211.

27. Baraldi, G., Perea, A., and Afonso, C. N. (2011). Dynamics of ions produced by laser ablation of ceramic Al_2O_3 and Al at 193 nm, *Appl. Phys. A*, **105**, pp. 75–79.

28. Amoruso, S., Berardi, V., Bruzzese, R., Capobianco, R., Velotta, R., and Armenante, M. (1996). High fluence laser ablation of aluminum targets: time-of-flight mass analysis of plasmas produced at wavelengths 532 and 355 nm, *Appl. Phys. A*, **62**, pp. 533–541.

29. Dreyfus, R. W. (1991). Cu0, Cu+, and Cu2 from excimer-ablated copper, *J. Appl. Phys.*, **69**, pp. 1721–1729.

30. Peláez, R. J., Afonso, C. N., Chen, J., Esposito, M., Lippert, T., Stender, D., and Wokaun, A. (2012). Relevance and formation mechanisms of negative ions upon ablation of Al_2O_3, *J. Phys. D: Appl. Phys.*, **45**, p. 285402.

31. Esposito, M., Bator, M., Döbeli, M., Lippert, T., Schneider, C. W., and Wokaun, A. (2011). Negative ions: the overlooked species in thin film growth by pulsed laser deposition, *Appl. Phys. Lett.*, **99**, p. 191501.

32. Kukreja, L. M., Verma, S., Pathrose, D. A., and Rao, B. T. (2014). Pulsed laser deposition of plasmonic-metal nanostructures, *J. Phys. D: Appl. Phys.*, **47**, p. 034015.

33. Resta, V., Afonso, C. N., Piscopiello, E., and Van Tendeloo, G. (2009). Role of substrate on nucleation and morphology of gold nanoparticles produced by pulsed laser deposition, *Phys. Rev. B*, **79**, p. 235409.

34. Cheung, J. C. (1994). Histoyry and fundamentals of pulsed laser deposition, in *Pulsed Laser Deposition of Thin Films*, eds. Chrisey, D. B., and Hubler, G. K. (John Wiley & Sons, New York).
35. Kools, J. C. S. (1994). Pulsed laser deposition of metals, in *Pulsed Laser Deposition of Thin Films*, eds. Chrisey, D. B., and Hubler, G. K. (John Wiley & Sons, New York).
36. Geysermans, P., Finocchi, F., Goniakowski, J., Hacquart, R., and Jupille, J. (2009). Combination of (100), (110) and (111) facets in MgO crystals shapes from dry to wet environment, *Phys. Chem. Chem. Phys.*, **11**, pp. 2228–2233.
37. Liu, H., and Dandy, D. S. (1995). Studies on nucleation process in diamond CVD: an overview of recent developments, *Diamond Relat. Mater.*, **4**, pp. 1173–1188.
38. Chen, J. S., Lau, S. P., Tay, B. K., Chen, G. Y., Sun, Z., Tan, Y. Y., Tan, G., and Chai, J. W. (2001). Surface energy of amorphous carbon films containing iron, *J. Appl. Phys.*, **89**, p. 7814.
39. Warrender, J. M., and Aziz, M. J. (2007). Kinetic energy effects on morphology evolution during pulsed laser deposition of metal-on-insulator films, *Phys. Rev. B*, **75**, p. 085433.
40. Warrender, J. M., and Aziz, M. J. (2007). Effect of deposition rate on morphology evolution of metal-on-insulator films grown by pulsed laser deposition, *Phys. Rev. B*, **76**, p. 045414.
41. Elofsson, V., Lü, B., Magnfält, D., Münger, E. P., and Sarakinos, K. (2014). Unravelling the physical mechanisms that determine microstructural evolution of ultrathin Volmer-Weber films, *J. Appl. Phys.*, **116**, p. 044302.
42. Donnelly, T., Krishnamurthy, S., Carney, K., McEvoy, N., and Lunney, J. G. (2007). Pulsed laser deposition of nanoparticle films of Au, *Appl. Surf. Sci.*, **254**, pp. 1303–1306.
43. Alonso, J. C., Diamant, R., Castillo, P., Acosta-García, M. C., Batina, N., and Haro-Poniatowski, E. (2009). Thin films of silver nanoparticles deposited in vacuum by pulsed laser ablation using a YAG:Nd laser, *Appl. Surf. Sci.*, **255**, pp. 4933–4937.
44. Dolbec, R., Irissou, E., Chaker, M., Guay, D., Rosei, F., and El Khakani, M. (2004). Growth dynamics of pulsed laser deposited Pt nanoparticles on highly oriented pyrolitic graphite substrates, *Phys. Rev. B*, **70**, pp. 1–4.
45. Resta, V., Gonzalo, J., Afonso, C. N., Piscopiello, E., and García López, J. (2011). Coverage induced regulation of Au nanoparticles during pulsed laser deposition, *J. Appl. Phys.*, **109**, pp. 1–7.

46. Barnes, J., Petford-Long, A. K., Doole, R. C., Serna, R., Gonzalo, J., Suárez-García, A., Afonso, C. N., and Hole, D. (2002). Structural studies of Ag nanocrystals embedded in amorphous Al_2O_3 grown by pulsed laser deposition, *Nanotechnology*, **13**, p. 305.

47. Perriere, J., Millon, E., and Fogarassy, E. (2006). *Recent Advances in Laser Processing of Materials* (Elsevier Science, Amsterdam).

48. Gonzalo, J., Perea, A., Babonneau, D., Afonso, C. N., Beer, N., Barnes, J.-P., Petford-Long, A. K., Hole, D. E., and Townsend, P. D. (2005). Competing processes during the production of metal nanoparticles by pulsed laser deposition, *Phys. Rev. B*, **71**, p. 125420.

49. Fähler, S., and Krebs, H.-U. (1996). Calculations and experiments of material removal and kinetic energy during pulsed laser ablation of metals, *Appl. Surf. Sci.*, **96–98**, pp. 61–65.

50. Cazzaniga, A., Ettlinger, R. B., Canulescu, S., Schou, J., and Pryds, N. (2014). Nanosecond laser ablation and deposition of silver, copper, zinc and tin, *Appl. Phys. A*, **117**, pp. 89–92.

51. Suárez-García, A., Barnes, J.-P., Serna, R., Petford-Long, A. K., Afonso, C. N., and Hole, D. (2003). The shallow implantation of bismuth during the growth of bismuth nanocrystals in Al_2O_3 by pulsed laser deposition, *Mater. Res. Soc. Symp. Proc.*, **780**, pp. 3–8.

52. Serna, R., Afonso, C. N., Ricolleau, C., Wang, Y., Zheng, Y., Gandais, M., and Vickridge, I. (2000). Artificially nanostructured $Cu:Al_2O_3$ films produced by pulsed laser deposition, *Appl. Phys. A*, **71**, pp. 583–586.

53. Ziegler, J. F., Biersack, J. P., and Ziegler, M. D. (2013). SRIM, http://www.srim.org/.

54. Fähler, S., Sturm, K., and Krebs, H.-U. (1999). Resputtering during the growth of pulsed-laser-deposited metallic films in vacuum and in an ambient gas, *Appl. Phys. Lett.*, **75**, p. 3766.

55. Thestrup, B., Toftmann, B., Schou, J., Doggett, B., and Lunney, J. G. (2002). Ion dynamics in laser ablation plumes from selected metals at 355 nm, *Appl. Surf. Sci.*, **197–198**, pp. 175–180.

56. Hidalgo, J. G., Serna, R., Haro-Poniatowski, E., and Afonso, C. N. (2004). Evidence for self-sputtering during pulsed laser deposition of Zn, *Appl. Phys. A*, **79**, pp. 915–918.

57. Resta, V., Peláez, R. J., and Afonso, C. N. (2014). Importance of ion bombardment during coverage of Au nanoparticles on their structural features and optical response, *J. Appl. Phys.*, **115**, p. 124303.

58. Xie, D., Wang, M. P., and Qi, W. H. (2004). A simplified model to calculate the surface-to-volume atomic ratio dependent cohesive energy of nanocrystals, *J. Phys. Condens. Matter*, **16**, pp. L401– L405.
59. Qi, W. H. H., Huang, B. Y. Y., Wang, M. P. P., Li, Z., and Yu, Z. M. M. (2007). Generalized bond-energy model for cohesive energy of small metallic particles, *Phys. Lett. A*, **370**, pp. 494–498.
60. Yamamura, Y., and Tawara, H. (1996). Energy dependence of ion-induced sputtering yields from monatomic solids at normal incidence, *At. Data Nucl. Data Tables*, **62**, pp. 149–253.
61. Serna, R., Suárez-García, A., Afonso, C. N., and Babonneau, D. (2006). Optical evidence for reactive processes when embedding Cu nanoparticles in Al_2O_3 by pulsed laser deposition, *Nanotechnology*, **17**, pp. 4588–4593.
62. Fähler, S., Kahl, S., Weisheit, M., Sturm, K., and Krebs, H. (2000). The interface of laser deposited Cu/Ag multilayers: evidence of the 'subsurface growth mode' during pulsed laser deposition, *Appl. Surf. Sci.*, **154–155**, pp. 419–423.
63. Eberl, C., Liese, T., Schlenkrich, F., Döring, F., Hofsäss, H., and Krebs, H.-U. (2013). Enhanced resputtering and asymmetric interface mixing in W/Si multilayers, *Appl. Phys. A*, **111**, pp. 431–437.
64. Robins, J., and Rhodin, T. (1964). Nucleation of metal crystals on ionic surfaces, *Surf. Sci.*, **2**, pp. 346–355.
65. Mativetsky, J. M., Fostner, S., Burke, S. A., and Grutter, P. (2008). The role of charge-induced defects in the growth of gold on an alkali halide surface, *Surf. Sci.*, **602**, pp. L21–L24.
66. Zenkevitch, A., Chevallier, J., and Khabelashvili, I. (1997). Nucleation and growth of pulsed laser deposited gold on sodium chloride (100), *Thin Solid Films*, **311**, pp. 119–123.
67. Resta, V., Caricato, A. P., Loiudice, A., Rizzo, A., Gigli, G., Taurino, A., Catalano, M., and Martino, M. (2013). Pulsed laser deposition of a dense and uniform Au nanoparticles layer for surface plasmon enhanced efficiency hybrid solar cells, *J. Nanopart. Res.*, **15**, p. 7.
68. Resta, V., Siegel, J., Bonse, J., Gonzalo, J., Afonso, C. N., Piscopiello, E., and Van Tenedeloo, G. (2006). Sharpening the shape distribution of gold nanoparticles by laser irradiation, *J. Appl. Phys.*, **100**, p. 084311.
69. Domingo, C., Resta, V., Sanchez-Cortes, S., Garcia-Ramos, J. V., and Gonzalo, J. (2007). Pulsed laser deposited au nanoparticles as substrates for surface-enhanced vibrational spectroscopy, *J. Phys. Chem. C*, **111**, pp. 8149–8152.

70. Højrup-Hansen, K., Ferrero, S., and Henry, C. R. (2004). Nucleation and growth kinetics of gold nanoparticles on MgO(1 0 0) studied by UHV-AFM, *Appl. Surf. Sci.*, **226**, pp. 167–172.
71. Zenkevitch, A., Chevallier, J., Khabelashvili, I., Zenkevitch, A., Chevallier, J., and Khabelashvili, I. (1997). Nucleation and growth of pulsed laser deposited gold on sodium chloride (100), *Thin Solid Films*, **311**, pp. 119–123.
72. Brune, H. (1998). Microscopic view of epitaxial metal growth: nucleation and aggregation, *Surf. Sci. Rep.*, **31**, pp. 125–229.
73. Irissou, E., Le Drogoff, B., Chaker, M., and Guay, D. (2003). Influence of the expansion dynamics of laser-produced gold plasmas on thin film structure grown in various atmospheres, *J. Appl. Phys.*, **94**, p. 4796.
74. Jubert, P.-O., Fruchart, O., and Meyer, C. (2003). Nucleation and surface diffusion in pulsed laser deposition of Fe on Mo(110), *Surf. Sci.*, **522**, pp. 8–16.
75. Afonso, C. N., Gonzalo, J., Serna, R., de Sande, J. C. G., Ricolleau, C., Grigis, C., Gandais, M., Hole, D. E., and Townsend, P. D. (1999). Vacuum versus gas environment for the synthesis of nanocomposite films by pulsed-laser deposition, *Appl. Phys. A*, **69**, pp. S201– S207.
76. Frank, F. C., and van der Merwe, J. H. (1949). One-dimensional dislocations. I. Static theory, *Proc. R. Soc. A Math. Phys. Eng. Sci.*, **198**, pp. 205–216.
77. Molina, L. M., and Hammer, B. (2004). Theoretical study of CO oxidation on Au nanoparticles supported by MgO(100), *Phys. Rev. B: Condens. Matter*, **69**, pp. 1–22.
78. Xie, D., Wang, M. P., and Qi, W. H. (2004). A simplified model to calculate the surface-to-volume atomic ratio dependent cohesive energy of nanocrystals, *J. Phys. Condens. Matter*, **16**, pp. L401– L405.
79. Venables, J. A. (1994). Atomic processes in crystal growth, *Surf. Sci.*, **299–300**, pp. 798–817.
80. Francis, A. J., and Salvador, P. A. (2007). Crystal orientation and surface morphology of face-centered-cubic metal thin films deposited upon single-crystal ceramic substrates using pulsed laser deposition, *J. Mater. Res.*, **22**, pp. 89–102.
81. Scott, R. C., Leedy, K. D., Bayraktaroglu, B., Look, D. C., Smith, D. J., Ding, D., Lu, X., and Zhang, Y.-H. (2011). Influence of substrate temperature and post-deposition annealing on material properties of Ga-doped ZnO prepared by pulsed laser deposition, *J. Electron. Mater.*, **40**, pp. 419–428.

82. Zhao, Y., Jiang, Y., and Fang, Y. (2007). The influence of substrate temperature on ZnO thin films prepared by PLD technique, *J. Cryst. Growth*, **307**, pp. 278–282.

83. Verma, S., Rao, B. T., Detty, A. P., Ganesan, V., Phase, D. M., Rai, S. K., Bose, A., Joshi, S. C., and Kukreja, L. M. (2015). Surface plasmon resonances of Ag-Au alloy nanoparticle films grown by sequential pulsed laser deposition at different compositions and temperatures, *J. Appl. Phys.*, **117**, p. 133105.

84. Gentili, D., Foschi, G., Valle, F., Cavallini, M., and Biscarini, F. (2012). Applications of dewetting in micro and nanotechnology, *Chem. Soc. Rev.*, **41**, p. 4430.

85. Wang, D., and Schaaf, P. (2013). Solid-state dewetting for fabrication of metallic nanoparticles and influences of nanostructured substrates and dealloying, *Phys. Status Solidi Appl. Mater. Sci.*, **210**, pp. 1544–1551.

86. Thompson, C. V. (2012). Solid-state dewetting of thin films, *Annu. Rev. Mater. Res.*, **42**, pp. 399–434.

87. Mizsei, J., and Lantto, V. (2001). In situ AFM, XRD and resistivity studies of the agglomeration of sputtered silver nanolayers, *J. Nanopart. Res.*, **3**, pp. 271–278.

88. Tanyeli, I., Nasser, H., Es, F., Bek, A., and Turan, R. (2013). Effect of surface type on structural and optical properties of Ag nanoparticles formed by dewetting, *Opt. Express*, **21**, pp. A798–A807.

89. Sadan, H., and Kaplan, W. D. (2006). Au–Sapphire (0001) solid–solid interfacial energy, *J. Mater. Sci.*, **41**, pp. 5099–5107.

90. Müller, C. M., Mornaghini, F. C. F., and Spolenak, R. (2008). Ordered arrays of faceted gold nanoparticles obtained by dewetting and nanosphere lithography, *Nanotechnology*, **19**, p. 485306.

91. Hirai, M., and Kumar, A. (2006). Wavelength tuning of surface plasmon resonance by annealing silver-copper nanoparticles, *J. Appl. Phys.*, **100**, p. 14309.

92. Bischof, J., Scherer, D., Herminghaus, S., and Leiderer, P. (1996). Dewetting modes of thin metallic films: nucleation of holes and spinodal dewetting, *Phys. Rev. Lett.*, **77**, pp. 1536–1539.

93. Trice, J., Thomas, D., Favazza, C., Sureshkumar, R., and Kalyanaraman, R. (2007). Pulsed-laser-induced dewetting in nanoscopic metal films: theory and experiments, *Phys. Rev. B: Condens. Matter*, **75**, pp. 1–15.

94. Fowlkes, J. D., Kondic, L., Diez, J., Wu, Y., and Rack, P. D. (2011). Self-assembly versus directed assembly of nanoparticles via pulsed laser

induced dewetting of patterned metal films, *Nano Lett.*, **11**, pp. 2478–2485.

95. Trice, J., Favazza, C., Thomas, D., Garcia, H., Kalyanaraman, R., and Sureshkumar, R. (2008). Novel self-organization mechanism in ultrathin liquid films: theory and experiment, *Phys. Rev. Lett.*, **101**, p. 017802.

96. Peláez, R. J., Afonso, C. N., Škereň, M., and Bulíř, J. (2015). Period dependence of laser induced patterns in metal films, *Nanotechnology*, **26**, p. 015302.

97. Ruffino, F., and Grimaldi, M. G. (2015). Controlled dewetting as fabrication and patterning strategy for metal nanostructures, *Phys. Status Solidi*, **212**, pp. 1662–1684.

98. Krishna, H., Sachan, R., Strader, J., Favazza, C., Khenner, M., and Kalyanaraman, R. (2010). Thickness-dependent spontaneous dewetting morphology of ultrathin Ag films, *Nanotechnology*, **21**, p. 155601.

99. Chen, C.-Y., Wang, J.-Y., Tsai, F.-J., Lu, Y.-C., Kiang, Y.-W., and Yang, C. C. (2009). Fabrication of sphere-like Au nanoparticles on substrate with laser irradiation and their polarized localized surface plasmon behaviors, *Opt. Express*, **17**, pp. 14186–14198.

100. Rack, P. D., Guan, Y., Fowlkes, J. D., Melechko, A. V., and Simpson, M. L. (2008). Pulsed laser dewetting of patterned thin metal films: a means of directed assembly, *Appl. Phys. Lett.*, **92**, p. 223108.

101. Habenicht, A., Olapinski, M., Burmeister, F., Leiderer, P., and Boneberg, J. (2005). Jumping nanodroplets, *Science*, **309**, pp. 2043–2045.

102. Henley, S. J., Carey, J. D., and Silva, S. R. P. (2005). Pulsed-laser-induced nanoscale island formation in thin metal-on-oxide films, *Phys. Rev. B*, **72**, p. 195408.

103. Ruffino, F., Pugliara, A., Carria, E., Bongiorno, C., Spinella, C., and Grimaldi, M. G. (2012). Formation of nanoparticles from laser irradiated Au thin film on SiO_2/Si: elucidating the Rayleigh-instability role, *Mater. Lett.*, **84**, pp. 27–30.

104. Hashimoto, S., Uwada, T., Hagiri, M., Takai, H., and Ueki, T. (2009). Gold nanoparticle-assisted laser surface modification of borosilicate glass substrates, *J. Phys. Chem. C*, **113**, pp. 20640–20647.

105. Rodríguez, C. E., Peláez, R. J., Afonso, C. N., Riedel, S., Leiderer, P., Jimenez-Rey, D., and Font, A. C. (2014). Plasmonic response and transformation mechanism upon single laser exposure of metal discontinuous films, *Appl. Surf. Sci.*, **302**, pp. 32–36.

106. Stolzenburg, H., Peretzki, P., Wang, N., Seibt, M., and Ihlemann, J. (2015). Implantation of plasmonic nanoparticles in SiO_2 by pulsed laser irradiation of gold films on SiO_x-coated fused silica and subsequent thermal annealing, *Appl. Surf. Sci.*, **374**, pp. 138–142.
107. Henley, S. J., Beliatis, M. J., Stolojan, V., and Silva, S. R. P. (2013). Laser implantation of plasmonic nanostructures into glass, *Nanoscale*, **5**, pp. 1054–1059.

Chapter 3

Nanosecond Laser Ablation of Solid Targets in a High-Pressure Atmosphere

Sebastiano Trusso,[a] Fortunato Neri,[b] and Paolo Maria Ossi[c]

[a]*IPCF-CNR, Instituto per i Processi Chimico-Fisici, V.le F. Stagno d'Alcontres 37, 98158 Messina, Italy*
[b]*Dipartimento di Scienze Matematiche e Informatiche, Scienze Fisiche e Scienze della Terra, V.le F. Stagno d'Alcontres 67, 98166 Messina, Italy*
[c]*Dipartimento di Energia & Centre for Nano Engineered Materials and Surfaces NEMAS, Politecnico di Milano, via Ponzio 34-3, 20133 Milano, Italy*
paolo.ossi@polimi.it

Pulsed laser ablation is an attractive technique to produce in a controlled way nanoparticles and thin films made of self-assembled nanoparticle arrays grown on suitable supports. When nanosecond laser pulses are used to irradiate a solid target, elemental or compound nanoparticles grow during the propagation through an ambient gas of the laser-generated plasma plume. Process parameters such as laser wavelength and energy density, target-to-substrate distance, and nature and pressure of the ambient gas, all affect the plasma expansion and thus nanoparticle size and kinetic energy, as well as the related distributions. At landing, particle size, energy, and mobility on the support influence film growth

Pulsed Laser Ablation: Advances and Applications in Nanoparticles and Nanostructuring Thin Films
Edited by Ion N. Mihailescu and Anna Paola Caricato
Copyright © 2018 Pan Stanford Publishing Pte. Ltd.
ISBN 978-981-4774-23-9 (Hardcover), 978-1-315-18523-1 (eBook)
www.panstanford.com

and its nanostructure, resulting in broad ranges of values of the physicochemical properties of the deposited films. After reviewing the peculiarities of plume expansion through a gas at high pressure, we discuss the synthesis of carbon-based films, exploring a wide range of gas pressures for different inert gases and the tailoring of nanostructures toward optimized sensing properties of WO_3 films made of self-assembled nanoparticles synthesized in reactive mixed gas atmospheres. We finally turn to noble metal nanoparticle arrays and their optical properties, specifically designed to produce substrates to be used in enhanced spectroscopies. Their application in surface-enhanced Raman spectroscopy to detect exiguous amounts of drugs of neurological interest in biological fluids is discussed.

3.1 Introduction

This contribution aims at offering a perspective on pulsed laser deposition (PLD) using nanosecond pulses in an ambient gas at high pressure. A major distinguishing feature of such a highly specific technique is the large yield of nanoparticles (NPs), one of the largest found among physical vapor deposition techniques. Indeed, during the last 15 years NPs of different materials over wide intervals of process conditions [1–3] were produced. The interest in NPs and films resulting from assembled NPs is steadily growing. Surface nanostructures can be engineered up to a significant extent and have been investigated due to the predicted and in part observed unusual transport (electronic, optical), magnetic, and chemical properties that characterize matter when its typical sizes are restricted to the nanometer scale. In particular, noble metal NPs are popular due to their attractive optical properties of interest in surface-enhanced vibrational spectroscopies, while their catalytic behavior finds application in the growth of nanotubes and nanorods [4]. In the following we go back to the mechanisms of NP formation active in a laser-generated plasma plume that expands through an ambient gas at high pressure and we discuss the delicate interplay of process parameters that govern such a formation. Some examples are addressed taking into consideration different kinds of materials. The application of noble metal NP arrays with ad hoc tailored surface nanostructures to detect exiguous amounts of drugs of clinical interest via surface-enhanced Raman spectroscopy (SERS) is critically surveyed.

3.2 Comparison between Some Basics of Laser Ablation in Vacuum and in a Gas at High Pressure

Similar to other plasma-assisted deposition techniques, PLD hinges on the attainment of a regime of nonthermodynamic interactions in the target volume subjected to laser irradiation. Such a regime consists of physicochemical processes leading to the formation and breaking of chemical bonds that involve assemblies of particles whose energy distributions are non-Maxwellian over the timescales associated with the considered processes. Pumping energy in a selective, nonthermal way into one degree of freedom of the target, namely the electrons, the latter are effectively decoupled by laser radiation from the remaining atomic/molecular system [5]. Absorption of light in the surface and subsurface volume of the target sensitively depends on its band structure and thus on its optical properties and on the combination of several laser characteristics, in particular wavelength, energy density, and pulse duration. Indeed, in any kind of material the deposited electromagnetic energy leads to ionization processes and to the concurrent generation of free electrons whose behavior resembles that of free electrons in metals. In particular, they can absorb via inverse bremsstrahlung enough laser energy to collisionally ionize neutrals of the matrix, producing plasmons, or unbound electrons, or excitons.

The coupling to lattice vibrations occurs in a few picoseconds, resulting in target heating and triggering a wealth of processes, including melting and ablation [6]. While in metals charge re-equilibration is immediate, in insulators and wide-bandgap semiconductors electron emission coincides with positive hole production; an electrostatically unstable region can result, due to the long time, of the order of picoseconds and sometimes longer, required to bulk electrons to fill in such holes. If the intense surface electric fields, of the order of 10^8 Vcm^{-1}, resulting as a consequence of charge separation, originate an electrostatic repulsion larger than the lattice binding energy, then a layer of surface ions is ejected from the target surface by a Coulomb explosion [7]. When the deposited laser power density is high, a condition frequently met in PLD experiments, hot carrier absorption becomes a relevant mechanism and the produced hot electron bath couples to the lattice with the electron–phonon collision timescale, resulting in superheating and phase explosion [8]. Associated with such mechanisms is the sudden release from the surface and the subsurface volume of the target of a consistent amount of matter. The ejecta are in the form of a number of species, such as neutral atoms/molecules, electrons, ions, and small clusters and constitute a hot, dense, partially ionized vapor cloud lying just above the target-irradiated area.

In the case of picosecond and nanosecond laser pulses the formation time of the initial vapor cloud is less than the pulse duration. So it further absorbs energy from the laser pulse tail. The relevant parameter here is the number density of the ablated species n_a whose critical value

$$n_{cr} = (4\pi c^2 \varepsilon_0 m_e e^{-2})\lambda^{-2} \tag{3.1}$$

determines the opacity of the plasma to the radiation and consequently its absorption efficiency. In Eq. 3.1, c and ε_0 are the light speed and the dielectric permittivity of vacuum, respectively; m_e and e the free electron mass and charge, respectively; and λ the laser wavelength. This peculiar mechanism of laser-sustained absorption is highly effective to drive, by inverse bremsstrahlung and by single-photon direct excitation, fast, full ionization of the original partly ionized, highly anisotropic plasma. The resulting dense plasma, with a typical equivalent electron temperature T_e between 2 eV and 10 eV, undergoes an initial isothermal expansion being accelerated by pressure gradients normal to the irradiated surface. A first highly collisional expansion stage occurs in the so-called Knudsen layer and results in a shifted center-of-mass Maxwellian velocity distribution of the species in the plasma plume. In the case of multiconstituent targets it is rather common to observe that the average velocity $\mathbf{v}_a$ of atoms with a different mass leaving the Knudsen layer is $\mathbf{v}_a \approx 10^4 \, \text{ms}^{-1}$, corresponding to kinetic energies between 10 and 10^2 eV. From the target surface also micrometric droplets are often ejected, mainly by hydrodynamic sputtering associated with repetitive, transient melting-solidification cycles of a nonideally flat surface that leads to swelling and detachment of surface asperities. Further plume expansion in vacuum leads to fast thermalization and neutralization.

When plasma expansion occurs in a fluid, typically a gas at different pressures, a complex phenomenology is observed, depending on gas pressure and on the ratio between the atomic masses of the ablated material and the ambient gas. An exhaustive, recent summary of the phenomenology observed at low gas pressure can be found in Ref. [9].

In this contribution we address the behavior of a plasma plume expanding through a gas at high pressure, with particular emphasis on the attainment of peculiar nanostructures of the deposited material, the possibility to control them, and their relation to specific film properties, focusing on optical properties. We mainly discuss the effect of plume propagation through inert gases. With respect to propagation in vacuum, plume expansion in a high-pressure medium is characterized by braking, meaningful plume-gas energy exchange, and spatial confinement, mainly in the direction normal to the target surface.

The ability of the ambient gas to confine the laser-generated plasma depends on the initial plasma energy and on the collision dynamics between the plasma species and gas atoms. Plasma energy is directly related to the laser energy density, or fluence (F), defined as the ratio between the laser pulse energy and the irradiated spot area on the target surface. Notably, only a fraction of the energy delivered by the laser pulse is transferred to the plasma, while a fraction is lost due to different mechanisms, including in particular thermal dissipation, laser light reflection at the target surface, and plasma shielding (see Eq. 3.1). Yet, over a range of F values, the higher the laser fluence, the higher the resulting plasma energy.

The ablated species experience increased scattering, both intraplume and by gas atoms: under such conditions shock waves are likely to be observed at the visible plume front. With respect to vacuum, the characteristic elliptical, forward-peaked plume shape converts to spherical. This effect is progressively more marked with increasing ambient gas pressure and mass. To observe comparable effects on plume dynamics, gas pressures that range between some tens and a few hundreds of pascals for massive gases like argon have roughly to be up-scaled by a factor of 10 for light gases like helium.

Already at intermediate gas pressure values, around 50 Pa, along with plume sharpening, plume emission is confined to the expansion front and a snowploughing effect develops. Both ablated species and ambient gas are compressed since early expansion times, in the range of fractions of a microsecond. With time elapsing, the inner plume pressure increases and it equals gas pressure, the plume is arrested, and its energy is dissipated in the high-pressure volume where most of plume-gas interaction occurs, leading to enhanced temperatures both of the gas and of the radiation. The plume moves backward, toward the target, marking a transition to a diffusional behavior of ablated species at long times, a marked interpenetration of plasma and gas being typical of this stage [10]. At very high gas pressure multiple shock waves travel through the quickly stopped plasma plume that experiences, besides confinement and backward propagation toward the target of a fraction of the ablated material, the development of regions of expansion and cooling. This peculiar phenomenology was reported for carbon plumes expanding at room temperature in Ar at $p \approx 4 \times 10^4$ Pa [11].

3.3 Nanoparticle Synthesis and Assembling upon Ablation in a High-Pressure Gas: Selected Examples

Carbon is one of the six most abundant elements in the solar system, and it enters several chemical reactions relevant for keeping in equilibrium the earth's biosphere. As an element it displays a range of structures and properties: a number of crystalline allotropes sum up to many noncrystalline C forms associated with the different possible types of bonding and different degrees of structural disorder. Such richness and the associated potentialities of application of C compounds, mainly with hydrogen, nitrogen, and oxygen, has conveyed since two centuries research interest in this element and its compounds both from the chemistry and the physics communities. The last C-based structures that fired strong activity were fullerene and its derivatives in the 1980s and the 1990s and more recently the graphene world, still expanding toward exploring applications of the related 2D sheetlike or flakelike carbon forms [12].

The strict dependence of the physical properties of C-based materials on the ratio between sp^2 (graphitelike) and sp^3 (diamond-like) bond coordination [13] helps in understanding their versatility. A number of sp^2-bonded carbons, ranging from microcrystalline graphite to glassy carbon, each characterized by a different degree of graphitic order are available. All noncrystalline carbons display a mixture of sp^3, sp^2, and even sp bonds, possibly in the presence of a fraction of hydrogen. Changing the relative abundance of the possible hybridizations results in a change of the density of states and of the energy gap. In the family of diamond-like carbon (DLC) films, besides the fraction of tetrahedrally bonded atoms, the major factors affecting both mechanical and optical properties are, in order, the degree of clustering of the sp^2 phase, its orientation, and anisotropy [13]. Hydrogen-free, dominantly sp^3-coordinated, tetrahedral amorphous carbon (ta-C) is characterized by a low friction coefficient and high hardness: its optical transparency in a wide spectral range and its absorbance in the UV are attractive for scratch-resistant thin coatings of lenses [14]. Over the last 30 years PLD has shown to be a versatile technique to synthesize carbon films in vacuum or at low background gas pressure, controlling bond hybridization, the degree of structural disorder, and the extent of clustering of sp^2-coordinated phase. DLC films with well-differentiated structural, mechanical, optical, and electronic properties were deposited [15]. The degree of the diamond-like character of the deposited films depends on process parameters such as vacuum, laser wavelength, and laser power density P. To obtain DLC a vacuum of at least 10^{-5} Pa and a minimum P value increasing with λ (e.g., $P \approx 3 \times 10^6$ W mm^{-2} at $\lambda = 248$ nm [16]) appear to be necessary. UV lasers are most suited since particulate emission from the target is strongly depressed and ta-C films are obtained at lower laser energy densities with respect to those required at longer wavelengths [17].

Much less attention was devoted to the synthesis in a high-pressure inert gas of C films, whose structure and surface nanostructuring depend on the conditions of NP formation in the expanding plasma plume, and of coalescence of deposited NPs onto the substrate. C films were deposited in helium and in argon at pressures between moderate (30 Pa) and high (2 kPa) changing P between 8.5 and 19 MW mm^{-2} using a nanosecond (pulse duration τ

= 20 ns) UV laser (λ = 248 nm). Increasing ambient gas pressure, the films display a hierarchy of nanostructures, from columnar, dense, with embedded nodules, to spongy, highly porous, discontinuous, brush-like, as shown in Fig. 3.1. The number density of nodules increases with increasing gas pressure, over the 30–70 Pa range, while their cap radius shrinks. At the intermediate gas pressure of 250 Pa, irrespective of the laser power density, low-density columnar growth coexists with nodule development. At such synthesis conditions nodules are rather big and they have a tendency toward sphericity and to loose intrapacking. Overall, these films resemble a nanopowder with a degree of intrafilm cohesion increasing with increasing P, while film–substrate adhesion is quite scarce for all P values. With respect to films synthesized in He, in films deposited in Ar over the same range of process conditions we observed three trends: the nodules are smaller, with a spherelike shape; their spatial distribution is more homogeneous; and their size distribution is narrower [18]. At high gas pressure, in the kPa range, highly porous and irregular films are deposited; while the peak thickness of such films can exceed 15 μm, at some points the substrate surface is barely covered. The film aspect is most similar to the outputs of ballistic models under conditions of no atomic mobility, characterized by frozen-in strongly off-equilibrium geometrical configurations; film–substrate adhesion is very poor [19].

Figure 3.1 Representative cross-sectional SEM views of C films deposited in inert gas at high pressure. (a) He at 60 Pa, P = 16 MW mm^{-2}: dense columns and nodules are visible; (b) He at 250 Pa, P = 8.5 MW mm^{-2}: evidence of loosely packed columns and nodules; (c) Ar at 2 kPa, P = 16 MW mm^{-2}: open, dendritic nanostructure.

Visible Raman spectroscopy indicates that all films deposited in the noble gas atmospheres described above are noncrystalline, sp^2 hybridized. In films prepared at lower gas pressure the deduced coherence length L_c, and thus the average size of the domains that coherently scatter radiation, is about 5 nm: these films can be classified as nanoglassy C.

Transmission electron microscopy (TEM) on the same films confirms the spectroscopic indications: the films are inhomogeneous, and they are made of noncrystalline aggregates with the average size in agreement with the L_c values deduced from Raman measurements. In samples deposited at high gas pressure TEM indicates the formation of agglomerates 10–20 nm in size, resulting from the mutual interconnection of smaller, low-energy NPs to form a loose, low-density, porous coating [20].

The role of an inert, high-pressure ambient gas filling the PLD chamber is to promote the synthesis of C NPs in the expanding plasma plume. The measured low fraction of sp^3-coordinated atoms in all films, independently of the laser power density P, is likely to be due to the strong scattering of plume species from He and Ar atoms and the subsequent cooling down of the plume to an equilibrium temperature between plume and ambient gas. The phenomenological counterparts of such effects are the increased visible fluorescence of the advancing plume front and the spatial confinement of the plume, together with its evident shrinking. With the laser parameters we adopted, the growth rate of a typical NP in the plume due to C atom agglomeration is considerably larger than the ablation rate of C atoms from the target surface [21]. Thus the plume is progressively depleted in C atoms with the increasing number/size of C NPs. In the frame of the ideal gas approximation, taking average parameter values over long times with respect to the highly nonlinear initial isothermal plume expansion regime, we calculated the average number of atoms N in a NP that reached its stationary size. Thus after the plume became collisionless at time t_f

$$N = (<n_a> \sigma_{a-a} <v> t_f)(<n_g> \sigma_{a-g} <v> t_f) \quad (3.2)$$

where n indicates the number density, σ refers to the elastic collision cross section, $<v>$ refers to the average particle velocity, and the subscripts a and g refer to ablated species and gas atoms, respectively. In our study, in the absence of plasma diagnostics we estimated the initial velocity of ablated C atoms $\mathbf{v} \approx 5 \times 10^5$ ms^{-1} and $t_f \approx 5 \times 10^{-6}$ s as based on measurements performed in experiments conducted in conditions similar to ours [21]. On these grounds, taking the spherical symmetry of the NPs, their estimated size ranges from about 5 nm ($N = 5 \times 10^3$) for expansions at 30 Pa to about 20 nm ($N = 5 \times 10^5$) for expansions at 2 kPa. The estimates are in agreement with the results of TEM measurements.

The same mechanisms of plume confinement, braking, and energy exchange between ablated species and gas atoms hold when the chemistry is less elementary, both on the gas and on the target side. We have taken into account the synthesis of compound materials in high-pressure gas mixtures, considering specifically WO_3, an n-type, wide-bandgap (2.5–5 eV) semiconductor used as a promising material for its selective susceptivity to a large number of toxic gases, such as NO_x, SO_2, H_2S, and NH_3 through relevant resistance changes [22]. Crystalline WO_3 films were prepared by PLD, pointing at the synthesis of samples with an open nanostructure associated to a high volumetric surface area and strong chemical reactivity, allowing the detection of very low concentrations of NH_3. Ammonia is extensively used in the production of fertilizers, plastics, and dyes and shows a threshold exposure limit as low as 25 ppm [23]. An UV laser (λ = 193 nm; τ = 25 ns) was focused on a rotating WO_3 target held at 873 K whose ablation was performed in a (1:1) mixed Ar/O_2 atmosphere at 10 Pa ambient pressure. The deposition conditions were selected after a thorough study of the dependence of composition, morphology, and structure of WO_x films on PLD deposition parameters [24, 25]. Representative images of film nanostructure, both top view and cross section, are displayed in Fig. 3.2a,b.

Figure 3.2 SEM views of the nanostructure of a WO_3 film deposited by PLD in high-pressure mixed gas ($Ar:O_2$) atmosphere. (a) Top view; (b) cross section with evidence of the columnar layered structure.

The irregular tilings that pave the film surface are rather loosely mutually interconnected, leaving a considerable open space at the junctions among adjacent tilings. The average size of such tiles ranges from 30 to 70 nm, with the most frequent value of about 50 nm. The cross section of the film indicates that on average, it is about 500 nm thick and it is made of aligned columns that grow slightly inclined (about 10°) with respect to the normal to the (100) Si substrate; for most columns the growth is regular up to the film surface, so they are similar to each other, with the most frequent average width of about 50 nm. From Fig. 3.2b in many columns a layered structure is discernible, each layer lying normal to the column growth axis. The film is highly compact near the interface with the substrate; then at a height of around 100 nm the columns begin to separate from each other and a void network develops among them up to the film surface. From Fig. 3.2b also an undulated surface can be discerned, corresponding to the top view in Fig. 3.2a. The deposition conditions are stationary, with an exiguous contamination degree, mainly Ti, as determined by energy-dispersive X-ray (EDX) analysis performed both along with scanning electron microscopy (SEM) and TEM observations. Thus the impingement ratio impurity/(W+O) at the growth surface is very low and consequently large, monoclinic V phase [26] crystalline WO_3 grains can grow. This is likely to explain the observed considerable column width, as compared to column height. Representative TEM pictures taken on the same sample are shown in Fig. 3.3. They indicate that the material has grown with a considerable degree of porosity: the film volume ratio porous:dense is about 80:20%, thus corroborating in a quantitative way the above qualitative SEM observations. The film was integrated in a metal-oxide semiconductor (MOS) device and tested against NH_3 at different temperatures. At 573 K the sensor response was reproducibly linear, with gas concentrations between 20 and 50 ppm, with a faster recovery time and maximum response than those of WO_3 sensors prepared by alternative methodologies [27].

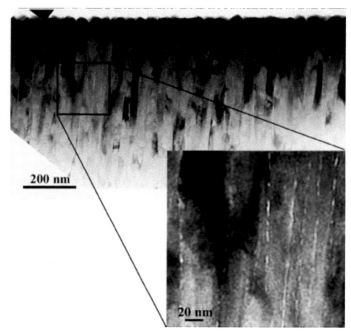

Figure 3.3 TEM pictures taken on the same sample as in Fig. 3.2. The columnar structure is evident as well as (see the magnified portion of the sample) the empty white-looking channels parallel to the column height.

3.4 Deposition of Noble Metal Nanoparticle Arrays for Application in Biomedical Sensing

The synthesis of NPs in laser-produced plasmas expanding through a high-pressure ambient gas, besides the spontaneous generation of a multiplicity of morphologies resulting from the mutual assembling of such NPs on a substrate, opens the way to the design of specific surface nanostructures for dedicated applications. This step requires the identification and control of process parameters relevant to NP synthesis and assembling. An exemplary case where tailoring NP growth and assembling are required to obtain meaningful applications is plasmonics, or surface plasmon polariton photonics, that hinges on the coupling of light to charges at metal surfaces. Although the basic ideas relevant to the field were formulated a century ago [28] modern plasmonics [29] required the development of nanofabrication techniques and the development of quantitative electromagnetic simulation that allow one to obtain optical components in the sub 100 nm size range. The investigation on the surface plasmon polariton excitations associated with light localization and guiding in such structures leads to the development of sensors, exploiting the possibility to locally enhance the intensity of electromagnetic fields to detect exiguous amounts of analyte via surface-enhanced spectroscopies, for example, SERS. Indeed, since about 40 years [30] it has been observed that noble metal metallic surfaces corrugated at the nanometer scale yield the most significant intensity enhancements of the Raman features of the investigated analyte. Major advances in SERS became feasible once associated with the tailoring of the metal surface nanostructure and roughness through the realization on a suitable support of uniform arrangements of identical NPs with a predesigned size. In the case of silver and gold NPs synthesized via PLD in a high-pressure inert gas like argon, we performed a detailed study to correlate process parameters and the features both of synthesized NPs in the expanding ablation plume and of their mutual interaction on the support that result in a variety of 2D architectures. We also tried to shed light on the more subtle influence that partly veiled deposition parameters have on the surface nanostructure of seemingly highly similar films, yet characterized by strongly different optical properties and, consequently, SERS behavior. Such efforts were directed at identifying suitable process conditions to synthesize SERS substrates useful in revealing exiguous amounts of drugs of clinical interest in patients affected by chronic neurological diseases, such as Parkinson's disease (PD) and epilepsy.

It is, thus, necessary to come back again to the discussion on PLD process parameters, in particular paying attention to laser fluence and the target mass ablated by a single laser pulse.

Laser fluence has a strong influence not only on plume energy but also on plume mass, that is, on the mass ablated from the target by a single laser pulse. The higher the F value, the larger the ablated mass per pulse. This feature plays an important role, in particular when the expanding plume interacts with an ambient gas. If the plume velocity is higher than the sound speed in the ambient gas, a shock wave develops, provided the gas mass compressed by the advancing plume front and surrounding it is comparable to the plume mass. Thus, plume mass M_p, gas density ρ_g, and plume velocity v_p determine in a complex way plume expansion dynamics and, in turn, the clustering processes in the gas phase. All these parameters enter Eq. 3.2 to determine the average number N of atoms in the gas phase–generated NPs. Since the atomic masses of Ag (107 amu) and Au (196 amu) are much larger than that of C (12 amu), scattering processes by Ar atoms are stronger in both cases and noble metal NPs are expected to land on a support with the residual kinetic energy much reduced with respect to C NPs so the developing surface nanostructures can be controlled by the deposition time or, equivalently, by the number of laser pulses. At a low laser pulse (LP) number, isolated, nearly spherical NPs are deposited; as long as the deposition proceeds NPs start to coalesce, giving rise to islands with elliptical shapes, which become more and more irregular, while at the same time increasing the average NP size; the subsequent stage consists of nearly percolated structures, and if the deposition lasts enough, semicontinous films are obtained [30, 31]. Since the optical properties of a NP array strongly depend on the surface morphology the fine control of all the above deposition parameters has to be achieved. As reported above the ablated mass per pulse M_p, is a function of the laser fluence. The ablation threshold F_t for Au and Ag using a nanosecond UV pulsed laser is just below $F_t = 1.0$ J cm^{-2}, so $F = 2.0$ J cm^{-2}, slightly above F_t, was adopted to grow Ag and Au thin films by PLD. Typically, a KrF UV laser ($\lambda = 248$ nm; $\pi = 25$ ns) delivered at room temperature a variable number of pulses, from 500 to 3×10^4, to elemental Ag or Au rotating targets to avoid extensive cratering. The reference vacuum was 10^{-4} Pa, and Ar pressure was changed in the deposition chamber between 10 and 10^2 Pa. In Fig. 3.4 are reported representative SEM pictures taken on Ag NP arrays deposited in Ar changing LP number and/or gas pressure. At 70 Pa Ar pressure moving from the very low LP value (500) to LP = 10^4 results in a densification of isolated NPs, showing an initial tendency to coalescence. The trend is unchanged at LP = 1.5×10^4, while at LP = 3×10^4 a definite evolution toward a deviation of the NP shape to elliptical and, in some instances, to an irregular shape is found. We observed that NP coalescence is strongly enhanced on lowering Ar pressure, as can be seen on comparing the nanostructures of the arrays prepared at the same LP number (1.5 × 10⁴) but different Ar pressures (10 Pa, Fig. 3.4c; 70 Pa, Fig. 3.4d). When we look at the optical properties of such nanostructures we notice the following trends: the surface plasmon resonance (SPR) consistently blue-shifts at a fixed LP number, on increasing the Ar pressure, as well as at fixed gas pressure, on lowering the LP number. In Fig. 3.4e,f both trends are shown. With Au NPs the same trends as reported for Ag NPs were observed.

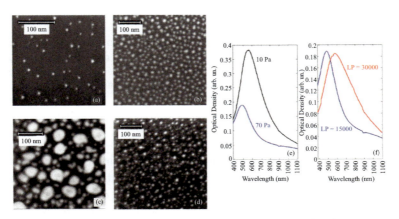

Figure 3.4 SEM pictures taken on Ag NP arrays deposited on (100) Si supports at different conditions. (a) Ar, 70 Pa; LP = 500; (b) Ar, 70 Pa; LP = 10⁴. (c) Ar, 10 Pa; LP = 15,000; (d) Ar, 70 Pa; LP = 15,000; (e) optical properties of two films deposited on a glass support with LP = 15,000, at low (10 Pa) and high (70 Pa) Ar pressure; evidence of a blue shift of the SPR with increasing gas pressure; (f) optical properties of two films deposited at the same Ar pressure (70 Pa), with LP = 15,000 and LP = 3 × 10⁴; evidence of a blue shift of the SPR with a lowering LP number.

The average size of NPs at their landing on the substrate as predicted by Eq. 3.2 were compared both for Ag and for Au plumes expanding in Ar, with the average size of NPs deposited at different gas pressures and the LP number low enough to avoid NP clustering. TEM observations [31] provided agreement between model predictions and experiment. Thus, we are able to control both NP size at landing, that lie in the range between 1 and 4 nm for Ag and between 2 and 6 nm for Au, and the ablation conditions that allow preserving NP integrity at landing on the support.

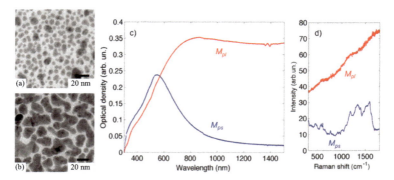

Figure 3.5 TEM pictures of the surface nanostructures of Ag films deposited on (100) Si supports with LP = 10^4, at the fluence of 2.8 J cm^{-2} with (a) low M_p (M_{ps} = 7.0 ng) and (b) high M_p (M_{pl} = 16.4 ng). (c) Surface plasmon resonance (SPR) peak position for films deposited on glass with M_{ps} and M_{pl}, respectively. (d) SERS spectra of rhodamine 6G taken from the same films as in (c), after dipping for 60 min in an aqueous solution at 10^{-6} M concentration. Exciting radiation: 632 nm He-Ne line.

Yet, the above trends do not provide a complete picture of the development of surface nanostructuring. We studied the relevance of M_p in a sequence of controlled depositions of Ag NP arrays, looking both at plasma dynamics and at film nanostructuring [32, 33]. Briefly, we kept the energy density of the KrF laser fixed at F = 2.8 J cm^{-2} and we fixed two values of the laser spot area, one being about three times that of the other. After careful measurements via stylus profilometry and reconstruction of the 3D crater profile we obtained M_{ps} = 7.0 ng for the smaller spot area and M_{pl} = 16.4 ng for the larger spot. At a constant LP number (10^4), with two Ar pressures (10 Pa; 70 Pa) islanded nanostructures were obtained, as exemplified in Fig. 3.5a,b, which refers to the films deposited at higher Ar pressure. Both of them are made of islands of irregular shapes and sizes, interconnected through a network of channels, again showing a degree of widths and lengths. Such nanostructures, highly magnified in the TEM images, appear qualitatively similar to each other, differing in the degree of surface coverage (49%, M_{ps}; 77%, M_{pl}) of the support, as estimated after conversion of the pictures to b/w tones. It is likely that the most frequently adopted SEM imaging of the surfaces under exam would not have revealed any meaningful difference between them. The optical properties of the two films are compared in Fig. 3.5c: in the film deposited with M_{ps} SPR is peaked around 590 nm and it is much narrower than for the film deposited with M_{pl}, where SPR is much broader, being centered around 820 nm. Besides this difference the SERS performance of the films, tested against the reference, easily detectable analyte rhodamine 6G (R6G), using aqueous solutions at the molar concentrations 10^{-4} and 10^{-6} is very different. The film deposited with M_{ps} provides a consistently better SERS response, while the one deposited with M_{pl} is less sensitive at the higher analyte concentration and nearly insensitive at the lower one, as shown in Fig. 3.5d.

Moving from the ability we gained to tailor the synthesis of NPs in the expanding laser-generated plasma and their mutual assembling on the support to produce different nanostructurings of the SERS substrates we produced Au NP arrays designed ad hoc to quantitatively assess small concentrations of drugs in various solutions, from more elementary like aqueous ones to realistic biological fluids of clinical interest. As an example we report on a recent study on apomorphine (APO, molecular weight 267.32 g mol^{-1}). APO is the less invasive drug used in the treatment of patients affected by severe PD, a neurodegenerative disease that affects about 0.4%–0.5% of the population and more that 1% of people aged over 65. PD is progressive and with increasing deterioration of the dopaminergic system, both fluctuations of the motor response to oral dopaminergic therapy and dyskinesias set on. APO is usually delivered by subcutaneous injection, and its concentration is adjusted on each single patient in relation to the motor condition and the possible collateral effects. To find the correct concentration and the rate of drug delivery during daytime is a procedure often difficult and time consuming. Presently a standardized way of the patient's self-rating of motor condition and hand tapping test is adopted, being mostly oriented to check the development of dyskinesias [34]. The established and affordable, though time consuming and costly, reference analytical technique high-performance liquid chromatography (HPLC) coupled with mass spectrometry (MS) is not adopted to quantitatively assess APO concentration. Compared to HPLC, SERS allows for a fast determination of drug concentration in biological fluids with minimal sample preparation. As such it is expected to improve checking the compliance of the patient to the treatment, which is relevant for keeping a good life quality in patients with chronic disease like PD who need daily drug intake.

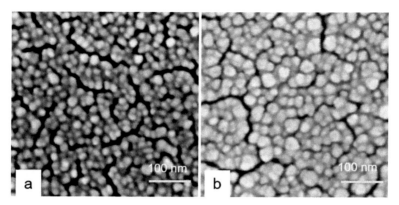

Figure 3.6 SEM micrographs (top views) of Au substrates deposited in Ar at 10^2 Pa on (100) Si supports with (a) LP$_s$ (LP = 10^4) and (b) LP$_l$ (LP = 2×10^4).

Au substrates were deposited in Ar at 10^2 Pa on Corning Glass 5079 or on (100) Si at the standard conditions described above, choosing as the LP number 10^4 (LP$_s$) and 2×10^4 (LP$_l$). Representative SEM pictures taken on both substrates are reported in Fig. 3.6a,b. Considerable uniformity of the nanostructure is evident throughout each of the two surfaces. An LP$_s$ substrate (Fig. 3.6a) consists of spheroidal, agglomerated NP$_s$, sized between 6 and 20 nm, that frequently give rise to islands with irregular shapes, separated from each other by channels whose length is between 30 and 200 nm and whose width is about 5–6 nm. An LP$_l$ substrate (Fig. 3.6b) is made of small, about 5 nm, NPs sticking together to give islands of sizes between 50 and 300 nm and shapes more and more irregular with increasing size. Inter-island channels are present, though they are less frequent than in an LP$_s$ substrate; the average channel width is about 5 nm and the length is up to 180 nm. It is noteworthy that the optical properties and thus the SERS of such family of films sensitively depends on the dimensions and spatial distribution of empty regions of the substrate, such as channels and their junctions [35] where hot spots in SERS take place. In Fig. 3.7 SERS spectra taken from solutions of APO at the concentrations of 6.6×10^{-5} M in blood (spectrum 2) and 3.3×10^{-6} M in blood serum (spectrum 1) are reported. The former spectrum was recorded using an LP$_s$ substrate, while an LP$_l$ substrate was chosen to take the latter. All the features of the SERS spectrum of APO, labeled by the symbol ∇ in Fig. 3.7, are recognizable in the blood serum sample with little or no shift in their wavenumbers, compared with a reference SERS spectrum collected from water solutions [36].

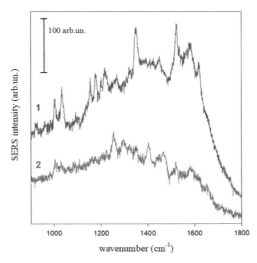

Figure 3.7 (1) SERS spectrum of apomorphine (APO) in blood serum (total APO concentration 3.3 × 10^{-6} M) taken on a Au substrate deposited on glass with LP$_l$ = 2 × 10^4 (Ar 10^2 Pa). (2) SERS spectrum of APO in blood plasma (total APO concentration 6.6 × 10^{-5} M) taken on a Au substrate with LP$_s$ = 10^4 (Ar 10^2 Pa). Exciting line: 785 nm.

Similar to what we observed in aqueous solutions, an LP$_l$ substrate (spectrum 1) definitely delivers better defined and more intense Raman features of APO with respect to those of spectrum 2 (substrate LP$_s$), although the latter was recorded from a solution at a drug concentration larger by more than 1 order of magnitude with respect to the former. The result demonstrates that to detect APO concentrations of clinical relevance, once other process parameters (such as gas nature, pressure, target-to-substrate distance, laser pulse energy, and wavelength) have been established, the fine-tuning of a few key parameters, for example, the LP number, is critical to improve the SERS response of a Au sensor with specifically designed nanometric architecture.

Since all the features of the SERS spectrum of APO are found in the blood serum sample (spectrum 1), detecting such a concentration in real samples is feasible, notwithstanding the complex matrix that characterizes these samples, where several molecular species contribute to the observed SERS spectrum. Indeed, the decrease by a factor of 10 of the detectable APO concentration in biological fluids with respect to aqueous solutions is likely to be due to the presence in the former of proteins (e.g., albumin), which bind APO to a relevant fraction [37]. We note that the easier APO detection in the case of spectrum 2 may be associated to the use of serum, a less complex medium than plasma.

The same conceptual approach has been adopted to investigate the concentration in different solutions of a popular antiepileptic drug, most employed in developing countries, such as carbamazepine (CBZ, molecular weight 236.27 gmol^{-1}). After taking reference Raman spectra of CBZ in MeOH solutions, SERS spectra were recorded at a CBZ concentration of 4.2×10^{-4} M using Au nanostructured substrates deposited in Ar at $p = 10^2$ Pa with LP = 2×10^4 using Corning Glass 5079 supports. Agreement was found between the position of each CBZ peak in the Raman spectrum and the position of the corresponding SERS feature [38]. We recently were able to detect CBZ features in human plasma at the concentrations of 2.12×10^{-3} M, 1.27×10^{-3} M, and 4.23×10^{-4} M, as well as to identify two major CBZ features in the SERS spectrum from the blood serum taken from a volunteer patient treated at the CBZ concentration of 5.04×10^{-5} M.

3.5 Conclusions

In conclusion laser ablation of a solid target in a gas at high pressure allows synthesizing NPs of different sizes in the expanding plasma plume, depending on the delicate interplay among laser characteristics (wavelength, energy per pulse, etc.), gas nature and pressure, and target (mass ablated per pulse), which govern the energetics and propagation of the plume. The NPs at their landing with reduced kinetic energy on a support suitable for the designed application undergo a limited surface diffusion until they come to rest, possibly after having experienced clustering with other NPs. The resulting surface nanostructure depends on the features of the constituent NPs and on the LP number, which provide matter pulses to the growing structure. A variety of self-assembled NP arrays are observed with largely differentiated morphologies. Such structures can find application in nonlinear vibrational spectroscopies, in particular as substrates for SERS. We have demonstrated how Au and Ag NP arrays can be used as SERS substrates to detect exiguous amounts of drugs used in the clinical treatment of neurological diseases.

Acknowledgments

The authors are grateful to several colleagues with whom they collaborated: A. Miotello, G. Radnoczi, M. Filipescu, N. Santo, M. Dinescu, E. Fazio, N. R. Agarwal, M. Tommasini, A. Lucotti, C. Zanchi, E. Ciusani, and U. de Grazia.

Mr. A. Mantegazza, Dipartimento di Energia, Politecnico di Milano is acknowledged for his advice with SEM observations. This research was partly supported by Polisocial Award 2014, project Controllare l'epilessia nei Paesi in via di sviluppo (Controlling Epilepsy in Developing Countries).

References

1. Wang, Y. L., Xu, W., Zhou, Y., Chu, L.-Z., and Fu, G.-S. (2007). Influence of pulse repetition rate on the average size of silicon nanoparticles deposited by laser ablation, *Laser Particle Beams*, **25**, pp. 9–13.
2. Irissou, E., Le Drogoff, B., Chaker, M., and Guay, D. (2003). Influence of the expansion dynamics of laser-produced gold plasmas on thin film structure grown in various atmospheres, *J. Appl. Phys.*, **94**, pp. 4796–4802.
3. Seal, K., Nelson, M. A., Ying, Z. C., Genov, D. A., Sarychev, A. K., and Shalaev, V. M. (2003). Growth, morphology, and optical and electrical properties of semicontinuous metallic films *Phys. Rev.*, **B67**, pp. 035318-13.
4. Bertoni, G., Cepek, C., Romanato, F., Casari, C. S., Li Bassi, A., Bottani, C. E., and Sancrotti, M. (2004). Growth of multi-wall and single-wall carbon nanotubes with in situ high vacuum catalyst deposition, *Carbon*, **42**, pp. 440–443.
5. Bäuerle, D. (2000). *Laser Processing and Chemistry*, (Springer, Berlin).
6. Lin, Z., Zhigilei, L.V., and Celli, V. (2008). Electron-phonon coupling and electron heat capacity of metals under conditions of strong electron-phonon nonequilibrium, *Phys. Rev. B*, **77**, pp. 075133–17.

7. Zink, J. C., Reif, J., and Matthias, E. (1992). Water adsorption on (111) surfaces of BaF2 and CaF2, *Phys. Rev. Lett.*, **68**, pp. 3595–3598.
8. Kelly, R., and Miotello, A. (1996) Comments on explosive mechanisms of laser sputtering, *Appl. Surf. Sci.*, **96**, pp. 205–215.
9. Geohegan, D. B., Puretzky, A. A., Rouleau, C., Jackson, J., Eres, G., Liu, Z., Styers-Barnett, D., Hu, H., Zhao, B., Ivanov, I., Xiao, K., and More K. (2010). In *Laser-Surface Interactions for New Materials Production*, eds. Miotello A., and Ossi P. M., p. 1 (Springer, Berlin).
10. Itina, T. E., Hermann, J., Delaporte, P., and Sentis, M. (2002). Laser-generated plasma plume expansion: Combined continuous-microscopic modeling, *Phys. Rev. E*, **66**, pp. 066406–12.
11. Geohegan, D. B., Puretzky, A. A., Hettich, R. L., Zheng, X-Y., Haufler, R. E., and Compton R. N. (1994). Gated ICCD Photography of the KrF–laser ablation of graphite into background gases, *Trans. Mat. Res. Soc. Jpn.*, **17**, p. 349.
12. Bianco, A., Cheng, H. M., Enoki, T., Gogotsi, Y., Hurt, R. H., Koratkar, N., Kyotani, T., Monthioux, M., Park, C. M., Tascon, J. M. D., and Zhang, J. (2013). All in the graphene family: a recommended nomenclature for two-dimensional carbon materials, Carbon, **65**, pp. 1–6.
13. Robertson, J. (2002) Diamond-like amorphous carbon, *Mater. Sci. Eng.*, **R27**, pp. 129–281.
14. Bonelli, M., Miotello, A., Mosaner, P., Casiraghi, C., and Ossi, P. M. (2003) Pulsed laser deposition of diamondlike carbon films on polycarbonate *J. Appl. Phys.*, **93**, pp. 859–867.
15. Bonelli, M., Fioravanti, A. P., Miotello, A., and Ossi, P. M. (2003). Structural and mechanical properties of ta-C films grown by pulsed laser deposition Europhys. Lett., **50**, pp. 501–506.
16. Voevodin, A. A., and Donley, M. S. (1996). Preparation of amorphous diamond-like carbon by pulsed laser deposition: a critical review, *Surf. Coat. Technol.*, **82**, pp. 199–213.
17. Haglund, R. F. (1998). In *Laser Adsorption and Desorption*, eds. Miller J. G., and Haglund R. F., p.15 (Academic Press, London).
18. Ossi, P. M., and Miotello, A. (2007). Control of cluster synthesis in nanoglassy carbon films, *J. Non-Cryst. Solids*, **353**, pp. 1860–1864.
19. Ossi, P. M., Bottani, C. E., and Miotello, A. (2005). Pulsed-laser deposition of carbon: from DLC to cluster-assembled films. *Thin Solid Films*, **482**, pp. 2–8.

20. Bolgiaghi, D., Miotello, A., Mosaner, P., Ossi, P. M., and Radnoczi, G. (2005). Pulsed laser deposition of glass-like cluster assembled carbon films *Carbon*, **43**, pp. 2122–2127.
21. Rode, A. V., Gamaly, E. G., and Luther-Davies, B. (2000). Formation of cluster-assembled carbon nano-foam by high-repetition-rate laser ablation, *Appl. Phys. A*, **70**, pp. 135–144.
22. Stoycheva, T., Annanouch, F. E., Garcia, I., Llobet, E., Blackman, C., Correig, X., and Vallejos, S. (2014). Micromachined gas sensors based on tungsten oxide nanoneedles directly integrated via aerosol assisted CVD, *Sensors Actuators B*, **198**, pp. 210–218.
23. www.cdc.gov/niosh/npg/npgd0028.html.
24. Filipescu, M., Ossi, P. M., Santo, N., and Dinescu, M. (2009). Radio-frequency assisted pulsed laser deposition of nanostructured WOx films, *Appl. Surf. Sci.*, **255**, pp. 9699–9702.
25. Dinescu, M., Filipescu, M., Ossi, P. M., and Santo, N. (2010). Nanoporous cluster-assembled WOx films prepared by radio-frequency assisted laser ablation, *Thin Solid Films*, **518**, pp. 4493–4498.
26. Cazzanelli, E., Vinegoni, C., Mariotto, G., Kuzmin, A., and Purans, J. (1999). Raman study of the phase transitions sequence in pure WO_3 at high temperature and in Hx WO_3 with variable hydrogen content, Solid State Ionics, **123**, pp. 67–74.
27. Palla-Papavlu, A., Filipescu, M., Schneider, C. V., Antohe, S., Ossi, P. M., Radnoczi, G., Dinescu M., Wokaum, A., and Lippert, T. (2016). Direct laser deposition of nanostructured tungsten oxide for sensing applications, *J. Phys. D: Appl. Phys.*, **49**, pp. 205101-8.
28. Editorial (2012). Surface plasmon resurrection, *Nat. Photonics*, **6**, p. 707.
29. Maier, S. A., Brongersma, M. L., Kik, P. G., Meltzer, S., Requicha, A. A. G., and Atwater, H. A. (2001). Plasmonics: a route to nanoscale optical devices., *Adv. Mater.*, **13**, pp. 1501–1505.
30. Fazio, E., Neri, F., Ossi, P. M., and Trusso, S. (2009). Growth process of nanostructured silver films pulsed laser ablated in high-pressure inert gas, *Appl. Surf. Sci.*, **255**, pp. 9676–9679.
31. Ossi, P. M., Neri, F., Santo, N., and Trusso, S. (2011). Noble metal nanoparticles produced by nanosecond laser ablation, *Appl. Phys. A*, **104**, pp. 829–837.
32. Spadaro, M. C., Fazio, E., Neri, F., Trusso, S., and Ossi, P. M. (2015). On the role of the ablated mass on the propagation of a laser-generated plasma in an ambient gas, *EPL*, **109**, pp. 25002-6.

33. Spadaro, M. C., Fazio, E., Neri, F., Trusso, S., and Ossi, P. M. (2014). On the influence of the mass ablated by a laser pulse on thin film morphology and optical properties, *Appl. Phys. A*, **117**, pp. 137–142.

34. Elia, A. E., Dollenz, C., Soliveri, P., and Albanese, A. (2012). Motor features and response to oral levodopa in patients with Parkinsons disease under continuous dopaminergic infusion or deep brain stimulation, *Eur. J. Neurol.*, **19**, pp. 76–83.

35. Agarwal, N. R., Neri, F., Trusso, S., and Ossi, P. M. (2013) Growth analysis of pulsed laser ablated films, *Plasmonics*, **8**, pp. 1707–1712.

36. Zanchi, C., Lucotti, A., Tommasini, M., Trusso, S., de Grazia, U., Ciusani, E., and Ossi, P. M. (2015). Au nanoparticle-based sensor for apomorphine detection in plasma, *Beilstein J. Nanotechnol.*, **6**, pp. 2224–2232.

37. Smith, R. V., Velagapudi, R. B., McLean, A. M., and Wilcox, R. E. (1985). Interactions of Apomorphine with serum and tissue proteins, *J. Med. Chem.*, **28**, pp. 613–620.

38. Zanchi, C., Lucotti, A., Tommasini, M., Trusso, S., de Grazia, U., Ciusani, E., and Ossi, P. M. (2016). Laser tailored nanoparticle arrays to detect molecules at dilute concentration, *Appl. Surf. Sci.*, **396**, pp. 1866–1874.

Chapter 4

Femtosecond Laser Ablation of Solid Targets in Vacuum and Low-Pressure Gas Atmosphere

Salvatore Amoruso

Dipartimento di Fisica "Ettore Pancini," Università degli Studi di Napoli Federico II
Complesso Universitario di Monte S. Angelo, Via Cintia, I-80126 Napoli, Italy
salvatore.amoruso@unina.it, amoruso@fisica.unina.it

The ejection of matter from a solid surface induced by irradiation with a high-intensity laser beam leads to a blow off of material moving rapidly away from the surface, a phenomenon termed "laser ablation." Studies on laser ablation and its uses in different fields (material analysis and processing, laser-produced plasma, film deposition, etc.) have kept the pace of the development of the different laser sources. The advent of ultrashort laser sources triggered the use of pulses of femtosecond duration, favoring a variety of new investigations and the development of femtosecond laser ablation of solid targets. This chapter summarizes diverse aspects of femtosecond laser ablation of a solid target in a high

---

Pulsed Laser Ablation: Advances and Applications in Nanoparticles and Nanostructuring Thin Films
Edited by Ion N. Mihailescu and Anna Paola Caricato
Copyright © 2018 Pan Stanford Publishing Pte. Ltd.
ISBN 978-981-4774-23-9 (Hardcover), 978-1-315-18523-1 (eBook)
www.panstanford.com

vacuum and in a low-pressure background gas, focusing on metals and semiconductors.

4.1 Introduction

There have been flourishing activities on laser ablation induced by nanosecond laser ablation (ns-LA) pulses, which are described in many books [1–6] and review articles [7–10]. Nowadays, ultrashort laser sources providing pulses of femtosecond (fs) duration are remarkable laboratory tools that have opened up a wide range of new investigations and femtosecond laser ablation (fs-LA) of solid targets is an emerging research field with a variety of possible applications [3, 11–15]. The main outcome of fs pulse duration is the minimization of heat diffusion effects, which is extremely important for precise microstructuring [16, 17]. The confinement of the energy deposition associated with the ultrashort laser pulse duration is paramount in applications where localized removal or limitation of target damage is important, such as surface micro- and nanostructuring, laser-induced forward transfer, and laser-induced breakdown spectroscopy (LIBS). The ablation blow off induced by fs pulses shows very peculiar characteristics that make it rather different from standard ns-LA, and the direct ejection of nanoparticles (NPs) is one of the most striking and intriguing features of fs-LA [18–20].

The fs-LA of metals and semiconductors typically occurs at laser pulse intensities of 10^{12}–10^{13} W/cm^2 (i.e., a laser fluence of 0.1–1 J/cm^2 for a typical pulse length of ~100 fs), for which a significant removal of target material occurs. The specific value of the laser pulse intensity threshold depends on target optical and thermal properties and on laser pulse wavelength and duration [5–7, 12]. Fs-LA differs in several ways from ns-LA. The laser pulse duration is much shorter than the characteristic time for energy transfer from electron to lattice τ_{el} (typically ~10 picoseconds [ps] for elemental metals); hence there is little hydrodynamic movement of the heated material during the laser pulse duration. Hence, for fs laser pulses the initial laser heating occurs almost at a solid density and can lead the matter to extreme temperature and pressure, generating novel material states that cannot be achieved by using longer pulses of comparable fluence [12, 16, 17]. Moreover, fs laser pulses are so short that interaction with the ejected particles does not occur, thus preventing complicated secondary laser–material interactions typical of ns-LA.

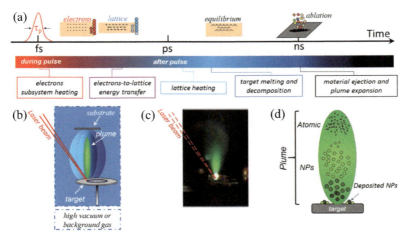

Figure 4.1 Description of the fs-LA process. (a) Sketch of the various temporal phases of the fs-LA process. (b) The laser pulse hits the target surface, producing an elongated plume that preferentially expands along the target normal, in high vacuum or in a low-pressure background gas. The ablated material can be collected on a suitable substrate. (c) Photo of the plume produced during laser ablation of a silver target in high vacuum. (d) Diagram showing the typical composition of the fs-LA plume with a faster atomic component preceding a slower NP population.

Figure 4.1a summarizes the series of processes occurring after irradiation of a metallic target with fs pulses. During the laser pulse, the electrons within the optical penetration depth absorb energy and are driven to a high temperature (the excited electrons can be assumed to thermalize almost instantaneously due to the short electron–electron interaction time), while the lattice remains relatively cold. In metals this nonequilibrium stage is normally described by two subequilibrium systems, hot electrons and cold lattice [21–23]. On a timescale of τ_{el}, the electron temperature drops through energy relaxation to the lattice and energy transport in deeper regions of the target by electron heat conduction. The evolution of the transient two-temperature system is described by the two-temperature model (TTM) [22, 23]. The system will then tend to reach equilibrium within a lapse of timeof $\sim$(3–5) τ_{el} [24, 25]. When the lattice temperature reaches high-enough values, melting may occur on a timescale between a few and a hundred ps. Finally, ablation of part of the heated region ensues. This stage can begin at a few tens of ps from the fs laser pulse and last up to several ns. Once ablation occurs, the ejected, partially ionized material forms a luminous ablation plume, which preferentially expands in a direction normal to the target surface, as depicted in Fig. 4.1b. The ablation plume propagates in high vacuum or in a low-pressure background gas, and the collection of the ablated material on a suitable substrate can allow deposition of thin films. Figure 4.1c shows a photo of the plume produced during fs-LA of a silver target in high vacuum.

The initial laser heating is nearly isochoric, and the relaxation of the hot, near-solid-density material proceeds in a different manner at different depths in the target. Material decomposition may occur through several different mechanisms, as will be described later, in Section 4.2. Near the surface the material is completely atomized, while deeper layers decompose into a mixture of vapor and liquid clusters through mechanical fragmentation and phase explosion [26–29]. This leads to the formation of a spatially structured fs ablation plume close to the target surface, whose evolution on a longer timescale gives rise to two distinct ablation components. Figure 4.2 illustrates this typical feature of fs-LA in the case of the ablation plume of a nickel target produced with a $\sim$300 fs laser pulse at a wavelength of 527 nm [29, 30]. In these investigations, the fs-LA plume characteristics were analyzed by means of intensified charge-coupled device (ICCD) fast imaging and optical emission spectroscopy of the ablated species self-emission [7, 29, 30]. Figure 4.2e reports a schematic diagram illustrating approximate timescales of the ablation process and of the subsequent optical emission from the ablated species, while Fig. 4.2f shows a sketch of the experimental setup typically exploited for the analysis of plume self-emission. The ICCD fast imaging provides 2D images of the plume emission, while optical emission spectroscopy (carried out by dispersing the emission with a spectrograph) allows registering the corresponding spectrum. Figures 4.2a and 4.2c show 2D images of the fs-LA plume emission registered at two different delays after the laser pulse. In each image, the intensity is normalized to its own maximum value. Figure 4.2a reports a typical image of the plume self-emission registered at a delay τ_D = 280 ns after the laser pulse [29, 30]. The plume image clearly evidences the layered structure of the ejected material. A faster, forward-directed component expands into vacuum at distances of several millimeters from the target surface on a timescale of hundreds of ns. This component is generally composed of neutral atoms, ions, and molecules (depending on the laser irradiation conditions and target properties), and its emission shows a characteristic line spectrum of the component species as that exemplified in Fig. 4.2b. This component is usually identified as an atomic plume. Analyses of the line emission intensities generally indicate an atomic plasma plume temperature of the order of 1 eV at early delays after the laser pulse, which successively reduces as a consequence of the expansion [31].

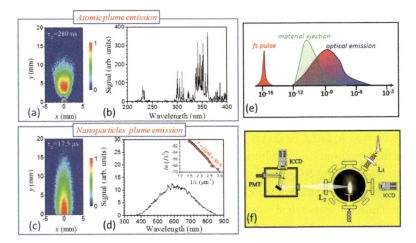

Figure 4.2 Peculiar characteristics of the plume generated during the fs-LA process. (a) ICCD image of a fs-LA nickel plume emission registered at τ_D = 280 ns after the laser pulse. (b) Typical optical emission spectrum of the atomic plume. (c) ICCD image of a fs-LA nickel plume emission registered at τ_D = 17 μs after the laser pulse, corresponding to the NP plume. (d) Typical optical emission spectrum of the NP plume. (e) Sketch with approximate timescales of ablation and subsequent visible emission from the ablation blow off. (f) Schematic of the experimental setup for ICCD imaging and optical emission spectroscopy of the fs-La plume. The ICCD fast imaging provides 2D images of the plume emission, while optical emission spectroscopy carried out by dispersing the emission with a spectrograph allows registering the corresponding spectrum.

The atomic plume is followed by a second, slower component that, for a short delay, is still confined very close to the target surface, as observed in Fig. 4.2a. A characteristic image of the slower component registered at a delay τ_D = 17.5 µs after the laser pulse is shown in Fig. 4.2c and indicates that its expansion occurs on a much longer timescale (several microseconds after the laser pulse arrival on the target surface) [29, 30]. This second component emits a broadband, structureless spectrum characteristic of NPs at high temperatures [18, 20]. Therefore, it is usually referred to as an NP plume. Once corrected for the instrumental spectral response and plotted as $\ln(I\lambda^5)$ versus λ, where I is the emission intensity registered at wavelength λ, the experimental data follow a decreasing linear dependence (see the inset of Fig. 4.2d) whose slope is inversely proportional to the NP temperature, which in this case is $T_{NP} \approx 2250$ K. The NP temperature at ~1 mm from the target surface is typically of a few thousand kelvin and successively reduces through radiative emission cooling during the late stages (starting from about 1 µs after the arrival of the laser pulse on the target) of the NP plume expansion into vacuum [32]. The NP plume is composed of small clusters and larger NPs [18–20, 29, 33, 34]. The NP characteristic broadband emission is a very useful method to spatially localize the NP plume and follow its propagation dynamics. Hence, differently from ns-LA, fs-LA of metals and semiconductors typically produces a fast-moving atomic plume, which can be partially ionized, and a slower-moving NP plume of hot NPs. The images of the atomic and NP plumes shown in Fig. 4.2 indicate that both components expand in a fairly symmetrical way about the normal to the target surface.

Typically, NPs constitute a consistent part of the ablated material, and the atomic plume can account for only 10%–20% of the total ablated mass [29, 35–37]. Some of the NPs generated in fs-LA can be deposited back onto the sample surface and form nanostructures [38, 39]. The effectiveness of NP back-deposition can depend on ambient gas pressure [36, 40, 41]. Moreover, after fs-LA the sample surface cools down at a very high rate (10^{13}–10^{15} K/s [42]) and rapidly resolidifies with the possibility to quench nonequilibrium states and give rise to the formation of diverse surface structures whose morphologies can be strongly influenced by the fs laser pulse parameters [38, 43–45]. Moreover, the ablated material produced during fs-LA can be collected on a suitable substrate and intriguing NP-assembled films can be elaborated [46–48].

Studies on the properties of the ablated material and of its expansion dynamics can improve the physical understanding of the overall fs-LA process, which is of paramount interest in the numerous research fields in which it is exploited. The diverse aspects of fs-LA mentioned earlier and evidenced by investigations of the fs-LA plume characteristics will be illustrated in the following sections.

4.2 Experimental and Theoretical Analyses of the Early Stage of Femtosecond Laser Ablation

Precise thermodynamic evolution of the decomposing target in the early stage (<1 ns) of the fs-LA process is not easily accessible to experiments, due to the very short timescale involved and the high density of the excited matter. Hence, theoretical investigations predominate [12, 26–29, 35, 49–51] and in some cases their predictions are compared with experimental findings (like ablation depth and particles velocities) obtained by studying the late stage of the process [29, 33, 50]. Experimental investigations of the modification induced by fs laser irradiation soon after the irradiation can be carried out by means of pump-probe approaches, in which an intense pump fs pulse excites the target and a subsequent fs probe pulse, at appropriate delay, probes the induced variation of the sample.

In a series of studies, the probe pulse was used to measure the variation of the optical surface reflectivity or to illuminate the sample surface for ultrafast time-resolved imaging and microscopy [52–55]. The results of these investigations, at fluences close to the ablation threshold, suggested that the early state of the excited matter relaxation leads to a nascent ablation material in the form of an inhomogeneous, two-phase gas/fluid mixture. A characteristic signature of such effect is the observation of Newton rings in the images of the irradiated surface registered at some delay after the ablation pulse (1–4 ns, depending on the target), which was explained as due to the interference between two sharp interfaces (the front and rear edges of the ablating layer) separated by an ablating material with a low average optical density [54, 56]. This effect was observed on different materials (Si, Al, Ti, Au, etc.), addressing it as a general feature of fs-LA of metals and semiconductors. However, the use of laser sources at visible and infrared wavelengths limits the spatial resolution and hinders sounding the properties of the decomposing material at atomic scale, while optical reflectivity measurements cannot provide insights into the relative proportions of gas and fluid components.

There have been some attempts to directly probe the state of aggregation of materials ejected shortly after fs laser heating of the target surface by means of X-ray probe beams. Glover et al. used a pump-probe approach in which the probing photons were time delayed X-rays (~80 ps and 400 eV) provided by a synchrotron radiation source [57, 58]. In this laser pump (800 nm, 200 fs, and 1 kHz) and X-ray-probe photoemission experiment, the early stage of laser ablation of a silicon target was analyzed by measuring Si 2p X-ray photoelectron energy. The photoelectrons allow probing only the ejecta since their escape depth (~1 nm) is shorter than the nominal ablation depth of silicon at the used fluence (>10 nm). Since the Si 2p photoemission peak position changes with the material phase (solid, liquid, and vapor), these measurements can allow following the transient evolution of the phases of the decomposing fluid material. The experimental results confirmed that, also in the high-fluence region (4–12 J/cm^2), the decomposing fluid is microscopically inhomogeneous, with a low average density, due to the vacuum expansion, accompanied by regions of high local density, thus indicating a predominance of a condensed (fluid) phase. The condensate phase was associated with NPs whose formation was explained as due to a high cooling rate associated with the initial material expansion that allows quenching the particular inhomogeneous fluid state of the matter [59].

Oguri et al. exploited ultrafast time-resolved X-ray absorption fine structure (XAFS) spectroscopy, with a time resolution of ~30 ps, to investigate the early stage of fs-LA of an Al target, in vacuum [60, 61]. XAFS spectroscopy is sensitive to the local electronic state and can provide information about the chemical bonding or the local structure of the material. This allowed differentiating between Al ions and Al neutrals of the vapor phase and liquid condensed phase in the form of NPs by means of the specific spectral signatures in the X-ray absorption spectrum. Figure 4.3 reports the spatial profiles of the three components registered at two different delays, τ_D = 500 ps (Fig. 4.3a) and τ_D = 10 ns (Fig. 4.3b), after the ablating pulse. These experimental results demonstrate the generation of a spatially structured nascent plume already in the early stage of the fs-LA process. The ablated material is composed of a faster atomic component moving in front of a hot liquid phase of slower NPs, already for τ_D = 500 ps. The atomic component presents a fraction of Al ions flying ahead of the Al neutrals, indicating that singly charged ions are slightly faster than neutral atoms. Fitting the spatial density profiles acquired at different delays τ_D with an exponential decaying function, Oguri et al. obtained the transient expansion lengths ℓ of the gaseous atomic component and liquid NPs up to a delay of 3 ns and derived their average velocities from the time-position plots of ℓ versus τ_D. The velocity of the atomic component was estimated to be $\sim 1.1 \times 10^4$ m/s, while the slower NPs moved at $\sim 6.9 \times 10^2$ m/s. The good consistency of these values with the expansion velocity of atoms ($\sim 1.2 \times 10^4$ m/s) and NPs (1.0×10^3 m/s) measured in a previous study of a fs-LA aluminum plume at a late stage of its expansion by time-gated optical emission spectroscopy [33] strongly supports the view that the plume structure forms in the first moments of the process and that the liquid component observed by XAFS spectroscopy successively develops into the NP plume observed at a later time (see Fig. 4.2, e.g.). XAFS spectroscopy also reveals that material ejection from the Al target surface starts within the first 30 ps and lasts for a few tens of ns and that the condensed liquid phase constitutes the major fraction of the ablated material, consistent with both the results of Glover et al. [57, 58] and other theoretical [29, 35, 62] and experimental [36] evidence.

The early stages of material decomposition have been extensively investigated theoretically, and two main approaches are generally used to describe the thermodynamic evolution of the hot target material, leading to ablation: hydrodynamic modeling (HM) [12, 50, 51, 58, 63] and molecular dynamics (MD) simulations [12, 26–29, 33, 35, 49, 49–51, 62–68]. In all cases, the TTM provides a good description of the thermal history of the electron and lattice subsystems [22–25]. HM couples the TTM with a set of hydrodynamic equations and an equation of the state of the material, solving for the evolution of the system in a self-consistent way. The thermodynamic history of the material is followed by analyzing the temporal variation of the fluid state of the matter in a continuum approach. In the MD approach, the MD part is responsible for handling the forces among the atoms at all times during the simulation, while the TTM part describes the coupling between electrons and phonons and is used to calculate the transfer and diffusion of the energy to the lattice, during and after irradiation. The interactions between the atoms are described in terms of an appropriate potential for the material under study. In some cases, a Lennard-Jones potential has been also used to obtain general characteristics of the process, which does not depend on a specific target material [26, 28, 69].

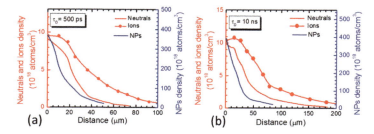

Figure 4.3 Density profiles of neutrals, singly charged ions, and nanoparticle populations measured at the early stage of fs-LA of an Al target by means of XAFS spectroscopy: (a) τ_D = 500 ps; (b) τ_D = 10 ns. The experimental data are from Ref. [60]. The ablation pulse has a duration of ~100 fs and a central wavelength of 790 nm, and the laser intensity is 8.5 × 10^{14} W/cm².

Typically, the comparison of model predictions with an experimental quantity can allow fixing unknown parameters or testing the reliability of a model in any specific case. For example, Colombier et al. studied fs-LA of copper and aluminum targets irradiated with 170 fs pulses at 800 nm, analyzing the variation of the ablation rate versus fluence, in the fluence range 1–35 J/cm², both by HM and experiments [50]. The HM predictions evidenced sharp fluence thresholds for ablation and saturation of the ablation depth at a high fluence, in fairly good agreement with the experimental findings [50]. Examples for the MD simulations can be found in Amoruso et al. for fs-LA of nickel [29] and in Nedyalkov et al. [70] for fs-LA of iron. In both cases, the comparison of the measured ablation depth [29, 70] and optical emission yield [29] dependence on fluence showed fairly good agreement with predictions of MD simulations.

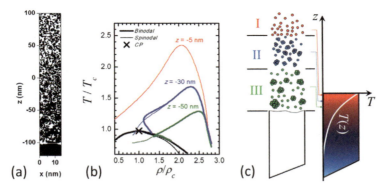

Figure 4.4 (a) Typical snapshot of a MD simulation showing the spatial structure of the nascent ablation plume. The position z = 0 marks the location of the pristine sample surface, and z < 0 refers to the inner part of the original target. The MD snapshot evidences that the target decomposition leads to a spatially structured plume with atoms flying ahead of larger chunks embedded into a vapor, forming a gas/fluid inhomogeneous mixture. (b) Example of thermodynamic paths in a temperature–density T–ρ phase diagram. The thick and thin solid black curves represent the bimodal and spinodal lines of the material, respectively. The cross marks the critical point (CP). The density and temperature are shown in reduced form, being normalized to the values of the critical point (ρ_c and T_c). The colored lines show the thermodynamic paths followed by different layers of material originally located at different depths z under the surface of the pristine target. The thick and thin lines for z = −30 nm and z = −50 nm represent the different paths followed by the fraction of material relaxing into the liquid and gas phase, respectively, at the end of the MD simulation. (c) Schematic picture of the fs-LA process: the left part sketches the original target state at the equilibrium between electrons and lattice, with a decreasing profile of the temperature $T(z)$ going inside the target. The different layers decompose according to their initial states, giving rise to a structured nascent plume characterized by the material in an atomic form (I) followed by a vapor/liquid mixed material phase resulting from fragmentation (II) and phase explosion (III) processes. The snapshot and the thermodynamic paths of panels (a) and (b) refer to nickel irradiated by a ~300 fs laser pulse at 527 nm at a fluence of ~1 J/cm².

Both for HM and MD simulations, analysis of the material state evolution through thermodynamic paths in the phase diagram allows one to follow the relaxation dynamics of the excited material and clarify mechanisms of target decomposition. One advantage of the MD simulations is the possibility to also get a direct visual inspection of the evolution of the simulated cell, which is not possible with the continuum description of HM. This allows obtaining a more direct view on the formation of clusters and aggregates in the nascent ablation plume and on their spatiotemporal evolution. Moreover, it permits one to get a straight clue on some striking aspects, like the variation of NP size dependence on experimental parameters or the spatial segregation of the ablating material according to their original position in the pristine sample [67, 71]. Important limitations of the MD simulations are the requirement of powerful calculation resources and the fact that only a very small portion of the sample can be generally modeled. Therefore, a realistic effect of the spatial profile of the laser beam intensity has been achieved in a limited number of cases by combining a number of simulations carried out at different values of the pulse fluence [68, 71].

Figure 4.4a reports an example of a typical MD snapshot showing the spatial structure of the nascent ablation plume at some delay after the laser pulse. The position $z = 0$ marks the location of the pristine sample surface, and $z < 0$ refers to the inner part of the original target. Figure 4.4b shows the temperature–density, T–ρ, phase diagram of nickel in reduced coordinates, that is, with variables normalized to the critical point (CP) (ρ_c = 2500 kg/m^3 and T_c = 9470 K) [63]. The colored lines show the evolution of layers of material initially located at three different depths underneath the original sample surface (i.e., z = −5, −30, and −50 nm), during fs-LA induced by 527 nm, ~300 fs laser pulses at a fluence of 1 J/cm^2. All the three layers experience a quasi-isochoric heating reaching a high-density fluid state at temperatures above the critical temperature T_c. Then, they relax following different paths, according to the initial state to which they were driven in the heating stage. In particular, the path corresponding to the material initially located closer to the surface (e.g., z = −5 nm) relaxes, passing above the CP and directly decomposing into an atomized gas. Instead, layers initially located at a larger depth below the initial material surface (e.g., z = −30 nm and z = −50 nm) undergo a different evolution characterized by a splitting of the thermodynamic path into two branches corresponding to a gaseous and a liquid phase of the material at the end of the simulation. The two branches are represented as thin and thick lines for gas and liquid phases, respectively. As for the uppermost layer at z = −30 nm, the gas branch proceeds into the unstable zone in the direction of decreasing density (on the left of the CP), while the liquid branch proceeds through the increase of the density and out of the unstable zone. Hence, separation of the two different phases occurs before the material enters the unstable region, and the decomposition takes place through a mechanical fragmentation process [26, 28, 62, 64]. In such a process the decomposition results into a homogeneous, clustered phase formed as a consequence of the rapid expansion of the supercritical fluid. Instead, the material of the layer at $z = -50$ nm, after a fast expansion, enters the metastable region located between the binodal and spinodal lines and then decomposes through a phase explosion mechanism [26, 28]. These complex decomposition dynamics lead to a spatially structured plume with atoms flying ahead of larger chunks embedded into a vapor forming a gas/fluid inhomogeneous mixture, as visually shown by the MD snapshot of Fig. 4.4a. The various mechanisms described above are schematically represented in Fig. 4.4c, which summarizes the current understanding of material decomposition in the fs-LA process. The left part of Fig. 4.4c shows the state of the target when an electron–lattice equilibrium is achieved, which is characterized by a profile of the temperature $T(z)$ decreasing toward the target inside. The different layers decompose according to their initial state, giving rise to a structured nascent plume characterized by material in an atomic form (I) followed by a vapor/liquid mixed material phase resulting from fragmentation (II) and phase explosion (III) processes. The faster atoms flying ahead of the ablated material, formed through mechanism I, and the larger clusters, produced by mechanisms II and III, eventually result in the atomic and NP plumes observed at larger distances and longer times after the laser pulse (see Figs. 4.2 and 4.3).

4.3 Experimental Analysis of Late Stages of Femtosecond Laser Ablation and Plume Propagation

The properties of the ablated material and of its expansion dynamics are of paramount interest in numerous research fields, such as pulsed laser deposition (PLD) of thin films, LIBS, ion beam generation, and NP synthesis and deposition. Direct analysis of the fs-LA plume expansion dynamics can also foster the physical understanding of the overall ablation process and allow relating the properties of the ejected plumes in the far field, for times of a few microseconds or more and distances from millimeters to centimeters from the target surface, to those near the target in the early stages of fs-LA. This section provides an overview of the features of fs-LA plume expansion in a high-vacuum environment and then illustrates the effects induced by a low-pressure background gas.

4.3.1 High-Vacuum Expansion

Different techniques can be exploited to investigate the propagation of the fs-LA plume in high-vacuum conditions [1, 7]. As already illustrated above (see Section 4.1 and Fig. 4.2), one of the most used approaches is the ICCD-based time-gated imaging technique [18, 29–34, 37]. Depending on the experimental configuration, it can allow either studying the spatial and temporal evolution of the various fs-LA plume components or obtaining information on the different plume species through spectral dispersion of the collected plume emission [72–75].

Time-gated imaging provides 2D snapshots of the 3D plume propagation and allows assessing several features of the plume expansion dynamics and spatiotemporal structure. Figure 4.5 summarizes typical behavior observed during the fs-LA plume in a vacuum through a series of sequences of 2D plume images registered at different values of delay τ_D and pulse fluence F. Each image is normalized to its maximum intensity, which is shown in the logarithmic scale. The data refer to fs-LA of a nickel target induced by ~300 fs laser pulses at 527 nm [30] but describe the general characteristics of the fs-LA plume structure.

Considering first the atomic plume, the left panels of Fig. 4.5 report two sequences of the atomic plume images acquired at short delays ($\tau_D < 1$ µs). They illustrate its temporal evolution at a fixed fluence F (upper panel) and the dependence on F at a fixed delay (lower panel). The atomic plume is characterized by an intense emission core and a less intense shoulder extending at a larger distance from the target surface, which is made visible by using a logarithmic intensity scale. The lower-intensity plume front can be easily overlooked on a linear-intensity scale, and its peculiar features will be described later. As shown by the temporal sequence of Fig. 4.5 (upper-left panel), the atomic plume gradually moves away from the target surface, at a typical average velocity of the order of 10^4 m/s, and completely detaches from the target surface at later times. As time goes by, its emission drops down rapidly due to density and temperature decrease consequent to expansion [31, 75]. The sequence of images in the lower-left panel illustrates the influence of the laser pulse fluence on the plume expansion velocity and shape. In particular, the atomic plume average velocity passes from $\sim 0.8 \times 10^4$ m/s at $F = 0.1$ J/cm^2 (image not shown) to $\sim 1.6 \times 10^4$ at $F = 0.6$ J/cm^2 and then reaches $\sim 2 \times 10^4$ m/s at $F = 0.9$ J/cm^2 [74]. Moreover, the lower-intensity component at the plume front tends to be more evident as the fluence increases.

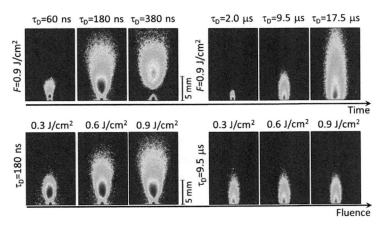

Figure 4.5 (Upper-left panel) 2D images of the early time expansion ($\tau_D < 1$ μs) of a fs-LA atomic plume at a laser fluence $F = 0.9$ J/cm^2 for three different time delays τ_D. (Upper-right panel) 2D images of the late time expansion ($\tau_D > 1$ μs) of a fs-LA NP plume at a laser fluence $F = 0.9$ J/cm^2 for three different time delays τ_D. (Lower-left panel) 2D images of a fs-LA atomic plume at $\tau_D = 180$ ns at three different values of the laser fluence F. (Lower-right panel) 2D images of a fs-LA NP plume at $\tau_D = 9.5$ μs at three different values of the laser fluence F. Each image is normalized to its own maximum intensity. The intensity is plotted in the logarithmic scale.

As for the NP plume, the images of Fig. 4.5 show that it expands at a much slower velocity and is still confined very close to the target surface at short delays. As already addressed above in Fig. 4.2, the full dynamic expansion of the NP plume occurs on a timescale of several μs, as illustrated in the upper-right panel of Fig. 4.5. The NP plume expansion is characterized by quite a different shape with respect to the atomic component and its emission always peaks at the target surface. The lower-right panel reports a sequence of NP plume images versus laser fluence F. It addresses a rather different dependence on F with respect to the atomic component: (i) the limited variation of the NP plume shape and extension at the three different values of the fluence suggests that the NP expansion velocity is almost independent of F; (ii) the average NP velocity is ~2×10^2 m/s. Atomic force microscopy (AFM) analysis of deposits shows that the NPs constituting the nickel NP plume of Fig. 4.5 have characteristic sizes of 20–50 nm [31, 76].

We now turn to the atomic plume internal spatial structure. Figure 4.6a reports two spectrally resolved images of the atomic plume emission in a spectral interval of ~60 nm centered at ~490 nm and registered at a laser fluence $F = 0.9$ J/cm^2, in the same condition as that in Fig. 4.5. The wavelength range was selected to record simultaneously typical optical emission lines of copper neutral and ionized atoms. The images were obtained by spectrally dispersing the atomic plume emission along the horizontal axis after imaging its light on the entrance slit of a spectrograph (see Fig. 4.2f). The spectrally resolved images of Fig. 4.6a show different emission lines of Cu excited neutrals (Cu*) and ions (Cu$^+$). The ions are located at the front of the plume, and their emission intensity is much less than that of neutrals. This suggests that ions constitute the main population of the atomic plume front but their fraction is minor with respect to the main neutral plume. Moreover, close to the target surface a structureless, broadband emission is present. At a longer delay, this emission is due to broadband radiation from the NP plume still located close to the target surface at short delays. Nevertheless, for a very short delay also another contribution of continuum light is likely, due to *bremsstrahlung* emission from the hot atomic plasma plume formed on the target at the very early moment after the laser pulse. Such an aspect of the atomic plume spatial structure was further investigated, at moderate (1–10 J/cm^2) [77] and high fluence ($F \leq 70$ J/cm^2) [78], in the case of a copper fs-LA plume produced by intense pulses provided by Ti:Sa laser sources (~50 fs and 800 nm). Figure 4.6b shows 2D spectrally resolved images of the Cu atomic plume registered at a laser fluence $F = 10$ J/cm (average intensity $I = 2 \times 10^{14}$ W/cm^2) that corresponds to about 20 times the ablation threshold at the experimental conditions of the experiment

[79]. The specific emission from neutrals and ions was obtained by using interferential optical filters separating their characteristic emissions. The left column of Fig. 4.6b refers to the corresponding overall plume emission, without spectral filtering. The intensity is plotted in a linear scale. Such studies confirmed that the most intense part of the plume emission is due to neutrals, while ions are characterized by a very faint emission signal, which is located at the very front of the plume. In Fig. 4.6b, it is also possible to observe some differences between expansion dynamics of ions and neutrals: the ionic plume expands with a higher velocity in the direction normal to the target, and its transverse dimension is larger than that of the neutrals. This, in turn, suggests that the generation of a fast ion population at the plume front followed by the neutral component of the atomic plasma plume is a rather general feature of fs-LA, which is more easily evidenced at a higher fluence. This aspect is not always observed because it can be easily overlooked in imaging analysis carried out without spectral filtering and can depend on the specific experimental conditions [77]. Anyway, all the imaging analyses of the fs-LA material blow off in the far field evidence a layered structure with a plume formed by two main components (atoms and NPs). This correlates well with the properties of a nascent ablation plume observed in the early stages of fs-LA and depicted by HM and MD simulations, as illustrated in the previous section.

Experimental analyses of the late stage of plume propagation in the far field indicate that clouds of ablated particles eventually expand in a fully 3D manner. At the late stage of the plume expansion, more conventional hydrodynamic descriptions are typically exploited to describe the dynamics of laser ablation plumes [80–82]. One of these is the adiabatic isentropic expansion model of Anisimov et al. [81, 82], which has been extensively applied to describe both the propagation and the angular profile of the deposition rate for ns-LA plumes [83–85]. This aspect can be particularly relevant for material analysis, production of particles beams, and thin-film deposition. The Anisimov model is based on a 3D, adiabatic, and isentropic self-similar solution of the gas dynamics equations, and it provides a gas dynamical description of the adiabatic phase of expansion of a laser ablation plume in a vacuum. For particles collected on a hemispherical surface, the Anisimov model predicts an angular distribution of the deposited/collected particles given by $Y(\theta) = Y_0 (1$

+ $\tan^2\theta)^{3/2}(1 + k^2\tan^2\theta)^{-3/2}$ [83], where θ is the angle measured with respect to the normal to the target surface, Y_0 is the peak intensity value at $\theta = 0$, and k is the asymptotic value of the longitudinal-to-transverse ratio of the plume in the plane of observation. The value of k is related to the angular width of the particle flux. Larger values of k correspond to a more forward-peaked expansion of the plume. The model was originally developed for the expansion of a neutral gas cloud, but successively it was successfully applied to describe the angular distribution of the overall plume and its ionized fraction for ns-LA in vacuum [83–85] and in a low-pressure background gas [86].

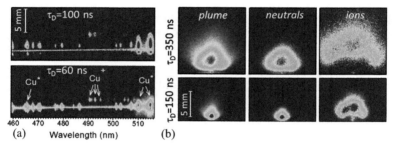

Figure 4.6 (a) 1D spectrally resolved images of a fs-LA copper plume produced by a ~300 fs laser pulse at 527 nm, at two different delays τ_D. A logarithmic intensity scale is used to compensate for the differences in the emission intensities of the various species. Each image is normalized to its own maximum intensity. In the images, the vertical position corresponds to the distance from the target surface along the atomic plume axis; thus the brightness of each pixel shows the emission intensity registered at that (wavelength and distance) coordinate. The slight tilt of the wavelength-dispersed emission with respect to the horizontal is caused by chromatic aberration. The signal detected below the position of the target surface is due to the reflection of the radiation by the target surface. (b) 2D images of a fs-LA copper plume emission at two different delays τ_D registered by spectrally resolved time-gated imaging of the atomic plume produced during fs-LA of a copper target with a ~50 fs laser pulse at 800 nm, for a fluence $F = 10$ J/cm^2. The intensity is plotted in a linear scale. In each image, the intensity is normalized to its own maximum value.

The Anisimov model was also shown to describe fairly well the angular distribution of the film thickness [87, 88] and of the ionized plume fraction produced by fs-LA of metallic targets in both bulk [77, 87–90] and thin film [91] form. Figure 4.7 reports an example of the angular variation of the film thickness (panel [a]) and ion flux (panel [b]) for different values of the laser fluence. One can observe that the ion profile is well described by the Anisimov model in both cases and k increases as the fluence value raises. This suggests that the atomic component of the fs-LA plume becomes more forward peaked for larger values of the laser fluence. The progressive narrowing of the ionic plume and the good agreement with the Anisimov formula have also been confirmed at much larger values of the laser pulse fluence (up to about 50 J/cm^2) by analyzing fs-LA of copper, with a ~50 fs pulse at 800 nm, over a rather large fluence range [77–79]. Instead, the angular distribution of the film thickness shows an opposite trend, becoming broader for a larger value of the laser fluence. As the NPs constitute the most part of the ablated material, the film thickness reflects the angular distribution of the collected NPs. Thus, the film thickness profile mainly illustrates the angular variation of the NP plume. Moreover, at the larger fluence value, the Anisimov model does not describe very well the film thickness profile. The difference with the atomic plume suggests that the observed changes in the NP plume angular distribution at larger fluences is mostly related to a mechanism that seems to influence only the NP dynamics. In this regard, MD simulations suggest that low-velocity NPs emitted during the fs-LA process can be pushed back and redeposited on the target due to the plasma pressure [66], supporting an interpretation of the change in the NP plume expansion with an increasing fluence as due to the effect of the atomic plasma pressure on the hydrodynamic evolution of the NP plume [87]. As a further remark, we note that at the low fluence values of Fig. 4.7, the ion plume angular width is broader (k for ions is less than k for NPs) than that of the NP plume. The opposite behavior has been reported for fs-LA of a silver target induced by ~500 fs pulses at 248 nm for $F \approx 2$ J/cm^2 [88], where $k = 4$ and $k = 2.4$ for ions and NPs, respectively, were observed. This indicates that in this last case the ionized part of the atomic plume is significantly narrower in angle than the overall plume, giving rise to deposit. The reason for this difference in behavior in the two studies discussed above is not yet completely clear at this stage. Nevertheless, the following comments are in order. One reason of discrepancy may be related to the higher fluence used in the case of Ag, which combined with its low reflectivity at 248 nm, could influence the original size of the ablated plume when the 3D expansion starts, which according to the Anisimov model affects the final angular width of the plume, as suggested in Ref. [88]. In addition, it is worth noting that the angular widths of ions and NPs for fs-LA of nickel show an opposite dependence on fluence. In particular, Donnelly et al. [87] observed a tendency to plateau for the NP plume angular width, while the width of the ion profile seems to continue to decrease as fluence raises, as also indicated by other investigations [78, 79]. Therefore, it may be possible that above a certain fluence value, depending on the material properties and laser pulse characteristics, the angular width of the ions reduces below that of the NP plume, as observed in the case of fs-LA of Ag with UV pulses [88].

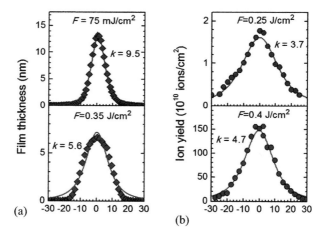

Figure 4.7 Examples of the angular variation of (a) film thickness and (b) ion flux. The data refer to fs-LA of a nickel target with ~300 fs, 527 nm laser pulses. F and k indicate the values of an average pulse fluence and the asymptotic value of the longitudinal-to-transverse ratio of the plume in the plane of observation, respectively. The curves are fits according to the Anisimov model.

4.3.2 Propagation in a Low-Pressure Background Gas

The use of a low-pressure ambient gas is well established in PLD with ns pulses, and it allows gaining control on energetic species and varying plume compositions, for example, in the fabrication of oxides [1–4]. The background gas acts as a moderator for the plume expansion during its propagation from the target surface to the collecting substrate [1–4, 92, 93]. Several investigations have addressed the influence of a background gas on the expansion dynamics of fs-LA plumes [36, 94–97]. Figure 4.8 reports a temporal sequence of 2D images of the plumes produced during fs-LA of an iron target in different ambient air pressures, from a high vacuum to 5×10^3 Pa [36]. The images for vacuum (first row) are representative of the plume expansion up to ~10 Pa and resemble what was discussed in the previous section for the typical plume produced during fs-LA in a vacuum. An atomic plume develops at earlier delays and is then followed by a NP plume at a later time. The 2D images of Fig. 4.8 clearly illustrate the different effects of the background gas pressure on the expansion dynamics of the atomic and NP plumes. At 3×10^2 Pa (second row), the atomic plume is progressively more confined: in the first stage of interaction ($\tau_D \approx 0.3$–0.9 μs) it shows a hemispherical shape, while at a longer delay ($\tau_D \approx 10$–20 μs) it is completely halted and assumes a nearly spherical form. Then, at a still longer delay the NP plume starts extending at distances larger than the stationary atomic plume, showing a rather different dynamical behavior. The NP plume is very much elongated in the direction normal to the target surface. Such features are also retained at higher pressure (third and fourth rows), with the atomic plume confinement and stopping gradually occurring earlier as pressure increases. The 2D images at $\tau_D = 22$ μs at $p = 10^3$ Pa and $\tau_D = 11$ μs at $p = 5 \times 10^3$ Pa clearly evidence the NP plume overpassing the confined atomic plume and address their different propagation dynamics. Then, at $p = 5 \times 10^3$ Pa also the NP plume front propagation is halted over distances of the order of about 5 mm for a delay of about 50 μs.

The different propagation dynamics of the two plumes were rationalized in Ref. [36] by considering the temporal variations of the plume front position: it passes progressively from a free expansion, at an early delay, to a braked propagation, eventually reaching a stopping regime later. These dynamics can be interpreted in the frame of an interaction model, taking into account the expansion of the plume and adjoint background gas experiencing the force due to the background gas pressure [93, 98, 99], and the different dimensionality of the propagation of the atomic and NP plume [36].

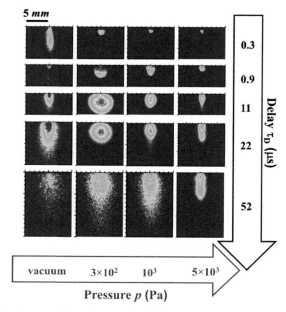

Figure 4.8 Temporal evolution of a fs-LA plume registered at different background gas pressure. The arrow on the right side of the 2D images reports the various delays τ_D after the laser pulse, while the horizontal arrow indicates the different pressures p of the background gas. Each 2D image is normalized to its own maximum intensity.

The stopping of the plume front propagation in an ambient gas can be associated with the physical condition where almost all the energy initially stored in each of the two plumes is converted in a soundwave propagating into the background gas at velocity c_g [100, 101]. The motion of the plume constituents later, after stopping, will move into the gas through diffusion [86]. The atomic plume approximately follows a hemispherical expansion; thus plume stopping takes place for a plume radius R_{st} such that $2/3 \pi \rho_g R_{st}^3 c_g^2 \approx E_A$, where E_A is the atomic plume energy. Instead, the NP plume propagation is almost 1D; hence the plume front halts at a distance z_{st} from the target surface such that $S z_{st} \rho_g c_g^2 \approx E_{NP}$, E_{NP} and S being the NP plume energy and the transverse section, respectively. From the previous relations, one can draw out as a dependence of the stopping distance versus pressure $R_{st} \propto p^{-1/3}$ for the atomic plume and $z_{st} \propto p^{-1}$ for the NP plume, in agreement with the experimental findings reported in Fig. 4.9.

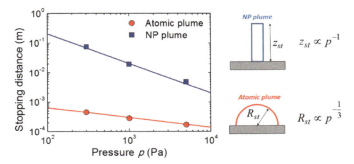

Figure 4.9 Variation of the plume stopping distance for the atomic and NP plumes. The symbols are experimental data from Ref. [36]. The solid lines are fits showing the two dependencies $R_{st} \propto p^{-1/3}$ (red line) and $z_{st} \propto p^{-1}$ (blue line), respectively. The right panels report two schematic diagrams of the atomic and NP plume shapes at stopping and the expected dependencies of the stopping distance on pressure, evidencing the different dimensionality of their expansion.

Dikawar et al. reported an extensive study on the atomic plume produced during fs-LA of brass in an Ar background, in a pressure range going from high-vacuum to atmospheric conditions [95, 97]. They observed that the atomic plume tends to be more spherically shaped in the pressure range 10–100 Pa. At higher pressure (>10^3 Pa), the stronger confinement of the background gas results in a torpedo-like form of the plume. Then, at still larger pressures (>10^4 Pa), the formation of a candle-like plume ensues as a consequence of the penetration of the background gas within the plume. Moreover, a dependence of the stopping distance for the atomic plume similar to that discussed above is reported up to atmospheric pressure [95]. Finally, Straw and Randholf investigated plasma-mediated chemical reactions occurring during fs-LA of a SiO_2 target in various reactive gases [96], addressing interesting aspects that are still scarcely investigated for fs-LA plasmas.

The confining effect of the background gas can also affect the deposition on a substrate located at a few centimeters in front of the target (see Fig. 4.1b). As an example, Fig. 4.10a shows the variation of the deposition rate as a function of the background gas pressure p registered during fs-LA of TiO_2 in ambient oxygen. The deposition rate is measured at a distance of 4 cm, and its value is normalized to the one registered in high-vacuum conditions. At low-pressure, the deposition rate is constant up to a pressure $p \approx 0.5$ Pa. Then, the influence of the background gas on the propagation of the ablated species leads to a reduction of the deposition rate, which sharply decreases and eventually reaches a plateau regime for p > 1 Pa. In this moderate-pressure regime, the deposition rate decreases to ~70% of that observed in the low-pressure regime. Figure 4.10b reports the atomic plume front position as a function of the delay τ_D at p = 0.5 Pa, that is, in the transition between low- and moderate-pressure regimes where a sharp reduction of the deposition rate is observed in Fig. 4.10a. It clearly shows that the atomic plume is gradually braked and then halted at a distance of ~2.8 cm from the target surface, thus suggesting that the observed ~30% reduction corresponds to the contribution of the atomic component. Then, at still larger pressure (p > 100 Pa) a high-pressure regime is reached where the deposition rate drastically reduces with p. In particular, at p ≈ 500 Pa the deposition rate reduces to ~10% of the low-pressure value. As Fig. 4.10b shows, at high pressure this occurs as a consequence of a strong confinement by the background gas on the NP plume, which is confined at a distance of ~1.8 cm from the target surface for p = 10^3 Pa.

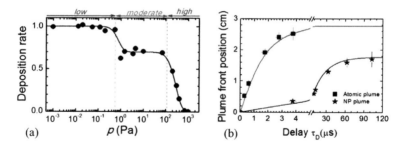

Figure 4.10 (a) Variation of the deposition rate with pressure p registered during fs-LA of TiO_2 in ambient oxygen: three different regimes—low, moderate, and high pressure—are identified. (b) Temporal variation of the plume front position for the atomic plume at p = 0.5 Pa (transition from the low to the moderate regime of panel [a]) and for the NP plume at p = 10^3 Pa (high-pressure regime) obtained by exploiting ICCD imaging. The vertical error bars account for uncertainties in the visual estimate of the luminous plume front position, while the horizontal error bars correspond to the acquisition gate width w. The horizontal axis is braked at ~5 μs to allow showing the two different temporal dynamics of the atomic and NP plumes in one plot. The curves are a guide to the eyes.

Finally, the presence of a background gas at relatively high pressure (e.g., kPa) can also lead to the suspension of ablated particles in the chamber for extended periods of time, which in some cases results in the formation of long, straight lines and webs of NPs, as observed by Tull et al. during fs-LA of silicon in H_2 and H_2S [103]. Moreover, the background gas pressure can influence the crystallinity of the NPs. Therefore, the background gas pressure and type can provide additional control on the structure, composition, and size distribution of the NPs generated during fs-LA.

4.4 Femtosecond Laser Ablation of Thin Films

Much of the research on fs-LA plumes deals with ablation of bulk solid targets, while experiments on fs-LA of a thin film supported by a transparent substrate are still very few and recent [91, 104–110]. Williams et al. investigated the angular distribution of the ionic plume generated in fs-LA, in a vacuum, of Ni films of variable thicknesses (up to 50 nm thick) on a sapphire substrate by exploiting ion probe diagnostics [91]. The experimental data evidenced a clear correlation between the width of the ion angular distribution and the film thickness, which was discussed in the frame of the Anisimov model of plume expansion. Amoruso et al. studied fs-LA of Au thin films on a transparent substrate both theoretically and experimentally, comparing it with the case of a bulk Au target [104, 105]. These studies address interesting differences both for the plumes' dynamics and the size distribution of the generated NPs. In particular, for the thin film it is observed that the atomic plume is less abundant, while the NP plume detaches from the target surface and expands at a larger average velocity. Moreover, the NPs produced by fs-LA of the Au thin film are smaller and characterized by a narrower size distribution. These experimental findings were rationalized by considering the very different temperature profile induced in the thin metallic film with respect to a bulk sample. The effects of the thin film characteristics, for example, thickness, on the NP size distribution were also investigated by Haustrup and O'Connor [106, 107]. They evidenced a relationship between the grain structure of the film and the generated NPs as well as an effect of thermoelastic stress related to the laser pulse wavelength.

Other studies considered the backward irradiation of the thin film through the transparent substrate [108–110], for example, as that typically exploited in the laser-induced forward transfer technique. In particular, Rouleau et al. investigated the transport of Pt NPs in an Ar background gas by Rayleigh scattering. NPs of ~200 nm size were observed to decelerate during the propagation in Ar at a pressure of less than ~5 × 10^3 Pa, in agreement with a linear drag model in the Epstein regime [110]. Moreover, both MD modeling and experiment demonstrate that the reduction in film thickness leads to a reduction of the NP sizes and a narrowing of the NP size distribution [109]. All these studies suggest that fs-LA of thin films, either with front or backward irradiation, provides a useful route for the synthesis of NPs with a narrow size distribution in the range of tens of nanometers.

4.5 Nanoparticles and Nanoparticle-Assembled Films

A variety of metal and semiconductor NPs have been produced by fs-LA [18–20, 34, 75, 76, 103–105]. As illustrated by the scanning electron microscopy (SEM) images reported in the upper row of Fig. 4.11, at a low number of laser pulses the collection of NPs on a suitable substrate leads to dispersed, less-than-one-layer deposits (Fig. 4.11a). This condition is typically exploited for the characterization of the NP size and shape characteristics [18–20]. Increasing the number of laser pulses N, more and more NPs are deposited, eventually leading to the elaboration of an assembly of individual NPs, indicated as NP-assembled films, at values of N of the order of 10^4–10^5. In this respect, it is worth noticing that Mirza et al. [111] studied the characteristics of the Ag NP deposited on fused quartz by using ns- and fs-LA for films with equivalent solid-density thicknesses of 1–7 nm. They observed that for ns pulses the NPs are well separated and roughly circular up to an equivalent thickness of ~3 nm but for higher thicknesses they coalesce and percolate. Instead, for fs pulses the NP films are formed by well-separated NP over the whole thickness range explored, addressing a striking peculiarity of fs-LA for NP-assembled film deposition.

The typical morphology of NP-assembled films of elemental metallic targets (e.g., Ag, Al, Au, Cu, Ni, and Pd) deposited in high vacuum by fs-LA is illustrated in Fig. 4.11c. Rather similar morphological characteristics are observed in the case of NP-assembled films of semiconductors (e.g., TiO$_2$, CdS, and ZnS). NP-assembled films of numerous materials have been produced by fs-LA, and their morphological, structural, and physical properties (e.g., optical, magnetic) have been characterized and discussed in a variety of cases [46, 47, 106, 107, 112–115]. Since most of the ablated material is in the form of NPs [29, 35–37, 116], fs-LA constitutes an effective and versatile technique for NP generation and associated elaboration of NP-assembled films on any suitable substrate.

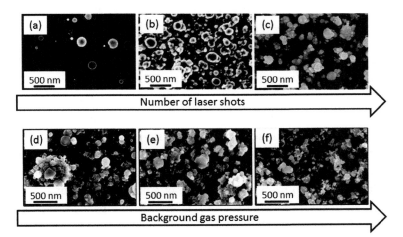

Figure 4.11 The upper panels show the variation of the morphology of the NP deposits on a substrate for a low (a), moderate (b), and large (c) number of laser pulses produced during fs-LA of a metallic target in high-vacuum conditions. The lower panels illustrate the variation of the morphology of TiO$_2$ NP-assembled films as a function of the background oxygen pressure deposited in the low- (d), moderate- (e), and high-pressure (f) regimes of Fig. 4.10a.

As illustrated above, the background gas pressure affects the deposition rate in the case of fs-LA of a TiO$_2$ target in oxygen ambient (see Fig. 4.10a). The lower row of Fig. 4.11b illustrates the corresponding changes in the morphology of NP-assembled films. In particular, panels (d–f) of Fig. 4.11 report typical SEM images of TiO$_2$ NP-assembled films deposited in the three different regimes observed in the pressure dependence of the deposition rate (see Fig. 4.10a). In the low-pressure regime (Fig. 4.11d), the film is rather porous and mainly composed of an assembly of globular structures decorated with a large number of smaller NPs whose typical size is in the range of 10–20 nm. At moderate pressure (Fig. 4.11e), the film morphology continues to resemble an agglomerated, colloidal-like nanostructure with NPs embedded in a glue-like body. In both regimes, a number of rather shapeless and compact clusters are also recognized in the deposited NP-assembled films. At high pressure (Fig. 4.11f), the morphology drastically changes. The film is a collection of spherically shaped NPs with typical sizes going from ~20 to ~200 nm. These NPs assemble into a packed-ball porous network, giving rise to a highly porous nanostructure of individual TiO_2 NPs.

4.6 Conclusions

This chapter aims at reviewing the current understanding of the process of fs-LA through the analysis of the ablated material and of its expansion dynamics. The diverse aspects of the fs-LA process have been illustrated, going from the interaction of the fs laser pulse with the target to the following relaxation dynamics of the excited target and from the formation of a nascent cloud of ablated material to the expansion (in a vacuum or into a background gas) of the ablated particles. Moreover, the generation of NPs and the deposition of NP-assembled films have been also discussed. The picture of fs-LA that emerges clearly indicates that the interaction of ultrashort laser pulses with solid metallic or semiconductor targets and the subsequent production of an ablation plume are particularly rich of fundamental aspects related to the variety of processes and mechanisms involved. All these aspects are particularly relevant in view of the number of applications (material processing, laser-produced plasma, synthesis of new material, etc.) based on fs-LA. In spite of the remarkable achievements reached so far on the comprehension of the diverse aspects of fs-LA, we expect that the research activities and related applications of fs-LA will continue to progress in the years ahead.

References

1. Chrisey, D. B., and Hubler, G. K. (1994). *Pulsed Laser Deposition of Thin Films* (Wiley-Interscience).
2. Miller, J. C., and Haglund, R. F. (1998). *Laser Ablation and Desorption* (Academic Press).
3. Phipps, C. (2007). *Laser Ablation and its Applications* (Springer).
4. Eason, R. (2006). *Pulsed Laser Deposition of Thin Films: Applications-Led Growth of Functional Materials* (Wiley-Interscience).
5. Bäuerle, D. W. (2011). *Laser Processing and Chemistry*, 4th ed. (Spinger).
6. Stafe, M., Marcu, A., and Puscas, N. (2014). *Pulsed Laser Ablation of Solids: Basics, Theory and Applications* (Springer).
7. Amoruso, S., Bruzzese, R., Spinelli, N., and Velotta, R. (1999). Characterization of laser-ablation plasmas, *J. Phys. B: At. Mol. Opt. Phys.*, **32**, pp. R131–R172.
8. Willmott, P. R., and Huber, J. R. (2000). Pulsed laser vaporization and deposition, *Rev. Mod. Phys.*, **72**, pp. 315–328.
9. Ashfold, M. N. R., Claeyssens, F., Fuge, G. M., and Henley, S. J. (2004). Pulsed laser ablation and deposition of thin films, *Chem. Soc. Rev.*, **33**, pp. 23–31.
10. De Giacomo, A., Dell'Aglio, M., Gaudiuso, R., Amoruso, S., and De Pascale, O. (2012). Effects of the background environment on formation, evolution and emission spectraof laser-induced plasmas, *Spectrochim. Acta B*, **78**, pp. 1–19.
11. Amoruso, S., Vitiello, M., and Wang, X. (2005). Femtosecond laser ablation and deposition, in *Pulsed Laser Deposition of Optoelectronic Films*, ed. Popescu, M. (INOE Publishing), pp. 41–80.
12. Gamaly, E. (2011). *Femtosecond Laser-Matter Interaction: Theory, Experiments and Applications* (Pan Stanford Publishing).
13. Sugioka, K., and Cheng, Y. (2013). *Ultrafast Laser Processing: From Micro- to Nanoscale* (Pan Stanford Publishing).
14. Harilal, S. S, Freeman, J. R., Diwakar, P. K., and Hassanein, A. (2014). Femtosecond laser ablation: fundamentals and applications, in *Laser-Induced Breakdown Spectroscopy: Theory and Applications*, eds. Musazzi, S., and Perini, U. (Springer), pp. 143–166.
15. Gerhard, C., Wieneke, S., and Viöl, W. (2015). *Laser Ablation: Fundamentals, Methods and Applications* (Nova).

16. Chichkov, B. N., Momma, C., Nolte, S., von Alvensleben, F., and Tünnermann, A. (1996). Femtosecond, picosecond and nanosecond laser ablation of solids, *Appl. Phys. A*, **63**, pp. 109–115.
17. Sugioka, K., Meunier, M., and Piqué, A. (2010). *Laser Precision Microfabrication* (Springer).
18. Amoruso, S., Bruzzese, R., Spinelli, N., Velotta, R., Vitiello, M., Wang, X., Ausanio, G., Iannotti, V., and Lanotte, L. (2004). Generation of silicon nanoparticles via femtosecondlaser ablation in vacuum, *Appl. Phys. Lett.*, **84**, pp. 4502–4504.
19. Eliezer, S., Eliaz, N., Grossman, E., Fisher, D., Gouzman, I., Henis, Z., Pecker, S., Horovitz, Y., Fraenkel, M., Maman, S., and Lereah, Y. (2004). Synthesis of nanoparticles with femtosecond laser pulses, *Phys. Rev. B*, **69**, pp. 144119-1–144119-6.
20. Amoruso, S., Ausanio, G., Bruzzese, R., Vitiello, M., and Wang, X. (2005). Femtosecond laser pulse irradiation of solid targets as a general route to nanoparticle formation in a vacuum, *Phys. Rev. B*, **71**, pp. 033406-1–033406-4.
21. Eesley, G. L. (1983). Observation of nonequilibrium electron heating in copper, *Phys. Rev. Lett.*, **51**, pp. 2140–2143.
22. Anisimov, S. I., Kapeliovich, B. L., and Perel'man, T. L. (1974). Electron emission from metal surfaces exposed to ultrashort laser pulses, *Sov. Phys. JETP*, **39**, pp. 375–377.
23. Anisimov, S. I., and Luk'yanchuk, B. S. (2002). Selected problems of laser ablation theory, *Phys. Usp.*, **45**, pp. 293–324.
24. Nolte, S., Momma, C., Jacobs, H., Tünnermann, A., Chichkov, B. N., Wellegehausen, B., and Welling, H. (1997). Ablation of metals by ultrashort laser pulses, *J. Opt. Soc. Am. B*, **14**, pp. 2716–2722.
25. Wellershoff, S.-S., Hohlfeld, J., Güdde, J., and Matthias, E. (1999). The role of electron–phonon coupling in femtosecond laser damage of metals, *Appl. Phys. A*, **69**, pp. S99–S107.
26. Perez, D., and Lewis, L. J. (2002). Ablation of solids under femtosecond laser pulses, *Phys. Rev. Lett.*, **89**, pp. 255504-1–255504-4.
27. Ivanov, D. S., and Zhigilei, L. V. (2003). Combined atomistic-continuum modeling of short-pulse laser melting and disintegration of metal films, *Phys. Rev. B*, **68**, 064114-1–064114-22.
28. Lewis, L. J., and Perez, D. (2010). Theory and simulation of laser ablation – from basic mechanisms to applications, in *Laser Precision Microfabrication*, eds. Sugioka, K., Meunier, M., and Piqué, A. (Springer), pp. 35–61.

29. Amoruso, S., Bruzzese, R., Wang, X., Nedialkov, N. N., and Atanasov, P. A. (2007). Femtosecond laser ablation of nickel in vacuum, *J. Phys. D: Appl. Phys.*, **40**, pp. 331–340.
30. Amoruso, S., Bruzzese, R., Pagano, C., and Wang, X. (2007). Features of plasma plume evolution and material removal efficiency during femtosecond laser ablation of nickel in high vacuum, *Appl. Phys. A*, **89**, pp. 1017–1024.
31. Verhoff, B., Harilal, S. S., Freeman, J. R., Diwakar, P. K., and Hassanein, A. (2012). Dynamics of femto- and nanosecond laser ablation plumes investigated using optical emission spectroscopy, *J. Appl. Phys.*, **112**, pp. 093303-1–093303-9.
32. Amoruso, S., Bruzzese, R., Spinelli, N., Velotta, R., Vitiello, M., and Wang, X. (2004). Emission of nanoparticles during ultrashort laser irradiation of silicon targets, *Europhys. Lett.*, **67**, pp. 404–410.
33. Amoruso, S., Bruzzese, R., Vitiello, M., Nedialkov, N. N., and Atanasov, P. A. (2005). Experimental and theoretical investigations of femtosecond laser ablation of aluminum in vacuum, *J. Appl. Phys.*, **98**, pp. 044907-1–044907-7.
34. Perriere, J., Boulmer-Leborgne, C., Benzerga, R., and Tricot, S. (2007). Nanoparticle formation by femtosecond laser ablation, *J. Phys. D: Appl. Phys.*, **40**, pp. 7069–7076.
35. Povarnitsyn, M. E., Itina, T. E., Sentis, M., Khishchenko, K. V., and Levashov, P. R. (2007). Material decomposition mechanisms in femtosecond laser interactions with metals, *Phys. Rev. B*, **75**, pp. 235414-1–235414-5.
36. Amoruso, S., Bruzzese, R., Wang, X., and Xia, J. (2008). Propagation of a femtosecond pulsed laser ablation plume into a background atmosphere, *Appl. Phys. Lett.*, **92**, pp. 041503-1–041503-3.
37. Noël, S., and Hermann, J. (2009). Reducing nanoparticles in metal ablation plumes produced by two delayed short laser pulses, *Appl. Phys. Lett.*, **94**, 053120-1–053120-3.
38. Pereira, A., Cros, A., Delaporte, P., Georgiou, S., Manousaki, A., Marine, W., and Sentis, M. (2004). Surface nanostructuring of metals by laser irradiation: effects of pulse duration, wavelength and gas atmosphere, *Appl. Phys. A*, **79**, pp. 1433–1437.
39. Vorobyev, A. Y., and Guo, C. (2005). Enhanced absorptance of gold following multipulse femtosecond laser ablation, *Phys. Rev. B*, **72**, pp. 195422-1–195422-5.

40. Vorobyev, A. Y., and Guo, C. (2007). Effects of nanostructure-covered femtosecond laser-induced periodic surface structures on optical absorptance of metals, *Appl. Phys. A*, **86**, pp. 235–241.

41. Pereira, A., Delaporte, P., Sentis, M., Marine, W., Thomann, A. L., and Boulmer-Leborgne, C. (2005). Optical and morphological investigation of backward-deposited layerinduced by laser ablation of steel in ambient air, *J. Appl. Phys.*, **98**, pp. 064902-1–064902-8.

42. Zavestovskaya, I. N., Kanavin, A. P., and Men'kova, N. A. (2008). Crystallization of metals under conditions of superfast cooling when materials are processed with ultrashort laser pulses, *J. Opt. Technol.*, **75**, pp. 353–358.

43. Vorobyev, A. Y., and Guo, C. (2013). Direct femtosecond laser surface nano/microstructuring and its applications, *Laser Photonics Rev.*, **7**, pp. 385–407.

44. Anoop, K. K., Rubano, A., Fittipaldi, R., Wang, X., Paparo, D., Vecchione, A., Marrucci, L., Bruzzese, R., and Amoruso, S. (2014). Femtosecond laser surface structuring of silicon using optical vortex beams generated by a q-plate, *Appl. Phys. Lett.*, **104**, pp. 241604-1–241604-4.

45. Anoop, K. K., Rubano, A., Fittipaldi, R., Wang, X., Paparo, D., Vecchione, A., Marrucci, L., Bruzzese, R., and Amoruso, S. (2014). Direct femtosecond laser ablation of copper with an optical vortex beam, *J. Appl. Phys.*, **116**, pp. 113102-1–113102-9.

46. Ausanio, G., Barone, A. C., Iannotti, V., Lanotte, L., Amoruso, S., Bruzzese, R., and Vitiello, M. (2004). Magnetic and morphological characteristics of nickel nanoparticles films produced by femtosecond laser ablation, *Appl. Phys. Lett.*, **85**, pp. 4103–4105.

47. Sanz, M., López-Arias, M., Marco, J. F., de Nalda, R., Amoruso, S., Ausanio, G., Lettieri, S., Bruzzese, R., Wang, X., and Castillejo, M. (2011). Ultrafast laser ablation and deposition of wide band gap semiconductors, *J. Phys. Chem. C*, **115**, pp. 3203–3211.

48. Amoruso, S., Tuzi, S., Pallotti, D. K., Aruta, C., Bruzzese, R., Chiarella, F., Fittipaldi, R., Lettieri, S., Maddalena, P., Sambri, A., Vecchione, A., and Wang, X. (2013). Structural characterization of nanoparticles-assembled titanium dioxide films produced by ultrafast laser ablation and deposition in background oxygen, *Appl. Surf. Sci.*, **270**, pp. 307–311.

49. Bulgakova, N. M. (2013). Fundamentals of ultrafast laser processing, in *Ultrafast Laser Processing: From Micro- to Nanoscale*, eds. Sugioka, K., and Cheng, Y. (Pan Stanford Publishing), pp. 99–182.

50. Colombier, J. P., Combis, P., Bonneau, F., Le Harzic, R., and Audouard, E. (2005). Hydrodynamic simulations of metal ablation by femtosecond laser irradiation, *Phys. Rev. B*, **71**, pp. 165406-1–165406-6.
51. Upadhyay, A. K., and Urbassek, H. M. (2006). Expansion flow and cluster distributions originating from ultrafast-laser-induced fragmentation of thin metal films: a molecular-dynamics study, *Phys. Rev. B*, **73**, pp. 035421-1–035421-7.
52. Shank, C. V., Yen, R., and Hirlimann, C. (1983). Time-resolved reflectivity measurements of femtosecond-optical-pulse-induced phase transitions in silicon, *Phys. Rev. Lett.*, **50**, pp. 454–457.
53. Downer, M. C., Fork, R. L., and Shank, C. V. (1985). Femtosecond imaging of melting and evaporation at a photoexcited silicon surface, *J. Opt. Soc. Am. B*, **2**, pp. 595–599.
54. Sokolowski-Tinten, K., Bialkowski, J., Cavalleri, A., von der Linde, D., Oparin, A., Meyer-ter-Vehn, J., and Anisimov, S. I. (1998). Transient states of matter during short pulse laser ablation, *Phys. Rev. Lett.*, **81**, pp. 224–227.
55. Kandyla, M., Shih, T., and Mazur, E. (2007). Femtosecond dynamics of the laser-induced solid-to-liquid phase transition in aluminum, *Phys. Rev. B*, **75**, pp. 214107-1–214107-7.
56. Inogamov, N. A., Petrov, Yu. V., Anisimov, S. I., Oparin, A. M., Shaposhnikov, N.V., von der Linde, D., and Meyer-ter-Vehn, J. (1999). Expansion of matter heated by an ultrashort laser pulse, *JETP Lett.*, **69**, pp. 310–316.
57. Glover, T. E., Ackerman, G. D., Lee, R. W., Padmore, H. A., and Young, D. A. (2004). Metal–insulator transitions in an expanding metallic fluid: particle formation during femtosecond laser ablation, *Chem. Phys.*, **299**, pp. 171–181.
58. Glover, T. E., Ackerman, G. D., Lee, R. W., and Young, D. A. (2004). Probing particle synthesis during femtosecond laser ablation: initial phase transition kinetics, *Appl. Phys. B*, **78**, pp. 995–1000.
59. Glover, T. E. (2003). Hydrodynamics of particle formation following femtosecond laser ablation, *J. Opt. Soc. Am. B*, **20**, pp. 125–131.
60. Oguri, K., Okano, Y., Nishikawa, T., and Nakano, H. (2007). Dynamical study of femtosecond-laser-ablated liquid-aluminum nanoparticles using spatiotemporally resolved X-ray-absorption fine-structure spectroscopy, *Phys. Rev. Lett.*, **99**, pp. 165003-1–165003-4.
61. Oguri, K., Okano, Y., Nishikawa, T., and Nakano, H. (2009). Dynamics of femtosecond laser ablation studied with time-resolved x-ray absorption fine structure imaging, *Phys. Rev. B*, **79**, pp. 144106-1–144106-10.

62. Lorazo, P., Lewis, L. J., and Meunier, M. (2003). Short-pulse laser ablation of solids: from phase explosion to fragmentation, *Phys. Rev. Lett.*, **91**, pp. 225502-1–225502-4.
63. Cheng, C., and Xu, X. (2005). Mechanisms of decomposition of metal during femtosecond laser ablation, *Phys. Rev. B*, **72**, pp. 165415-1–165415-15.
64. Lorazo, P., Lewis, L. J., and Meunier, M. (2006). Thermodynamic pathways to melting, ablation, and solidification in absorbing solids under pulsed laser irradiation, *Phys. Rev. B*, **73**, pp. 134108-1–134108-22.
65. Colombier, J. P., Combis, P., Stoian, R., and Audouard, E. (2007). High shock release in ultrafast laser irradiated metals: scenario for material ejection, *Phys. Rev. B*, **75**, pp. 104105-1–104105-11.
66. Zhigilei, L. V., Lin, Z., and Ivanov, D. S. (2009). Atomistic modeling of short pulse laser ablation of metals: connections between melting, spallation, and phase explosion, *J. Phys. Chem. C*, **113**, pp. 11892–11906.
67. Tsakiris, N., Anoop, K. K., Ausanio, G., Gill-Comeau, M., Bruzzese, R., Amoruso, S. and Lewis, L. J. (2014). Ultrashort laser ablation of bulk copper targets: dynamics and size distribution of the generated nanoparticles, *J. Appl. Phys.*, **115**, pp. 243301-1–243301-11.
68. Wu, C., and Zhigilei, L. V. (2014). Microscopic mechanisms of laser spallation and ablation of metal targets from large-scale molecular dynamics simulations, *Appl. Phys. A*, **114**, pp. 11–32.
69. Bouilly, D., Perez, D., and Lewis, L. J. (2007). Damage in materials following ablation by ultrashort laser pulses: a molecular-dynamics study, *Phys. Rev. B*, **76**, pp. 184119-1–184119-9.
70. Nedialkov, N. N., Imamova, S. E., Atanasov, P. A., Heusel, G., Breitling, D., Ruf, A., Hügelb, H., Dausinger, F., and Berger, P. (2004). Laser ablation of iron by ultrashort laser pulses, *Thin Solid Films*, **453–454**, pp. 496–500.
71. Amoruso, S., Bruzzese, R., Wang, X. Nedialkov, N. N., and Atanasov, P. A. (2007). An analysis of the dependence on photon energy of the process of nanoparticle generation by femtosecond laser ablation in a vacuum, *Nanotechnology*, **18**, pp. 145612-1–145612-6.
72. Siegel, J., Epurescu, G., Perea, A., Gordillo-Vázquez, F. J., Gonzalo, J., and Afonso, C. N. (2004). Temporally and spectrally resolved imaging of laser-induced plasmas, *Opt. Lett.*, **29**, pp. 2228–2230.

73. Noël, S., Hermann, J., and Itina, T. (2007). Investigation of nanoparticle generation during femtosecond laser ablation of metals, *Appl. Surf. Sci.*, **253**, pp. 6310–6315.
74. Wang, X., Amoruso, S., and Xia, J. (2009). Temporally and spectrally resolved analysis of a copper plasma plume produced by ultrafast laser ablation, *Appl. Surf. Sci.*, **255**, pp. 5211–5214.
75. Albert, O., Roger, S., Glinec, Y., Loulergue, J. C., Etchepare, J., Boulmer-Leborgne, C., Perrière, J., and Millon, E. (2003). Time-resolved spectroscopy measurements of a titanium plasma induced by nanosecond and femtosecond lasers, *Appl. Phys. A*, **76**, pp. 319–323.
76. Amoruso, S., Ausanio, G., Bruzzese, R., Lanotte, L., Scardi, P., Vitiello, M., and Wang, X. (2006). Synthesis of nanocrystal films via femtosecond laser ablation in vacuum, *J. Phys. Condens. Matter*, **18**, pp. L49–L53.
77. Anoop, K. K., Ni, X., Wang, X., Amoruso, S., and Bruzzese, R. (2014). Fast ion generation in femtosecond laser ablation of a metallic target at moderate laser intensity, *Laser Phys.*, **24**, pp. 105902-1–105902-1.
78. Anoop, K. K., Polek, M. P., Bruzzese, R., Amoruso, S., and Harilal, S. S. (2015). Multidiagnostic analysis of ion dynamics in ultrafast laser ablation of metals over a large fluence range, *J. Appl. Phys.*, **117**, pp. 083108-1–083108-9.
79. Ni, X., Anoop, K. K., Wang, X., Paparo, D., Amoruso, S., and Bruzzese, R. (2014). Dynamics of femtosecond laser-produced plasma ions, *Appl. Phys. A*, **117**, pp. 111–115.
80. Singh, R. K., and Narayan, J. (1990). Pulsed laser evaporation technique for deposition of thin films: physics and theoretical model, *Phys. Rev. B*, **41**, pp. 8843–8859.
81. Anisimov, S. I., Bauerle, D., and Luk'yanchuk, B. S. (1993). Gas dynamics and film profile in pulsed-laser deposition of materials, *Phys. Rev. B*, **48**, pp. 12076–12081.
82. Anisimov, S. I., Luk'yanchuk, B. S., and Luches, A. (1996). *Appl. Surf. Sci.*, **96–98**, pp. 24–32.
83. Hansen, T. N., Schou, J., and Lunney, J. G. (1999). Langmuir probe study of plasma expansion in pulsed laser ablation, *Appl. Phys. A*, **69**, pp. S601–S604.
84. Toftmann, B., Schou, J., and Lunney, J. G. (2003). Dynamics of the plume produced by nanosecond ultraviolet laser ablation of metals, *Phys. Rev. B*, **67**, pp. 104101-1–104101-5.

85. Toftmann, B., and Schou, J. (2013). Time-resolved and integrated angular distributions of plume ions from silver at low and medium laser fluence, *Appl. Phys. A*, **112**, pp. 197–202.

86. Amoruso, S., Toftmann, B., and Schou, J. (2004). Thermalization of a UV laser ablation plume in a background gas: from a directed to a diffusionlike flow, *Phys. Rev. E*, **69**, pp. 056403-1–056403-6.

87. Donnelly, T., Lunney, J. G., Amoruso, S., Bruzzese, R., Wang, X., and Ni, X. (2010). Dynamics of the plumes produced by ultrafast laser ablation of metals, *J. Appl. Phys.*, **108**, pp. 043309-1–043309-13.

88. Toftmann, B., Doggett, B., Budtz-Jørgensen, C., Schou, J., and Lunney, J. G. (2013). Femtosecond ultraviolet laser ablation of silver and comparison with nanosecond ablation, *J. Appl. Phys.*, **113**, pp. 083304-1–083304-7.

89. Donnelly, T., Lunney, J. G., Amoruso, S., Bruzzese, R., Wang, X., and Ni, X. (2009). Double pulse ultrafast laser ablation of nickel in vacuum, *J. Appl. Phys.*, **106**, pp. 013304-1–013304-5.

90. Donnelly, T., Lunney, J. G., Amoruso, S., Bruzzese, R., Wang, X., and Ni, X. (2010). Angular distributions of plume components in ultrafast laser ablation of metal targets, *Appl. Phys. A*, **100**, 569–574.

91. Williams, G. O., Favre, S., and O'Connor, G. M. (2009). Directional ion emission from thin films under femtosecond laser irradiation, *Appl. Phys. Lett.*, **94**, pp. 101503-1–101503-3.

92. Geohegan, D. B. (1992). Fast intensified CCD photography of YBa2Cu3O7-x laser ablation in vacuum and ambient oxygen, *Appl. Phys. Lett.*, **60**, pp. 2732–2734.

93. Sambri, A., Amoruso, S., Wang, X., Radovic', M., Miletto Granozio, F., and Bruzzese, R. (2007). Substrate heating influence on plume propagation during pulsed laser deposition of complex oxides, *Appl. Phys. Lett.*, **91**, pp. 151501-1–151501-3.

94. Al-Shboul, K. F., Harilal, S. S., and Hassanein, A. (2012). Spatio-temporal mapping of ablated species in ultrafast laser-produced graphite plasmas, *Appl. Phys. Lett.*, **100**, pp. 221106-1–221106-4.

95. Diwakar, P. K., Harilal, S. S., Hassanein, A., and Phillips, M. C. (2014). Expansion dynamics of ultrafast laser produced plasmas in the presence of ambient argon, *J. Appl. Phys.*, **116**, pp. 133301-1–133301-1

96. Straw, M., and Randolph, R. (2014). Direct spatiotemporal analysis of femtosecond laser-induced plasma-mediated chemical reactions, *Laser Phys. Lett.*, **11**, pp. 035601-1–035601-7.

97. Diwakar, P. K., Harilal, S. S., Phillips, M. C., and Hassanein, A. (2015). Characterization of ultrafast laser-ablation plasma plumes at various Ar ambient pressures, *J. Appl. Phys.*, **118**, pp. 043305-1–043305-8.
98. Amoruso, S., Schou, J., and Lunney, J. G. (2008). Influence of the atomic mass of the background gas on laser ablation plume propagation, *Appl. Phys. A*, **92**, pp. 907–911.
99. Amoruso, S., Schou, J., and Lunney, J. G. (2010). Energy balance of a laser ablation plume expanding in a background gas, *Appl. Phys. A*, **101**, pp. 209–214.
100. Amoruso, S., Bruzzese, R., Spinelli, N., Velotta, R., Vitiello, M., and Wang, X. (2003). Dynamics of laser-ablated MgB_2 plasma expanding in argon probed by optical emission spectroscopy, *Phys. Rev. B*, **67**, pp. 224503-1–224503-11.
101. Schou, J., Amoruso, S., and Lunney, J. G. (2007). Plume dynamics, in *Laser Ablation and its Applications*, ed. Phipps, C. (Springer), pp. 67–95.
102. Pallotti, D. K., Ni, X., Fittipaldi, R., Wang, X., Lettieri, S., Vecchione, A., and Amoruso, S. (2015). Laser ablation and deposition of titanium dioxide with ultrashort pulses at 527 nm, *Appl. Phys. B*, **119**, pp. 445–452.
103. Tull, B. R., Carey, J. E., Sheehy, M. A., Friend, C., and Mazur, E. (2006). Formation of silicon nanoparticles and web-like aggregates by femtosecond laser ablation in a background gas, *Appl. Phys. A*, **83**, pp. 341–346.
104. Amoruso, S., Nedyalkov, N. N., Wang, X., Ausanio, G., Bruzzese, R., and Atanasov, P. A. (2011). Ultrafast laser ablation of gold thin film targets, *J. Appl. Phys.*, **110**, pp. 124303-1–124303-4.
105. Amoruso, S., Nedyalkov, N. N., Wang, X., Ausanio, G., Bruzzese, R., and Atanasov, P. A. (2014). Ultrashort-pulse laser ablation of gold thin film targets: theory and experiment, *Thin Solid Films*, **550**, pp. 190–198.
106. Haustrup, N., and O'Connor, G. M. (2012). Impact of wavelength dependent thermo-elastic laser ablation mechanism on the generation of nanoparticles from thin gold films, *Appl. Phys. Lett.*, **101**, pp. 263107-1–263107-5.
107. Haustrup, N., and O'Connor, G. M. (2012). Confinement of laser-material interactions by metal film thickness for nanoparticle generation, *J. Nanosci. Nanotechnol.*, **12**, pp. 8656–8661.
108. Murphy, R. D., Abere, M. J., Schrider, K. J., Torralva, B., and Yalisove, S. M. (2013). Nanoparticle size and morphology control using ultrafast laser induced forward transfer of Ni thin films, *Appl. Phys. Lett.*, **103**, pp. 093113-1–093113-5.

109. Rouleau, C. M., Shih, C.-Y., Wu, C., Zhigilei, L. V., Puretzky, A. A., and Geohegan, D. B. (2014). Nanoparticle generation and transport resulting from femtosecond laser ablation of ultrathin metal films: time-resolved measurements and molecular dynamics simulations, *Appl. Phys. Lett.*, **104**, pp. 193106-1–193106-5.

110. Rouleau, C. M., Puretzky, A. A., and Geohegan, D. B. (2014). Slowing of femtosecond laser-generated nanoparticles in a background gas, *Appl. Phys. Lett.*, **105**, pp. 213108-1–213108-4.

111. Mirza, I., O'Connell, G., Wang, J. J., and Lunney, J. G. (2014). Comparison of nanosecond and femtosecond pulsed laser deposition of silver nanoparticle films, *Nanotechnology*, **25**, pp. 265301-1–265301-10.

112. Trelenberg, T. W., Dinh, L. N., Stuart, B. C., and Balooch, M. (2004). Femtosecond pulsed laser ablation of metal alloy and semiconductor targets, *Appl. Surf. Sci.*, **229**, 268–274.

113. Trelenberg, T. W., Dinh, L. N., Saw, C. K., Stuart, B. C., and Balooch, M. (2004). Femtosecond pulsed laser ablation of GaAs, *Appl. Surf. Sci.*, **221**, pp. 364–369.

114. Ausanio, G., Barone, A. C., Iannotti, V., Scardi, P., D'Incau, M., Amoruso, S. Vitiello, M., and Lanotte, L. (2006). Morphology, structure and magnetic properties of (Tb0.3Dy0.7Fe2)100−xFex nanogranular films produced by ultrashort pulsed laser deposition, *Nanotechnology*, **17**, pp. 536–542.

115. Ausanio, G., Campana, C., Iannotti, V., Amoruso, S., Wang, X., and Lanotte, L. (2010). Elastomagnetic and elastoresistive effects in CoFe films produced by femtosecond pulsed laser deposition, *IEEE Trans. Magn.*, **46**, pp. 479–482.

116. O'Connell, G., Donnelly, T., and Lunney, J. G. (2014). Nanoparticle plume dynamics in femtosecond laser ablation of gold, *Appl. Phys. A*, **117**, pp. 289–293.

Chapter 5

Short-Pulse Laser Near-Field Ablation of Solid Targets under Liquids

M. Ulmeanu,[a] P. Petkov,[b] F. Jipa,[c] E. Brousseau,[b] and M. N. R. Ashfold[a]

[a]*School of Chemistry, University of Bristol, UK*
[b]*Cardiff School of Engineering, Cardiff University, UK*
[c]*National Institute for Laser, Plasma and Radiation Physics, Romania*
magdalena.ulmeanu@bristol.ac.uk

5.1 Introduction

The relentless demand for ever more complex nanopatterned surfaces and large periodic arrays, fabricated by low-cost and effective lithographic methods, has stimulated a huge upsurge in interest in near-field optics, that is, in phenomena associated with nonpropagating and highly localized electromagnetic fields and their interaction with matter. The ability to localize the optical energy to length scales smaller than the diffraction limit, defined as $\lambda/2$ (λ being the wavelength of light), has made the scanning near-field optical microscopy (SNOM) technique an attractive field

---

Pulsed Laser Ablation: Advances and Applications in Nanoparticles and Nanostructuring Thin Films
Edited by Ion N. Mihailescu and Anna Paola Caricato
Copyright © 2018 Pan Stanford Publishing Pte. Ltd.
ISBN 978-981-4774-23-9 (Hardcover), 978-1-315-18523-1 (eBook)
www.panstanford.com

for near-field optics studies. Various types of probe tips have led to the development of many types of SNOM, for example, single-mode optical fibers [1], tip–sample spacing stabilized by shear-force control [2], and modulation of the scattered electric field from the end of a sharp silicon tip [3]. The SNOM technique requires scanning of the samples, which can disturb the optical near field by multiscattering or by strong interaction between the surface and the probes, which complicates interpretation of the scanned images. Nonoptically probing near-field microscopy has been introduced in an effort to overcome this problem [4].

Optical near-field effects can be generated in the vicinity of metal or dielectric nanoparticles, where the localized electromagnetic energy leads to an increase of the intensity of the incident field. Spherical particles were self-assembled on surfaces, and subwavelength structures were obtained due to particle-enhanced laser irradiation [5, 6]. The patterning mechanism was found to involve near-field optical resonance effects induced by the particles on the surface. Numerical calculations were presented by solving the electromagnetic boundary problem. Thus near-field laser ablation (NF-LA) was introduced as a lithographic method that does not require a complex system to pattern a large surface area. Developing a sufficient understanding of such near-field optical effects to allow their use in nanopatterning remains a key challenge in this research field.

One method for imaging the complex 2D near-field intensity distribution beneath a scattering colloidal particle has recently been demonstrated [7]. Compared with the NF-LA method, where the near field is visualized via local ablation of the substrate in the shape of a crater corresponding to the peak in the field distribution, the method described in Ref. [7] succeeded in imprinting the whole complex near-field distribution on a photosensitive substrate. In this work, a single microsphere was placed on a thin film of photosensitive phase-change material $Ge_2Sb_5Te_5$, exposed to a single, short laser pulse, and the spatial intensity modulation of the near field was recorded in the film as a pattern of different material phases. The spatial electric field distribution at the substrate plane was simulated, taking into account both scattering from the sphere and reflected plane waves from the substrate, thereby describing an infinite series of multiple scattering events at the sphere and the substrate. The optical contrast allows visualization of the high dynamic range of the field distributions and suggests potential applications in near-field optical lithography.

Another recent milestone study in the arena of material processing based on near-field optical lithography employed optical trapping methods to hold a colloidal particle in a liquid medium and thereby enable direct-write nanopatterning [8, 9]. In these approaches, a laser beam was focused in the near field by a dielectric microbead, positioned using a Bessel beam optical trap, or a self-propelled Janus spherical motor. The trapped particle could be translated across the surface to write the nanoscale patterns directly, and potential applications in material processing, microscopy, and biosensing at the submicrometer scale were discussed.

Laser-initiated liquid-assisted colloidal (LILAC) lithography, pioneered by the present authors, exploits many of the virtues of both of these methods [10, 11]. The optical near field in the vicinity of an array of colloidal microspheres self-assembled onto a substrate of interest (e.g., Si or GaAs in the work reported to date) and immersed in a range of different liquids exhibits a rich, user-controllable, spatial modulation that is capable of carving complex 2D and 3D patterns on the substrate surface. Herein, we present the LILAC lithography technique and illustrate its potential applicability for a range of applications in microelectronics, nanophotonics, and nanomedicine.

5.2 Working Principle of the LILAC Lithography Technique

In LILAC lithographic processing, a mask of colloidal particles is assembled in an ordered layer structure on a substrate surface. The immobilized colloidal particles are then immersed into a liquid medium. Figures 5.1a and 5.1b afford a schematic comparison of the NF-LA and LILAC techniques and the surface patterns they can yield.

In NF-LA, colloidal particles self-assembled onto a (in this case) Si substrate are irradiated with a single laser pulse in air. The near-field ablation pattern caused by a single particle takes the form of a crater in the silicon (Si) substrate, as shown in the scanning electron microscopy (SEM) image shown in Fig. 5.1a. This pattern corresponds to the peak in the field distribution, which lies directly beneath the particle. The SEM image shown in Fig. 5.1b illustrates the totally different surface pattern that can be obtained by immersing the colloidal-particle-covered substrate in a liquid. The black/darker rings in this image correspond to regions where the laser fluence F was locally enhanced (either by near-field effects or by the scattered light field of the colloidal particle), while the bright rings show regions exposed (or shadowed) to a lower F.

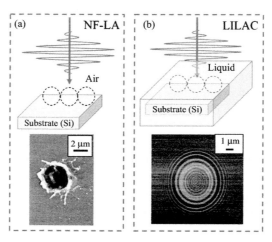

Figure 5.1 Schematic representations of (a) the near-field laser ablation (NF-LA) technique and (b) laser-initiated liquid-assisted colloidal (LILAC) lithography using a liquid with a refractive index higher than that of the particle. (Below) SEM images of the surface patterns formed by single-pulse laser irradiation of an isolated colloidal particle on a Si substrate in the near-field regime.

The following sections detail each step in the implementation of LILAC lithography and a theoretical and experimental approach to this novel surface patterning technique.

5.2.1 Preparing the Si Substrates

The Si substrates were cut from n-type Si wafers (crystal GmbH) and pretreated in an oxygen plasma (Femto Diener, 80 W power, and 0.5 mbar O_2 pressure) for 10 min to remove contaminants and render the surface hydrophilic. We assumed bulk values for the optical constants of these Si substrates (i.e., n = 5.61 and k = 3.014 at λ = 355 nm) [12]. To clean the Si wafers, they were first sonicated in a solvent like isopropanol. The Si surface was then treated by a low-pressure oxygen plasma technology that not only eliminates organic contaminations but also increases the hydrophilicity, which plays a key role for the next step—the spin coating of the colloidal particles. The Si wafers were then placed in a vacuum chamber, which was evacuated using a vacuum pump. O_2 at 0.1–0.5 mbar pressure was then fed into the chamber, and the Si substrates were plasma treated for 10–20 minutes. By the end of this process, the surface of the wafer was chemical clean, residue-free, and highly hydrophilic.

5.2.2 Preparing the Colloidal Mask

All chemical materials in this study were obtained commercially and used as supplied. Aqueous suspensions of monodisperse (coefficient of variation [CV] of 10%–15%) colloidal particles with different mean radii, R ranging from 5 µm to 175 nm (10% solid concentration, with a use-selected refractive index $n_{colloid}$ in the range of 1.3–1.7 measured at λ = 589 nm) were sourced from Bangs Laboratories and Sigma Aldrich.

One key parameter when seeking large areas of ordered nanostructures is the ordering of the colloidal particles on the surface at the start of the nanopatterning process. Spin coating is a reliable, relatively inexpensive fabrication technique for producing regular arrays of colloidal particles [13]. The spin-coating process can be divided into three stages: (i) a 15 µl suspension containing the colloidal particles is dropped onto the substrate maintained in a stable position; (ii) the substrate is rotated with a user-selected angular velocity, expressed in revolutions per minute (rpm), with the step sequence shown in Table 5.1; and (iii) the evaporation stage that accompanies the spinning determines the nature of the resulting film.

Table 5.1 Step sequence (duration and angular velocity) used in the spin-coating process

Step number	Time (sec)	Velocity (rpm)
1	30	300
2	30	400
3	30	1500
4	30	2000

Figure 5.2 shows low- and high-magnification SEM images of monolayer colloidal-crystal films formed by spin-coating, which will be employed in the subsequent investigations.

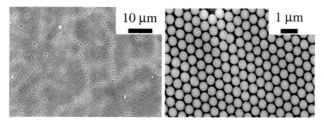

Figure 5.2 SEM images of self-assembled colloidal particle monolayers on Si substrates, formed using a spin-coating technique.

5.2.3 Laser Processing Parameters

Samples were irradiated with a series of single laser pulses from a super-rapid laser source from Lumera Lasers GMbh, diode-pumped master oscillator power amplifier (MOPA) Nd:YVO$_4$ picosecond (8 ps pulse duration) source, maximum power 2 W at λ = 355 nm. The laser beam was controlled by WaveRunner software (Nutfield Technology) through a Galvo scanner utilizing a telecentric lens with focusing distance f = 103 mm. The maximum field size of the current setup is 50 mm × 50 mm. An appropriate scanning pattern was established in order to create single-pulse individually exposed areas equally spaced, keeping the repetition frequency at 10 kHz. Power levels P from 5 mW to 100 mW were used, with a determined range of fluence F varying from 0.28 to 5.66 J/cm^2. The focused beam on the target surface had a Gaussian spatial beam profile, as illustrated in Fig. 5.3.

5.2.4 Focusing the Laser Beam through the Liquids

Several phenomena need consideration when the laser beam penetrates a layer of liquid en route to the substrate surface. First, the focal length of the focusing objective changes due to the refraction by the liquid layer [14].

To achieve the optimum focal spot dimension on the substrate surface, the focusing objective has to be moved by a distance

$\Delta f = h(1 - 1/n_{liquid})$, where f is the focal length, h is the thickness of the liquid layer, and n_{liquid} is the refractive index of the liquid. Possible attenuation of the laser beam intensity when it passes through the liquid is another factor to bear in mind. This attenuation arises from absorption and/or scattering of photons by the molecules of the liquid [15]. Finally, we need to consider the effective transmission of the light T_m into the substrate sample at normal incidence. This is given by $T_m = (4n_{liquid}n_s)/(n_{liquid} + n_s)^2 + k_s^2$, where n_s and k_s are the real and imaginary parts of the complex refractive index $n_s + ik_s$ of the Si substrate at the wavelength of interest. In the case of a Si substrate, this implies that only $X\%$ of the light intensity incident on the surface actually enters the substrate, and the other $(100 - X)\%$ is reflected.

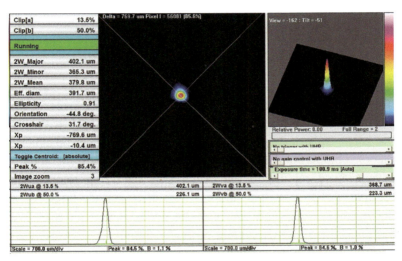

Figure 5.3 Gaussian beam profile of the picosecond 355 nm laser pulses used in the present experiments.

Laser irradiation of the colloidal particles spin-coated onto the Si substrates was performed under four different liquids (volume ~1 cm³, contained in an open trough of dimension 25 × 25 × 20 mm³), the refractive indices n_{liquid} of which are listed in Table 5.2. Data for methanol, carbon tetrachloride, and toluene at $\lambda = 532$ nm are taken from Ref. [16], while that for $\lambda = 355$ nm are obtained from an interpolation of the dispersion equation provided in this reference. For acetone, the data are taken from Ref. [17]. Attenuation of the incident laser radiation by these various different solvents will be negligible, given the thin liquid samples employed and the smallness of their absorption coefficients $\alpha(\text{cm}^{-1}) < 0.04$ at $\lambda = 355$ nm [18].

Table 5.2 Liquids used in the present LILAC lithography studies, together with their respective chemical formulae and refractive indices n_{liquid} (at $\lambda = 532$ nm and $\lambda = 355$ nm)

Liquid	Chemical formula	n_{liquid} at $\lambda = 532$ nm	n_{liquid} at $\lambda = 355$ nm
Methanol	CH_3OH	1.33	1.34
Acetone	CH_3COCH_3	1.36	1.37
Carbon tetrachloride	CCl_4	1.46	1.48
Toluene	$C_6H_5CH_3$	1.49	1.52

5.2.5 Finite-Difference Time Domain Simulations

The field lines of the energy flux (the Poynting vector) around a small particle were modeled by classical Mie theory. Mie's paper from 1908 [19] gave the first outline as to how to compute light scattering by small spherical particles using Maxwell's electromagnetic theory. Computational methods based on Mie scattering theory have evolved rapidly, to the extent that scattering calculations for spherical spheres 1 or more orders of magnitude larger than the incident wavelength can now be undertaken with ease. We have performed numerical Mie theory analysis using RSoft Design group's Full Wave 9.1 software, based on the finite-difference time domain (FDTD) method. The algorithm to model the optical near field around a single particle solves Maxwell's curl equations as a function of discrete time and space.

Figure 5.4 shows illustrative simulations for a colloidal particle with $R = 1.5$ μm ($n_{\text{colloid}} = 1.44$) placed in air or immersed in a liquid medium with a refractive index ranging from $n_{\text{liquid}} = 1.35$ to 1.50. The laser wavelength is $\lambda = 400$ nm, and the polarization is linear. The displayed quantity is the cross-sectional view of the Poynting vector magnitude, defined as the directional energy flux per unit area in the vicinity of the colloidal particle, S_z for the different media, as a function of distance Z beyond the particle. These model data illustrate that the focal point for the colloidal particle in air is located at the position $Z/R = 1.1$, that is, close to the particle surface. The field enhancement decays exponentially, meaning that for efficient ablation/patterning, the particle surface should be maintained within the near-field distance.

As Fig. 5.4 also shows, the focusing and field enhancement properties are dependent upon the refractive index of the surrounding medium. We identify two different regimes: (i) when $n_{colloid} \leq n_{liquid}$ (Fig. 5.4a–f) and (ii) when $n_{colloid} > n_{liquid}$ (Fig. 5.4g–i). In all cases the field enhancement is decaying more slowly than if $n = 1$, resulting in a significantly increased depth of focus. If $n_{colloid} < n_{liquid}$, we do not observe any field intensification under the colloidal particle, only at and beyond its edge.

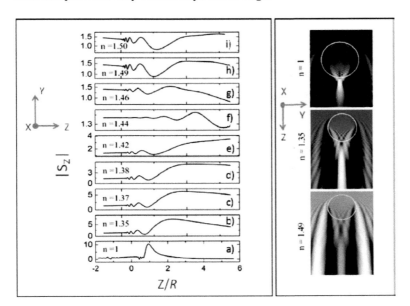

Figure 5.4 Cross-sectional views (along Z, at X = 0, left) illustrating the magnitude of the Poynting vector $|S_z|$ in the vicinity of an isolated R = 1.5 μm silica particle ($n_{colloid}$ = 1.44, indicated by the open white circle in the right image) immersed in different liquid media. The right-hand panels are false grayscale plots showing the near-field enhancement in liquids with n_{liquid} = 1.0 (top), 1.35 (middle), and 1.49 (bottom). The incident laser beam (λ = 400 nm) is linearly polarized along the Y axis and propagates in the Z direction.

5.3 Experimental Demonstrations

Figure 5.5 shows a selection of SEM images that illustrate the potential of LILAC lithography in comparison with NF-LA. The Si substrate in each case was covered with a monolayer of $R = 1.5$ μm silica colloidal particles. The low-resolution image (Fig. 5.5a) shows a sequence of processed areas, each of which is the result of a single-shot exposure. As reported previously, the colloidal particles in the irradiated area are efficiently removed—a process termed "dry laser cleaning" [20]. The other panels show higher-resolution views of (Fig. 5.5b) an unirradiated region of the hexagonal closed-packed (hcp) monolayer and selected areas of Si substrates after single-shot laser processing in (Fig. 5.5c) air, (Fig. 5.5d) CH$_3$OH, and (Fig. 5.5e) C$_6$H$_5$CH$_3$.

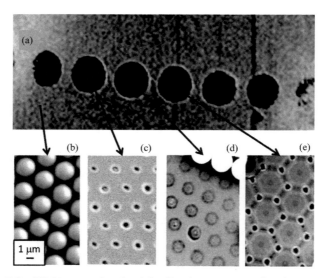

Figure 5.5 SEM images showing (a) a Si substrate covered with a monolayer of $R = 1.5$ μm silica colloidal particles after LILAC lithography treatment, (b) an unirradiated region of the monolayer-covered surface, and detailed views of the Si surface after single-shot laser processing in (c) air ($n = 1$), (d) CH$_3$OH ($n = 1.34$), and (e) C$_6$H$_5$CH$_3$ ($n = 1.52$).

These images illustrate the regularity of the 2D patterning of the Si substrate achieved using the LILAC lithography technique and the obvious change in the patterning depending upon whether n_{liquid} is less than panel (d), or greater than panel (e), $n_{colloid}$. The features in panel (d) were formed by single-shot irradiation with $F = 0.1$ mJ cm^{-2} and are centered directly beneath the original positions of the colloidal particles in the hcp monolayer on the Si surface. As panel (e) illustrates, the detail of the nanostructuring is very different in the case that $n_{liquid} > n_{colloid}$. This latter sample, prepared by single-pulse laser processing with $F = 0.2$ mJ cm^{-2}, shows clear features in the interstices between the original colloidal particles and weaker features directly under the original site of each particle.

One particularly noteworthy feature of the data shown in Fig. 5.5e is the finding that the scale of the nanopatterning achieved using the (comparatively low-cost) LILAC technique is within the diffraction limit. This is illustrated more clearly in Fig. 5.6, which shows two examples of this subdiffraction limited patterning of a Si substrate surface. The SEM image in Fig. 5.6a shows holes with full-width at half-maximum (FWHM) of ~80 nm and depth ~150 nm, as determined from the accompanying atomic force microscopy (AFM) line scan, while the SEM images Fig. 5.6b show local bumps with base diameters of ~300 nm and maximum heights of ~100 nm.

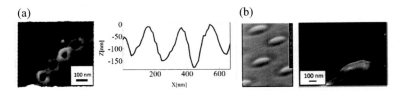

Figure 5.6 Examples of subdiffraction limited patterning of a Si substrate surface. (a) SEM image showing holes with full-width at half-maximum (FWHM) of ~80 nm and depth of ~150 nm, as determined from the accompanying AFM line scan. (b) SEM images of local bumps with base diameters of ~300 nm and maximum heights of ~100 nm.

During LILAC processing, we have also discerned evidence for laser-induced bubble formation (Fig. 5.7). Absorption in regions of highest light intensity near the substrate-liquid interface results in local vaporization (or dissociation), generating gas bubbles at the interface. The image in Fig. 5.7 illustrates the complex patterning that can be obtained from superposed contributions from diffraction on a gas bubble and from the array of colloidal particles. This exciting but still very preliminary result provides further motivation for an in-depth study of the liquid medium (its absorptivity, refractive index, viscosity, vapor pressure, photostability, etc.) on the etching and patterning of surfaces using the LILAC method. Many different commercially available liquids have been used in previous immersion lithography studies, several of which display the preferred combination of a high refractive index yet low absorbance at ultraviolet (UV) wavelengths [21].

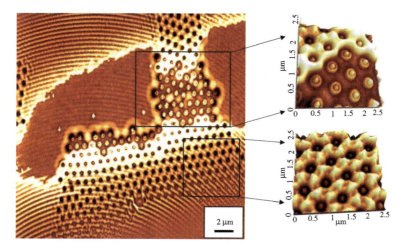

Figure 5.7 AFM image illustrating the complex surface structuring caused by gas bubble diffraction patterns (surface-wave-like structures) overlapping with structures created by the LILAC lithographic technique, with particular regions shown on an expanded scale on the right. The substrate in this case is GaAs (radius of colloidal particle, R = 350 nm and $n_{colloid}$ = 1.44), in acetone (n_{liquid} = 1.33), and the excitation wavelength λ = 355 nm.

5.4 Conclusions

Many aspects of the proposed work will have direct benefits for academic research in wide areas of science, engineering, and other areas beyond. The new opportunities afforded by creating a new low-cost nanoscale lithographical concept and method will have broad impacts in many academic communities—including but not restricted to laser–material processing, near-field optics (the distinctive near-field optical effects based on the superposition of waves scattered and focused by colloids will advance studies in this area), liquid manipulation, optical trapping, colloidal particles (new shape and material designs), soft materials, and nanopatterning and imprinting.

Acknowledgments

The authors acknowledge support from the Research Executive Agency FP7-PEOPLE-2013-IEF-625403, through the Marie Curie Intra European Fellowship.

References

1. Pohl, D. W., Denk, W., and Lanz, M. (1984). Optical stethoscopy: image recording with resolution λ/20, *Appl. Phys. Lett.*, **44**, pp. 651–653.
2. Betzig, E., Trautman, J. K., Harris, T. D., Weiner, J. S., and Kostelak, R. L. (1991). Breaking the diffraction barrier: optical microscopy on a nanometric scale, *Science*, **251**, pp. 1468–1470.
3. Zenhausern, F., O'Boyle, M. P., and Wickramasinghe, H. K. (1994). Apertureless near-field optical microscope, *Appl. Phys. Lett.*, **65**, pp. 1623–1625.
4. Kawata, Y., Egami, C., Nakamura, O., Sugihara, O., Okamoto, N., Tsuchimori, M., and Watanabe, O. (1999). Non-optically probing near-field microscopy, *Opt. Commun.*, **161**, pp. 6–12.
5. Lu, Y. F., Zhang, L., Song, W. D., Zheng, Y. W., and Luk'yanchuk, B. S. (2000). Laser writing of a subwavelength structure on silicon (100) surfaces with particle-enhanced optical irradiation, *JETP Lett.*, **72**, pp. 457–459.
6. Mosbacher, M., Münzer, H.-J., Zimmermann, J., Solis, J., Boneberg, J., and Leiderer, P. (2001). Optical field enhancement effects in laser-assisted particle removal, *Appl. Phys. A*, **72**, pp. 41–44.
7. Kühler, P., García de Abajo, F. J., Solis, J., Mosbacher, M., Leiderer, P., Afonso, C. N., and Siegel, J. (2009). Imprinting the optical near field of microstructures with nanometer resolution, *Small*, **5**, pp. 1825–1829.
8. McLeod, E., and Arnold, C. B. (2008). Subwavelength direct-write nanopatterning using optically trapped microspheres, *Nat. Nanotechnol.*, **3**, pp. 413–417.
9. Li, J., Gao, W., Dong, R., Pei, A., Sattayasamitsathit, S., and Wang, J. (2014). Nanomotor lithography, *Nat. Commun.*, **5**, pp. 1–7.

10. Ulmeanu, M., Petkov, P., Ursescu, D., Maraloiu, V., Jipa, F., Brousseau, E., and Ashfold, M. N. R. (2015). Pattern formation on silicon by laser-initiated liquid-assisted colloidal (LILAC) lithography, *Nanotechnology*, **26**, p. 455303.

11. Ulmeanu, M., Grubb, M. P., Jipa, F., Quignon, B., and Ashfold, M. N. R. (2015). 3-D patterning of silicon by laser-initiated, liquid-assisted colloidal (LILAC) lithography, *J. Colloid Interface Sci.*, **447**, pp. 258–262.

12. Palik, E. D. (1988). *Handbook of Optical Constants of Solids* (Academic, New York).

13. Chen, J., Dong, P., Di, D., Wang, C., Wang, H., Wang, J., and Wu, X. (2013). Controllable fabrication of 2D colloidal-crystal films with polystyrene nanospheres of various diameters by spin-coating, *Appl. Surf. Sci.*, **270**, pp. 6–15.

14. Menendez-Manjon, A., Wagener, P., and Barcikowski, S. (2011). Transfer-matrix method for efficient ablation by pulsed laser ablation and nanoparticle generation in liquids, *J. Phys. Chem. C*, **115**, pp. 5108–5114.

15. Wang, C., Huo, H., Johnson, M., Shen, M., and Mazur, E. (2010). The thresholds of surface nano/micro-morphology modifications with femtosecond laser pulse irradiations, *Nanotechnology*, **21**, p. 075304.

16. Moutzouris, K., Papamichael, M., Betsis, S. C., Stavrakas, I., Hloupis, G., and Triantis, D. (2014). Refractive, dispersive and thermo-optic properties of twelve organic solvents in the visible and near-infrared, *Appl. Phys. B*, **116**, pp. 617–622.

17. Rheims, J., Koser, J., and Wriedt, T. (1997). Refractive-index measurements in the near-IR using an Abbe refractometer, *Meas. Sci. Technol.*, **8**, pp. 601–605.

18. Duley, W. W. (2005). *UV Lasers: Effects and Applications in Material Science* (Cambridge University Press, UK).

19. Mie, G. (1908). Beitraege zur Optic Trueber Medien, Speziell kolloidaler Metalloesungen, *Ann. Phys.*, **330**, pp. 377–445.

20. Zheng, Y. W., Luk'yanchuk, B. S., Lu, Y. F., Song, W. D., and Mai, Z. H. (2001). Dry laser cleaning of particles from solid substrates: experiments and theory, *J. Appl. Phys.*, **90**, pp. 2135–2142.

21. Sanders, D. P. (2010). Advances in patterning materials for 193 nm immersion lithography, *Chem. Rev.*, **110**, pp. 321–360.

Chapter 6

MAPLE Deposition of Nanomaterials

Enikö György[a,b] and Anna Paola Caricato[c]

[a]*Consejo Superior de Investigaciones Cientificas, Instituto de Ciencia de Materiales de Barcelona, Campus UAB, Cerdanyola del Valles 08193, Spain*
[b]*National Institute for Lasers, Plasma and Radiation Physics, P. O. Box MG 36, 77125 Bucharest, Romania*
[c]*Department of Mathematics and Physics "E. De Giorgi," University of Salento, Via Arnesano, I-73100 Lecce, Italy*
egyorgy@icmab.es, annapaola.caricato@unisalento.it

In this chapter, experimental and numerical simulation results concerning the immobilization of nanoparticles and nanocomposites onto solid substrate surfaces by the matrix-assisted pulsed laser evaporation (MAPLE) technique are reviewed. The influence of the principal experimental parameters on the laser-transferred material's morphological characteristics and chemical composition and on the congruent transfer of the starting nanoentities are outlined and discussed both for ultraviolet (UV)- and infrared (IR)-MAPLE processes. Moreover, recent advances concerning nanoparticle deposition by resonant IR-MAPLE (RIR-MAPLE) and inverse-MAPLE are summarized.

Pulsed Laser Ablation: Advances and Applications in Nanoparticles and Nanostructuring Thin Films
Edited by Ion N. Mihailescu and Anna Paola Caricato
Copyright © 2018 Pan Stanford Publishing Pte. Ltd.
ISBN 978-981-4774-23-9 (Hardcover), 978-1-315-18523-1 (eBook)
www.panstanford.com

6.1 Introduction

Nanoparticles (NPs) are defined as particles with external dimensions or internal features within the nanometer range (typically <100 nm). They are known to have novel functional, optical, electronic, magnetic, and thermal properties, not observed in the bulk form of the same material [1, 2]. Due to their attractive functional properties, NPs are expected to have a considerable impact on many industrial activities, processes, and products. Energy production and storage, electronics and optoelectronics, chemical and biological sensing, environmental pollution monitoring and mitigation, and medical and pharmaceutical industries are among the most important applications fields where nanomaterials endow new functionalities [3]. At the nanometer scale the materials' functional properties are dependent on size and shape, besides chemical composition. As a consequence, the properties of the nanomaterials can be tuned and adapted to specific applications [2, 3]. However, in many technological fields the new properties cannot be fully exploited in the powder-like form of the nanomaterials. Immobilization of NPs onto solid substrate surfaces is a way to bridge this gap, for most of the application fields.

Therefore, there has been an increasing interest during the last few years for the development of new deposition techniques for the immobilization of nanomaterials onto solid substrates in the form of both nanostructures and continuous thin films. Low-temperature techniques that could ensure the immobilization of nanomaterials onto flexible substrates are of particular interest [4]. Nanometer-thick coatings for large-area photovoltaics or self-cleaning applications are advantageous since they are also cost effective, requiring only reduced amounts of materials to cover extended surface areas. However, the most important aspect in the design of new NP-containing products is the possibility to degrade or recover them from the environment. Due to their size, nanomaterials interact with biological structures. Therefore, there are many possible health risks associated with the presence of nanowaste in the ecosystem. The main advantage of NPs immobilized onto solid substrates is that they can be easily removed from the environment and recycled.

Conventional solvent-based chemical deposition methods are usually simple and cost effective. However, they do not ensure good control over the deposited layers' thickness, uniformity, and adherence to the substrate surface. Furthermore, deposition of multilayers is complicated due to solvents' incompatibilities. Moreover, each deposition method is limited to a specific class of materials, often implying the use of toxic chemical substances, pretreated substrates, and/or high temperatures.

The development of versatile methods for the immobilization of nanoentities onto solid substrates that ensure the reproduction of the geometrical characteristics, size, and shape, as well as chemical composition of the initial, native nanomaterials, would be advantageous from the application point of view. Additional requirements are uniform substrate coverage, thickness control, good adherence to the substrate surface, possibility of multilayers growth, avoidance of the use of hazardous chemical products, and high processing temperatures.

Laser-based methods could represent an alternative to conventional techniques since they ensure good control over the deposited material's thickness, adherence to the substrate's surface, and easy fabrication of multilayers [5]. Moreover, the materials can be deposited on any kind of substrate materials, including flexible polymers. In particular, pulsed laser deposition (PLD) is known to be a versatile technique for the synthesis and deposition of metallic and ceramic NPs. However, one of the major drawbacks of PLD is that the chemical composition of the NPs can differ significantly from that of the initial starting materials used for the preparation of the targets. Another inconvenience of PLD is that it does not guarantee good control over the deposited NP's dimensions. Usually, the particles generated and deposited by PLD have a wide size distribution.

A more recent laser-based technique, called matrix-assisted pulsed laser evaporation (MAPLE) [6, 7], was developed initially for the processing and immobilization of organic and bio-organic materials. MAPLE is to certain extent a derivative of PLD. The main characteristic of MAPLE is that the material of interest is dissolved or dispersed in a solvent matrix, aimed to protect the material from the direct action of the incident laser radiation. To this aim, solvent matrixes with high absorption at the wavelength of the laser radiation are chosen, with the purpose of shielding the materials of interest, preventing photoinduced decomposition or formation of structural and chemical defects. The concentration of the material in the MAPLE targets is low, typically in the range of 0.5–10 wt%.

The solution is cooled in liquid nitrogen until solidification and kept frozen during the laser transfer process. Moreover, the laser pulse intensity is 1–2 orders of magnitude lower as compared to that characteristic for PLD.

The MAPLE deposition hardware does not substantially differ from the ones commonly used in PLD. Excimer or Nd:YAG lasers (third harmonic at 355 nm) are mostly used, since ultraviolet (UV) radiation couples with almost any target material. Moreover, infrared (IR) laser wavelengths can be also used in order to further minimize the interaction with the solute material (generally Er:YAG laser; 2.9 µm). Nanosecond pulses are used to deliver enough energy to a solvent layer to let it evaporate. The main difference with respect to PLD systems is the target holder, since the target has to be kept at a very low temperature during the ablation process. It requires that a liquid nitrogen reservoir be connected to the target holder. It is usually manufactured from high-conductivity oxygen-free copper, crossed by a stem of the same material supporting the target holder. The target should rotate (3–10 Hz, typically), like in PLD, to allow smooth erosion of the frozen solution. Feedthroughs and connectors have to be accurately designed, with properly chosen gaskets, to allow rotation at low temperature without seizing problems.

Besides organic and bio-organic materials, the possibility to transfer by MAPLE inorganic nanoentities has been demonstrated [8–30]. It was shown that under certain experimental conditions, the MAPLE technique could solve the main drawbacks of PLD. Through the appropriate selection of process parameters, laser wavelength, nature of solvents, laser pulse intensity, and pressure of the ambient reactive atmosphere, the chemical composition and size of the laser-transferred particles can be controlled. Table 6.1 provides an overview of the nanoentities and nanocomposite materials deposited by MAPLE, together with the main process characteristics, the nature of the matrix solvent, and the laser wavelength used for the irradiation of the targets. The MAPLE-deposited nanoentities and nanocomposites are quite challenging for many technological and ordinary life applications like, for example, energy storage, electronic devices, mechanical systems, biochemical sensors [9, 16], gas sensors [18, 20], nanocrystal quantum dot (QD) lasers, nonlinear optics, photovoltaics technology [10], tribology [11], catalysis and photocatalysis [25, 26], biomedical applications [21, 22], and solar cells [27, 28].

Table 6.1 List of nanomaterials deposited by the MAPLE technique

Material	Solvent matrix	Laser source	Ref.
Single-walled carbon nanotubes (SWCNTs)	-Chloroform ($CHCl_3$)	193 nm ArF* excimer laser	[8]
Single-walled carbon nanotubes (SWCNTs)	-Toluene (C_7H_8)	248 nm KrF* excimer laser	[9]
Multiwalled carbon nanotubes (MWCNTs)	-Toluene (C_7H_8)	248 nm KrF* excimer laser	[10]
CdSe/ZnS quantum dots	-Toluene (C_7H_8)	248 nm KrF* excimer laser	[11, 12]
Carbon nanopearls	-Acetone (C_3H_6O) -Dimethylsulfoxide (C_2H_6OS) -Dimethylformadine (C_3H_7NO) -Toluene (C_7H_8) -Methanol (CH_3OH) -Ethyl acetate ($C_4H_8O_2$)	248 nm KrF* excimer laser	[13]
Carbon nanopearls/gold composites	-Toluene (C_7H_8)	248 nm KrF* excimer laser and magnetron sputtering (Au target)	[14]
Multiwalled carbon nanotubes (MWCNTs)/ PMMA polymer composites	-Toluene (C_7H_8)	248 nm KrF* excimer laser	[15]
Single-walled carbon nanotubes (SWCNTs)/ PMMA polymer composites	-Toluene (C_7H_8)	248 nm KrF* excimer laser	[16]
SnO_2 nanoparticles	-Toluene (C_7H_8)	248 nm KrF* excimer laser	[17]
TiO_2 nanoparticles/nanorods	-Toluene (C_7H_8)	248 nm KrF* excimer laser	[18–20]

(Continued)

Table 6.1 (Continued)

Material	Solvent matrix	Laser source	Ref.
Fe$_2$O$_3$/dextran composites	-Distilled water	248 nm KrF* excimer laser	[21, 22]
Graphene oxide platelets	-Distilled water	248 nm KrF* excimer laser	[23]
TiO$_2$ nanoparticles/graphene oxide platelet composites	-Distilled water	248 nm KrF* excimer laser	[24, 25]
TiO$_2$ nanoparticles/graphene oxide platelet composites	-Distilled water	248 nm KrF* excimer laser, 2.94 μm Er:YAG	[26]
CdSe/MEH-CN-PPv polymer composites	-Benzyl alcohol (C$_6$H$_5$CH$_2$OH): distilled water -Toluene (C$_7$H$_8$): distilled water	2.94 μm Er:YAG	[27]
CdSe CdSe/PCPDTBT polymer composites	-Chlorobenzene (C$_6$H$_5$Cl) or trichlorobenzene (C$_6$H$_3$Cl$_3$): phenol (C$_6$H$_5$OH): distilled water	2.94 μm Er:YAG	[28]
TiO$_2$ nanoparticles	-Butyl alcohol (C$_4$H$_9$OH) -Distilled water	2.94 μm Er:YAG	[29]
TiO$_2$, ZnO nanoparticles TiO$_2$/PMMA composites	-Ethanol (C$_2$H$_5$OH): water -Pentanol (C$_5$H$_{11}$OH)	2.94 μm Er:YAG	[30]

In particular, interesting gas-sensing results were achieved by using a thin layer of TiO_2 NPs in the anatase phase (~10 nm size) deposited by MAPLE onto interdigitated alumina substrates [18, 19]. The details about the target preparation and deposition procedure are reported in Refs. [18, 19]. The deposited NPs well reproduced the complex morphology of alumina grains, with a uniform distribution. Rough substrates improve the performance of the sensors by increasing the active area of the sensing films. Electrical tests performed in a controlled atmosphere in the presence of ethanol and acetone vapors put in evidence a high value of the sensor response, even at very low concentrations (20–200 ppm in dry air) of both vapors [31]. MAPLE-deposited TiO_2 nanorods in the brookite phase exhibited optically activated enhancement of the response toward 1 ppm of NO_2 oxidizing gas mixed in dry air upon irradiating the sensing layer with UV light with low energy close to the TiO_2-sensing layer bandgap width [20]. The response is expressed as the ratio $R = I_{gas}/I_{air}$, where I_{gas} and I_{air} are the current in the presence of toxic gas and dry air, respectively. A ratio $R_{light}/R_{dark} = 1.6$ was inferred at a working temperature of the sensing layer of 300°C. The starting materials (TiO_2 nanorods in the brookite phase with a mean size of 5×50 nm^2 and doleate/oleyl amine as the capping layer) were dissolved in pure toluene (0.016 wt% TiO_2) and once frozen at liquid nitrogen temperature (LNT) (77 K) were irradiated with a KrF* excimer laser ($\tau = 20$ ns, $v = 10$ Hz, and $F = 150, 250$, and 350 mJ/cm^2). Several kinds of substrates were used to fully characterize the deposited layers: (100) single-crystal Si wafers, silica slides, Cu carbon-coated grids, and alumina interdigitated slabs [20].

TiO_2/Au/reduced graphene oxide (rGO), as well as nitrogen-doped TiO_2/Au/rGO nanocomposite thin films were deposited by the UV-MAPLE technique in order to evaluate their photocatalytic activity. For this purpose, titanium dioxide and TiO_2/Au/rGO nanocomposite thin films were grown in controlled O_2 or N_2 atmospheres (2–20 Pa). A KrF* laser beam ($\tau \approx 25$ ns, $v = 10$ Hz, and $F = 0.40$ J/cm^2) was used for the irradiation of the targets consisting of TiO_2 NPs or mixtures of TiO_2 NPs, Au NPs, and GO platelets in aqueous solutions. Anatase phase TiO_2 NPs with an average diameter of ~20 nm, GO sheets with about 1 mm^2 surface area, and Au NPs with an average diameter of around 100 nm were used as base materials for the preparation of MAPLE targets. Pure TiO_2 and $TiO_2/Au/GO$ nanocomposite thin films were prepared by dispersing in distilled water 1 wt% TiO_2 NPs and 1 wt% TiO_2 NPs, 1 wt% Au NPs, and 5 wt% GO sheets, respectively. BK7 glass plates with a 1 × 1 cm² surface area were used as substrates, placed parallel to the target at a separation distance of 4 cm. During depositions the substrates were kept at a constant temperature of 50°C.

The photocatalytic activity of the $TiO_2/Au/rGO$ nanocomposite materials synthesized in an O_2 atmosphere was improved by around 170% as compared to pure TiO_2. Moreover, the gradual incorporation of nitrogen into the rGO structure was achieved through the control of the ambient N_2 gas pressure, resulting in materials with enhanced photocatalytic efficiency, about 260% higher as compared to pure TiO_2 [25].

Very recently, this work was implemented by studying also the wetting and electrical properties of the MAPLE-deposited TiO_2 and TiO_2/GO layers with the addition of Ag NPs [32]. The synthesized ternary $TiO_2/rGO/Ag$ coatings were highly hydrophobic and had low electrical resistance.

The use of IR laser beams resonant with vibration frequencies of the solvent is a powerful strategy to further reduce the photochemical and/or photothermal damages of the solute material. The working principle of the resonant infrared–MAPLE (RIR-MAPLE) technique is outlined in Section 6.3.

RIR-MAPLE-deposited hybrid nanocomposite thin films—like CdSe colloidal quantum dot/poly[2-methoxy-5-(2′-ethylhexyloxy)-1,4-(1-cyanovinylene)phenylene] (MEH-CN-PPV) and CdSe poly[2,6-(4,4-bis-(2-ethylhexyl)-4H-cyclopenta[2,1-b;3,4-b0]dithiophene)-alt-4,7-(2,1,3 benzothiadiazole)] (PCPDTBT)—demonstrated the ability to finely tune the NP distribution within organic films, not ordinarily achievable using traditional solution-based deposition techniques, like spin coating and drop casting. The well-controlled nanocomposite films have significant implications for the development of optoelectronic devices based on these materials, increasing the performances of optoelectronic devices [27, 28]. The target preparation methods were made according to the procedures described in Refs. [27, 28, 33]. The depositions were conducted with an Er:YAG laser at $F = 1.8$–2 J/cm², $v = 2$ Hz, and a 7 cm target-to-substrate distance. The thickness of the films was within the 80–100 nm range.

Hybrid organic solar cells based on nanocomposite films deposited by the RIR-MAPLE technique were fabricated as well [28]. This was the first report on a polymer-NP hybrid organic solar cell fabricated by a MAPLE-related technique.

For the fabrication and characterization of hybrid organic solar cells, a PEDOT:PSS solution was spin-coated on prepatterned indium tin oxide (ITO)-coated glass substrates to yield a 40 nm thick film. Then, the active layer (80–100 nm thick) was deposited by RIR-MAPLE. Next, the deposited films were annealed using a hotplate inside a glove box at 140°C for 10 min. Finally, Al contacts (150 nm thick) were deposited by thermal evaporation in vacuum. All solar cells were encapsulated (using epoxy) and tested in the ambient atmosphere under a simulated illumination.

An important aspect of the study was to determine the optimal CdSe NP loading in the blended film. The 80% CdSe loading device yielded the best device performance and has a power conversion efficiency (PCE) of around 0.4%. While the hybrid organic solar cells fabricated by RIR-MAPLE clearly exhibited the photovoltaic effect, the measured PCEs of the devices were lower than those typically exhibited by spin cast devices [34, 35]. One possible reason for the lower PCE, despite the reduced phase segregation in the hybrid nanocomposite thin films, is the presence of the surfactant sodium dodecyl sulfate (SDS) (0.005 wt% in the CdSe NP emulsion) in the device. The 0.005 wt% SDS is necessary to stabilize the CdSe target emulsion; yet, SDS is an insulating molecule, and if some portion of this material is deposited in the corresponding solar cells, the device performance could be degraded. It will be necessary to reduce the amount of surfactant in the NP target emulsion to improve the solar cell performance.

6.2 Ultraviolet Matrix-Assisted Pulsed Laser Evaporation

As underlined earlier, with an appropriate selection of process parameters (like laser wavelength, nature of solvents, laser pulse intensity, and pressure of the ambient reactive atmosphere) it is possible to control the chemical composition, crystalline nature, and size and shape of the laser-transferred NPs.

For example, MAPLE deposition of TiO_2 NPs for gas sensing was accomplished by Caricato et al. [18]. First, TiO_2 colloidal NPs (size 10 nm) in the anatase phase were prepared by standard procedures [36]. Then they were diluted in deionized water with a concentration of 0.2 wt% and put in an ultrasonic bath for 10 minutes to prevent aggregation. Afterward, the solution was frozen at LNT and quickly placed into a vacuum chamber on a rotating target holder, cooled with liquid nitrogen to guarantee a low and constant temperature (−160°C). The frozen target was irradiated in vacuum (5×10^{-4} Pa) with an ArF* (τ = 20 ns, 10 Hz, and F = 0.55 J/cm^2) excimer laser source. The number of subsequent laser pulses, applied to deposit a single film on a substrate placed in front of the target at the distance of 36 mm, was 6500. The films were deposited on different substrates: silica, Si(100), and interdigitated alumina (Al_2O_3) slabs, for the different characterizations.

High-resolution scanning electron microscopy (SEM) images of the TiO_2 NP films deposited on Si substrates showed that the NPs preserved the starting dimensions, although a tendency to form aggregates was noticed. By comparison with TiO_2 NP films deposited by the spin-coating technique, starting from a solution with the same TiO_2 NP concentration (i.e., 0.2 wt%), a much more uniform coverage of the substrate was observed for the MAPLE-deposited film.

A very interesting result is that a uniform film of NPs was also obtained on rough Al_2O_3 substrates used for gas-sensing measurements, following the morphology of the alumina grains (Fig. 6.1). The preservation of the anatase crystal phase was evidenced by X-ray diffraction (XRD) spectra, where the characteristic peaks of the anatase phase at 2θ = 25°, corresponding to the reflection by the (101) crystallographic plane, are well evident.

The optical energy gap that resulted was of about 3.6 eV, as compared to the bulk anatase TiO_2 value of 3.18 eV, and close to the value reported in the literature for anatase TiO_2 NP Langmuir–Blodgett films [37].

As can be observed in Table 6.1, in UV-MAPLE toluene is chosen the most frequently as organic solvent, due to its high absorption in the UV spectral range, at the 248 nm wavelength, emitted by the KrF* excimer laser source. However, the obtained results demonstrate that the laser radiation is absorbed also by the nanoentities embedded in the organic solvents. Despite the highly absorbing solvent, at laser fluences exceeding a threshold value, the morphological features of the initial NPs used for the preparation of the MAPLE are not maintained during the laser interaction and transfer processes.

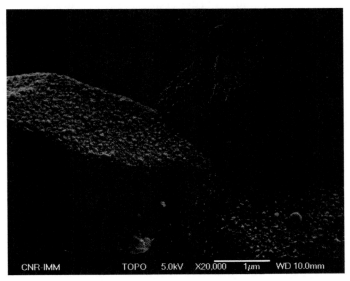

Figure 6.1 High-resolution SEM micrograph of a TiO_2 nanoparticle thin film MAPLE-deposited onto a rough alumina substrate. Reprinted from Ref. [18], Copyright (2007), with permission from Elsevier.

In Refs. [19, 38] the influence of the laser fluence on the preservation of the shape and phase of TiO_2 nanorods was investigated. In this study, size-tunable brookite (orthorhombic) TiO_2 nanorods, covered with an oleate/oleylamine capping layer, were chemically synthesized by a colloidal nonhydrolytic sol-gel route [39]. Such TiO_2 nanorods, with average dimensions of 3–4 nm × 20–50 nm, were dispersed in toluene to form the MAPLE target irradiated with a KrF* excimer laser at decreasing laser fluences (0.350, 0.150, 0.100, 0.050, and 0.025 J/cm^2) and a repetition rate of 10 Hz. The target-to-substrate distance was set at 40 mm and a number (6000) of subsequent laser pulses was applied.

At the highest deposition fluence (0.35 J/cm^2) the film (average thickness of ~150 nm) was composed of individually distinguishable

TiO$_2$ nanorods and crystalline spherical NPs having an average diameter of ~13 nm. The occurrence of TiO$_2$ spheres and the onset of traces of the rutile phase, both absent in the starting solution (Fig. 6.2), confirmed that nanorod melting/coalescence processes take place, driving their transformation into the most thermodynamically stable spherical shapes. These features are consistent with the size-dependent relative stability order of the TiO$_2$ phases at the nanoscale [40]. When decreasing the laser fluence from 0.150 to 0.025 J/cm^2, a decreasing fraction of sphere-like NPs was observed (Fig. 6.2), with no relevant differences for 0.150 and 0.100 J/cm^2, while 0.050 J/cm^2 was estimated as the threshold fluence for the onset of sphere-shaped NP formation. Only at 0.025 J/cm^2 a congruent transfer of nanorods with intact size/shape and phase from the solid state to the vapor phase and subsequent deposition onto a suitable substrate have been secured.

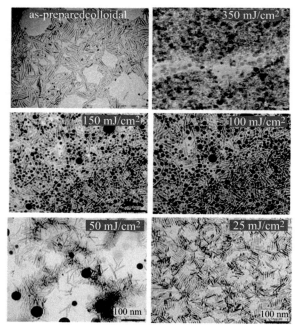

Figure 6.2 Bright-field TEM image of the as-prepared colloidal TiO$_2$ nanorods and a low-magnification TEM image of the MAPLE film deposited at decreasing laser fluences (0.350, 0.150, 0.100, 0.050, and 0.025 J/cm^2). Reprinted from Ref. [19], Copyright (2014), with permission from Elsevier.

The observed shape and size changes of the TiO$_2$ nanorods can be explained considering size-dependent melting point depression of nanostructures and transfer of thermal energy to the solute following laser irradiation of the MAPLE target despite the low laser fluence values (up to 0.35 J/cm^2) and the quite high melting temperature (~1850°C) of TiO$_2$ bulk. In this case, 600°C could be considered as an upper estimation of the melting point of the deposited nanorods. Moreover, the brookite-to-rutile phase transformation for TiO$_2$ NPs has been reported to occur in the temperature range 500°C–600°C [41]. It is worth recalling that the majority of the involved structures are brookite phase and only the largest spheres may have undergone a brookite-to-rutile phase transition.

The crucial role of heating effects inside the MAPLE target is demonstrated by the fact that the lowest considered fluence value (0.025 J/cm^2) is below the toluene ablation threshold at 248 nm (0.060 J/cm^2) [42]. This aspect is particularly intriguing because laser ablation of organics-/polymer-based MAPLE target is driven by the solvent ablation. The observed ablation threshold lower than the solvent one could be related to changes in the thermal transport regime and/or to effects induced by the low-dimensional solute. In fact, the thermal conductivity of TiO$_2$-based nanofluid is reported to enhance (up to 33% and more for nanorods than for nanospheres) in respect with the solvent alone [43, 44].

Also in the case of MAPLE-deposited CdSe/ZnS core-shell QDs, the shape and dimension were preserved only for low fluence values [11, 12]. In Fig. 6.3 we show transmission electron microscopy (TEM) images of a reference drop-cast sample prepared by the deposition onto a SiO$_2$ glass substrate of a drop from the solution, CdSe/ZnS core-shell QDs in toluene, used as targets in UV-MAPLE experiments (Fig. 6.3a) and CdSe/ZnS core-shell QDs immobilized onto a similar SiO$_2$ glass substrate by UV-MAPLE at (Fig. 6.3b) 0.25 and (Fig. 6.3c) 0.45 J/cm^2 laser fluences, respectively. The dimensions, of around 5 nm in diameter, and the spherical shape of the NPs were preserved after the laser transfer only at low laser fluences (Fig. 6.3b). In the case of the sample deposited at a 0.45 J/cm^2 laser fluence, besides the spherical particles, elongated whisker-like nanostructures can be identified with a high, about 2.5, aspect ratio (Fig. 6.3c).

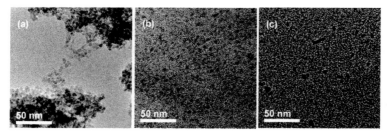

Figure 6.3 TEM images of CdSe/ZnS core-shell QDs. (a) A reference drop-cast sample and (b, c) QDs deposited on SiO$_2$ glass substrates by UV-MAPLE using toluene as a solvent matrix, at (b) 0.25 and (c) 0.45 J/cm^2 laser fluences. Reproduced from Ref. [12] with permission from Wiley.

The reference sample contains particles with two different crystalline structures. The interplanar distances measured by high-resolution transmission electron microscopy (HRTEM) can be assigned to the lattice plane reflection of the ZnS hexagonal wurtzite and ZnS F 43m (zinc-blende) cubic phases corresponding to the ZnS shell of the QDs.

The crystalline structure of the laser-transferred particles was unchanged as compared to that of the reference nonirradiated QDs only for low laser fluences. The measured interplanar distances correspond to the lattice plane reflections of the hexagonal and cubic zinc-blende ZnS phases (Fig. 6.4a,b). Conversely, in the case of the sample deposited at a higher, 0.45 J/cm^2, laser fluence the interplanar distance of the quantum rods (Fig. 6.4c) of 0.25 nm can be assigned to the {102} lattice plane reflection of the hexagonal CdSe phase.

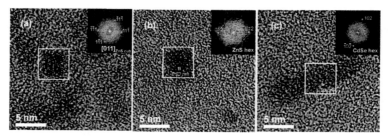

Figure 6.4 HRTEM images and fast Fourier transformation patterns of CdSe/ZnS core-shell QDs deposited on SiO$_2$ glass substrates by UV-MAPLE using toluene as a solvent matrix at (a,b) 0.25 and (c) 0.45 J/cm^2 laser fluences. Reproduced from Ref. [12] with permission from Wiley.

A similar trend in the evolution of the shell structure of multiwalled carbon nanotubes (MWCNTs) was observed with a gradual increase of the laser fluence (Fig. 6.5).

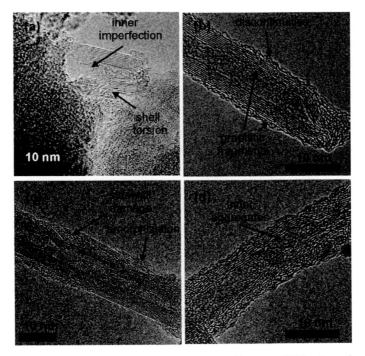

Figure 6.5 HRTEM images of MWCNTs deposited by UV-MAPLE using toluene as the solvent matrix at (a) 0.10, (b) 0.20, (c) 0.35, and (d) 0.45 J/cm^2 incident laser fluence values. Reproduced from Ref. [10] with permission from Springer.

At the lowest, 0.10 J/cm^2 laser fluence, the nanotubes transferred to the substrate surface preserve the initial shell structure (Fig. 6.5a). The nonirradiated MWCNTs used for the preparation of the MAPLE targets present some structural imperfections such as shell bending and variations of the diameter along the main tube axis. Therefore, the torsion of the shells observed in the HRTEM image was most likely already present in the initial MWCNTs. With the increase of the laser fluence a progressive increment in damage of the MWCNTs' shells was observed.

First, the gradual amorphization of the outer shells takes place (Fig. 6.5b–d). Moreover, at high laser fluences small, nanometer-size fragments can be identified in the inner cavity of the nanotubes, which most probably are constituted of graphitic fragments detached from the inner walls of the MWCNTs. At the highest incident laser fluence, 0.45 J/cm^2, the nanotubes are covered with an almost continuous amorphous outer shell (Fig. 6.5d). Both the amorphization of the outer shell as well as inner fragmentation lead to the gradual reduction of the number of shells with the increase of the incident laser fluence value.

Moreover, at high laser fluence values the deposited material consists of particulates forming separate islands, without covering completely the substrate surface, most probably due to the melting followed by the vaporization of the QDs. With the decrease of the laser fluence the surface morphology changes, being characterized by a continuous coverage over the substrate surface and the presence of circular features with diameters from about 1 µm to a few micrometers (Fig. 6.6).

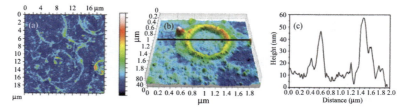

Figure 6.6 Top view and tilted atomic force microscopy (AFM) micrographs as well as surface profile across the line marked in the AFM micrograph for CdSe/ZnS core-shell QDs deposited on a SiO$_2$ glass substrate by UV-MAPLE using toluene as a solvent matrix at a 0.25 J/cm^2 laser fluence. Reproduced from Ref. [12] with permission from Wiley.

The observed surface features were attributed to photothermal ablation mechanisms [10–12]. Indeed, according to numerical simulations, at incident laser fluence values beyond 0.20 J/cm^2 the surface temperature of frozen toluene, with an initial temperature of 77 K, surpasses the sublimation threshold, reaching the boiling point of toluene, 383 K (110°C), at atmospheric pressure (Fig. 6.7a). Under vacuum conditions the boiling point is lower than at atmospheric pressure. As a consequence, explosive boiling could take place at laser fluences exceeding 0.20 J/cm^2.

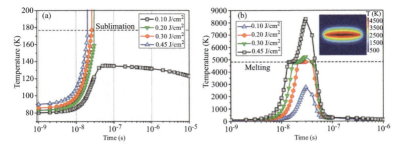

Figure 6.7 Numerical simulation of temperature evolution of (a) frozen toluene at 77 K, as well as (b) a single MWCNT immersed in frozen toluene irradiated at 0.10, 0.20, and 0.45 J/cm² incident laser fluence values, a 248 nm wavelength, and a 25 ns pulse duration, as a function of time. (Inset) Temperature map of a MWCNT irradiated with a 0.20 J/cm² laser fluence at 40 ns after the beginning of the laser pulse. Reproduced from Ref. [10] with permission from Springer.

Previous numerical simulation results suggested that explosive decomposition of the MAPLE target leads to the expulsion of liquid droplets with dimensions comparable to the optical penetration depth of the matrix [45]. For a toluene matrix the optical penetration depth was estimated to be 8.9 nm from the inverse of the linear optical absorption coefficient at a 248 nm wavelength.

In Ref. [45] the numerical simulation of laser interactions with polymers in a toluene matrix explains the formation of "deflated balloon" structures present of the MAPLE deposited films' surface as a result of liquid droplet ejections. The droplets consist of mixtures of liquid toluene solvent, polymer, and vapor phase solvent molecules. Following the deposition of the droplet on to a substrate surface, the solvent evaporates, leaving behind the characteristic circular structures. The described ablation mechanism can allow for the transport of individual nanoentities as well as larger aggregates.

Similar surface features were observed also in the case of carbon nanopearl films deposited by UV-MAPLE using a highly absorbing solvent such as acetone, dimethyl formamide, ethyl acetate, or toluene [13]. Conversely, the effect was minimal in the case of solvents with low absorption at the incident laser radiation, such as dimethyl sulfoxide and methanol.

On the other hand, the temperature at the irradiated frozen toluene surface at a low, 0.10 J/cm², laser fluence remains below the value of the sublimation temperature at atmospheric pressure

(Fig. 6.7a). Nevertheless, the deposition of MWCNTs was also observed at 0.10 J/cm^2, most probably due to the lower sublimation temperature of toluene under vacuum conditions.

The temperature evolution of MWCNTs in a frozen toluene matrix during laser irradiation was calculated [10]. The results of the numerical calculations are presented in Fig. 6.7b. As can be observed, photothermal absorption mechanisms lead to the increase of the carbon nanotubes' temperature to several thousands of degrees during the laser pulse duration, for the entire investigated laser fluence range, 0.10–0.45 J/cm^2. As a consequence, MWCNTs could induce the sublimation of the surrounding toluene via heat transfer even at the lowest, 0.10 J/cm^2, laser fluence. During this process, the toluene solvent could carry the MWCNTs toward the substrate surface. This process is very similar to that reported above for TiO$_2$ nanorods deposition.

The melting threshold intensity was determined to be 12 MW/cm^2, corresponding to a fluence of 0.24 J/cm^2, by direct laser irradiation of CNTs with a KrF* excimer laser source [46]. Indeed, according to numerical simulations, the melting temperature of MWCNTs is surpassed for laser fluence values higher than 0.20 J/cm^2 (Fig. 6.7b). Thus, parts of the MWCNTs, especially those situated in the outmost surface layer of the frozen composite MAPLE target, are most probably damaged by laser fluences higher than this threshold value. HRTEM analyses exhibit the amorphization of MWCNTs outer shells at laser fluences above this threshold (Fig. 6.5c,d). The earlier-described photothermal explosive boiling ablation mechanisms allow for the transport not only of nanotubes but also of larger structural elements as interconnected nanotube bundles, reducing the thermal effects that could take place under direct laser irradiation.

However, besides individual nanotubes and nanotube bundles, a flat, coating-like material was identified on the sample's surface (Fig. 6.8a). The presence of this undesired coating could be attributed to the photolysis of toluene under multipulse laser irradiation. Indeed, laser photolysis of toluene was reported to yield gaseous hydrocarbons and low-volatile aromatic hydrocarbons, along with solid material consisting of graphitic carbon and polymer-like C/H compounds [47].

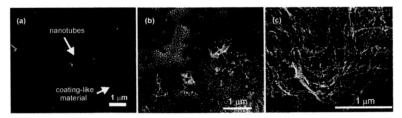

Figure 6.8 FE-SEM images of MWCNTs deposited by UV-MAPLE using toluene as a solvent matrix (a) as-deposited and (c) after postdeposition thermal annealing treatment in air at 300°C for 5 h and (c) UV-ozone treatment for 2 h. Reproduced from Ref. [10] with permission from Springer.

The observed coating-like material could not be eliminated by low-temperature postdeposition annealing. Thermal annealing in air up to 200°C does not cause any surface morphological change. As can be observed in Fig. 6.8b, thermal annealing at 300°C for 5 h induces clusterization, resulting in the formation of hundreds of nanometer-size particles. Moreover, the polymer-like residue was not removed. Conversely, UV-ozone treatment leads to the complete elimination of the polymeric material (Fig. 6.8c).

A polymeric material on the film surface can also originate from the capping layer of the NPs themselves, if characterized by a low vapor pressure, as reported for the deposition of SnO_2 colloidal NPs 3.6 ± 0.6 nm in diameter in the cassiterite phase. These NPs were prepared according to standard procedures [36] and presented one monolayer of a trioctylphosphine capping layer. For immobilization by MAPLE, they were diluted in toluene (0.2 wt%) and frozen at LNT. The frozen target was irradiated in a vacuum chamber with a KrF* laser beam (τ = 20 ns, ν = 10 Hz, and F = 0.35 J/cm^2). Each film was deposited with 6000 laser pulses. SEM inspection showed that the as-deposited film consists of uniformly distributed elongated structures (Fig. 6.9). The average film thickness was evaluated to be 150 ± 50 nm.

The Fourier transform infrared (FTIR) spectrum showed different absorption bands, which were ascribed predominantly to the NP capping layer, which is necessary to avoid precipitation. However, the use of this capping layer determines, as a consequence, its presence also in the toluene solution with an estimated concentration of 10% in volume. Trioctylphosphine has a vapor pressure of 120 Pa at 20°C, which is much lower than that of toluene at the same temperature

(2900 Pa). The consequence is that it is not effectively pumped out during the MAPLE process. Accordingly, it reaches the substrate, contributing to the composition of the deposited film. No traces of the capping layer were present on the film after annealing at 400°C. In fact, on the corresponding FTIR spectrum only one peak was clearly visible at 667 cm^{-1}. This peak is assigned to the vibration of the antisymmetric O-Sn-O bridging bond [48]. The average dimension of the resulting NP was 4 ± 1 nm, in accordance with the NP dimension of the starting solution [36]. Some larger NPs, with dimensions of 10–20 nm, were also noticed. They can be the result of either NP coalescence caused by the postdeposition annealing or the existence of these agglomerates in the starting solution itself.

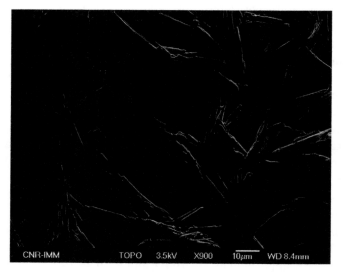

Figure 6.9 SEM micrograph of a MAPLE-deposited SnO$_2$ nanoparticle film on a Si substrate. From Ref. [17]. © IOP Publishing. Reproduced with permission. All rights reserved.

UV-visible absorption spectra were recorded to characterize the optical absorbance of the nanocrystalline SnO$_2$ film. The optical energy gap E_g was determined to be 4.24 eV (direct allowed transition). This value is higher than the one of 3.6 eV reported in literature [49] for bulk SnO$_2$. The observed blue shift of the absorption edge is attributed to the small dimension of the nanocrystalline particles. The optical bandgap of the nanocrystalline particles depends on the particle radius, due to quantum confinement of electrons and holes, as reported by different authors [50, 51].

Besides the codeposition of undesired polymeric materials, other "side effects" were also observed in the case of toluene or chloroform matrixes, as carbon impurities [14], C-Cl absorption bands in the composition of the material deposited on the substrate surface [52], or changes in the film's surface morphology [53]. All these negative aspects must be taken into account since, due to their high absorption in the UV spectral region, toluene or chloroform are the most commonly used solvents in UV-MAPLE experiments. However, substitution of toluene or chloroform with other solvents, such as methanol, acetone, or dimethyl sulfoxide, was demonstrated to produce a much lower deposition rate on the substrate surface due to their poorer absorbance and reduced vapor pressure as compared to toluene or chloroform [13].

To avoid the negative effects of solvents, the toluene matrix was substituted with distilled water in further experiments. This is an opposite situation to the traditional MAPLE: nanoentities with high absorption at the laser wavelength, TiO_2 NPs, and GO platelets were embedded in a nontoxic distilled water solvent matrix, transparent to the laser radiation [54]. The experiments were performed in low-pressure oxygen or nitrogen atmospheres.

As can be observed in Fig. 6.10a, the TiO_2 thin film deposited on the substrate surface is composed of large, spherical particles with diameters of hundreds of nanometers, up to 1 μm. Their round shape and larger dimensions, as compared to the initial TiO_2 NPs of around 20 nm used for the preparation of the MAPLE targets, suggest that they are formed in the liquid phase by the coalescence of the molten NPs. Conversely, the TiO_2/GO composite film (Fig. 6.10b) consists of irregular shape aggregates, individual sheets, and spherical particles.

The incident UV laser radiation is also absorbed by the GO platelets, causing their partial reduction, as confirmed by X-ray photoelectron spectroscopy results. In Fig. 6.11 the C 1s spectrum of initial GO platelets used for the preparation of the MAPLE targets (Fig. 6.11a) as well as the C 1s spectrum corresponding to the TiO_2/GO composite thin film obtained by the UV-MAPLE process

(Fig. 6.11b) are presented. The C 1s lines were deconvoluted in four lines: (I) C=C bonds centered at 284.8 eV binding energy and oxygen-containing functional groups and (II–IV) situated at high binding energy [55]. The C=C bonds correspond to graphite-like sp^2 C of the conjugated honeycomb lattice. The deconvoluted lines centered at higher binding energy values are attributed to C–O bonds, including contributions from the C–O–C epoxy (line II) and C–OH hydroxyl (line III), as well as C=O bonds (line IV) of the C=O carbonyl and O–C=O carboxyl groups [55]. As can be observed, the intensity of the lines corresponding to oxygen functional groups is significantly diminished in the case of the spectrum corresponding to the TiO_2/GO composite thin film as compared to that of lines corresponding to C–O and C=O bonds in the spectrum of the initial, nonirradiated GO platelets. The diminishment of the lines' intensity points to the significant decrease of the number of the functional oxygens during laser processing and immobilization and the formation of rGO graphene-like platelets. This is an important result since reduction of GO by conventional methods is time consuming and usually implies high temperatures of around 1000°C or the use of toxic chemical substances [56]. Moreover, the reduction process and thus the chemical composition of the laser-immobilized rGO platelets can be easily controlled through the ambient oxygen pressure [25].

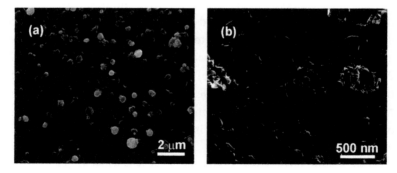

Figure 6.10 Field emission SEM images of (a) TiO_2 and (b) TiO_2/GO nanocomposite thin films deposited by UV-MAPLE using distilled water as a solvent matrix. Reproduced from Ref. [24] with permission from Wiley.

An additional advantage of the UV-MAPLE process lies in the fact that it allows for the simultaneous reduction and anion doping of the GO platelets. Indeed, upon the substitution of the low-pressure oxygen with nitrogen in the irradiation chamber, the controlled nitrogen doping of the GO platelets was achieved [25]. The N inclusion into the GO structure is reflected by the increase of the II–IV peaks' intensities of the C 1s spectrum (Fig. 6.12a), as compared to the samples obtained in low-pressure oxygen (Fig. 6.11b). Indeed, the C–N and C=N bonds are situated at the higher energy side of the C 1s spectrum, overlapping with the peaks corresponding to the oxygen functional groups [57]. Correspondingly, the N1s spectrum of the nanocomposite thin film deposited in nitrogen atmosphere (Fig. 6.12b) can be deconvoluted in two lines, assigned to nitrogen-containing functional groups, including C–N and C=N bonds of (I) pyrrolic-N and (II) quaternary-N [57].

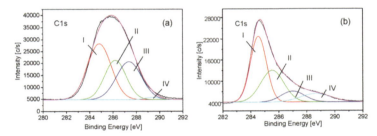

Figure 6.11 C 1s X-ray photoelectron spectroscopy spectra of (a) nonirradiated graphene oxide GO and (b) TiO$_2$/GO nanocomposite thin film obtained by UV-MAPLE using distilled water as a solvent matrix in 10 Pa O$_2$. Reproduced from Ref. [24] with permission from Wiley.

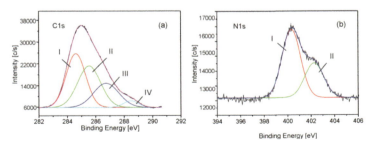

Figure 6.12 (a) C 1s and (b) N1s X-ray photoelectron spectroscopy spectra of TiO$_2$/GO nanocomposite thin film obtained by UV-MAPLE using distilled water as a solvent matrix in 10 Pa N$_2$. Reproduced from Ref. [24] with permission from Wiley.

6.3 Infrared Matrix-Assisted Pulsed Laser Evaporation

The main disadvantage of UV-MAPLE technique is that the shape and dimensions of the NPs are not maintained during the laser processing and transfer when using distilled water as a solvent material. The structure and morphology of the nanoentities could not be preserved in the case of the toluene matrix either, with the exception of low laser fluence values, resulting, in turn, in co-deposition of polymer-like residues and low deposition rates. This is due to the fact that despite the selection of organic solvents with high absorption, the laser radiation is absorbed also by the material of interest, resulting in structural and morphological changes. As can be observed in Figs. 6.2, 6.3b, 6.4c, 6.5b–d, and 6.10a, the NPs deposited on the substrate's surface have different structures, shapes, and dimensions as compared to the base materials used for the preparation of the MAPLE targets, even in case of an organic solvent matrix with high absorption at the wavelength of the incident laser radiation.

One strategy to overcome this drawback is to replace the UV laser source. To this aim, an Er:YAG laser emitting in the IR spectral range at the wavelength of 2940 nm and 350 μs pulse duration was used. This laser source has the advantage that the laser pulse energy is below the bandgap values of the materials to be immobilized on the substrate surface, transition metal oxides, and graphene oxide. Moreover, its emission wavelength is resonant with the O–H vibrational modes of the water matrix. The process is called resonant IR-MAPLE (RIR-MAPLE). It was observed that the amount of material immobilized on the substrate surface is significantly higher during the RIR-MAPLE process as compared to the UV-MAPLE process, keeping other parameters, such as oxygen pressure or target-to-substrate distance, identical. Besides, the sample obtained by RIR-MAPLE is constituted of tens of nanometer-size NPs (Fig. 6.13b), very similar to the nonirradiated material (Fig. 6.13a), in contrast to the same material immobilized by UV-MAPLE (Fig. 6.10a). These results are in good agreement with other studies reported in the literature concerning RIR-MAPLE deposition of CdSe/polymer composites [28] or TiO_2 and ZnO NPs [29, 30]. The NP films deposited by RIR-

MAPLE using low NP concentrations when preparing the MAPLE targets are continuous and uniform, characterized by a much lower surface roughness due to the absence of the large aggregates typically formed during UV-MAPLE [26, 28–30]. On the contrary, at high NP concentration values, phase separation and clusterization of the NPs can take place in the MAPLE target dispersions before freezing in liquid nitrogen, leading to thin films characterized by large surface roughness and the presence of micron-size aggregates [27, 29].

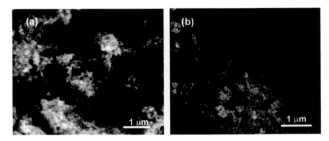

Figure 6.13 Field emission SEM images of (a) a TiO$_2$/GO reference drop-cast sample and (b) TiO$_2$/GO nanocomposite thin films deposited by the IR-MAPLE technique using distilled water as the solvent matrix. Reprinted with permission from Ref. [26]. Copyright (2014) American Chemical Society.

Furthermore, a decrease in the number of oxygen-containing functional groups during RIR-MAPLE transfer of GO platelets, conducting to the formation of rGO structures, was observed (Fig. 6.14), similar to the UV-MAPLE process (Fig. 6.11b).

The chemical composition of the TiO$_2$ particles remains identical to that of the initial material, both during UV- and RIR-MAPLE processes. The temperature evolution of the MAPLE targets constituted by TiO$_2$ NPs, GO platelets, and distilled water as the matrix was estimated by numerical simulations during both the IR- and UV-MAPLE processes (Fig. 6.15).

The temperature variation of the frozen, 77 K initial temperature, distilled water matrix under IR laser irradiation is presented in Fig. 6.15a. The absorption coefficient of water ice at the wavelength of 2940 nm was considered comparable to that of liquid water at room temperature [58]. The energy absorbed by the water ice matrix is not thermally confined in the irradiated volume determined by the laser spot area and the optical penetration depth, due to the long, 350 µs laser pulses. For water ice the thermal diffusion time was estimated at 1 μs [59]. Moreover, the absorption of the IR laser radiation in the GO platelets and TiO$_2$ NPs was considered negligible. Indeed, the incident IR photon energy, 0.4 eV, is much below the bandgaps of both GO and TiO$_2$. On the other hand, according to the phase diagram, water sublimation starts at around 220 K at an ambient pressure of 2 Pa [60].

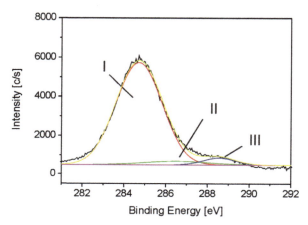

Figure 6.14 C 1s X-ray photoelectron spectroscopy spectrum of a TiO$_2$/GO nanocomposite thin film obtained by the IR-MAPLE technique using distilled water as the solvent matrix in 10 Pa O$_2$. Reprinted with permission from Ref. [26]. Copyright (2014) American Chemical Society.

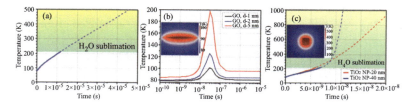

Figure 6.15 Numerical simulations of temperature evolution (a) of a distilled water matrix under IR laser irradiation and (b) GO platelets with 1, 2, and 5 nm thicknesses under UV laser irradiation. (Inset) Temperature distribution in a cross section of a 1 nm thick GO platelet at t = 30 ns. (c) TiO$_2$ nanoparticles with 20 and 40 nm diameters under UV laser irradiation. (Inset) Temperature distribution in a cross section of a TiO$_2$ nanoparticle 20 nm in diameter during the action of the laser pulse, at t = 12 ns. Reprinted with permission from Ref. [26]. Copyright (2014) American Chemical Society.

Numerical simulations show that during IR laser irradiation, the sublimation of the water matrix starts within the first 10 μs of the laser pulse (Fig. 6.15a). The water vapors transfer the GO platelets and TiO_2 NPs to the substrate surface. Consequently, initial material deposition could be attributed to laser-induced sublimation of the water matrix. Furthermore, during the laser pulse, the water matrix will continue to absorb the incident laser radiation and will heat the surrounding GO platelets and TiO_2 NPs. The reduction of GO platelets (Fig. 6.14) proceeds by thermal mechanisms during the IR-MAPLE process, since the thermal decomposition of GO is reported to take place at around 200°C [61]. On the other hand, a temperature rise of thousands of degrees is estimated by numerical simulations due to the very long laser pulse duration (Fig. 6.15a). However, due to the reduced, about 575 K, spinodal temperature (around 90% of the critical temperature) of water, phase explosion could take place, the liquid droplets transporting the TiO_2 NPs and GO platelets toward the substrate surface. Indeed, shadow graph and dark-field Schlieren images taken during the ablation process support the ejection of liquid droplets, suggesting the onset of the explosive boiling regime in the first 30 μs of the laser pulse [62].

In the case of the UV-MAPLE process, the absorption of laser radiation by the water matrix is considered negligible. Conversely, the UV photon energy is absorbed by the GO platelets and TiO_2 NPs. In Fig. 6.15b we present the maximum temperature reached during the UV laser pulse by the GO platelets with different thicknesses. The temperature of the GO platelets thicker than 5 nm surpasses the water sublimation threshold temperature. The neighbor thermally affected water volume is a few micrometers in diameter. Thicker GO aggregates attain higher maximum temperatures, and the thermal cycles are longer.

Similarly, the temperature evolution of the TiO_2 NPs depends on their dimensions. The temperature increases rapidly during the laser pulse (Fig. 6.15c), reaching hundreds of kelvins in just a few nanoseconds. The estimated thermally affected water volume around TiO_2 NPs is of the order of hundreds of nanometers. Initially, heat transfer to the surrounding water matrix could induce a rapid temperature increase, leading to the onset of ablation and to the transport of the GO and TiO_2 molecules toward the substrate surface. Indeed, the water sublimation temperature is already reached after the first 7 ns of the laser pulse (Fig. 6.15c). On the other hand, the morphology of the TiO_2 particles deposited on the substrate surface (Fig. 6.10a) indicates that their temperature reached the melting threshold, around 2000 K, during laser irradiation since they were molten and merged into larger particles as compared to the initial NPs. As a consequence, the surrounding water matrix could reach its spinodal temperature. The TiO_2 NPs and GO platelets could be ejected as part of the water droplets generated in the process of the explosive boiling of the overheated water matrix.

6.4 Inverse Matrix-Assisted Pulsed Laser Evaporation

Steiner et al. [63] introduced a variation of MAPLE, where it is possible to simultaneously synthesize and deposit well-dispersed NPs or nanoporous films through an inversion of the conventional MAPLE process. This inversion aims to direct the laser energy preferentially into the solute molecules to induce a reaction, rather than using a strongly absorbing solvent to shield them. Unlike the conventional MAPLE deposition of NPs, which involves tailoring low fluences and preferentially absorbing solvents to avoid altering the properties of prefabricated NPs, the inverse MAPLE technique relies on the extreme energetic conditions induced by the laser. This provides the inverse MAPLE technique with the potential to produce novel nonequilibrium alloys and NPs with controlled sizes and unique thermal histories.

In the inverse MAPLE process, as presented in Ref. [63], a solution or fine dispersion of metal-based acetate precursors in deionized water, generally of 1–2 wt%, is used as starting material. The solution is flash-frozen with liquid nitrogen to form a solid MAPLE target. A pulsed KrF* excimer laser (τ = 25 ns, v = 5 Hz, and F = 0.25–1.00 J/cm^2) was used for irradiation.

The inverse MAPLE process is initiated by the decomposition of the metal-based acetate precursors, which deliver the metallic ions required to form NPs. Metal-based acetate compounds are an incorporation of metallic ions into a coordination complex with one or more organic acetate ligands (CH_3COO^-). The KrF* UV light contains sufficient energy per photon to photochemically or photothermally decompose any of the bonds within an acetate molecule during a single photonic event. Size distributions of NPs deposited by the inverse MAPLE technique do not exhibit any discernible dependence on fluence or deposition pressure, indicating that NPs are formed within the target prior to ejection [64]. Following decomposition of the acetate precursors, the released metal ions in the target will begin to coalesce and form NPs. Once NPs have begun to form, in areas of high acetate concentration, they significantly increase the local absorption, introducing elevated temperatures and additional thermal energy within the target for diffusion. The temperatures reached by the irradiated NPs not only increase diffusion in surrounding regions but can exceed the explosive boiling point of the water matrix and are, therefore, capable of decomposing neighboring acetate precursors. The acetate with the lower decomposition temperature is expected to provide an initial seeding of NPs, which then serve as the dominant absorption species in the target and are able to decompose any surrounding acetates. NP formation, as well as most acetate decomposition, is expected to occur within the first several hundred microns of the target. When a surface volume containing fully formed NPs absorbs sufficient energy during a laser pulse to induce an energetic phase change in the surrounding matrix, all or part of the volume will be ejected, carrying NPs along with the expanding solvent into the vacuum. Below the threshold for energetic ejection, NPs are still able to heat and evaporate portions of the volatile matrix, increasing the local density of NPs and the probability of ejection within a subsequent pulse. As a result, a dense network of NPs is expected to form at the surface of the target, with different regions simultaneously ejected and repopulated during the duration of each laser pulse.

Support for this theory is provided through the analysis of deposited NPs and by novel characterization of MAPLE targets after irradiation via cryostage SEM.

The formation of Pd NPs by inverse MAPLE from a palladium acetate solution was studied as a function of carrier solvent, laser-pulse number, metal precursor concentration, and postdeposition thermal heating [65]. Structural and compositional analyses demonstrated that the solvent critically determines the size, morphology, and size distribution of the resulting NPs. A proper choice of the physical properties of the solvent allows sharpening the size distribution and increasing the uniformity of the NP thin-film coverage over substrates. Solvents with a low dynamical viscosity coefficient and high volatility are responsible for the formation of relatively larger NPs. In fact, for diethyl-ether-derived samples, a bimodal distribution of NP sizes spanning from ~1 nm up to 20 nm was obtained. Conversely, by using acetone, a monomodal distribution of sizes in the 1–6 nm range (mean diameter of 1.5 ± 0.7 nm) and a more uniform and densely packed surface coverage (NP coverage was twice as dense as the one obtained with diethyl ether) resulted. The cumulative effects of laser-pulse number and solute concentration seem to be less influential than the type of solvent carrier used.

6.5 Conclusions

Recent investigations demonstrate that the MAPLE technique could represent an alternative to conventional NP immobilization techniques. Through the appropriate selection of process parameters, laser wavelength, nature of solvents, laser pulse intensity, and pressure of the ambient inert or reactive atmosphere, the chemical composition and geometrical characteristics of the laser-transferred particles can be controlled. It was found that both UV- and IR-MAPLE processes are effective for immobilization of nanoentities onto solid substrates. Employing solvents with high absorption at the incident laser radiation under the UV-MAPLE process ensures the preservation of the nanoentities' shapes, dimensions, and chemical compositions at reduced laser intensity values. Important side effects can be eliminated by the substitution of organic solvents with distilled water. However, the dimensions and shape of oxide NPs are maintained only during the RIR-MAPLE process when distilled water is used as the solvent material. Moreover, through inverse MAPLE, starting from solution or dispersion of metal-based acetate precursors, metallic NPs can be produced, with well-controlled morphology and size distribution. The inverse MAPLE process is determined primarily by the physical properties of the solvent materials used for the preparation of the MAPLE targets.

Acknowledgments

The authors are profoundly grateful to Prof. A. Luches for his help and fruitful discussions.

References

1. Lövestam, G., Rauscher, H., Roebben, G., Klüttgen B.S., Gibson, N., Putaud, J.P., and Stamm, H. (2010). Considerations on a definition of nanomaterial for regulatory purposes, *JRC Reference Reports* (Joint Research Centre of the European Commission).
2. Vajtai, R. (2013). *Handbook of Nanomaterials* (Springer-Verlag, Berlin, Heidelberg).
3. Stark, W. J., Stoessel, P. R., Wohlleben, W., and Hafner, A. (2015). Industrial applications of nanoparticles, *Chem. Soc. Rev.*, **44**, pp. 5793–5805.
4. Koo, J. H., Seo, J., and Lee, T. (2012). Nanomaterials on flexible substrates to explore innovative functions: from energy harvesting to bio-integrated electronics, *Thin Solid Films*, **524**, pp. 1–19.
5. Eason, R. (2007). *Pulsed Laser Deposition of Thin Films: Applications-Led Growth of Functional Materials* (Wiley, Hoboken, New Jersey).
6. Bubb, D. M., Ringeisen, B. R., Callahan, J. H., Galicia, M., Vertes, A., Horwitz, J. S., McGill, R. A., Houser, E. J., Wu, P. K., Pique, A., and Chrisey, D. B. (2001). Vapor deposition of intact polyethylene glycol thin films, *Appl. Phys. A*, **73**, pp. 121–123.
7. Ringeisen, B. R., Callahan, J., Wu, P. K., Pique, A., Spargo, B., McGill, R. A., Bucaro, M., Kim, H., Bubb, D. M., and Chrisey, D. B. (2001). Novel laser-based deposition of active protein thin films, *Langmuir*, **17**, pp. 3472–3479.
8. Wu, P. K., Ringeisen, B. R., Krizman, D. B., Frondoza, C. G., Brooks, M., Bubb, D. M., Auyeung, R. C. Y., Pique, A., Spargo, B., McGill, R. A., and Chrisey, D. B. (2003). Laser transfer of biomaterials: matrix-assisted pulsed laser evaporation (MAPLE) and MAPLE direct write, *Rev. Sci. Instrum.*, **74**, pp. 2546–2557.
9. Pérez del Pino, A., György, E., Cabana, L., Ballesteros, B., and Tobias, G. (2012). Deposition of functionalized single wall carbon nanotubes through matrix assisted pulsed laser evaporation, *Carbon*, **50**, pp. 4450–4458.

10. György, E., Pérez del Pino, Á., Roqueta, J., Ballesteros, B., Cabana, L., and Tobias, G. (2013). Effect of laser radiation on multi-wall carbon nanotubes: study of shell structure and immobilization process, *J. Nanopart. Res.*, **15** pp. 1852 (1–11).
11. György, E., Pérez del Pino, A., Roqueta, J., Ballesteros, B., Miguel, A. S., Maycock, C., and Oliva, A. G. (2011). Synthesis and laser immobilization onto solid substrates of CdSe/ZnS core-shell quantum dots, *J. Phys. Chem. C*, **115**, pp. 15210–15216.
12. György, E., Pérez del Pino, A., Roqueta, J., Miguel, A. S., Maycock, C., and Oliva, A. G. (2012). Synthesis and characterization of CdSe/ZnS core-shell quantum dots immobilized on solid substrates through laser irradiation, *Phys. Status Solidi A*, **209**, pp. 2201–2207.
13. Hunter, C. N., Check, M. H., Bultman, J. E., and Voevodin, A. A. (2008). Development of matrix assisted pulsed laser evaporation (MAPLE) for deposition of disperse films of carbon nanoparticles, *Surf. Coat. Technol.*, **203**, pp. 300–306.
14. Hunter, C. N., Check, M. H., Muratore, C., and Voevodin, A. A. (2010). Electrostatic quadrupole plasma mass spectrometer measurements during thin film depositions using simultaneous matrix assisted pulsed laser evaporation and magnetron sputtering, *J. Vac. Sci. Technol. A*, **28**, pp. 419–424.
15. Sellinger, A. T., Leveugle, E. M., Gogick, K., Zhigilei, L. V., and Fitz-Gerald, J. M. (2006). Laser processing of polymer nanocomposite thin films, *J. Vac. Sci. Technol. A*, **24**, pp. 1618–1622.
16. Sellinger, A. T., Martin, A. H., and Fitz-Gerald, J. M. (2008). Effect of substrate temperature on poly(methyl methacrylate) nanocomposite thin films deposited by matrix assisted pulsed laser evaporation, *Thin Solid Films*, **516**, pp. 6033–6040.
17. Caricato, A. P., Epifani, M., Martino, M., Romano, F., Rella, R., Taurino, A., Tunno, T., and Valerini, D. (2009). MAPLE deposition and characterization of SnO_2 colloidal nanoparticle thin films, *J. Phys. D: Appl. Phys.*, **42**, pp. 095105–095110.
18. Rella, R., Spadavecchia, J., Manera, M. G., Capone, S., Taurino, A., Martino, M., Caricato, A. P., and Tunno, T. (2007). Acetone and ethanol solid-state gas sensors based on TiO_2 nanoparticles thin film deposited by matrix assisted pulsed laser evaporation, *Sens. Actuators B*, **127**, pp. 426–431.
19. Caricato, A. P., Arima, V., Catalano, M., Cesaria, M., Cozzoli, P. D., Martino, M., Taurino, A., Rella, R., Scarfiello, R., Tunno, T., and Zacheo, A. (2014). MAPLE deposition of nanomaterials, *Appl. Surf. Sci.*, **302**, pp. 92–98.

20. Manera, M. G., Taurino, A., Catalano, M., Rella, R., Caricato, A. P., Buonsanti, R., Cozzoli, P. D., and Martino, M. (2012). Enhancement of the optically activated NO_2 gas sensing response of brookite TiO_2 nanorods/nanoparticles thin films deposited by matrix-assisted pulsed-laser evaporation, *Sens. Actuators B*, **616**, pp. 869–879.
21. Predoi, D., Ciobanu, C. S., Radu, M., Costache, M., Dinischiotu, A., Popescu, C., Axente, E., Mihailescu, I. N., and György, E. (2012). Hybrid dextran-iron oxide thin films deposited by laser techniques for biomedical applications, *Mater. Sci. Eng. C*, **32**, pp. 296–302.
22. Ciobanu, C. S., Iconaru, S. L., György, E., Radu, M., Costache, M., Dinischiotu, A., Le Coustumer, P., Lafdi, K., and Predoi, D. (2012). Biomedical properties and preparation of iron oxide-dextran nanostructures by MAPLE technique, *Chem. Cent. J.*, **6**, pp. 17 (1–12).
23. Pérez del Pino, Á., György, E., Logofatu, C., and Duta, A. (2013). Study of the deposition of graphene oxide by matrix-assisted pulsed laser evaporation, *J. Phys. D: Appl. Phys.*, **50**, p. 505309.
24. György, E., Pérez del Pino, Á., Logofatu, C., Cazan, C., and Duta, A. (2014). Simultaneous laser-induced reduction and nitrogen doping of graphene oxide in titanium oxide/graphene oxide composites, *J. Am. Ceram. Soc.*, **97**, pp. 2718–2724.
25. Datcu, A., Duta, L., Pérez del Pino, A., Logofatu, C., Luculescu, C., Duta, A., Perniu, D., and György, E. (2015). One-step preparation of nitrogen doped titanium oxide/Au/reduced graphene oxide composite thin films for photocatalytic applications, *RSC Adv.*, **5**, pp. 49771–49779.
26. O'Malley, S. M., Tomko, J., Pérez del Pino, A., Logofatu, C., and György, E. (2014). Resonant infrared and ultraviolet matrix-assisted pulsed laser evaporation of titanium oxide/graphene oxide composites: a comparative study, *J. Phys. Chem. C*, **118**, pp. 27911–27919.
27. Pate, R., Lantz, K. R., and Stiff-Roberts, A. D. (2009). Resonant infrared matrix-assisted pulsed laser evaporation of CdSe colloidal quantum dot/poly[2-methoxy-5-(2'-ethylhexyloxy)-1,4-(1-cyano vinylene) phenylene] hybrid nanocomposite thin films, *Thin Solid Films*, **517**, pp. 6798–6802.
28. Ge, W., Atewologun, A., and Stiff-Roberts, A. D. (2015). Hybrid nanocomposite thin films deposited by emulsion-based resonant infrared matrix-assisted pulsed laser evaporation for photovoltaic applications, *Org. Electron.*, **22**, pp. 98–107.
29. Mayo, D. C., Paul, O., Airuoyo, I. J., Pan, Z., Schriver, K. E., Avanesyan, S. M., Park, H. K., Mu, R. R., and Haglund, R. F. (2013). Resonant infrared

matrix-assisted pulsed laser evaporation of TiO$_2$ nanoparticle films, *Appl. Phys. A*, **110**, pp. 923–928.

30. Singaravelu, S., Mayo, D. C., Park, H. K., Schriver, K. E., Klopf, J. M., Kelley, M. J., and Haglund R. F. Jr. (2014). Fabrication and performance of polymer–nanocomposite anti-reflective thin films deposited by RIR-MAPLE, *Appl. Phys. A*, **117**, pp. 1415–1423.

31. Rella, R., Spadavecchia, J., Manera, M. G., Capone, S., Taurino, A., Martino, M., Caricato, A. P., and Tunno, T. (2007). Acetone and ethanol solid state gas sensors based on TiO$_2$ nanoparticles thin film deposited by matrix assisted pulsed laser evaporation, *Sens. Actuators B*, **127**, pp. 426–431.

32. György, E., Perez del Pino, A., Datcu, A., Duta, L., Logofatu, C., Iordache, I., and Duta A. (2016). Titanium oxide–graphene oxide–silver ternary composite thin films grown by matrix assisted pulsed laser evaporation, *Ceram. Int.*, **42** pp. 16191–16187.

33. Ge, W., Nyikayaramba, G., and Stiff-Roberts, A. D. (2014). Bulk heterojunction PCPDTBT:PC$_{71}$BM organic solar cells deposited by emulsion-based, resonant infrared matrix-assisted pulsed laser evaporation, *Appl. Phys. Lett.*, **104**, p. 223901.

34. Zhou, Y., Eck, M., Veit, C., Zimmermann, B., Rauscher, F., Niyamakom, P., Yilmaz, S., Dumsch, I., Allard, S., Scherf, U., and Krüger, M. (2011). Efficiency enhancement for bulk-heterojunction hybrid solar cells based on acid treated CdSe quantum dots and low band gap PCPDTBT, *Sol. Energy Mater. Sol. Cells*, **95**, pp. 1232–1237.

35. Zhou, R. J., Stalder, R., Xie, D. P., Cao, W. R., Zheng, Y., Yang, Y. X., Plaisant, M., Holloway, P. H., Schanze, K. S., Reynolds, J. R., and Xue, J. G. (2013). Enhancing the efficiency of solution-processed polymer: colloidal nanocrystal hybrid photovoltaic cells using ethanedithiol treatment, *ACS Nano*, **7**, pp. 4846–4854.

36. Epifani, M., Arbiol, J., Díaz, R., Perálvarez, M. J., Siciliano, P., and Rella, R. (2005). Synthesis of SnO$_2$ and ZnO colloidal nanocrystals from the decomposition of Tin (II) 2-ethylhexanoate and Zinc (II) 2-ethylhexanoate, *Chem. Mater.*, **17**, pp. 6468–6472.

37. Coutinho, J. P. G., and Barbosa, M. T. C. M. (2006). Characterization of TiO$_2$ nanoparticles in Langmuir-Blodgett films, *J. Fluoresc.*, **16**, pp. 387–392.

38. Caricato, A. P., Belviso, M. R., Catalano, M., Cesaria, M., Cozzoli, P. D., Luches, A., Manera, M. G., Martino, M., Rella, R., and Taurino, A. (2011). Study of titania nanorod films deposited by matrix-assisted pulsed laser evaporation as a function of laser fluence, *Appl. Phys. A*, **105**, pp. 605–610.

39. Buonsanti, R., Grillo, V., Carlino, E., Giannini, C., Kipp, T., Cingolani, R., and Cozzoli, P. D. (2008). Nonhydrolytic synthesis of high-quality anisotropically shaped brookite TiO_2 nanocrystals, *J. Am. Chem. Soc.*, **130**, pp. 11223–11233.

40. Zhang, H., and Banfield, J. F. (2000). Understanding polymorphic phase transformation behavior during growth of nanocrystalline aggregates: insights from TiO_2, *J. Phys. Chem. B*, **104**, pp. 3481–3487.

41. Li, J. G., and Ishigaki, T. (2004). Brookite → rutile phase transformation of TiO_2 studied with monodispersed particle, *Acta Mater.*, **52**, pp. 5143–5150.

42. Kokkinaki, O., and Georgiou, S. (2007). Laser ablation of cryogenic films: implications to matrix-assisted pulsed laser deposition of biopolymers and dedicated applications in nanotechnology, *Digest J. Nanomater. Biostruct.*, **2**, pp. 221–241.

43. Murshed, S. M. S., Leong, K. C., and Yang, C. (2005). Enhanced thermal conductivity of TiO_2—water based nanofluids, *Int. J. Therm. Sci.*, **44**, pp. 367–373.

44. Pak, B. C., and Cho, Y. I. (1998). Hydrodynamic and heat transfer study of dispersed fluids with submicron metallic oxide particles, *Exp. Heat Transfer*, **11**, pp. 151–170.

45. Leveugle, E., Sellinger, A., Fitz-Gerald, J. M., and Zhigilei, L. V. (2007). Making molecular balloons in laser-induced explosive boiling of polymer solutions, *Phys. Rev. Lett.*, **98**, p. 216101.

46. Ohsumi, K., Honda, T., Kim, W. S., Oh, C. B., Murakami, K., Abo, S., Wakaya, F., Takai, M., Nakata, S., Hosono, A., and Okuda, S. (2007). KrF laser surface treatment of carbon nanotube cathodes with and without reactive ion etching, *J. Vac. Sci. Technol. B*, **25**, pp. 557–560.

47. Pola, J., Urbanova, M., Bastl, Z., Plzak, Z., Subrt, J., Vorlicek, V., Gregora, I., Crowley, C., and Taylor, R. (1997). Laser photolysis of liquid benzene and toluene: graphitic and polymeric carbon formation at ambient temperature, *Carbon*, **35**, pp. 605–611.

48. Gu, F., Wang, S. F., Song, C. F., Lu, M. K., Qi, Y. X., Zhou, G. J., Xu, D., and Yuan, D. R. (2003). Synthesis and luminescence properties of SnO_2 nanoparticles, *Chem. Phys. Lett.*, **372**, pp. 451–454.

49. Suda, Y., Kawasaki, H., Namba, J., Iwatsuji, K., Doi, K., and Wada, K. (2003). Properties of palladium doped tin oxide thin films for gas sensors grown by PLD method combined with sputtering process, *Surf. Coat. Technol.*, **174–175**, pp. 1293–1296.

50. Lee, E. J. H., Ribeiro, C., Giraldi, T. R., Longo, E., Leite, E. R., and Varela, J. A. (2004). Photoluminescence in quantum-confined SnO_2 nanocrystals: evidence of free exciton decay, *Appl. Phys. Lett.*, **84**, pp. 1745–1747.

51. Das, S., Kar, S., and Chaudhuri, S. (2006). Optical properties of SnO_2 nanoparticles and nanorods synthesized by solvothermal process, *J. Appl. Phys.*, **99**, pp. 114303 (1–7).

52. Bubb, D. M., Wu, P. K., Horwitz, J. S., Callahan, J. H., Galicia, M., Vertes, A., McGill, R. A., Houser, E. J., Ringeisen, B. R., and Chrisey, D. B. (2002). The effect of the matrix on film properties in matrix-assisted pulsed laser evaporation, *J. Appl. Phys.*, **91**, pp. 2055–2058.

53. Caricato, A. P., Arima, V., Cesaria, M., Martino, M., Tunno, T., Rinaldi, R., and Zacheo, A. (2013). Solvent-related effects in MAPLE mechanism, *Appl. Phys. B*, **113**, pp. 463–471.

54. Warren, S. (1984). Optical constants of ice from the ultraviolet to the microwave, *Appl. Opt.*, **23**, pp. 1206–1225.

55. Stankovich, S., Dikinm, D. A., Piner, R. D., Kohlhaas, K. M., Kleinhammes, A., Jia, Y., Wu, Y., Nguyen, S. T., and Ruoff, R. S. (2007). Synthesis of graphene-based nanosheets via chemical reduction of exfoliated graphite oxide, *Carbon*, **45**, pp. 1558–1565.

56. Pei, S., and Cheng, H. M. (2012). The reduction of graphene oxide, *Carbon*, **50**, pp. 3210–3228.

57. Sun, L., Wang, L., Tian, C., Tan, T., Xie, Y., Shi, K., Li, M., and Fu, H. (2012). Nitrogen-doped graphene with high nitrogen level via a one-step hydrothermal reaction of graphene oxide with urea for superior capacitive energy storage, *RSC Adv.*, **2**, pp. 4498–4506.

58. Pirkl, A., Soltwisch, J., Draude, F., and Dreisewerd, K. (2012). Infrared matrix-assisted laser desorption/ionization orthogonal-time-of-flight mass spectrometry employing a cooling stage and water ice as a matrix, *Anal. Chem.*, **84**, pp. 5669–5676.

59. Bubb, D. M., Johnson, S. L., Collins, B., and Haglund, R. F. Jr. (2010). Thermal confinement and temperature-dependent absorption in resonant infrared ablation of frozen liquid targets, *J. Phys. Chem. C*, **114**, pp. 5611–5616.

60. Feistel, R., and Wagner, W. (2006). A new equation of state for H_2O ice Ih. *J. Phys. Chem. Ref. Data*, **35**, pp. 1021–1047.

61. Hu, Z., Chen, Y., Hou, Q., Yin, R., Liu, F., and Chen, H. (2012). Characterization of graphite oxide after heat treatment, *New J. Chem.*, **36**, pp. 1373–1377.

62. Nahen, K., and Vogel, A. (2002). Plume dynamics and shielding by the ablation plume during Er:YAG laser ablation, *J. Biomed. Opt.*, **7**, pp. 165–178.
63. Steiner, M. A., and Fitz-Gerald, J. M. (2015). Dynamics of the inverse MAPLE nanoparticle deposition process, *Appl. Phys. A*, **119**, pp. 629–638.
64. Allmond, C., Sellinger, A., Gogick, K., and Fitz-Gerald, J. (2007). Photochemical synthesis and deposition of noble metal nanoparticles, *Appl. Phys. A*, **86**, pp. 477–480.
65. Cesaria, M., Caricato, A. P., Taurino, A., Resta, V., Belviso, M. R., Cozzoli, P. D., and Martino, M. (2015). Matrix-assisted pulsed laser evaporation deposition of Pd nanoparticles: the role of solvent, *Sci. Adv. Mater.*, **7**, pp. 1–13.

Chapter 7

Thin Films and Nanoparticles by Pulsed Laser Deposition: Wetting, Adherence, and Nanostructuring

Carmen Ristoscu and Ion N. Mihailescu

Laser Department, National Institute for Lasers, Plasma and Radiation Physics,
Atomistilor 409, Magurele, Ilfov, 00175, Romania
carmen.ristoscu@inflpr.ro, ion.mihailescu@inflpr.ro

Lasers characterized by spectral purity and spatial and temporal coherence are unique energy sources that ensure the maximum incident intensity on the surface of any sample. Ablation was used for the first time back in 1962 for material expulsion by visible lasers. The generated plasma not only controls the complex interaction phenomena between the laser radiation and various media but also can be exploited for improving laser radiation coupling and eventually the efficient processing of materials. After 1988, pulsed laser deposition emerged as a versatile and reliable technology for fabricating high-quality nanostructured layers. Laser ablation has been accordingly extensively used as an efficient tool for the preparation of nanomaterials in the form of thin films and particles.

Pulsed Laser Ablation: Advances and Applications in Nanoparticles and Nanostructuring Thin Films
Edited by Ion N. Mihailescu and Anna Paola Caricato
Copyright © 2018 Pan Stanford Publishing Pte. Ltd.
ISBN 978-981-4774-23-9 (Hardcover), 978-1-315-18523-1 (eBook)
www.panstanford.com

The wetting, adherence, and nanostructuring properties of the synthesized coatings are described and discussed in this context.

7.1 Introduction

The synthesis of materials and/or devices with new properties by controlled manipulation of their micro- or nanostructures became an emerging interdisciplinary field in solid-state physics, chemistry, biology, and materials science. Nanotechnology aims for designing, characterization, production, and application of structures, devices, and systems, controlling shape, size, and composition at nanoscale. Synthesis methods are classified after applied strategy (bottom-up or top-down approach), nature of the process (physical, chemical, biological, etc.), energy source (laser, plasma, ion sputtering, electron beam, microwave, hydrothermal, freeze drying, high-energy ball milling, combustion, flame, or supercritical) or media (gas, liquid, or solid).

The European Commission adopted on October 18, 2011, the following definition of a nanomaterial [1].

A natural, incidental or manufactured material containing particles, in an unbound state or as an aggregate or as an agglomerate and where, for 50% or more of the particles in the number size distribution, one or more external dimensions is in the size range 1–100 nm. In specific cases and where warranted by concerns for the environment, health, safety or competitiveness the number size distribution threshold of 50% may be replaced by a threshold between 1 and 50%.

Nanomaterials have accordingly structural features in between atomic and bulk [2]. This is a consequence of the low dimensional size of the materials, which entails (i) a large fraction of surface atoms, (ii) high surface energy, (iii) spatial confinement, and (iv) reduced imperfections, not existing in the corresponding bulk materials. Due to small dimensions, nanomaterials have extremely large surface-area-to-volume ratios, resulting in surface-dependent material properties.

However, it is generally accepted that nanomaterials exhibit some disadvantages. We thus mention the instability of the too-fine particles, which could act as strong explosive points due to the high surface area in direct contact with oxygen; impurity; biological harmfulness; difficulty in synthesis, storage, and application; and difficulty in recycling and disposal.

Physical or chemical vapor deposition, ion implantation, and laser technologies are mostly used to synthesize solid surfaces at a nanometer scale. Surfaces displaying improved corrosion resistance, hardness, wear resistance, or protective coatings stand for representative examples of today's technologies, where the properties of a thin layer are improved by the presence of a nanometer-size microstructure on the surface [3]. The interaction of the laser radiation with matter is the major concern for the synthesis of nanomaterials with laser beams. The use of pulsed laser deposition (PLD) for manufacturing new materials involves the appropriate selection of laser parameters in respect to the photophysical properties of the target material. The absorbed laser energy can react with the matter, initiating a thermal or/and chemical process. Commonly, the mechanisms that are initiated as a result of laser–matter interaction are classified by the dedicated literature [4] as photochemical, photothermal, and photophysical. In particular, the following aspects should be considered:

- The physical state of the material
- The type of material, that is, conductor, insulator, or semiconductor
- Laser beam parameters: wavelength, pulse width, fluence, beam diameter, and polarization
- Impurities, defects, and crystalline structure of the target material
- Relaxation, thermalization, and initial excitation times

In PLD, the absorbed energy, transformed into chemical and/or thermal energy, breaks intermolecular bonds. This process seldom results in the production of smaller flakes or particles with a wide size distribution, which is considered one of the disadvantages of the top-down method. For this reason, the bottom-up approach is considered simpler and more precise for the synthesis of nanoparticles (NPs) less than 100 nm and the top-down approach is preferred for the synthesis of thin films (TFs) and particles larger than 100 nm.

Laser ablation has been therefore extensively used as an efficient tool for the preparation of nanostructured films [5–18]. The wetting, adherence, and nanostructuring of the coatings stand for essential parameters and processes involved in NP and TF production and envisaged applications. The wetting controls the interaction between the nano-objects and the substrate, which is basically governed by adherence, while nanostructuring allows for fabrication of designed products with appropriate wettability and adherence performances. These and related phenomena represent the main subject of this chapter.

7.2 Wetting

7.2.1 Definitions

Wettability is a main property of the surface governed not only by the geometrical structure but also by the chemical composition. Surface wettability is an important property of surface engineering, characterized by a static contact angle (θ) between the liquid and solid surface at the triple-phase contact line. The *static contact angle* of liquid on a solid surface is a basic parameter to describe the solid surface wettability (Fig. 7.1) and is related to the interfacial energies via Young's equation $\gamma_{lv}\cos\theta = \gamma_{sv} - \gamma_{sl}$. Here γ_{lv}, γ_{sv}, and γ_{sl} are the surface energies corresponding to liquid–vapor, solid–vapor, and solid–liquid interfaces, respectively.

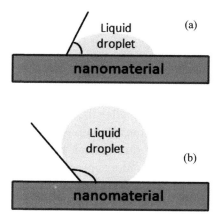

Figure 7.1 Hydrophilic (a) versus hydrophobic (b) surfaces.

Surfaces with static contact angles $\theta < 90°$ are considered to be hydrophilic and generally have good wettability properties. Surfaces with contact angles $\theta = 90°–150°$ are hydrophobic, and surfaces with $\theta > 150°$ and sliding angles less than $10°$ are considered to be superhydrophobic and are often used for liquid-repelling applications [19] or everyday garments and medical clothing [20]. On the other hand, a superhydrophilic surface generally displays a water contact angle lower than $5°$. Such two extreme cases and their corresponding wettability patterning surfaces have attracted great interest due to their importance in both theoretical research and practical applications (Fig. 7.2) [21].

Figure 7.2 Some natural cases, such as a lotus leaf, a mosquito eye, a rose petal, spider silk, butterfly wings, gecko feet, a desert beetle, and a water strider, and their potential applications for biomimetic superantiwetting surfaces, such as self-cleaning, anti-icing/antifogging, microdroplet manipulation, fog/water collection, water/oil separation, antibioadhesion, microtemplate for patterning, and friction reduction. © 2015 Huang JY, Lai YK. Published in Ref. [21] under CC BY 3.0 license. Available from: http://dx.doi.org/10.5772/60826.

When the surface morphology is uniform in different directions, such a surface is considered to be isotropic and usually the wettability of such a surface is also *isotropic*. Therefore, the apparent contact angle measured from different directions will be very similar. In engineering applications perfect isotropy is impossible to achieve. However, on anisotropic surfaces where properties of surface will change in different directions [22], the wettability of the surface is often *anisotropic*. The apparent contact angle is usually smaller in the direction perpendicular to the surface anisotropy ($\theta\perp$) and larger in the parallel direction ($\theta_{||}$), as described by Ma et al. [23].

The *surface energy* (for details see also Chapter 1 of this book) can be calculated making use of different theoretical models that take into consideration the known wetting properties of the selected materials with particular "model liquids." Accordingly, to determine the surface energy of the materials, one has to measure first the contact angles at the surfaces wetted by specific model liquids with predefined properties (in particular a well-defined surface tension) and then their polar and dispersive components. For contact angles with water measured for similar materials, Kalin and Polajnar observed a clear trend in increasing contact angles in the same order as the decreasing the surface energy [24].

Two wetting *models* developed independently by Cassie and Wenzel [25, 26] are generally employed to correlate theoretically the surface roughness with the apparent contact angles (measured on the average plane of the surface, which can be significantly different from the real contact angle on the basis of the local orientation of the surface). The contact angle θ in a noncomposite state is described by the noncomposite wetting model (*Wenzel model*) as follows: $\cos\theta = \chi_w \cos\theta_e$, where χ_w is Wenzel roughness factor and θ_e is an intrinsic angle.

As for the contact angle in a composite state, θ is described by the composite wetting model (*Cassie model*) as follows: $\cos\theta = \chi_c f \cos\theta_e - \chi_c + 1$, where χ_c is the fraction of the droplet's base area contacting with the solid, called the Cassie roughness factor. f is the ratio of the actual area to the projected area of the droplet solid contact. It generally is equal to 1.0.

On a hydrophilic rough surface, the surface is fully wetted, that is, the droplet is in the *Wenzel state*. On a hydrophobic rough surface, the surface is only partially wetted, that is, the droplet is in the *Cassie–Baxter state* [26, 27]. Fundamental research into wetting phenomena is consequently required to validate the equations that describe the apparent contact angle on hydrophilic and hydrophobic rough surfaces [28].

Wettability in general can be affected by physical roughness and chemical heterogeneity. Therefore, it is important to stress that both factors are important and physicochemical properties are to be considered especially in the analysis of anisotropic wetting. Wetting and dewetting are fundamental phenomena involved in key technological processes, such as micro- and nanofabrication by spinodal dewetting, wetting on gradient surfaces, nanolithography, induced wetting by capillary or electrocapillarity, solid lubrication, chemical or light induced wetting, painting, nano- and macroparticle-coated liquid surfaces, surface polishing on a nanolevel, flotation, gas/oil recovery, and surface self-cleansing [29]. An important part of each material is its surface. Surfaces are the first to come into contact with other materials and are also in contact with the atmosphere or environment in which the material is exposed. The interactions at the surfaces affect significantly the efficiency of the bulk material and, therefore, of the whole system [24].

7.2.2 Case Examples

Hydrophilic and hydrophobic surfaces can be investigated numerically by the Boltzmann lattice method [22]. For ideal systems the calculation of the adhesive forces is possible using analytical approaches for van der Waals and polar forces. But there is still a lack of knowledge about how to consider impacts like roughness and irregular shape of the solids during the calculation [30].

As mentioned before, wettability can be affected by physical roughness and chemical heterogeneity. Therefore, it is important to note that both factors are relevant and physicochemical properties should be considered when analyzing the anisotropic wetting. Although the wetting properties on the irregularly microstructured surface were theoretically simulated in respect to the relationship between surface fractal feather and heterogeneous nucleation [31], the results were not further verified by dedicated experiments or actual applications.

Yande Liang et al. produced/designed a thermodynamic model to demonstrate the influence of roughness on anisotropic wetting by analyzing geometry morphology of machined surfaces (2D model surfaces) [32]. They also designed experiments on machined surfaces having a wide range of roughness of hydrophilic and hydrophobic materials. The droplet's anisotropy found on machined surfaces increased with the mean slope of the roughness profile. It indicates that roughness on anisotropic wetting on hydrophilic materials has a stronger effect than that on hydrophobic materials. Furthermore, the contact angles predicted by their model were consistent with the experimentally ones, that is, the effect of roughness on anisotropic wetting on hydrophilic materials was stronger than that on hydrophobic materials, which, in its turn, is characterized by the static contact angle and droplet's distortion.

Kalin and Polajnar observed that steel and most diamond-like carbon (DLC) coatings have a similar surface energy; usually DLC coatings have reduced values [24]. Ceramics have a high surface energy because of the very large polar component. Polymers also exhibit a similarly elevated surface energy, but this is because of their high dispersive component, since they have a much lower polar component.

The effect of liquid droplet size on wetting behavior of a laser-textured SiC surface was studied [33]. The wetting state remained stable on a smooth surface, while transforming from hydrophilic to hydrophobic on a textured surface with a droplet size of 0.04 µl. It remained unchanged after approaching a droplet size of 1 µl. This phenomenon, in accordance with Ref. [33] is associated with the ratio of the base diameter and the groove width.

In Ref. [34] we made an exhaustive introduction of the wetting phenomenon and correlations between well-known Young, Cassie, and Wenzel approaches. The contact angle measurement is thoroughly explained, and relevant examples are given. A discussion on hydrophobic or hydrophilic nanostructures, with a special focus on ZnO, SiO_x, TiO_2, and DLC materials onto textile (polyester, polyamide, cotton/polyester, and poly[lactic acid]) or metallic substrates for medical purposes, is made.

A one-step PLD procedure to achieve either hydrophobic or hydrophilic ZnO structures (TFs or NPs), without any complementary postdeposition treatments of the surface, was recently proposed

[20]. Depending on the number of applied laser pulses, well-separated NPs (for 10 pulses) or compact TFs (for 100 pulses) were obtained. Changing the ambient gas nature and pressure inside the PLD chamber, hydrophilic or hydrophobic surfaces could be synthesized. The expected properties of the textiles coated with ZnO were investigated at room temperature (RT) by static contact angle measurements. For TF, a contact angle of 157° was measured, which renders to these films the attribute of superhydrophobic (Fig. 7.3). A model was proposed in order to explain the major difference of the wetting behavior for TFs and NPs. Moreover, the capability of these nanostructures to completely inhibit fungal development and neutralize bacteria was found to be a direct consequence of their wetting behavior.

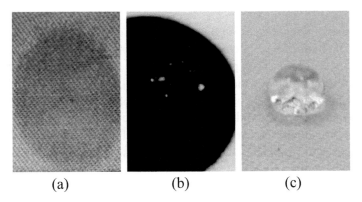

Figure 7.3 Textile material partially coated with ZnO nanostructures: (a) hydrophilic TF deposited in a 13 Pa oxygen flux, (b) hydrophobic TF deposited in vacuum, and (c) hydrophobic NP deposited in vacuum. © 2015 Duta L, Popescu AC, Zgura I, Preda N, Mihailescu IN. Published in Ref. [34] under CC BY 3.0 license. Available from: http://dx.doi.org/10.5772/60808.

We studied also the intercalation of a hydrophobin nanolayer between the substrate and a ZnO film, which can boost the oxide efficiency against microorganisms with a higher natural resistance [35]. When used alone, the hydrophobin had no effect on both *Candida albicans* colonies and six strains of filamentous fungi. In the case of simple finishing with ZnO, the reduction rate was of 50% and 70%, respectively, of the colonies in 24 h.

Apart from the usual method of measuring the contact angle, Moldovan and Enachescu proposed scanning polarization force microscopy (SPFM) as one of the most advanced noncontact techniques having nanometer spatial resolution, which can be successfully used, together with suitable theoretical models, to determine the potential energy, disjoining pressure, and spreading coefficient from the dependence of the contact angle on droplet height, for micro- and nanodroplets [36]. Accordingly, the contact angle was directly measured from the acquired topography images of the droplets. SPFM overcomes the difficulties inherent in classical atomic force microscopy (AFM) techniques (e.g., difficulty to maintain a stable feedback on liquid surfaces) and offers a direct way for the measurement of the microscopic contact angle.

7.3 Adherence

The use of nanostructured films extends to electronic, engineering, optical, biomedical, nuclear, and space applications. The degree of adherence is a key parameter because it is closely related to interatomic and intermolecular forces [37]. TFs (generally below 1 µm and occasionally of the order of tens of nanometers) could be fragile, and, consequently, they need to be supported by a substrate. Next, the adherence determines the durability and longevity of TF devices and governs the growth's kinetics. The wear properties are also connected to the adherence of layers. Protective coatings are vital for improving corrosion resistance, especially in the case of medical implants. In a PLD process, the adherence is influenced by substrate type, cleaning and temperature, deposition rate, target material, and gas nature and pressure. Moreover, studies have proved that the adherence is strongly correlated to wettability, described in the previous section.

7.3.1 Basic Mechanisms

The recent literature on adherence properties contains studies of three main adherence mechanisms: mechanical coupling, molecular bonding, and thermodynamic adhesion [38]. The mechanical coupling, or interlocking (hook-and-eye) adhesion mechanism, is based on the adhesive keying into the surface of the substrate. This is similar to glue on wood, meaning that the glue goes into the rough irregularities on the surface of the wood (Fig. 7.4) [39].

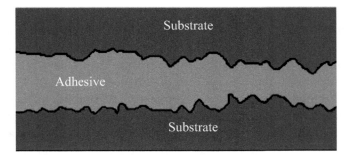

Figure 7.4 Illustration of mechanical coupling between two substrates. Reprinted from Ref. [38], Copyright (2009), with permission from Elsevier.

Molecular bonding is the most widely accepted mechanism for explaining adhesion between two surfaces in close contact. It involves intermolecular forces between adhesive and substrate, that is, dipole–dipole interactions, van der Waals forces, and chemical interactions (i.e., ionic, covalent, and metallic bonding). This mechanism describes the strength of the adhesive joints by interfacial forces and also by the presence of polar groups [40].

Significant research has been reported taking into account the thermodynamic adsorption model of adhesion. The unique advantage of the thermodynamic mechanism over the other mechanisms is that it does not require molecular interaction for good adhesion but only an equilibrium process at the interface [41].

7.3.2 Investigations and Examples

Any consideration of adherence mechanisms requires information about the physical and chemical properties of the adhering surfaces and the delamination surfaces in cases where adhesion has failed in use or as a result of mechanical testing. There are a number of surface characterization techniques utilized for investigating properties related to adhesion mechanisms and adhesion strength. These include time-of-flight secondary ion mass spectrometry (ToF-SIMS), X-ray photoelectron spectroscopy (XPS), AFM, secondary electron microscopy (SEM), attenuated total reflectance infrared (ATR-IR) spectroscopy, and other microscopy techniques, plus methods sensitive to surface energy, such as optical contact angle analysis. There have been numerous studies that have looked at surface properties, such as roughness, polarity, chemical composition, and surface free energy, to describe and explain adhesion phenomena at a surface or interface using the above-mentioned techniques. A pull-off test was carried out to prove and measure the tensile strength of the studied materials.

The bonding strength of calcium phosphate coatings on metallic substrates can be evaluated using several techniques, such as the standard tensile adhesion test [42], interfacial indentation test [43], tensile adhesion strength (TAS) [44], and indentation method [45]. However, there are limitations of these techniques to accurately measuring the adhesion strength, such as a probability of penetration of glue into the coating layer and a dependence of coating failure on the flaw distribution at the edge of specimen [42]. The hydroxyapatite (HA) coatings with the densest structure (i.e., the lowest porosity and a predominantly amorphous phase) have a higher TAS than those of lower density [44, 46].

In Ref. [47] we synthesized novel ovine (SHA)- and bovine (BHA)-derived HA TFs on titanium substrates by PLD for a new generation of implants. We chose these new coating materials as they could be considered as a prospective competitor to synthetic HA used for implantology applications.

The bonding strength (tensile) at the biofunctional film-substrate interface is considered a critical factor in fabricating high-quality implants and long-term stability of these devices in situ [48–50]. Both adhesive fracture (at the interface), in the center of the tested area, and cohesive fracture (near the interface or in the film volume), at the tested area edges, were involved in the film mechanical failure.

The adherence values recorded for the HA films (43 ± 7 MPa) are generally similar to the ones often reported in literature for these type of PLD films [51]. Significantly higher pull-out adherence values have been obtained in the case of SHA (66 ± 8 MPa) and BHA:Li (75 ± 4 MPa) (Table 7.1). The decreased value of adherence recorded for the BHA:CIG (49 ± 10 MPa) structure could be attributed to the intrinsic friability of the glassy doping phase under mechanical stress.

We note that solely pull-out adherence values of the order of 40 MPa are considered acceptable for these types of implant coatings [50, 52, 53]. Therefore, the excellent value of adherence obtained for BHA:Li films should be emphasized. The increased values of adherence obtained for BHA films could be explained in terms of intentional (Li) and unintentional (Na, Mg, and F) doping of these structures. Cai et al. [54] have observed similar effects with Mg and F doping.

Table 7.1 Pull-out bonding strength values for the HA films prepared by PLD

Sample	Pull-out adherence value (MPa)
SHA	66 ± 8
BHA:Li	75 ± 4
BHA:CIG	49 ± 10
HA films	43 ± 7

We consider that these improved performances should allow for obtaining implant coatings with a shorter osseointegration time and better osseoconductive characteristics.

In Ref. [55] is reported the synthesis of 700 nm thick bioglass (BG) TFs by magnetron sputtering (MS) from 45S5 BG targets in the argon atmosphere. For some samples of further denoted BGG, we introduced a ~70 nm thick mixed glass–Ti buffer layer ($BG_{1-x}Ti_x$ [x = 0 – 1]) with a gradient of composition by cosputtering. For bonding strength comparison, we synthesized by radio frequency–magnetron sputtering (RF-MS), under identical experimental conditions, 700 nm BG/Ti "abrupt coatings" (BGA) without the intermediate buffer layer between the substrate and the film.

As known, the improvement of the mechanical and biological properties of films can be achieved by the transformation of BG into glass ceramics via heat treatments [56, 57]. We have chosen for our study an annealing temperature of 650°C in order to induce the partial crystallization of the combeite ($Na_2CaSi_2O_6$) phase.

As most implants have complex geometries, which imply the involvement of the biomechanical forces (shear, tension, etc.) and bending moments, a straightforward method for evaluating the BG/Ti interface behavior at tensile forces is the pull-out test. For the BGA structure, the bonding failure occurred at a mean value of 29 ± 7 MPa. This is a rather low bonding strength value but similar to those reported in literature [58, 59]. This effect is mainly due to the significant difference between the thermal expansion coefficients of the BG film (~17 × 10^{-6}/°C) and the titanium alloy substrate (9.2–9.6 × 10^{-6}/°C), at high temperatures [58, 60, 61].

A 1.7 times higher bonding strength (50 ± 6 MPa) was obtained in the case of the BGG structure. This result confirms that the co-sputtering of BG structures with graded buffer layers is a proper solution for preparing adherent BG coatings.

The osteoblasts spread over the surface, adopting typical polyhedral shapes with numerous focal adhesion points and protrusions infiltrating deep into the films. An ideal coating should provide a proper mechanical support while exhibiting an enhanced bioactivity. We consider, therefore, that the combeite and wollastonite phases forming after thermal treatment played in our case a double positive role: they are highly bioactive and provide a strong mechanical support quite suitable for implant and prosthesis applications.

7.4 Nanostructuring

7.4.1 Definitions

Nanostructured materials have attracted an increased interest in previous years due to a remarkable combination of mechanical, electrical, optical, and magnetic properties. Some prominent examples follow, as also mentioned in Ref. [62]:

- Nanophase ceramics are of special interest because they exhibit large ductility at elevated temperatures in respect to the coarse-grained ceramics.
- Gas tight materials, dense parts, or porous coatings were fabricated from nanosized metallic powders. Cold welding properties, together with the ductility, make them suitable for metal–metal bonding, mainly in the electronics industry.
- Nanostructured semiconductors demonstrate a variety of nonlinear optical properties. Semiconductor Q-particles also present quantum confinement effects, which may result is particular properties, like luminescence in silicon powders and silicon germanium quantum dots exploited in infrared optoelectronic devices. One should mention that nanostructured semiconductors are also used as window layers in solar cells.

- One should stress that single nanosized magnetic particles are in fact monodomains. Therefore, in magnetic nanophase materials the grains correspond to domains, while boundaries correspond with disordered walls. Small particles have individual atomic structures with discrete electronic states, responsible for particular properties in addition to the superparamagnetism behavior. Magnetic nanocomposites were applied in ferrofluids, high-density information storage, and magnetic refrigeration.
- Nanostructured metal clusters and colloids of mono- or plurimetallic composition have a large potential for utilization in catalytic applications. They may act as precursors for new types of heterogeneous catalysts and have been shown to provide substantial advantages in chemical transformations and electrocatalysis in respect with activity, selectivity, and lifetime.
- Nanostructured metal-oxide TFs are recognized in the manufacturing of gas sensors (volatile compounds and aromatic hydrocarbons) with superior sensitivity and selectivity. They were, therefore, applied for rechargeable batteries' fabrication to be used in cars or customer goods.
- Polymer-based composites with a high content of inorganic particles, resulting in a large dielectric constant, are interesting materials for the next generation of photonic bandgap structures.

7.4.2 Imaging of Nanostructures

7.4.2.1 Conventional imaging

The simplest diagnostic of a nanostructure is by imaging. The most used methods are SEM, AFM, and transmission electron microscopy (TEM). They evolved/advanced following the need to better visualize/optically characterize the produced nanostructures.

AFM is used for morphological examination of organic, biological, and inorganic surfaces, providing topographical information with a minimal preparation of the sample. Special attention has to be paid to the sample inclination under microscope. Its vibration is reflected in the poor quality of the recorded image. Depending on the working mode, AFM could prove a destructive technique, especially when in contact mode.

Resolutions of 2 nm and of 0.27 nm are currently available for SEM and TEM instruments, respectively. In both cases, a preliminary preparation of the samples is required. In TEM, the samples need special preparation: milling, thinning, and the right positioning under the microscope. The nonconductive probes have to be metalized in order to obtain SEM images with an improved contrast.

7.4.2.2 Differential evanescent light intensity imaging

The modeling of photon extraction from traveling waves in nanostructured materials is not yet accurately described. The differential evanescent light intensity (DELI) imaging method, which offers an excellent background along with a good spatial contract for very small objects near the surface structures, was proposed [63–70]. DELI has the unique advantage that the electromagnetic field that propagates into the waveguide does not interfere with the extracted field orthogonal to surface. Accordingly, DELI is an optical microscopy method based on the capture of the propagating field redirected by the nanolayer from the evanescent fields in optical waveguides.

Experimentally, the light intensity pattern extracted from the deposited particles at the waveguide surface from a light beam propagating through the transparent substrate reaches a charge-coupled device (CCD) camera and is captured as 2D images. The physical principle is similar to the one of scanning tunneling optical microscope (STOM), where generated evanescent waves illuminate the sample with a light beam, exhibiting an angle of incidence larger than the critical angle for total internal reflection. DELI proved easier than AFM and SEM and is suitable for evaluating large areas for its nanometer profile.

In our DELI phenomenological analysis, we assumed that the evanescent field produced in air in the close neighborhood of the waveguide is perturbed by the nanostructured deposited material molecules. The molecules get polarized in the presence of the electric component of the evanescent field, and the dipole emission model can be applied [71]. Our proposed DELI photon extraction model allows us to obtain a relation between the nanolayer thickness, h, and the normalized integrated optical density (IOD):

$$IOD_2(h_2)/IOD_1(h_1) \approx h_2/h_1 \qquad (7.1)$$

As a result, by calibrating h_1 we can estimate the absolute thickness h_2 of any other layer. h_1 can be inferred by SEM, AFM, profilometry, or other independent techniques.

As the thickness is directly related to the photon extraction efficiency, γ, we can estimate also this parameter.

DELI was applied to measure the thickness of PLD films of ZnO [69] or TiO_2 [70]. Using the photon extraction model and depending on the number of laser pulses and substrate temperature applied for the deposition of ZnO films, thicknesses in the range 1–105 nm were inferred (Table 7.2).

Table 7.2 Thickness of ZnO films versus the number of applied laser pulses and substrate temperature

Number of laser pulses	Thickness at 27°C (nm)	Thickness at 100°C (nm)
100	5.5	1
200	9.7	3.7
400	7.3	14.4
600	23.3	17.9
800	93.3	25.1
1000	105	100

Accordingly, the values for γ were of 0.01235 nm^{-1} for films deposited at 27°C and of 0.01146 nm^{-1} for films at 100°C, respectively.

In the same manner, for TiO_2 films thickness measurements in the range of 3–100 nm are collected in Table 7.3.

Table 7.3 Thickness of TiO_2 films versus the number of applied laser pulses and substrate temperature

Number of laser pulses	Thickness at 27°C (nm)	Thickness at 100°C (nm)
100	3	23
200	6	34
400	15	86
1000	100	100

The photon extraction parameter, γ, was 0.03183 nm^{-1} at 27°C and 0.01121 nm^{-1} at 100°C.

This evanescent light method can be applied also for photodeposited nanostructures (a-Se) [63, 64], sputtered a-Au layers [68], and polyethylene nanolayers obtained by matrix-assisted pulsed laser evaporation (MAPLE) [65–67].

7.4.3 Nanostructuring with Advanced PLD Techniques

We studied the effect of temporally shaped pulse upon the properties of aluminum nitride thin layers synthesized by PLD with 200 fs laser pulses generated by a Ti-sapphire laser source [72]. We showed that the film morphology and structure can be gradually modified when applying monopulses of different duration (AlN-1) or passing to a sequence of two pulses of different intensities (AlN-2 and AlN-3).

For AlN-1 samples (Fig. 7.5a), three classes of surface particulates could be distinguished: particulates smaller than 100 nm; medium-size particulates, up to 1 µm; and large particulates, up to 2 µm. The large particulates were rather rare. The typical surfaces of the AlN-2 film (Fig. 7.5b) also showed large crystallites, ranging up to 1.5 µm, with a rather high density. In the case of the AlN-3 samples (Fig. 7.5c), the particulates could be grouped into three classes according to their average size: particulates around 100 nm; particulates around 500 nm; and particulates larger than 1.5 µm, up to 2.5 µm. The large particulates showed well-defined facets. When using temporally shaped pulses, the amount of metallic Al increased in films; even the total energy was identical in all cases. The shaping was accompanied by the appearance of an amorphous phase and of few traces of crystalline cubic AlN.

The measured average particulate density was rather similar in the three cases, namely (5 ± 0.8) × 10^8 cm^{-2} for the AlN-1 samples, (4.8 ± 0.7) × 10^8 cm^{-2} for the AlN-2 samples, and (5.6 ± 0.8) × 10^8 cm^{-2} for the AlN-3 samples, with an about 15% counting error in each case. The particulate average size resulting from the histogram analysis and curve fitting was 390 ± 5 nm in the case of AlN-1 samples, 230 ± 3 nm for AlN-2 samples, and 310 ± 4 nm in the case of AlN-3 samples.

This nanostructuring technique was applied also for SiC films. The SiC film deposited with Ti-sapphire common pulses shows a high density of droplets (in average 62 μm^{-2}), reaching up to 400 nm in size [73] (Fig. 7.6). Comparatively, a striking reduction of the droplets' density can be observed for the film obtained with time-tailored laser pulses, where an average droplet density of 8.6 μm^{-2} has been determined from the SEM images. Also, in this latter case, a significant decrease of the droplet size is evident, the largest of them being smaller than 200 nm.

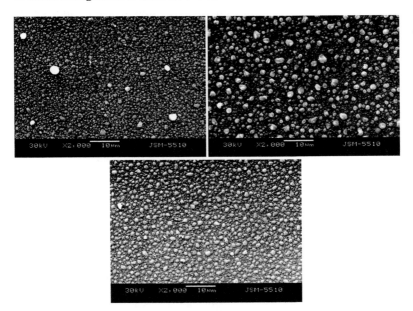

Figure 7.5 SEM images showing the surface morphology of samples AlN-1 (a), AlN-2 (b), and AlN-3 (c).

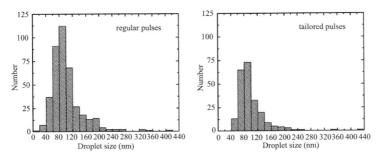

Figure 7.6 Histograms recorded on SiC films obtained with regular or shaped fs pulses.

The material morphology and structure can be gradually modified when applying the shaping of the ultrashort fs laser pulses into two successive pulses under the same temporal envelope as the initial laser pulse or temporally shaped pulse trains with picosecond separation (monopulses of different duration or a sequence of two pulses of different intensities) [74]. By optimization of the temporal shaping of the pulses besides the other laser parameters (wavelength, energy, beam homogeneity, and fluence), one could choose an appropriate regime to eliminate excessive photothermal and photomechanical effects and obtain films with the desired crystalline phase and number and dimension of grains/particulates or controlled porosity.

In the search for optimal coatings exhibiting the desired mechanical, structural, electrical, tribological, or biological properties, material libraries have been deposited by the *combinatorial-PLD* (*c-PLD*) technique [75, 76]. c-PLD was applied to synthesize on titanium implants Ag-doped C (C:Ag) films for a new generation of multifunctional coated implants with increased resistance to microbial colonization and high biocompatibility.

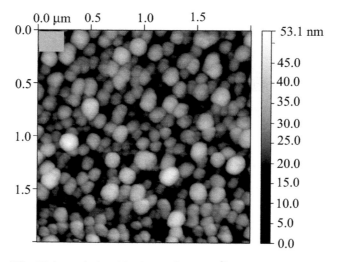

Figure 7.7 High-resolution AFM image (2 × 2 µm²).

The C-rich areas are mainly composed of fine nanometric grains (with a diameter ranging from 20 to 30 nm). Average roughness (RMS) values lower than ~5 and ~9 nm are observed. When the doping Ag concentration is increasing, the surface topology becomes rougher. It consists of a homogeneous envelope of closely packed submicron particles (0.12–0.2 μm), as evidenced by high-resolution AFM analysis (Fig. 7.7).

7.4.4 Applications

In the following we present some examples of nanomaterials in the form of metallic NPs, grown by PLD, for surface-enhanced Raman spectroscopy (SERS) applications, metal-oxide TFs with potential applications in gas sensing, nanostructured metal oxides as electrocatalysts, and ceramics.

7.4.4.1 Metal oxides for gas sensing

Stankova et al. obtained, by PLD, nanostructured optical waveguide films of TiO_2 and WO_{3-x}, for sensing applications, having high transparence and root mean square (RMS) roughness values between 4 and 8 nm [77]. Cross-sectional SEM images presented columnar growth morphology with well-resolved nanocolumns having diameters of approximately up to 100 nm. By m-line spectroscopy, WO_{3-x} and TiO_2 layers revealed high selectivity and sensitivity to 0.1% NH_3 and to 0.3% CO_2, respectively. The sensitivity decreased 1 order of magnitude when the gases were diluted in N_2. The testing system reversibility, response, and reproducibility and the fast response and recover time are the main advantages of the gas-sensing layers observed at RT operation conditions.

We reported on new features of nanostructured coatings used in trace gas detection applications based on the modification of optical parameters [78].

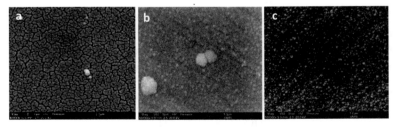

Figure 7.8 SEM microimages of Pd:SnO_2 structures deposited in 10 Pa O_2 at (a) RT, (b) 350°C, and (c) 500°C.

Figure 7.8 shows the morphologies for the three types of Pd:SnO$_2$ samples deposited in 10 Pa O$_2$ at (Fig. 7.8a) RT, (Fig. 7.8b) 350°C, and (Fig. 7.8c) 500°C. The sample deposited at RT exhibited a rough surface with cracks, while the two others were rather smooth and densely packed. A quasi-ordered nanostructure can be noticed in the sample deposited at 500°C.

AFM studies (Fig. 7.9) sustained these observations. The surface of the sample deposited at RT was porous, with large particles (>100 nm). By contrast, samples deposited at 350°C and 500°C were denser, and their particle dimensions were down to ~100 nm and ~80 nm, respectively. An even slight increase in the SnO$_2$ grain size can lead to significant changes in the film structure, which is responsible for both the catalytic and gas-sensing properties of metal oxides [79].

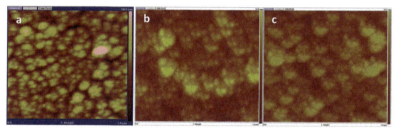

Figure 7.9 AFM microimages of Pd:SnO$_2$ structures deposited in 10 Pa O$_2$ at (a) RT, (b) 350°C, and (c) 500°C.

Twin films deposited on quartz were investigated using the m-line technique to monitor the variation of optical properties when exposed to butane diluted in N$_2$ down to 100 ppm concentration. Our results demonstrated that the gas-sensing properties were mainly a function of the structure and surface morphology of the samples, which in turn depended on substrate temperature. Moreover, it was evident that the higher the substrate temperature during PLD, the higher the gas sensitivity of the obtained sensors. Our investigations did not evidence within measurement limits any remnant contamination of the sensing elements.

7.4.4.2 Fuel cell elements

Wang et al. prepared TiO$_2$ TFs with various morphologies by PLD at RT onto a conductive microfibrous carbon paper substrate [80]. Carbon paper is the type of substrate employed in energy storage and conversion devices. TiO$_2$ films deposited under vacuum and in the presence of a mild pressure of oxygen are very smooth and dense. Instead, TiO$_2$ films deposited in the presence of a helium atmosphere are porous structures and vertically aligned. An increase in the helium pressure leads to the formation of forest-like vertically aligned nanostructures. They observed that regardless of the film morphology, all TiO$_2$ films (amorphous rutile phase) demonstrated excellent catalyst-supporting properties to Pt, chosen as a model catalyst. Laser-synthesized TiO$_2$ films display enhanced catalytic activity as compared to bare Pt.

The impact of a nanostructured Ni-yttria-stabilized zirconia (Ni-YSZ) anode on low-temperature solid oxide fuel cell (LT-SOFC) performance is investigated in [81]. Anode-supported SOFCs based on TF (~1 µm) electrolytes (TF-SOFCs) with and without the nanostructured Ni-YSZ (grain size = ~100 nm) anode were fabricated. The cell performance at low temperatures of the nanostructured Ni-YSZ anode significantly increased as compared to that of the cell without it.

7.4.4.3 Nanoparticles for SERS

For NPs of noble metals, such as Au, Ag, and Cu, the strong surface plasmon resonances (SPR) are in UV–vis range. Associated to the SPR of such individual nanostructures there is a strong amplification of a local electromagnetic field near the corrugated metal surface. This amplification induces the so-called electromagnetic enhancement of the Raman scattering signal, and with chemical enhancement mechanism, it is operative in SERS [82–84].

A rough silver surface was obtained by ultrashort PLD using a Nd:glass laser (λ = 527 nm) delivering pulses of 250 fs [85]. The authors observed that SERS activity results, investigated by Raman scattering of rhodamine B adsorbed onto rough silver substrates, are strongly related to the effective surface area (ESA) of the silver surfaces. Moreover, composition, velocity, excitation temperature, and density of short and ultrashort plasmas are quite similar, showing that film properties are strongly related to the plasma features.

We also detected traces of toluene Nd SERS [86] when investigating/analyzing niobium rods chemically modified by fs laser pulses. As a result, the formation of Nb oxides on the surface, as well as of Nb hydrides, was observed. After ethanol cleaning, a strong SERS signal originating from the toluene residual traces was evidenced. We hypothesized that the islands on the Nb surface act as hot spots for toluene molecules, leading to a boosted SERS effect, while the rippled structure of the surrounding dark material does not contribute significantly to enhanced scattering. Further, it was observed that a laser-irradiated Nb surface is able to provide a SERS enhancement of $\sim 1.3 \times 10^3$ times for rhodamine 6G solutions. Thus, for the first time it was shown that a Nb/Nb oxide surface could exhibit SERS functionality with potential applications in biological/biochemical screening or for sensing of dangerous environmental substances.

Nedyalkov et al. present the fabrication of nanostructured Au films by PLD in air at atmospheric pressure using a Nd:YAG (Lotis) laser system operated at 10 Hz, at third harmonic, at wavelength of λ = 355 nm and pulse duration of 15 ns [87]. The layer consists of microsized droplets on a nanostructured film, composed of aggregates of NPs a few tens of nanometers in size. The deposited film has a porous structure that shows characteristic optical spectra related to the excitation of plasmons and can be directly used in practical applications as SERS and sensor devices.

7.5 Conclusions

PLD proved to be a reliable prospective alternative for fabricating TFs and NPs for key top technologies, including biomedicine. The new special commandments allowed the production of performant nanostructures that surpass many of the up-to-date limits. They can be manufactured from low-cost renewable resources that can replace the present materials. In this context, there were formulated some mandatory requirements for TF and NP wetting in conjunction with adherence and nanostructuring level.

Presently, there are well-established technologies to synthesize, characterize, and eventually use the newly designed PLD nanostructures with tailored wettability, adherence, and degree of nanostructuring. The field is in continuous expansion, and herewith significant references were reviewed.

Acknowledgments

The authors acknowledge the support of UEFISCDI under the contracts ID304/2011 and PCCA 244/2014.

References

1. https://en.wikipedia.org/wiki/Nanomaterials#cite_note-21
2. https://nccr.iitm.ac.in/2011.pdf
3. Gleiter, H. (2000). Nanostructured materials: basic concepts and microstructure, *Acta Mater.*, **48**(1), pp. 1–29
4. Habiba, K., Makarov, V. I., Weiner, B. R., and Morel, G. (2010). Fabrication of nanomaterials by pulsed laser synthesis, in *Manufacturing Nanostructures*, eds. Ahmed, W., and Ali, N. (One Central Press), pp. 263–292.
5. Smith, H. M., and Turner, A. F. (1965). Vacuum deposited thin films using a ruby laser, *Appl. Opt.*, **4**(1), pp. 147–148.
6. Miller, J. C. (1994). *Laser Ablation Principles and Applications* (Springer-Verlag, Berlin).
7. Chrisey, D. B., and Hubler, G. K. (1994). *Pulsed Laser Deposition of Thin Films* (Wiley-Interscience).
8. Von Allmen, M., and Blatter, A. (1995). *Laser-Beam Interactions with Materials*, 2nd ed. (Springer, Berlin, Heidelberg, New York, Barcelona, Budapest, Hong Kong, London, Milan, Paris, Tokyo).
9. Belouet, C. (1996). Thin film growth by the pulsed laser assisted deposition technique, *Appl. Surf. Sci.*, **96–98**, pp. 630–642
10. Miller, J. C., and Haglund, R. F. (1998). *Laser Ablation and Desorption* (Academic Press).
11. Eason, R. (2006). *Pulsed Laser Deposition of Thin Films: Applications-Led Growth of Functional Materials* (Wiley-Interscience).
12. Bäuerle, D. W. (2011). *Laser Processing and Chemistry*, 4th ed. (Spinger).
13. Mihailescu, I. N., and Gyorgy, E. (1999). Pulsed laser deposition: an overview, in 4th International Commission for Optics (ICO) Book *International Trends in Optics and Photonics*, ed. Asakura, T. (ICO President), pp. 201–214.
14. Mihailescu, I. N., and Hermann, J. (2010). Laser plasma interactions, in *Laser Processing of Materials: Fundamentals, Applications, and Developments*, ed. Schaaf, P., Springer Series in Materials Science (Springer, Heidelberg, Dordrecht, London, New York), pp. 51–90.

15. Mihailescu, I. N., Ristoscu, C., Bigi, A., and Mayer, I. (2010). Advanced biomimetic implants based on nanostructured coatings synthesized by pulsed laser technologies, in *Laser-Surface Interactions for New Materials Production: Tailoring Structure and Properties*, eds. Miotello, A., and Ossi, P. M., Springer Series in Materials Science (Springer), Vol. 130, pp. 235–260.

16. Ristoscu, C., and Mihailescu, I. N. (2013). Biomimetic coatings by pulsed laser deposition, in *Laser Technology in Biomimetics: Basics and Applications*, eds. Belegratis, M., and Schmidt, V., Springer Series in Biological and Medical Physics, Biomedical Engineering (Springer-Verlag, Berlin, Heidelberg), pp. 163–191.

17. Popescu, A. C., Ulmeanu, M., Ristoscu, C., and Mihailescu, I. N. (2015). Deposition and surface modification of thin solid structures by high intensity pulsed laser irradiation, in *Laser Surface Engineering: Processes and Applications*, eds. Lawrence, J., and Waugh, D., Woodhead Publishing Series in Electronic and Optical Materials (Woodhead), pp. 287–313.

18. Sima, F., Ristoscu, C., Duta, L., Gallet, O., Anselme, K., and Mihailescu, I. N. (2016). Laser thin films deposition and characterization for biomedical applications, in *Laser Surface Modification of Biomaterials, Techniques and Applications*, 1st ed., ed. Vilar, R. (Woodhead), pp. 77–125.

19. Chen, F., Zhang, D., Yang, Q., Wang, X., Dai, B., Li, X., Hao, X., Ding, Y., Si, J., and Hou, X. (2011). Anisotropic wetting on microstrips surface fabricated by femtosecond laser, *Langmuir*, **27**(1), pp. 359–365.

20. Popescu, A. C., Dorcioman, G., Duta, L., Mihailescu, I. N., Stan, G. E., Pasuk, I., Zgura, I., Beica, T., Enculescu, I., Ianculescu, A., and Dumitrescu, I. (2011). Radical modification of the wetting behavior of textiles coated with ZnO thin films and nanoparticles when changing the ambient pressure in the pulsed laser deposition process, *J. Appl. Phys.*, **110**(6), p. 064321.

21. Huang, J.-Y., and Lai, Y.-K. (2015). TiO$_2$-based surfaces with special wettability: from nature to biomimetic application, in *Wetting and Wettability*, ed. Aliofkhazraei, M. (InTech), pp. 47–84.

22. Kubiak, K. J., and Mathia, T. G. (2014). Anisotropic wetting of hydrophobic and hydrophilic surfaces: modelling by Lattice Boltzmann method, *Procedia Eng.*, **79**, pp. 45–48.

23. Ma, C., Bai, S., Peng, X., and Meng, Y. (2013). Anisotropic wettability of laser micro-grooved SiC surfaces, *Appl. Surf. Sci.*, **284**, pp. 930–935.

24. Kalin, M., and Polajnar, M. (2014). The wetting of steel, DLC coatings, ceramics and polymers with oils and water: the importance and correlations of surface energy, surface tension, contact angle and spreading, *Appl. Surf. Sci.*, **293**, pp. 97–108.
25. Cassie, A. B. D. (1948). Contact angles, *Discuss. Faraday Soc.*, **3**, pp. 11–16.
26. Wenzel, R. N. (1936). Resistance of solid surfaces to wetting by water, *J. Phys. Chem.*, **28**, pp. 988–994.
27. Cassie, A., and Baxter, S. (1944). Wettability of porous surfaces, *Trans. Faraday Sci.*, **30**, pp. 546–551.
28. Yu, D. I., Kwak, H. J., Doh, S. W., Kang, H. C., Ahn, H. S., Kiyofumi, M., Park, H. S., and Kim, M. H. (2015). Wetting and evaporation phenomena of water droplets on textured Surfaces, *Int. J. Heat Mass Transfer*, **90**, pp. 191–200.
29. Nikolov, A., and Wasan, D. (2014). Wetting–dewetting films: the role of structural forces, *Adv. Colloid Interface Sci.*, **206**, pp. 207–221.
30. Fritzsche, J., and Peuker, U. A. (2015). Wetting and adhesive forces on rough surfaces: an experimental and theoretical study, *Procedia Eng.*, **102**, pp. 45–53.
31. Wang, M., Zhang, Y. Zheng, H. Y., Lin, X., and Huang, W. D. (2012). Investigation of the heterogeneous nucleation on fractal surfaces, *J. Mater. Sci. Technol.*, **28**, pp. 1169–1174.
32. Liang, Y., Shu, L., Natsu, W., and He, F. (2015). Anisotropic wetting characteristics versus roughness on machined surfaces of hydrophilic and hydrophobic materials, *Appl. Surf. Sci.*, **331**, pp. 41–49.
33. Wang, R., and Bai, S. (2015). Effect of droplet size on wetting behavior on laser textured SiC surface, *Appl. Surf. Sci.*, **353**, pp. 564–567.
34. Duta, L., Popescu, A. C., Zgura, I., Preda N., and Mihailescu, I. N. (2015). Wettability of nanostructured surfaces, in *Wetting and Wettability*, ed. Aliofkhazraei, M. (InTech), pp. 207–252.
35. Popescu, A. C., Stan, G. E., Duta, L., Dorcioman, G., Iordache, O., Dumitrescu, I., Pasuk, I., and Mihailescu, I. N. (2013). Influence of a hydrophobin underlayer on the structuring and antimicrobial properties of ZnO films, *J. Mater. Sci.*, **48**(23), pp. 8329–8336.
36. Moldovan, A., and Enachescu, M. (2015). Wetting properties at nanometer scale, in *Wetting and Wettability*, ed. Aliofkhazraei, M. (InTech), pp. 15–45.
37. Mittal, K. L. (1976). Adhesion measurement of thin films, *Electrocomponent Sci. Technol.*, **3**, pp. 21–42.

38. Awaja, F., Gilbert, M., Kelly, G., Fox, B., and Pigram, P. J. (2009). Adhesion of polymers, *Prog. Polym. Sci.*, **34**, pp. 948–968.
39. Wake, W. C. (1982). *Adhesion and the Formulation of Adhesives*, 2nd ed. (Applied Science, Essex).
40. Sharpe, L. H. (1993). In *The Interfacial Interactions in Polymeric Composites*, ed, Akovali, G., NATO ASI Series (Kluwer Press, Dordrecht), Vol. 230, pp. 1–20.
41. Lipatov, Yu. S. (1995). *Polymer Reinforcement* (ChemTec, Toronto).
42. Tsui, Y., Doyle, C., and Clyne, T. (1998). Plasma sprayed hydroxyapatite coatings on titanium substrates. Part 1: mechanical properties and residual stress levels, *Biomaterials*, **19**, pp. 2015–2030.
43. Lo, W., Grant, D. M., Ball, M. D., Welsh, B. S., Howdle, S. M., Antonov, E. N., Bagratashvili, V. N., and Popov, V. K. (2000). Physical, chemical, and biological characterization of pulsed laser deposited and plasma sputtered hydroxyapatite thin films on titanium alloy, *J. Biomed. Mater. Res.*, **50**, pp. 536–545.
44. Yang, Y.-C., and Chang, E. (2001). Influence of residual stress on bonding strength and fracture of plasma-sprayed hydroxyapatite coatings on Ti-6Al-4V substrate, *Biomaterials*, **22**, pp. 1827–1836.
45. Takeuchi, S., Ito, M., and Takeda, K. (1990). Modelling of residual stress in plasma-sprayed coatings: effect of substrate temperature, *Surf. Coat. Technol.*, **43**, pp. 426–435.
46. Kweh, S., Khor, K., and Cheang, P. (2002). An in vitro investigation of plasma sprayed hydroxyapatite (HA) coatings produced with flame-spheroidized feedstock, *Biomaterials*, **23**, pp. 775–785.
47. Duta, L., Oktar, F. N., Stan, G. E., Popescu-Pelin, G., Serban, N., Luculescu, C., and Mihailescu, I. N. (2013). Novel doped hydroxyapatite thin films obtained by pulsed laser deposition, *Appl. Surf. Sci.*, **265**, pp. 41–49.
48. Sima, L. E., Stan, G. E., Morosanu, C. O., Melinescu, A., Ianculescu, A., Melinte, R., Neamtu, J., and Petrescu, S. M. (2010). Differentiation of mesenchymal stem cells onto highly adherent radio frequency-sputtered carbonated hydroxylapatite thin films, *J. Biomed. Mater. Res. A*, **95A**, pp. 1203–1214.
49. Stan, G. E., Pasuk, I., Husanu, M. A., Enculescu, I., Pina, S., Lemos, A. F., Tulyaganov, D. U., Mabrouk, K. E. L., and Ferreira, J. M. F. (2011). Highly adherent bioactive glass thin films synthesized by magnetron sputtering at low temperature, *J. Mater. Sci. – Mater. Med.*, **22**, pp. 2693–2710.

50. Draft International Standards [ISO/DIS] (1999). Implants for surgery – hydroxyapatite ceramic: part 1 and 2, p. 13779.
51. Bao, Q., Chen, C., Wang, D., Ji, Q., and Lei, T. (2005). Pulsed laser deposition and its current research status in preparing hydroxyapatite thin films, *Appl. Surf. Sci.*, **252**, pp. 1538–1544.
52. American Society for Testing and Materials [ASTM] (2009). Standard specification for composition of ceramic hydroxylapatite for surgical implants, F 1185-03, pp. 514–515.
53. Food and Drug Administration [FDA] (1997). Calcium phosphate (Ca–P) coating draft guidance for preparation of FDA submissions for orthopedic and dental endooseous implants, pp. 1–14.
54. Cai, Y., Zhang, S., Zeng, X., Qian, M., Sun, D., and Weng, W. (2011). Interfacial study of magnesium-containing fluoridated hydroxyapatite coatings, *Thin Solid Films*, **519**, pp. 4629–4633.
55. Stan, G. E., Popescu, A. C., Mihailescu, I. N., Marcov, D. A., Mustata, R. C., Sima, L. E., Petrescu, S. M., Ianculescu, A., Trusca, R., and Morosanu, C. O. (2010). On the bioactivity of adherent bioglass thin films synthesized by magnetron sputtering techniques, *Thin Solid Films*, **518**, pp. 5955–5964.
56. Filho, O. P., LaTorre, G. P., and Hench L. L. (1996). Effect of crystallization on apatite-layer formation of bioactive glass 45S5, *J. Biomed. Mater. Res.*, **30**, pp. 509–514.
57. El Batal, H. A., Azooz, M. A., Khalil, E. M. A., Soltan Monem, A., and Hamdy, Y. M. (2003). Characterization of some bioglass–ceramics, *Mater. Chem. Phys.*, **80**, pp. 599–609.
58. Goller, G. (2004). The effect of bond coat on mechanical properties of plasma sprayed bioglass-titanium coatings, *Ceram. Int.*, **30**, pp. 351–355.
59. Mardare, C. C., Mardare, A. I., Fernandes, J. R. F., Joanni, E., Pina, S. C. A., Fernandes, M. H. V., and Correia, R. N. (2003). Deposition of bioactive glass-ceramic thin-films by RF magnetron sputtering, *J. Eur. Ceram. Soc.*, **23**, pp. 1027–1030.
60. Hench, L. (1993). In *An Introduction to Bioceramics*, ed. Wilson, J. (World Scientific), 386 p.
61. Peddi, L., Brow, R. K., and Brown, R. F. (2008). Bioactive borate glass coatings for titanium alloys, *J. Mater. Sci. - Mater. Med.*, **19**, pp. 3145–3152.
62. http://ltp.epfl.ch/files/content/sites/ltp/files/shared/Teaching/Master/03-IntroductionToNanomaterials/LectureSupportAll.pdf

63. Socol, G., Axente, E., Oane, M., Voicu, L., Petris, A., Vlad, V., Mihailescu, I. N., Mirchin, N., Margolin, R., Naot D., and Peled, A. (2007). Nanoscopic deposited structures analyzed by an evanescent optical method, *Appl. Surf. Sci.*, **253**, pp. 6535–6538.

64. Socol, G., Axente, E., Oane, M., Voicu, L., Dinescu, A., Petris A., Vlad, V., Mihailescu, I. N., Mirchin, N., Margolin, R., Naot D., and Peled, A. (2007). Using differential evanescent light intensity for evaluating profiles and growth rates in KrF laser photodeposited nanostructures, *J. Mater. Sci. – Mater. Electron.*, **18**, pp. S207–S211.

65. Mirchin, N., Gankin, M., Gorodetsky, U., Popescu, S. A., Lapsker, I., Peled, A., Duta, L., Dorcioman, G., Popescu, A., and Mihailescu, I. N. (2010). Estimation of polyethylene nanothin layer morphology by differential evanescent light intensity imaging, *J. Nanophoton.*, **4**, p. 041760.

66. Lapsker, I., Mirchin, N., Gorodetsky, U., Popescu, S. A., Peled, A., Duta, L., Dorcioman, G., Popescu, A. C., and Mihailescu, I. N. (2010). Morphology of polyethylene nanolayers: a study by evanescent light microscopy, *J. Mater. Sci.*, **45**, pp. 6332–6338.

67. Popescu, S. A., Apter, B., Mirchin, N., Gorodetsky, U., Lapsker, I., Peled, A., Duta, L., Dorcioman, G., Popescu, A., and Mihailescu, I. N. (2011). Study of polyethylene nanolayers by evanescent light microscopy, *Appl. Phys. A*, **104**(3), pp. 997–1002.

68. Mirchin, N., Apter, B., Lapsker, I., Fogel, V., Gorodetsky, U., Popescu, S. A., Peled, A., Popescu-Pelin, G., Dorcioman, G., Duta, L., Popescu, A. C., and Mihailescu, I. N. (2012). Measuring nanolayer profiles of various materials by evanescent light technique, *J. Nanosci. Nanotechnol.*, **12**(4), pp. 2668–2671.

69. Mirchin, N., Peled, A., Duta, L., Popescu, A. C., Dorcioman, G., and Mihailescu, I. N. (2013). Nanoprofiles evaluation of ZnO thin films by an evanescent light method, *Microsc. Res. Tech.*, **76**(10), pp. 992–996.

70. Mirchin, N., Peled, A., Azoulay, J., Duta, L., Dorcioman, G., Popescu, A. C., and Mihailescu, I. N. (2014). Nanoprofiles of TiO_2 films deposited by PLD using an evanescent light method, *World J. Eng.*, **11**(2), pp. 111–116.

71. de Fornel, F. (2000). *Evanescent Waves: From Newtonian Optics to Atomic Optics* (Springer, Berlin).

72. Ristoscu, C., Ghica, C., Papadopoulou, E. L., Socol, G., Gray, D., Mironov, B., Mihailescu, I. N., and Fotakis, C. (2011). Modification of AlN thin films morphology and structure by temporally shaping of fs laser pulses used for deposition, *Thin Solid Films*, **519**, pp. 6381–6387.

73. Ristoscu, C., Socol, G., Ghica, C., Mihailescu, I. N., Gray, D., Klini, A., Manousaki, A., Anglos, D., and Fotakis, C. (2006). Femtosecond pulse shaping for phase and morphology control in PLD: synthesis of cubic SiC, *Appl. Surf. Sci.*, **252**(13), pp. 4857–4862.

74. Ristoscu, C., and Mihailescu, I. N. (2011). Effect of pulse laser duration and shape on PLD thin films morphology and structure, in *Lasers - Applications in Science and Industry*, ed. Jakubczak, K. (InTech), pp. 53–74.

75. Craciun, D., Socol, G., Stefan, N., Miroiu, M., Mihailescu, I. N., Galca, A.-C., and Craciun, V. (2009). Chemical composition of ZrC thin films grown by pulsed laser deposition, *Appl. Surf. Sci.*, **255**, pp. 5288–5291.

76. Craciun, D., Socol, G., Stefan, N., Miroiu, M., Mihailescu, I. N., Galca, A. C., and Craciun, V. (2009). Structural investigations of ITO-ZnO films grown by the combinatorial pulsed laser deposition technique, *Appl. Surf. Sci.*, **255**(10), pp. 5288–5291.

77. Stankova, N. E., Dimitrov, I. G., Atanasov, P. A., Sakano, T., Yata, Y., and Obara, M. (2010). Nanostructured optical waveguide films of TiO_2 and WO_{3-x} for photonic gas sensors, *Thin Solid Films*, **518**, pp. 4597–4602.

78. Ristoscu, C., Mihailescu, I. N., Caiteanu, D., Mihailescu, C. N., Mazingue, Th., Escoubas, L., Perrone, A., and Du, H. (2008). Nanostructured thin optical sensors for detection of gas traces, in *Functionalized Nanoscale Materials, Devices, & Systems*, ed. Vaseashta, A., and Mihailescu, I. N. (Springer Science + Business Media B.V.), pp. 27–50.

79. Golovanov, V., Korotcenkov, G., Brinzari, V., Cornet, A., Morante, J., Arbiol, J., and Rossyniol, E. (2002). CO–water interaction with SnO_2 gas sensors: role of orientation effects, in *Proceeding of the 16th International Conference on Transducers*, EUROSENSORS-XVI, Prague, Czech Republic, pp. 926–929 (CD).

80. Wang, Y., Tabet-Aoul, A., and Mohamedi, M. (2015). Laser synthesis of hierarchically organized nanostructured TiO_2 films on microfibrous carbon paper substrate: characterization and electrocatalyst supporting properties, *J. Power Sources*, **299**, pp. 149–155.

81. Park, J. H., Han, S. M., Yoon, K. J., Kim, H., Hong, J., Kim, B.-K., Lee, J.-H., and Son, J.-W. (2016). Impact of nanostructured anode on low-temperature performance of thin-film-based anode-supported solid oxide fuel cells, *J. Power Sources*, **315**, pp. 324–330.

82. Campion, A., and Kambhampati, P. (1998). Surface-enhanced Raman scattering, *Chem. Soc. Rev.*, **27**, pp. 241–250.

83. Moskovits, M. (1985). Surface-enhanced spectroscopy, *Rev. Mod. Phys.*, **57**, pp. 783–826.

84. Otto, A., Mrozek, I., Grabhorn, H., and Akemann, W. (1992). Surface-enhanced Raman scattering, *J. Phys. Condens. Matter*, **4**, pp. 1143–1212.
85. De Bonis, A., Galasso, A., Ibris, N., Sansone, M., Santagata, A., and Teghil, R. (2012). Ultra-short pulsed laser deposition of thin silver films for surface enhanced Raman scattering, *Surf. Coat. Technol.*, **207**, pp. 279–285.
86. Ivanov, V. G., Vlakhov, E. S., Stan, G. E., Zamfirescu, M., Albu, C., Mihailescu, N., Negut, I., Luculescu, C., Socol, M., Ristoscu, C., and Mihailescu, I. N. (2015). Surface-enhanced Raman scattering activity of niobium surface after irradiation with femtosecond laser pulses, *J. Appl. Phys.*, **118**, p. 203104.
87. Nedyalkov, N., Nikolov, A., Atanasov, P., Alexandrov, M., Terakawa, M., and Shimizu, H. (2014). Nanostructured Au film produced by pulsed laser deposition in air at atmospheric pressure, *Opt. Laser Technol.*, **64**, pp. 41–45.

Chapter 8

Core-Shell Nanoparticles for Energy Storage Applications

Manish Kothakonda,[a] Briley Bourgeois,[a] Brian C. Riggs,[a] Venkata Sreenivas Puli,[a] Ravinder Elupula,[b] Muhammad Ejaz,[b] Shiva Adireddy,[a] Scott M. Grayson,[b] and Douglas B. Chrisey[a]

[a]*Department of Physics and Engineering Physics, Tulane University, New Orleans, LA 70118, USA*
[b]*Department of Chemistry, Tulane University, New Orleans, LA 70118, USA*
mkothako@tulane.edu, manish.hcu.phy@gmail.com

Core-shell nanomaterials and nanostructures have become an important research area in the field of science and engineering after only a few decades. These materials, with different sizes and different shapes (spherical, starlike, or tubular) of core-shell thickness and with different surface morphologies, possess different properties. In this chapter, we will mainly focus on core-shell nanoparticles that are used for energy storage applications, their synthesis, characterization, and properties. While gaining deeper insight into the area, we will explore a detailed overview of different techniques used for the preparation of various ceramic core-shell nanostructures with tunable sizes and tailored structures.

Pulsed Laser Ablation: Advances and Applications in Nanoparticles and Nanostructuring Thin Films
Edited by Ion N. Mihailescu and Anna Paola Caricato
Copyright © 2018 Pan Stanford Publishing Pte. Ltd.
ISBN 978-981-4774-23-9 (Hardcover), 978-1-315-18523-1 (eBook)
www.panstanford.com

8.1 Introduction

The development of nanostructured materials is one of the primary research goals in reaching advanced multifunctional properties and achieving better performance for novel applications. This has brought increased research attention to many potential methods for combining two or more disparate material properties into one material. Nanostructures surrounded by thin surface layers have shown the ability to substantially change the functionality and properties of the original nanostructure, such as the thermal stability; catalytic activity; and optical, magnetic, and electronic properties. In the late 1980s, these types of heterostructures were realized in semiconductor nanoparticles [1-3]. In the early 1990s, the term "core-shell" was coined by researchers [4, 5] and has been the subject of increasing research efforts in the material science community.

Core-shell nanostructures are developed with the goal of combining two materials and thus two properties within one structure. For example, nanoparticles of iron oxide coated with a silica layer demonstrate magnetic properties from the core iron oxide and optical luminescent properties from the shell of silica [6]. Magnetic core-shell nanoparticles have been of particular interest because of the shell layer's biocompatibility and chemical stability, making it suitable for biomedical applications [7, 8], such as targeted drug delivery [9, 10]. Furthermore, the shell structure has been used to change the surface charge and reactivity of the whole particle. For applications in printed electronics, metallic nanoparticles are coated with a shell material that provides exceptional colloidal stability and can be optically transparent and chemically inert [11].

In many cases for various applications, as shown in Fig. 8.1, core-shell nanostructures have emerged as a low-cost alternative to traditional materials due to the design and geometry of the thin shell of active material [12-14]. A good example of this is the use of cost-effective oxides coated in a layer of phosphates to make a variety of pigments [15]. Core-shell particles also introduce tunability in a variety of ways, including controlling the shell thickness. For example, by increasing the gold layer thickness on silica nanoparticles, the absorption band can be adjusted from the visible to the infrared region, matching the surface plasmon resonance band with the desired wavelength of light [16]. Similar ideas were demonstrated through the preparation of core-shell nanoparticles that exhibited the quantum confinement effect or novel electronic, magnetic, and optical properties. One example is forming gold nanoparticles coated with two shells: a bottom shell of dense silica and a top layer of CdSe quantum dots. This multilayered approach allowed control of CdSe quantum dots in many properties [17].

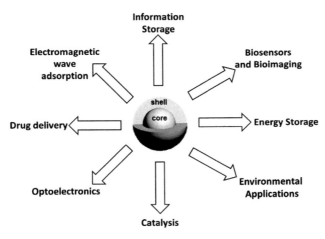

Figure 8.1 Applications of core-shell nanoparticles in various areas of modern technology. Reproduced from Ref. [20] with permission of The Royal Society of Chemistry.

Another core-shell system can be categorized as one in which a particular combination of the core and shell materials results in an entirely new property rather than a combination of the properties of the individual materials. Core-shell nanostructures with ferroelectric shells and ferromagnetic cores have been reported over the last few years. These novel properties were determined to be caused by the strain-mediated interaction between core and shell phases [18, 19].

There are a significant number of synthesis methods for core-shell particles that have been developed over the past 20 years. This chapter will examine the synthesis routes, methods of characterization, and resulting properties of core-shell nanoparticles.

As a case study, we will focus on core-shell nanoparticles for energy storage application [20–27].

8.1.1 Nanoparticle Property Selection

The end goal for capacitive energy storage is to increase the total energy density. The dielectric energy density of a material can be reduced down to two material properties: the dielectric constant/permittivity and the breakdown field. As shown in Eq. 8.1 the energy density (E_d) is related to dielectric permittivity and the square of the breakdown voltage (E_b), where ε_0 and ε_r are relative permittivity and permittivity of free space, respectively:

$$\text{Energy density }(E_d) = \frac{\text{Stored energy}}{\text{Volume}} \left(\frac{J}{cm^3}\right) \text{ or } \left(\frac{Watt-s}{cm^3}\right)$$

$$= E_d = \frac{1}{2}\varepsilon_0 \varepsilon_r E_b^2 = \frac{1}{2}CV_m^2 \frac{1}{Ad} \quad (8.1)$$

To improve the energy storage property of capacitors, it is important to increase both the dielectric constant and the breakdown field. However, the most common dielectric materials, such as organic polymers and inorganic ceramics, do not meet the current energy storage requirements of advanced capacitors. Conventional ceramic dielectric materials have high permittivities but suffer from low breakdown fields, making them only applicable for low-voltage devices. Organic polymers, such as polystyrene (PS), polypropylene (PP), and poly(ethylene terephthalate) (PET), which have high breakdown fields but also low dielectric constants, require significant fields in order to store suitable energy due to their low polarization. Despite their detriments, both materials have a significant role in modern electronics and electrical power systems due to their rapid charge and discharge capabilities [28–30]. Core-shell nanostructures have the potential to possess the dual properties of the high dielectric constant of ceramic nanoparticles and a high breakdown strength, leading to high energy storage densities [31–40]. Combined with the excellent processability of polymers, these core-shell nanostructures are an attractive option for future capacitor materials.

There are a variety of different types of core-shell structures that have been explored in order to develop high-energy-density materials. These are summarized in Fig. 8.2, including a metallic shell on a different metal core, glass coating on a metal core, polymer shell on a metal core, nonmetallic shell on a nonmetallic core, and polymer shell on a different polymer core. In the present study, we will be focused on metal core and polymer shell nanoparticles [41–48].

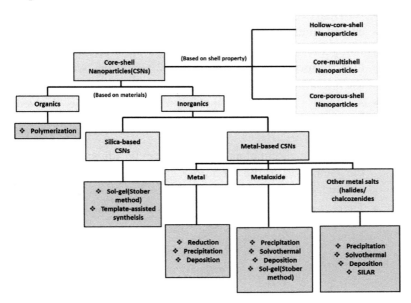

Figure 8.2 General classification of core-shell nanoparticles on the basis of material type and properties of their shells. Reproduced from Ref. [20] with permission of The Royal Society of Chemistry.

8.1.2 Nanoparticle Synthesis

The synthesis of metal core and polymer shell nanoparticles for energy storage involves two main strategies: the first one is the "grafting from" strategy and the other one is "grafting to." These two are explained in detail next [49–54].

8.1.2.1 Core-shell nanoparticles prepared by the grafting-from route

This strategy relies on the formation of nanocomposites by the in situ polymerization of monomers on initiator-functionalized nanoparticle surfaces. The key to this approach is the introduction of a sufficient quantity of initiating sites on the nanoparticle surfaces [55–58]. There are two feasible and robust grafting techniques, atom transfer radical polymerization (ATRP) and reversible addition-fragmentation chain transfer (RAFT) polymerization (explained in Fig. 8.3), which provide many advantages:

- Nanoparticle aggregation is prevented by the shell layer coated on the nanoparticle surfaces.
- Using the shell layer as a matrix, nanocomposites can be directly formed from core-shell nanoparticles, which allow the preparation of high-quality, highly filled nanocomposites.
- A strong nanoparticle/matrix interface can be derived as a result of robustly bonded polymer chains onto the nanoparticle surfaces.
- The concentration of nanoparticles can be adjusted by tuning the feed ratio of the monomer and initiator functionalized nanoparticles.
- There is a broad range of monomers that can be polymerized using the grafting route [58–64].

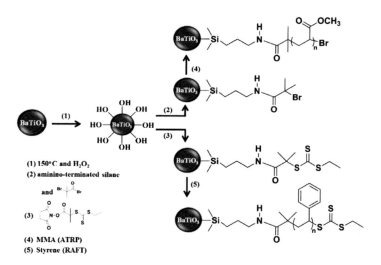

Figure 8.3 Illustration of the synthesis routes for high-k poly(methyl methacrylate) (PMMA) and polystyrene (PS) nanocomposites by ATRP and RAFT polymerization. Reproduced from Ref. [56] with permission of The Royal Society of Chemistry.

8.1.2.2 Core-shell nanoparticles prepared by the grafting-to route

This strategy relies on grafting the preprepared polymer chains on the nanoparticles surface. These polymer chains are grafted onto the nanoparticle surface by a reaction between the polymer end groups and functional groups. The grafting-to technique allows us better control of the molecular composition and the molecular weight of the polymer chains than the grafting-from technique. Click chemistry, as illustrated in Fig. 8.4, is the versatile method used to link the reaction partners, which has not only the advantage of high yield of the end product but also solvent insensitivity and moderate reaction conditions. Therefore, click reactions are an inventive method for the grafting of polymer chains on nanoparticle surfaces [50, 58, 65–68].

Tchoul and coworkers have prepared a series of core-shell PS@TiO$_2$ nanocomposites by a Cu(I)-catalyzed alkyne-azide click reaction. The dielectric constant of the PS core-shell (100 kg·mol^{-1}) nanocomposites with 27 vol% TiO$_2$ was reported to be 6.4 at 100 Hz, while the dielectric loss was reported to be as low as 0.625%. These nanocomposites will have applications in thin-film transistors as gate dielectrics to achieve high carrier mobility and low leakage current. The only disadvantage reported in the process is that the catalyst CuBr cannot be easily removed from the nanocomposites because of strong complexation with azides and triazoles, which may show fixed frequency-dependent dielectric properties and high dielectric losses [65, 69].

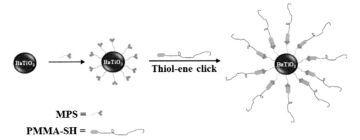

Figure 8.4 Illustrations of the synthesis process for PMMA@BaTiO$_3$ nanocomposites by thiol-ene click reactionsReprinted with permission from Ref. [70]. Copyright (2014) American Chemical Society.

The reported polymer matrices include poly(vinylidene fluoride) (PVDF) and its copolymers, poly(vinylidene-fluoride-co-tri-fluoroethylene) (PVDF-TrFE) [70], poly(vinylidene fluoride-co-hexafluoropropylene) (PVDF-HFP), PVDF-HFP ethylene, poly(vinylidene-fluoride-co-tri-fluoroethylene-chlorotrifluoroethylene) (PVDF-TrFE-CTFE) [70, 71], poly(methyl methacrylate) (PMMA) [56, 72], polycarbonate (PC) [38, 73], epoxy resin, sulfonate styrene-b-(ethylene-ranbutylene)-b-styrene (S-SEBS) [74], PP, and PS [65].

Many polymer nanocomposites have been prepared successfully in the past, yet there are some drawbacks for high-dielectric-constant nanoparticles. Firstly, the agglomeration of nanoparticles is a difficult task to elude due to the high surface energy of inorganic nanoparticles. The agglomeration will not only lower the breakdown strength but also give rise to inadmissible porosity for a low dielectric constant and increase in dielectric losses [75].

A second major drawback is the weak interface adhesion between the fillers and the matrix, which is due to the significant difference of intrinsic surface properties between the inorganic filler and the organic polymer. It is known that active interface adhesion between the fillers and the matrix is critical to increasing the breakdown strength and energy storage density. Therefore, several methods have been used to tailor the polymer/filler interface: (i) nanoparticles modified with a coupling agent, (ii) end-functionalized polymer chains grafted onto the nanoparticle surfaces, and (iii) polymerization of the monomer onto nanoparticle surfaces by in situ.

A few examples of ceramic-polymer nanocomposites used for energy storage are a PVDF ferroelectric polymer and, more recently, its copolymers, such as those with hexafluoropropylene (PVDF-HFP), chlorotrifluoroethylene (PVDF-CTFE), and trifluoroethylene (PVDF-TrFE). These composites are noted for their high dielectric constants and low dielectric losses, with a high dielectric breakdown strength, when compared to other polymers and are widely explored in electrical energy storage capacitors [76, 77].

Jiang and coworkers have also worked on core-shell biodegradable polylactic acid (PLA)-encapsulated high-k $BaTiO_3$ nanoparticles, which in turn showed enhanced dielectric properties and also energy storage capabilities. Surface-initiated ring-opening polymerization (ROP) was used to prepare core-shell structured BT@PLA nanoparticles displaying high potential for environmentally friendly dielectric and energy storage applications. The dielectric constants for the nanocomposites with BT, PD@BT, and PDA@PLA@BT are observed to be 7.52, 8.10, and 8.74, respectively.

Huang and Jiang have used RAFT and ATRP techniques to prepare core-shell high-k PMMA@BaTiO$_3$ and PS@BaTiO$_3$ nanocomposites, respectively. Dispersion of the observed nanoparticles was homogeneous in both cases. The dielectric constant of PS@BaTiO$_3$ nanocomposites with 48 vol% BaTiO$_3$ increased from 2.8, which was observed for pure PS, to 24, while the dielectric loss did not change much in either case. Stability of dielectric constant and dielectric loss was observed over a broad range of frequencies, which is a very effective demand for functional devices that operate over wide ranges of frequencies [56, 57].

Perry et al. [38, 39] prepared PFBPA BaTiO$_3$: P(VDF-HFP)-based polymer nanocomposites with a high energy storage density by treating the BT nanoparticles with coupling agents. The study revealed that controlled variation of the volume fractions of nanoparticles could maximize the permittivity. The study reported that at 50%–60% of the nanoparticle volume fraction the maximum permittivity was observed.

Marks and coworkers fabricated a series of PP nanocomposites via in situ olefin polymerization. Good nanoparticle dispersion, high dielectric constants, high breakdown strengths, low dielectric losses, and high energy storage densities were achieved in these PP nanocomposites [60, 66, 77, 78].

Jiang et al. designed a novel core-double-shell structure to improve further the dielectric properties of polymer nanocomposites. The first core-shell is hyperbranched aromatic polyamide (HBP), grafted from the surface of nanoparticles. The second layer of the core is PMMA, grafted from the terminal groups of HBP via ATRP. The core-double-shell nanocomposite possesses the advantage of the two layers of polymers with a high energy storage density, a high dielectric constant, and a low dielectric loss in comparison with PMMA/BaTiO$_3$ nanoparticles prepared by conventional methods. They have also introduced core-satellite nanoassemblies comprising BaTiO$_3$ and Ag nanoparticles (Ag@BaTiO$_3$). This group has developed a new strategy for developing polymer composites with high breakdown strengths, high dielectric constants, and low dielectric losses. The dielectric constant was relatively enhanced by the usage of PVDF as a polymer matrix. The PVDF–Ag@BaTiO$_3$ nanocomposites make use of the high dielectric constant of BaTiO$_3$ nanoparticles, further well-known Coulomb blockade, and quantum confinement effects of ultratiny Ag nanoparticles, which exhibit exceptionally enhanced energy density and energy storage efficiency in comparison with the BaTiO$_3$ nanoparticles [28].

Substantial amounts of energy storage density (14.4 J/cm^3, at an applied field of 10^5 V/cm) and low energy loss were reported in percolative metallic aluminum-PP nanocomposites [63].

8.2 Experimental Section

Glycidyl methacrylate (GMA, 97+%, Fluka) was passed through neutral alumina to purify. Copper (I) chloride (CuCl, 99.99%, Aldrich); p-toluene sulfonyl chloride (PTSC, 97%, Sigma-Aldrich); 2-(4-chlorosulfonylphenyl) ethyl trichlorosilane (CTCS, 50% in methylene chloride, Gelest); anhydrous anisole (99.7%, Sigma-Aldrich); 4,4'-dinonyl-2,2'dipyridine (DNDP, 97%, Aldrich); and barium titanate (BaTiO$_3$, ≥99%, Aldrich) nanopowder in the cubic crystalline phase were used as starting materials.

8.2.1 Nanoparticle Synthesis

Figure 8.5 summarizes various techniques for nanoparticle preparation. Preparation of nanoparticles is categorized into two major techniques. The first one is the top-down approach, where an external force is applied to a solid that leads to its breakup into smaller particles. The bottom-up approach begins at the atomic scale and builds the nanostructures until the desired sizes and shapes are achieved [79–81].

The top-down approach is further subdivided into dry and wet grinding. In the dry-grinding method, the solid substance is grounded as a result of compression, shock, or friction using well-known methods such as a jet mill, a hammer mill, a shearing mill, a roller mill, a shock shearing mill, a ball mill, and a tumbling mill. On the other hand, wet grinding of a solid substrate is carried out using a vibratory ball mill, a tumbling ball mill, a planetary ball mill, a centrifugal fluid mill, an agitating beads mill, a flow conduit beads mill, an annular gap beads mill, or a wet jet mill. Comparing the dry and wet methods, the wet method is more suitable for obtaining highly dispersed nanoparticles. Also, the mechanochemical method and the mechanical alloying method are also known top-down methods [81–85].

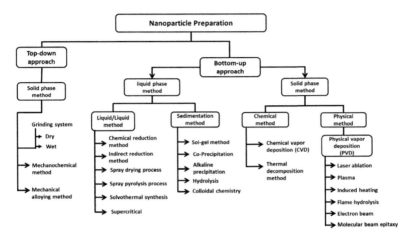

Figure 8.5 Typical synthetic methods for nanoparticles for the top-down and bottom-up approaches.

The bottom-up approach is divided into solid-phase and liquid-phase methods. The chemical vapor deposition (CVD) method can produce ultrafine particles of less than 1 μm by the chemical reaction occurring in the solid phase, whereas the physical vapor deposition (PVD) method uses cooling of the evaporated material to produce nanoparticles. Although the gaseous phase methods minimize the occurrence of organic impurities in the particles compared to the liquid-phase methods, they necessitate the use of complicated vacuum equipment whose disadvantages are the high costs involved and low productivity. The thermal decomposition method is particularly productive in the metal-oxide nanoparticle preparation [86–94].

Liquid-phase methods allow us to synthesize a wide variety of nanoparticles [95–98]. Liquid-phase methods are subdivided into liquid/liquid and sedimentation methods. A typical example of a liquid/liquid method is a chemical reduction of metal ions, which possesses an advantage in the facile fabrication of nanoparticles of various shapes, such as nanorods, nanowires, nanoprisms, nanoplates, and hollow nanoparticles [99, 100].

The solvothermal method is a conventional method for the preparation of inorganic nanomaterials. Solvothermal synthesis is performed in stainless steel autoclaves, with Teflon liners, which can withhold high-pressure systems. High pressure is developed in the autoclaves with the rise in temperature. This method not only has an advantage of controlled morphology of nanoparticles but also possesses high crystallinity and better surface functionalization [101].

Other well-known methods are spray drying, spray pyrolysis, and the supercritical method. The general technique in the sedimentation method is a sol-gel process, which has been used extensively for the fabrication of metal-oxide nanoparticles [102–106].

Although various techniques have been summarized in Fig. 8.5, there are some features to consider that are common to all the methods. The synthesis of nanoparticles requires the use of a device or a process that fulfils the following:

- Control over particle size, size distribution, shape, crystal structure, and chemical composition
- Reproducibility
- Scalability and cost-effectiveness

8.2.2 Nanoparticles Synthesis by Pulsed Laser Ablation

Pulsed laser ablation has drawn much attention due to its high potential in laser-based material processing, including nanocrystal growth, thin solid film preparation, surface cleaning, and microelectronic device fabrication. Ejection of macroscopic amounts of species from the solid material substrate by interaction with ultrashort laser pulses (10^{-13} to 10^{-8} s) is defined to be laser ablation. Laser ablation can be carried out in vacuum, gas, and liquid mediums. Generation of nanostructures by laser ablation is usually referred to as a bottom-up process due to nucleation, growth, and assembly of clusters of ablated species [107–118].

The advantage of preparation of nanostructures by laser ablation is it can be applied to almost every solid material. Changes in the laser parameters, such as wavelength, laser energy, and pulse duration, control the size and shape of the nanostructures. Also, laser ablation is not a time-consuming process, causing limited thermal damage to the base material, which is necessary for precise material processing. The prepared size-controlled nanoparticles are further grafted using one of the techniques for the formation of core-shell nanostructures [113, 114, 119–121].

8.2.3 Synthesis of BaTiO$_3$ Nanoparticles by the Solvothermal Method

The solvothermal method is commonly employed in the synthesis of perovskite nanomaterials. The solvothermal method allows for a high level of control over the material phase, structure, shape, and size. The characteristics of the solvothermal method include excellent control over the morphology of nanomaterials, fairly uniform dispersion of nanoparticle size, low-temperature processing, and long reaction times. A typical solvothermal reaction is performed in a stainless steel autoclave. The autoclave serves as a strong reaction chamber to hold the pressurized experiments. There is another material lining the interior of the autoclave. Usually, this lining is made of Teflon to provide a low reactivity surface and limit heterogeneous nucleation in the reaction.

Solvothermal reactions are performed using several liquid solutions. The critical elements making up these solutions are typically some organic surfactant acting as a capping agent, precursor powders dissolved in some liquid, a bulk solution in which the reaction occurs, and a basic solution to alter the pH of the reaction solution. These chemicals are mixed inside of the reaction vessel. The reaction vessel is placed inside of the stainless steel autoclave, and the autoclave is heated in an oven. The temperature of the reaction can be varied to change the formation of the nanoparticles. A typical reaction temperature is 150°C–250°C. The reaction is performed over a relatively long period lasting for about 24–72 h.

The growth of the nanoparticles is governed, overall, by the preferred energy state of the monomer, barium and titanium atoms in this case. In a solvothermal reaction, several factors affect the monomer as it moves throughout the bulk of the reaction liquid. The monomers themselves move under a random Brownian motion throughout the bulk liquid. This motion is affected by the fluid properties of the liquid; the interaction, or lack thereof, between the molecules of the liquid and the monomer; and the elevated temperature of the reaction vessel. The monomer proceeds to move throughout the bulk solution until it is energetically favorable for the monomer to precipitate out of solution. This is known as nucleation. The experimental properties of the solvothermal method are set up so that the particles will exhibit homogeneous nucleation, or the nucleation of particles directly into the solution, as opposed to heterogeneous nucleation, or the nucleation of particles against an existing surface. This is the primary reason for choosing the slick surface of Teflon as the lining of the reaction vessel.

After nucleation, the particles begin to grow. This process is controlled through the random motion of the solution, the specifics of the chemical reaction, and the amount of available monomer for the reaction to take place. The onset of nucleation, with respect to the concentration of available monomer, begins when the concentration of the solution reaches a supersaturated point. This critical concentration signifies that the solution is so saturated that the monomer would prefer to fall out of the solution. After this occurs, another monomer available in the solution interacts with the precipitated monomer to form a crystalline nanoparticle. Because less energy is needed for the monomer to heterogeneously nucleate on the surface of the particle, growth of the nanoparticles can occur at a concentration lower than the critical concentration point. The concentrations of the monomer can be used to control the size of the nanoparticles. An example of this is shown in Fig. 8.6 [122].

The rates at which the particles grow are determined by either a diffusion- or a reaction-limited regime [122]. The reaction regime occurs when monomer supply is plentiful and the particles grow as quickly as the new monomer can be absorbed. The diffusion-limited regime occurs when particles absorb the monomer as quickly as the monomer arrives at the particle surface. Ideally, this difference in required energy allows for the separation of the nucleation and growth processes [123]. A lack of separation between the nucleation and growth regime leads to a wider dispersion of particle sizes due to nucleates formed in different time periods growing at similar rates. The long reaction time of the solvothermal method helps to narrow this dispersion. Figure 8.7 shows the concentration profile under a typical nanoparticle synthesis. C_{min} denotes the minimum concentration needed for the solution to become supersaturated and homogenous nucleation to occur. Homogenous nucleation stops after the concentration dips below the minimum concentration level. Nanoparticle growth will continue until the concentration reaches the saturation point, denoted as C_s [124].

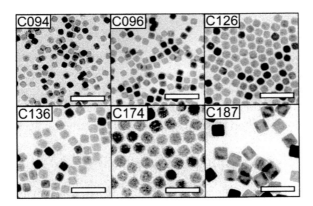

Figure 8.6 An example of nanoparticle size control (scale bar is 50 nm).

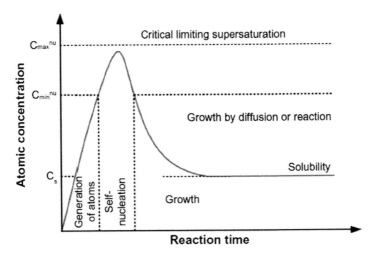

Figure 8.7 Graphical representation of the LaMer diagram depicting the concentration controlled nucleation and growth of nanoparticles. Reprinted with permission from Ref. [124]. Copyright (1950) American Chemical Society.

Different capping agents are of interest while using the solvothermal method. Commonly used organic capping agents, such as oleic acid and oleylamine, can have preferential bonding to certain faces of the nanoparticle structure. Taking advantage of this preferential bonding allows for shape control over nanoparticles. The capping agents also serve as a limiting factor for the size of the nanoparticles. The capping agents bond to the surface of the nanoparticles during growth and prevent new monomers from reacting with the nanoparticles, hence the name "capping agent." This also helps to prevent agglomeration of nanoparticles and Ostwald ripening [125].

Successful implementation of the solvothermal method in the growth of nanoparticles can achieve monodisperse particle growth and shape control of various particles. For these reasons and the adjustability of the process, it is advantageous to use the solvothermal method in the growth of $BaTiO_3$ nanoparticles.

8.2.4 Polymerization of Nanoparticles

As we discussed earlier, the two main strategies of polymerization are grafting-from and grafting-to strategies. These two techniques are elaborated below.

8.2.4.1 Synthesis of PGMA-BaTiO$_3$ core-shell nanostructures by grafting-from

In this section, we will learn about the synthesis procedure for the preparation of core-shell-structured hybrid $BaTiO_3$-poly(glycidyl methacrylate) (PGMA) nanoparticles, using surface-initiated atom transfer radical polymerization (SI-ATRP). The two-step process is shown in the Fig. 8.8. Firstly initiating sites are immobilized onto $BaTiO_3$ nanoparticles and in the next part, SI-ATRP of GMA from initiator-immobilized nanoparticles.

For the immobilization of ATRP initiator onto $BaTiO_3$, 1 g of powdered $BaTiO_3$ was dispersed in 40 mL of anhydrous toluene and sonicated for 20 min. 1.6 g of 50% CTCS in CH_2CL_2 was dissolved in 12 mL of anhydrous toluene and added to the above solution under argon gas and stirred vigorously at room temperature for 18 h. The modified $BaTiO_3$ powder was collected by repeated dispersion into the solvent by sonication and recollected by centrifugation using toluene and tetrahydrofuran (THF), each three times. In the end, the purified initiator-functionalized BaTiO$_3$ nanoparticles were dried under vacuum at room temperature and were directly employed in the polymer grafting.

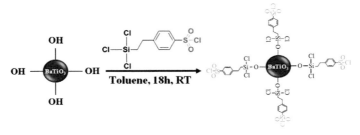

Figure 8.8 Synthesis of BaTiO$_3$-PGMA core-shell nanocomposites by SI-ATRP of GMA from BaTiO$_3$ nanoparticles. Reproduced from Ref. [126] with permission from Wiley.

In the surface-induced ATRP of GMA from BaTiO$_3$, 500 mg of functionalized nanoparticles were dispersed in a mixture of 30 g of anisole and 15 g of GMA in a 100 mL Schlenk flask. After the purge of 30 min., CuCl was added to the suspension under Ar flow, followed by the addition of DNDP and PTSC as a free initiator. The nanoparticles grafted by PGMA were isolated from the free polymer by repeated dispersion in THF. CuCl present in trace amounts was removed by additional washes of THF and associated solvent mixtures. The particles removed were finally dried at room temperature.

Also, ungrafted BaTiO$_3$ nanocomposite films were prepared by a conventional solution mixing process using BaTiO$_3$ nanoparticles and a PGMA polymer. Mainly, four different compositions were prepared (80PGMA-20 BaTiO$_3$, 75PGMA-25BaTiO$_3$, 50PGMA-50BaTiO$_3$, and 75BaTiO$_3$-25PGMA). The prepared nanoparticles and PGMA powders were dissolved in 10 mL of N,N-dimethylfomamide (DMF) and vigorously stirred to homogenize the material. The films were obtained by drop-casting the solution onto the copper substrate and curing at 80°C overnight for the formation of the thick, uniform films. However, the compatibility difference between the polymer and ungrafted nanoparticles prevented the production of uniform films.

8.2.4.2 Synthesis of PVDF-HFP-GMA-BaTiO₃ core-shell nanostructures by grafting-to

In this section, we will learn about the synthesis of the grafting-to method in a three-step process, as shown in Fig. 8.9. Firstly, the BaTiO$_3$ nanoparticles are prepared and functionalized with –OH groups at their surface. Secondly, PGMA is introduced onto the PVDF-HFP chains via in situ ATRP and the as-prepared polymer is denoted as PVDF-HFP-GMA. In the final step, the BaTiO$_3$ nanoparticles are dispersed into the preprepared polymer for the formation of PVDF-HFP-GMA-BaTiO$_3$ core-shell nanostructures. These three steps are elaborated below.

As we know, BaTiO$_3$ nanoparticles are prepared by the conventional solvothermal method. The prepared BaTiO$_3$ nanoparticles are cleaned by deionized water and dried in vacuum at 80°C for 12 h. The dried nanoparticles are treated with a H$_2$O$_2$ aqueous solution, added to toluene, and further sonicated for 30 min, 20 g of γ-APS was added, and the mixture was heated to 80°C for 24 h under a N$_2$ atmosphere for surface functionalization. Then, the functionalized nanoparticles were recovered by centrifugation at 9000 rpm for 5 min. The obtained nanoparticles were washed with toluene twice and dried under vacuum at 80°C for 12 h.

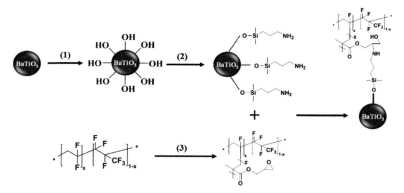

(1) H$_2$O$_2$, 105°C, 4h
(2) APS, Toluene
(3) CuBr, PMDETA, GMA

Figure 8.9 Schematic illustration of the preparation process for the PVDF-HFP-GMA/BT nanocomposites. Reproduced from Ref. [127] with permission of The Royal Society of Chemistry.

In the second step, functionalization of PVDF-HFP with GMA was carried out. Firstly, 20 g of PVDF-HFP was dissolved in 300 mL of DMF in a round-bottomed flask at 50°C, and 20 g of GMA and 1.435 g of CuBr were added to the dissolved PVDF-HFP solution and sealed. The oxygen present in the flask was carefully removed by evacuating and back-filling with N_2 gas three times. Then, 1.733 g of PMDETA was added by a syringe. This mixture was stirred for 24 h at 60°C. The products were precipitated in methanol, and the mixture was filtered to obtain PVDF-HFP-GMA. The obtained PVDF-HFP-GMA was dried under vacuum.

In the final step, PVDF-HFP-GMA-BaTiO$_3$ nanocomposites were prepared by the blending method. Firstly, the required amount of BaTiO$_3$-APS nanoparticles were mixed with DMF and sonicated for 30 min. Then the PVDF-HFP-GMA polymer was dissolved in DMF at 50°C. Then, the two prepared solutions were mixed and stirred with a magnetic stirrer at 80°C for 30 min. The resulting mixture was precipitated in methanol and was filtered to get polymer nanocomposites. The obtained core-shell nanocomposites were dried under vacuum at 80°C for 12 h. The dried nanocomposites were compressed into films at 180°C under pressure. Finally, nanocomposites containing different BaTiO$_3$ fractions (10%, 20%, 30%, 40%, and 50%) were prepared and were denoted by GMA-BT-10, GMA-BT-20, GMA-BT-30, GMA-BT-40, and GMA-BT-50, respectively. Different techniques were used to characterize all the different compositions of core-shell nanocomposites.

8.3 Materials Characterization

Size exclusion chromatography was carried out with three-column series from Polymer Laboratories, on the Waters model 1500 series pump (Milford, MA) consisting of PLgel 5 μm Mixed D (300 × 7.5 mm^2), PLgel 5 μm 500 Å (300 × 7.5 mm^2), and PLgel 5 μm 50 Å (300 × 7.5 mm^2) columns. The system Model 2487 was fitted with a differential refractometer detector. THF was used as the mobile phase (1 mL/min. flow rate). PS was used as a linear standard in the system. Collected data was processed using the Precision Acquire software. Fourier transform infrared (FTIR) spectroscopy was done using NEXUS 670 FT-IR SEP. The mixture of analyte and KBr was ground into a fine powder using mortar and pestle. Further, the powder was made into a pellet by application of pressure. Thermogravimetric analysis (TGA) was performed using the instrument TGA 2950 Thermogravimetric Analyzer under a nitrogen atmosphere with a heating rate of 20°C/min. The acquired data were then processed using the TA Instruments Universal Analysis software. Transmission electron microscopy (TEM) was performed using FEI G2 F30 Tecnai TEM, which operated at 200 kV. The dielectric properties were measured by an HP4294A LCR meter by using copper foil as a counterelectrode and silver paint as the electrode for thick films in a frequency range of 1–100 kHz with an applied electric field of 100 mV. The breakdown voltage of the core-shell capacitor was measured at room temperature using a Trek (10 kV) high-voltage amplifier.

8.4 Experimental Observation

The sizes of the commercially purchased nanoparticles were exhibited to be 50–100 nm. The literature reported by various groups indicates that the hydroxyl groups are already present on the surface of pristine $BaTiO_3$ nanoparticles and further exploited for the functionalization. The commercially available CTCS silane coupling agent was immobilized directly onto the nanoparticles without the use of H_2O_2, in a single step. CTCS usually produces ATRP-initiating sites in which hydrolysis of chlorosilyl occurs to form Si-O-Si covalent bonds with inorganic surfaces such as glass. However, the best initiating group for ATRP is the chlorosulfonylphenyl group ($-Ph-SO_2Cl$) as Cl is easily preoccupied with the $u^lX/2L$ complex, with the produced $-Ph-SO_2*$ radical initiating radical polymerization. The FTIR spectrum of $BaTiO_3$ shows only bands at 3422 cm^{-1}, 1442 cm^{-1}, and 596 cm^{-1}, a broad absorption band representing some –OH groups that are present on the surface of $BaTiO_3$, the stretching vibration of $-CO_3^{2-}$ from the residual $BaCO_3$ in the $BaTiO_3$, and the vibration mode of a metal-dioxo bridge, respectively. The absorption bands at 1000 cm^{-1} and 1015 cm^{-1} in the FTIR spectrum of $BaTiO_3$-CTCS (Fig. 8.10b) confirm the Si–O– particle and the Ti–O–Si group, respectively. Also, they suggest that the silane was successfully introduced onto the surface of $BaTiO_3$. Grafted CTCS onto the surface of $BaTiO_3$

was indicated by an absorption band at 1337 cm^{-1}. However, strong metal-dioxo bridge absorbance was detected due to absorption at ~544 cm^{-1}. The observed peak at 1200 cm^{-1} indicates sulfonyl chloride groups.

The FTIR spectrum of BaTiO$_3$-PGMA (Fig. 8.10c) shows that absorption bands at 1730 cm^{-1} and 907c^{-1} are due to the strong carbonyl group and epoxy ring, respectively, indicating successful grafting of PGMA to nanoparticles.

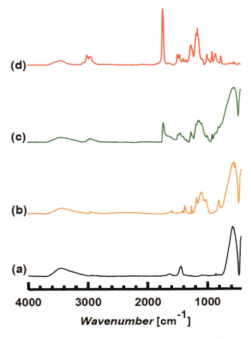

Figure 8.10 FTIR spectra of (a) as-received BaTiO$_3$, (b) BaTiO$_3$-CTCS, (c) BaTiO$_3$-PGMA, and (d) pure PGMA. Reproduced from Ref. [126] with permission from Wiley.

Further analysis was carried out by TGA (Fig. 8.11), which indeed confirmed the immobilization of ATRP sites onto BaTiO$_3$. Due to adsorbed moisture and surface hydroxyl groups, the BaTiO$_3$ nanoparticles show about 0.70% mass loss before reaching 300°C. The additional mass loss of 1.3% in Fig. 8.11b at 600°C relative to BaTiO$_3$ provides further evidence that ATRP-initiating sites were immobilized onto nanoparticles successfully. The results confirm that CTCS can be directly immobilized onto pristine $BaTiO_3$ nanoparticles without any prior oxidative treatment.

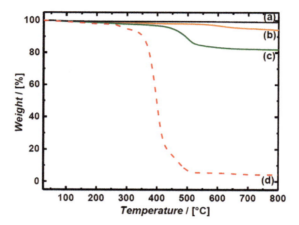

Figure 8.11 TGA curves of (a) as-received $BaTiO_3$, (b) $BaTiO_3$-CTCS, (c) $BaTiO_3$-PGMA, and (d) pure PGMA. Reproduced from Ref. [126] with permission from Wiley.

TGA is successful in providing quantitative data about the mass of grafted PGMA relative to the mass of nanoparticles. The $BaTiO_3$-PGMA nanoparticles exhibited substantial weight losses at around 450°C, as shown in Fig. 8.11c.

The gel permeation chromatography (GPC) analysis of the PGMA grafting reaction gave an average molecular weight estimated to be 12,000 and a polydispersity index of around 1.34. The free polymer molecular weight suggests that the grafted polymer chains had approximately 84 repeating units. A relatively low dispersity index confirms that polymerization was highly controlled.

Morphological characterization of $BaTiO_3$-PGMA nanoparticles was analyzed by TEM to observe the surface morphology of the nanoparticles. After polymerization, $BaTiO_3$-PGMA nanoparticles exhibit limited aggregation, as shown in the Fig. 8.12a. The PGMA shell that encapsulates $BaTiO_3$ nanoparticles exists in a uniform continuous polymer film of a thickness of ~20 nm. These results clearly confirm that PGMA has been successfully grafted onto the $BaTiO_3$ nanoparticles.

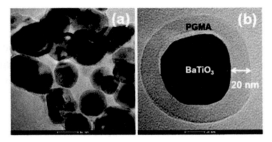

Figure 8.12 TEM images of (a) BaTiO$_3$-PGMA (low magnification) and (b) BaTiO$_3$-PGMA (high magnification). Reproduced from Ref. [126] with permission from Wiley.

8.4.1 Dielectric Properties and Leakage Current Behavior

The polycrystalline electroceramics complex impedance spectroscopy is a well-known technique to describe electrical properties. This method contributes to the in-depth analysis of dielectric materials, like conductivity and dielectric constant. A Cole-Cole plot is widely used to communicate the frequency response of the system in the present context.

To calculate dielectric permittivity, the frequency-dependent capacitance properties of pristine PGMA and BaTiO$_3$-PGMA nanoparticles are measured. The dielectric permittivity of pristine PGMA and BaTiO$_3$-PGMA nanoparticles decreased as the frequency was increased from 100 Hz to 100 kHz. Core-shell nanoparticles showed a significantly higher dielectric constant as compared to a pristine PGMA film (PGMA-$\varepsilon \approx 5.3$, and BaTiO$_3$-PGMA $\varepsilon \approx 54$) at room temperature. A relatively high dielectric permittivity ε of ~1100 in the core-shell-structured BaTiO$_3$-PGMA nanocomposite is attributed to the well-dispersed ceramic filler BaTiO$_3$ nanoparticles.

Energy storage properties were calculated for core-shell-structured BaTiO$_3$-PGMA nanocomposites for energy storage capacitor applications. The energy densities of pure PGMA and BaTiO$_3$-PGMA nanocomposites were analyzed using the linear dielectric capacitor equation at the maximum dielectric breakdown strength. The energy densities for pure PGMA and BaTiO$_3$-PGMA were calculated to be 5.86 J/cm^3 and ~21.51 J/cm^3, respectively. This shows the addition of grafted BaTiO$_3$ fillers to PGMA-enhanced dielectric permittivity and showed an ultrahigh energy storage density.

Dielectric loss (tan δ) is another crucial parameter in electrical capacitor applications. The dielectric loss parameter determines the performance of the capacitor. It is also defined as the loss proportional to the dissipated energy in the form of heat. Figure 8.13 shows that the dielectric losses of PGMA and BaTiO$_3$-PGMA are reported to be 0.030 and 0.039, respectively. Low dielectric losses in BaTiO$_3$-PGMA nanocomposite films might be due to well-dispersed BaTiO$_3$ nanoparticles as well as the insulating PGMA shell around each nanoparticle.

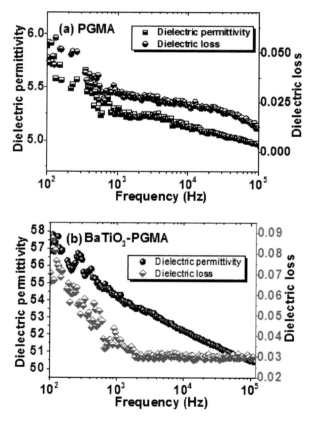

Figure 8.13 Frequency-dependent dielectric properties at 1 kHZ–100 kHz: (a) PGMA and (b) BaTiO$_3$-PGMA nanocomposite. Reproduced from Ref. [126] with permission from Wiley.

Pristine PGMA films have decreased dielectric losses as the frequency is increased from 1 kHz to 100 kHz. Further, the BaTiO$_3$-PGMA nanocomposite films followed a nearly constant frequency-independent behavior from 2 kHz to 100 kHz. The report suggests that the BaTiO$_3$ nanoparticles have shown more dielectric losses compared to PGMA and BaTiO$_3$-PGMA nanocomposites [126]. The low dielectric loss in the BaTiO$_3$-PGMA nanocomposite films might be due to the insulating PGMA layer around each nanoparticle and the enhanced dispersion of the BaTiO$_3$ nanoparticles. As mentioned earlier, the energy densities of pure PGMA and BaTiO$_3$-PGMA nanocomposites were calculated using linear dielectric capacitor equations at the dielectric breakdown strength maximum. These results are promising when compared to reported surface-modified BaTiO$_3$-PS nanocomposites or PMMA-grafted-BaTiO$_3$ nanoparticles for energy storage [126]. Core-satellite Ag@BaTiO$_3$ nanoassemblies of polymer composites have reported a higher dielectric constant ε of ~45 but failed to achieve higher energy densities [127–129]. Guo et al. [60] reported well-dispersed PP-grafted BaTiO$_3$, TiO$_2$, and ZrO$_2$ nanocomposites by in situ polymerization techniques. These nanocomposites exhibited very low leakage current densities and a low dielectric constant of ~6.1, with an energy density of ~9.4 J/cm^3, at a dielectric breakdown strength of around 6 MV/cm. Also, Fredin et al. [78] also reported highly dispersed PP-grafted BaTiO$_3$, ZrO$_2$, SrTiO$_3$, Mgo, and Ba$_{0.5}$Sr$_{0.5}$TiO$_3$ nanocomposites with an energy density as high as 15.6 J/cm^3. Recently Adireddy et al. [130] also reported flexible freestanding Ba$_{0.5}$Sr$_{0.5}$TiO$_3$-PVDF nanocomposite films with a dielectric constant of ~27 with an energy density of ~9.7 J/cm^3 at a dielectric breakdown strength of around 2.85 MV/cm. Xie et al. reported a high dielectric constant of 20–55 and a low dielectric loss, with an energy density of ~0.03 J/cm^3, at 1.2 MV/cm.

It can be concluded from the above-mentioned literature that the present BaTiO$_3$-PGMA nanocomposites are promising materials for energy storage applications because of their facile synthesis, low cost, and rapid response to the applied electric field.

Variation of AC electrical conductivity (σ_{ac}) of PGMA and BaTiO$_3$-PGMA nanocomposites as a function of frequency at room temperature is shown in Fig. 8.14. The AC electrical conductivity was calculated from dielectric properties using the relation $\sigma_{ac} = \varepsilon_0 \varepsilon_r \omega \tan(\delta)$. The BaTiO$_3$-PGMA nanocomposite has shown a higher σ_{ac} than pristine PGMA capacitors. It is observed that the conductivity increases with an increase in frequency in both PGMA and BaTiO$_3$-PGMA nanocomposites. The continuous solid line denotes the fit of experimental data to a power law.

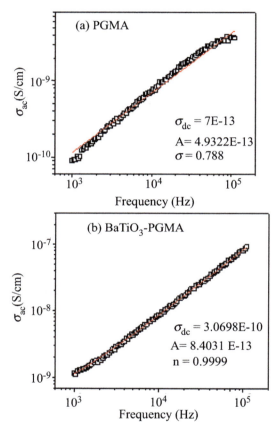

Figure 8.14 Variation of AC conductivity with the frequency of (a) a PGMA nanocomposite and (b) a BaTiO$_3$-PGMA nanocomposite. Reproduced from Ref. [126] with permission from Wiley.

The universal power law explains the frequency-dependent behavior, that is, $\sigma_{ac} = \sigma_0 + A\omega^n$, where σ_0 is the frequency-independent conductivity, ω is the angular frequency, A is the pre-exponential factor, and n is the power law exponent ($0 < n < 1$). The frequency exponent (n) and σ_{ac} for PGMA were calculated to be $n \approx 0.78$ and 7×10^{-13}, respectively.

Figure 8.15 shows the current-electric field (I-E) curves of PGMA and BaTiO$_3$-PGMA nanocomposites at room temperature. A low leakage current I of 10^{-11}–10^{-9} A was observed for PGMA and BaTiO$_3$-PGMA nanocomposites before the electrical breakdown of the capacitor films. The low leakage current means that the capacitor can store more energy, whereas a high leakage current means more power is consumed, and, therefore, less energy can be stored.

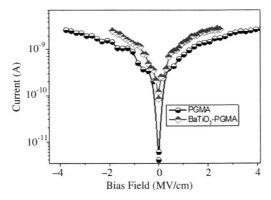

Figure 8.15 Electric field–dependent leakage current behavior of PGMA and BaTiO$_3$-PGMA nanocomposites. Reproduced from Ref. [126] with permission from Wiley.

8.5 Conclusions

Core-shell nanostructures have emerged as valuable and versatile structures for energy storage because of their additive or complementarily enhanced properties compared to their one-component counterparts. Recent developments in the preparation of core-shell nanostructures and their energy storage applications, including ATRP, RAFT, and click reactions, have been summarized in this chapter. Advanced techniques used for the characterization of core-shell nanostructures have also been briefly illustrated. The key structural factors affecting the energy storage activity of the nanostructured materials, namely dielectric constant, nanoparticle dispersion, dielectric loss, and energy density, were identified.

In particular, $BaTiO_3$-PGMA core-shell nanostructures were examined in detailed. $BaTiO_3$-PGMA core-shell nanostructures with a well-defined 20 nm PGMA coating were successfully prepared by SI-ATRP of GMA from the surface of $BaTiO_3$ nanoparticles. A series of electronic measurements were made to compare PGMA films and composites of the grafted $BaTiO_3$ nanoparticles. Improved dielectric properties were obtained due to well-dispersed $BaTiO_3$ nanoparticles as well as the insulating PGMA shell around each particle. Also, low leakage currents of the order of 10^{-11} to 10^{-9} A were observed in both pure and nanocomposite films. Also, high energy densities and high energy storage efficiencies were obtained for the core-shell nanostructures. AC conductivity measurements showed good agreement with the universal power law. The significant improvement observed for polymer-grafted $BaTiO_3$ nanocomposites highlights the promise of grafted nanomaterials for optimizing future high-energy-storage-capacitor materials.

These advancements in core-shell nanostructures are expected to provide a solid and stable platform for the development of energy storage benign protocols in the near future.

References

1. Spanhel, L., Weller, H., and Henglein, A. (1987). Photochemistry of semiconductor colloids. 22. Electron ejection from illuminated cadmium sulfide into attached titanium and zinc oxide particles, *J. Am. Chem. Soc.*, **109**, pp. 6632–6635.

2. Youn, H. C., Baral, S., and Fendler, J. H. (1988). Dihexadecyl phosphate, vesicle-stabilized and in situ generated mixed cadmium sulfide and zinc sulfide semiconductor particles: preparation and utilization for photosensitized charge separation and hydrogen generation, *J. Phys. Chem.*, **92**, pp. 6320–6327.

3. Henglein, A. (1989). Small-particle research: physicochemical properties of extremely small colloidal metal and semiconductor particles, *Chem. Rev.*, **89**, pp. 1861–1873.

4. Hoener, C. F., Allan, K. A., Bard, A. J., Campion, A., Fox, M. A., Mallouk, T. E., Webber, S. E., and White, J. M. (1992). Demonstration of a shell-core structure in layered CdSe-ZnSe small particles by X-ray photoelectron and Auger spectroscopies, *J. Phys. Chem.*, **96**, pp. 3812–3817.

5. Zhou, H. S., Sasahara, H., Honma, I., Komiyama, H., and Haus, J. W. (1994). Coated semiconductor nanoparticles: the CdS/PbS system's photoluminescence properties, *Chem. Mater.*, **6**, pp. 1534–1541.

6. Chen, Z., Wang, Z. L., Zhan, P., Zhang, J. H., Zhang, W. Y., Wang, H. T., and Ming, N. B. (2004). Preparation of metallodielectric composite particles with multishell structure, *Langmuir*, **20**, pp. 3042–3046.

7. Wagner, J., Autenrieth, T., and Hempelmann, R. (2002). Core shell particles consisting of cobalt ferrite and silica as model ferrofluids [$CoFe_2O_4$–SiO_2 core shell particles], *J. Magn. Magn. Mater.*, **252**, pp. 4–6.

8. Xie, X., Zhang, X., Zhang, H., Chen, D., and Fei, W. (2004). Preparation and application of surface-coated superparamagnetic nanobeads in the isolation of genomic DNA, *J. Magn. Magn. Mater.*, **277**, pp. 16–23.

9. West, J. L., and Halas, N. J. (2000). Applications of nanotechnology to biotechnology, *Curr. Opin. Biotechnol.*, **11**, pp. 215–217.

10. Sparnacci, K., Laus, M., Tondelli, L., Magnani, L., and Bernardi, C. (2002). Core–shell microspheres by dispersion polymerization as drug delivery systems, *Macromol. Chem. Phys.*, **203**, pp. 1364–1369.

11. Liz-Marzán, L. M., Giersig, M., and Mulvaney, P. (1996). Synthesis of nanosized gold–silica core–shell particles, *Langmuir*, **12**, pp. 4329–4335.

12. Tzika, P. A., Boyce, M. C., and Parks, D. M. (2000). Micromechanics of deformation in particle-toughened polyamides, *J. Mech. Phys. Solids*, **48**, pp. 1893–1929.

13. Duan, H. L., Jiao, Y., Yi, X., Huang, Z. P., and Wang, J. (2006). Solutions of inhomogeneity problems with graded shells and application to core-shell nanoparticles and composites, *J. Mech. Phys. Solids*, **54**, pp. 1401–1425.

14. Zhang, Q., Lee, I., Joo, J. B., Zaera, F., and Yin, Y. (2013). Core–shell nanostructured catalysts, *Acc. Chem. Res.*, **46**, pp. 1816–1824.

15. Ahmed, N. M. (2008). Modified zinc oxide-phosphate core-shell pigments in solvent-based paints, *Anti-Corros. Methods Mater.*, **55**, pp. 333–340.

16. Oldenburg, S. J., Averitt, R. D., Westcott, S. L., and Halas, N. J. (1998). Nanoengineering of optical resonances, *Chem. Phys. Lett.*, **288**, pp. 243–247.

17. Liu, N., Prall, B. S., and Klimov, V. I. (2006). Hybrid gold/silica/nanocrystal-quantum-dot superstructures: synthesis and analysis of semiconductor–metal interactions, *J. Am. Chem. Soc.*, **128**, pp. 15362–15363.

18. Zeng, H., Sun, S. H., Li, J., Wang, Z. L., and Liu, J. P. (2004). Tailoring magnetic properties of core/shell nanoparticles, *Appl. Phys. Lett.*, **85**, pp. 792–794.
19. Zhang, H. P., Yang, L. C., Fu, L. J., Cao, Q., Sun, D. L., Wu, Y. P., and Holze, R. (2009). Core-shell structured electrode materials for lithium ion batteries, *J. Solid State Electrochem.*, **13**, pp. 1521–1527.
20. Gawande, M. B., Goswami, A., Asefa, T., Guo, H., Biradar, A. V., Peng, D.-L., Zboril, R., and Varma, R. S. (2015). Core–shell nanoparticles: synthesis and applications in catalysis and electrocatalysis, *Chem. Soc. Rev.*, **44**, pp. 7540–7590.
21. Salgueiriño-Maceira, V., Correa-Duarte, M. A., Spasova, M., Liz-Marzán, L. M., and Farle, M. (2006). Composite silica spheres with magnetic and luminescent functionalities, *Adv. Funct. Mater.*, **16**, pp. 509–514.
22. Riggs, B. C., Adireddy, S., Rehm, C. H., Puli, V. S., Elupula, R., and Chrisey, D. B. (2015). Polymer nanocomposites for energy storage applications, *Mater. Today: Proc.*, **2**, pp. 3853–3863.
23. Adireddy, S., Puli, V. S., Sklare, S. C., Lou, T. J., Riggs, B. C., Elupula, R., Grayson, S. M., and Chrisey, D. B. (2014). PVDF–BaSrTiO$_3$ nanocomposites for flexible electrical energy storage devices, *Emerging Mater. Res.*, **3**, pp. 265–270.
24. Puli, V. S., Pradhan, D. K., Riggs, B. C., Adireddy, S., Katiyar, R. S., and Chrisey, D. B. (2014). Synthesis and characterization of lead-free ternary component BST–BCT–BZT ceramic capacitors, *J. Adv. Dielectrics*, **04**, p. 1450014.
25. Coondoo, I., Panwar, N., Puli, V. S., and Katiyar, R. S. (2011). Ferroelectric and piezoelectric studies on Mo-substituted SrBi$_2$Ta$_2$O$_9$ ferroelectric ceramics, *Integr. Ferroelectr.*, **124**, pp. 1–9.
26. Puli, V. S., Pradhan, D. K., Riggs, B. C., Chrisey, D. B., and Katiyar, R. S. (2014). Investigations on structure, ferroelectric, piezoelectric and energy storage properties of barium calcium titanate (BCT) ceramics, *J. Alloys Compd.*, **584**, pp. 369–373.
27. Puli, V. S., Pradhan, D. K., Chrisey, D. B., Tomozawa, M., Sharma, G. L., Scott, J. F., and Katiyar, R. S. (2012). Structure, dielectric, ferroelectric, and energy density properties of (1 − x)BZT–xBCT ceramic capacitors for energy storage applications, *J. Mater. Sci.*, **48**, pp. 2151–2157.
28. Xie, L., Huang, X., Huang, Y., Yang, K., and Jiang, P. (2013). Core@double-shell structured BaTiO$_3$–polymer nanocomposites with high dielectric constant and low dielectric loss for energy storage application, *J. Phys. Chem. C*, **117**, pp. 22525–22537.

29. Rabuffi, M., and Picci, G. (2002). Status quo and future prospects for metallized polypropylene energy storage capacitors, *IEEE Trans. Plasma Sci.*, **30**, pp. 1939–1942.
30. Ávila, H. A., Ramajo, L. A., Góes, M. S., Reboredo, M. M., Castro, M. S., and Parra, R. (2013). Dielectric behavior of epoxy/BaTiO$_3$ composites using nanostructured ceramic fibers obtained by electrospinning, *ACS Appl. Mater. Interfaces*, **5**, pp. 505–510.
31. Wang, Q., and Zhu, L. (2011). Polymer nanocomposites for electrical energy storage, *J. Polym. Sci., Part B: Polym. Phys.*, **49**, pp. 1421–1429.
32. Li, J., Seok, S. I., Chu, B., Dogan, F., Zhang, Q., and Wang, Q. (2009). Nanocomposites of ferroelectric polymers with TiO$_2$ nanoparticles exhibiting significantly enhanced electrical energy density, *Adv. Mater.*, **21**, pp. 217–221.
33. Dang, Z.-M., Yuan, J.-K., Zha, J.-W., Zhou, T., Li, S.-T., and Hu, G.-H. (2012). Fundamentals, processes and applications of high-permittivity polymer–matrix composites, *Prog. Mater. Sci.*, **57**, pp. 660–723.
34. Wen, F., Xu, Z., Xia, W., Wei, X., and Zhang, Z. (2013). High energy density nanocomposites based on poly(vinylidene fluoride-chlorotrifluoroethylene) and barium titanate, *Polym. Eng. Sci.*, **53**, pp. 897–904.
35. Beier, C. W., Sanders, J. M., and Brutchey, R. L. (2013). Improved breakdown strength and energy density in thin-film polyimide nanocomposites with small barium strontium titanate nanocrystal fillers, *J. Phys. Chem. C*, **117**, pp. 6958–6965.
36. Tomer, V., Polizos, G., Manias, E., and Randall, C. A. (2010). Epoxy-based nanocomposites for electrical energy storage. I: effects of montmorillonite and barium titanate nanofillers, *J. Appl. Phys.*, **108**, p. 074116.
37. Xie, L., Huang, X., Li, B.-W., Zhi, C., Tanaka, T., and Jiang, P. (2013). Core–satellite Ag@BaTiO$_3$ nanoassemblies for fabrication of polymer nanocomposites with high discharged energy density, high breakdown strength and low dielectric loss, *Phys. Chem. Chem. Phys.*, **15**, pp. 17560–17569.
38. Kim, P., Jones, S. C., Hotchkiss, P. J., Haddock, J. N., Kippelen, B., Marder, S. R., and Perry, J. W. (2007). Phosphonic acid-modified barium titanate polymer nanocomposites with high permittivity and dielectric strength, *Adv. Mater.*, **19**, pp. 1001–1005.
39. Kim, P., Doss, N. M., Tillotson, J. P., Hotchkiss, P. J., Pan, M.-J., Marder, S. R., Li, J., Calame, J. P., and Perry, J. W. (2009). High energy density

nanocomposites based on surface-modified BaTiO$_3$ and a ferroelectric polymer, *ACS Nano*, **3**, pp. 2581–2592.

40. Puli, V. S., Pradhan, D. K., Kumar, A., Katiyar, R. S., Su, X., Busta, C. M., Tomozawa, M., and Chrisey, D. B. (2012). Structure and dielectric properties of BaO–B2O$_3$–ZnO–[(BaZr$_{0.2}$Ti$_{0.80}$)O$_3$]$_{0.85}$ − [(Ba$_{0.70}$Ca$_{0.30}$)TiO$_3$]$_{0.15}$ glass–ceramics for energy storage, *J. Mater. Sci.: Mater. Electron.*, **23**, pp. 2005–2009.

41. Fu, Q., Yang, F., and Bao, X. (2013). Interface-confined oxide nanostructures for catalytic oxidation reactions, *Acc. Chem. Res.*, **46**, pp. 1692–1701.

42. O'Mahony, G. E., Ford, A., and Maguire, A. R. (2013). Asymmetric oxidation of sulphides, *J. Sulfur Chem.*, **34**, pp. 301–341.

43. Kaczorowska, K., Kolarska, Z., Mitka, K., and Kowalski, P. (2005). Oxidation of sulfides to sulfoxides. Part 2: oxidation by hydrogen peroxide, *Tetrahedron*, **61**, pp. 8315–8327.

44. Qiao, Z.-A., Zhang, P., Chai, S.-H., Chi, M., Veith, G. M., Gallego, N. C., Kidder, M., and Dai, S. (2014). Lab-in-a-shell: encapsulating metal clusters for size sieving catalysis, *J. Am. Chem. Soc.*, **136**, pp. 11260–11263.

45. Shokouhimehr, M., Shin, K.-Y., Lee, J. S., Hackett, M. J., Jun, S. W., Oh, M. H., Jang, J., and Hyeon, T. (2014). Magnetically recyclable core–shell nanocatalysts for efficient heterogeneous oxidation of alcohols, *J. Mater. Chem. A*, **2**, pp. 7593–7599.

46. Enache, D. I., Edwards, J. K., Landon, P., Solsona-Espriu, B., Carley, A. F., Herzing, A. A., Watanabe, M., Kiely, C. J., Knight, D. W., and Hutchings, G. J. (2006). Solvent-free oxidation of primary alcohols to aldehydes using Au-Pd/TiO$_2$ catalysts, *Science*, **311**, pp. 362–365.

47. Guo, X., Fu, Q., Ning, Y., Wei, M., Li, M., Zhang, S., Jiang, Z., and Bao, X. (2012). Ferrous centers confined on core–shell nanostructures for low-temperature CO oxidation, *J. Am. Chem. Soc.*, **134**, pp. 12350–12353.

48. Wu, C., Lim, Z.-Y., Zhou, C., Guo Wang, W., Zhou, S., Yin, H., and Zhu, Y. (2013). A soft-templated method to synthesize sintering-resistant Au-mesoporous-silica core–shell nanocatalysts with sub-5 nm single-cores, *Chem. Commun.*, **49**, pp. 3215–3217.

49. Paniagua, S. A., Kim, Y., Henry, K., Kumar, R., Perry, J. W., and Marder, S. R. (2014). Surface-initiated polymerization from barium titanate nanoparticles for hybrid dielectric capacitors, *ACS Appl. Mater. Interfaces*, **6**, pp. 3477–3482.

50. Maliakal, A., Katz, H., Cotts, P. M., Subramoney, S., and Mirau, P. (2005). Inorganic oxide core, polymer shell nanocomposite as a high K gate

dielectric for flexible electronics applications, *J. Am. Chem. Soc.*, **127**, pp. 14655–14662.

51. Tang, H., Wang, P., Zheng, P., and Liu, X. (2016). Core-shell structured BaTiO$_3$@polymer hybrid nanofiller for poly(arylene ether nitrile) nanocomposites with enhanced dielectric properties and high thermal stability, *Compos. Sci. Technol.*, **123**, pp. 134–142.

52. Liang, X., and Nazar, L. F. (2016). In situ reactive assembly of scalable core–shell sulfur–MnO$_2$ composite cathodes, *ACS Nano*, **10**(4), 4192–4198.

53. Shin, J., Lee, K. Y., Yeo, T., and Choi, W. (2016). Facile one-pot transformation of iron oxides from Fe$_2$O$_3$ nanoparticles to nanostructured Fe$_3$O$_4$@C core-shell composites via combustion waves, *Sci. Rep.*, **6**, p. 21792.

54. Peng, L., Zhang, H., Fang, L., Bai, Y., and Wang, Y. (2016). Designed functional systems for high-performance lithium-ion batteries anode: from solid to hollow, and to core–shell NiCo$_2$O$_4$ nanoparticles encapsulated in ultrathin carbon nanosheets, *ACS Appl. Mater. Interfaces*, **8**, pp. 4745–4753.

55. Yu, S., Qin, F., and Wang, G. (2016). Improving the dielectric properties of poly(vinylidene fluoride) composites by using poly(vinyl pyrrolidone)-encapsulated polyaniline nanorods, *J. Mater. Chem. C*, **4**, pp. 1504–1510.

56. Xie, L., Huang, X., Wu, C., and Jiang, P. (2011). Core-shell structured poly(methyl methacrylate)/BaTiO$_3$ nanocomposites prepared by in situ atom transfer radical polymerization: a route to high dielectric constant materials with the inherent low loss of the base polymer, *J. Mater. Chem.*, **21**, pp. 5897–5906.

57. Yang, K., Huang, X., Xie, L., Wu, C., Jiang, P., and Tanaka, T. (2012). Core–shell structured polystyrene/BaTiO$_3$ hybrid nanodielectrics prepared by in situ RAFT polymerization: a route to high dielectric constant and low loss materials with weak frequency dependence, *Macromol. Rapid Commun.*, **33**, pp. 1921–1926.

58. Huang, X., and Jiang, P. (2015). Core–shell structured high-k polymer nanocomposites for energy storage and dielectric applications, *Adv. Mater.*, **27**, pp. 546–554.

59. Yang, K., Huang, X., Huang, Y., Xie, L., and Jiang, P. (2013). Fluoropolymer@BaTiO$_3$ hybrid nanoparticles prepared via RAFT polymerization: toward ferroelectric polymer nanocomposites with high dielectric constant and low dielectric loss for energy storage application, *Chem. Mater.*, **25**, pp. 2327–2338.

60. Guo, N., DiBenedetto, S. A., Tewari, P., Lanagan, M. T., Ratner, M. A., and Marks, T. J. (2010). Nanoparticle, size, shape, and interfacial effects on leakage current density, permittivity, and breakdown strength of metal oxide–polyolefin nanocomposites: experiment and theory. *Chem. Mater.*, **22**, pp. 1567–1578.

61. Li, H., Bian, Z., Zhu, J., Zhang, D., Li, G., Huo, Y., Li, H., and Lu, Y. (2007). Mesoporous titania spheres with tunable chamber stucture and enhanced photocatalytic activity, *J. Am. Chem. Soc.*, **129**, pp. 8406–8407.

62. Fredin, L. A., Li, Z., Ratner, M. A., Lanagan, M. T., and Marks, T. J. (2012). Enhanced energy storage and suppressed dielectric loss in oxide core-shell–polyolefin nanocomposites by moderating internal surface area and increasing shell thickness, *Adv. Mater.*, **24**, pp. 5946–5953.

63. Fredin, L. A., Li, Z., Lanagan, M. T., Ratner, M. A., and Marks, T. J. (2013). Substantial recoverable energy storage in percolative metallic aluminum-polypropylene nanocomposites, *Adv. Funct. Mater.*, **23**, pp. 3560–3569.

64. Fredin, L. A., Li, Z., Lanagan, M. T., Ratner, M. A., and Marks, T. J. (2013). Sustainable high capacitance at high frequencies: metallic aluminum-polypropylene nanocomposites, *ACS Nano*, **7**, pp. 396–407.

65. Tchoul, M. N., Fillery, S. P., Koerner, H., Drummy, L. F., Oyerokun, F. T., Mirau, P. A., Durstock, M. F., and Vaia, R. A. (2010). Assemblies of titanium dioxide-polystyrene hybrid nanoparticles for dielectric applications, *Chem. Mater.*, **22**, pp. 1749–1759.

66. Li, Z., Fredin, L. A., Tewari, P., DiBenedetto, S. A., Lanagan, M. T., Ratner, M. A., and Marks, T. J. (2010). In situ catalytic encapsulation of core-shell nanoparticles having variable shell thickness: dielectric and energy storage properties of high-permittivity metal oxide nanocomposites, *Chem. Mater.*, **22**, pp. 5154–5164.

67. Jung, H. M., Kang, J.-H., Yang, S. Y., Won, J. C., and Kim, Y. S. (2010). Barium titanate nanoparticles with diblock copolymer shielding layers for high-energy density nanocomposites, *Chem. Mater.*, **22**, pp. 450–456.

68. Sreerama, S. G., Elupula, R., Laurent, B. A., Zhang, B., and Grayson, S. M. (2014). Use of MALDI-ToF MS to elucidate the structure of oligomeric impurities formed during "click" cyclization of polystyrene, *React. Funct. Polym.*, **80**, pp. 83–94.

69. Elupula, R., Laurent, B. A., and Grayson, S. M. (2006). Advances in the synthesis of cyclic polymers, In *Materials Science and Technology* (Wiley-VCH Verlag GmbH & Co. KGaA).

70. Yang, K., Huang, X., Zhu, M., Xie, L., Tanaka, T., and Jiang, P. (2014). Combining RAFT polymerization and thiol–ene click reaction for core-shell structured polymer@BaTiO$_3$ nanodielectrics with high dielectric constant, low dielectric loss, and high energy storage capability, *Appl. Mater. Interfaces*, **6**, 1812–1822.
71. Li, J., Claude, J., Norena-Franco, L. E., Seok, S. I., and. Wang, Q. (2008). Electrical energy storage in ferroelectric polymer nanocomposites containing surface-functionalized BaTiO$_3$ nanoparticles, *Chem. Mater.*, **20**, pp. 6304–6306.
72. Chu, B., Neese, B., Lin, M., Lu, S.-G., and Zhang, Q. M. (2008). Enhancement of dielectric energy density in the poly(vinylidene fluoride)-based terpolymer/copolymer blends,, *Appl. Phys. Lett.*, **93**, p. 152903.
73. Stefanescu, E. A., Tan, X., Lin, Z., Bowler, N., and Kessler, M. R. (2010). Multifunctional PMMA-ceramic composites as structural dielectrics, *Polymer*, **51**, pp. 5823–5832.
74. Ramesh, S., Shutzberg, B. A., Huang, C., Jie, G., and Giannelis, E. P. (2003). Dielectric nanocomposites for integral thin film capacitors: materials design, fabrication and integration issues, *IEEE Trans. Adv. Packag.*, **26**, pp. 17–24.
75. Yang, T.-I., and Kofinas, P. (2007). Dielectric properties of polymer nanoparticle composites, *Polymer*, **48**, pp. 791–798.
76. Huang, X., Zhang, X., and Jiang, H. (2014). Energy storage via polyvinylidene fluoride dielectric on the counterelectrode of dye-sensitized solar cells, *J. Power Sources*, **248**, pp. 434–438.
77. Chu, B., Zhou, X., Neese, B., Zhang, Q. M., and Bauer, F. (2006). Relaxor ferroelectric polymer–poly(vinylidene fluoride-trifluoroethylene-chlorofluoroethylene) terpolymer high electric energy density and field dependent dielectric response, *Ferroelectrics*, **331**, pp. 35–42.
78. Guo, N., DiBenedetto, S. A., Kwon, D.-K., Wang, L., Russell, M. T., Lanagan, M. T., Facchetti, A., and Marks, T. J. (2007). Supported metallocene catalysis for in situ synthesis of high energy density metal oxide nanocomposites, *J. Am. Chem. Soc.*, **129**, pp. 766–767.
79. Fredin, L. A., Li, Z., Ratner, M. A., Lanagan, M. T., and Marks, T. J. (2012). Energy storage: enhanced energy storage and suppressed dielectric loss in oxide core–shell–polyolefin nanocomposites by moderating internal surface area and increasing shell, *Adv. Mater.*, **24**, pp. 5945–5945.
80. Wang, Y., Wei, C., Cong, H., Yang, Q., Wu, Y., Su, B., Zhao, Y., Wang, J., and Jiang, L. (2016). Hybrid top-down/bottom-up strategy using

superwettability for the fabrication of patterned colloidal assembly, *ACS Appl. Mater. Interfaces*, **8**, pp. 4985–4993.

81. Singh, A. V., and Mehta, K. K. (2016). Top-down versus bottom-up nanoengineering routes to design advanced oropharmacological products, *Curr. Pharm. Des.*, **22**, pp. 1534–1545.

82. Gregorczyk, K., and Knez, M. (2016). Hybrid nanomaterials through molecular and atomic layer deposition: Top down, bottom up, and in-between approaches to new materials, *Prog. Mater. Sci.*, **75**, pp. 1–37.

83. Yang, L., Chu, D., Wang, L., Ge, G., and Sun, H. (2016). Facile synthesis of porous flower-like SrCO3 nanostructures by integrating bottom-up and top-down routes, *Mater. Lett.*, **167**, pp. 4–8.

84. Meng, Y., Yu, T., Zhang, S., and Deng, C. (2016). Top-down synthesis of muscle-inspired alluaudite Na$_2$+2xFe$_2$-x(SO$_4$)$_3$/SWNT spindle as a high-rate and high-potential cathode for sodium-ion batteries, *J. Mater. Chem. A*, **4**, pp. 1624–1631.

85. Li, P., Yan, X., He, Z., Ji, J., Hu, J., Li, G., Lian, K., and Zhang, W. (2016). α-Fe2O3 concave and hollow nanocrystals: top-down etching synthesis and their comparative photocatalytic activities, *CrystEngComm*, **18**, pp. 1752–1759.

86. Engels, J. F., Roose, J., Zhai, D. S., Yip, K. M., Lee, M. S., Tang, B. Z., and Renneberg, R. (2016). Aggregation-induced emissive nanoparticles for fluorescence signaling in a low cost paper-based immunoassay, *Colloids Surf. B*, **143**, pp. 440–446.

87. Wais, U., Jackson, A. W., Zuo, Y., Xiang, Y., He, T., and Zhang, H. (2016). Drug nanoparticles by emulsion-freeze-drying via the employment of branched block copolymer nanoparticles, *J. Controlled Release*, **222**, pp. 141–150.

88. Cross, C. E., Hemminger, J. C., and Penner, R. M. (2007). Physical vapor deposition of one-dimensional nanoparticle arrays on graphite: seeding the electrodeposition of gold nanowires, *Langmuir*, **23**, pp. 10372–10379.

89. Wang, X. M., Zuo, J., Keil, P., and Grundmeier, G. (2007). Comparing the growth of PVD silver nanoparticles on ultra thin fluorocarbon plasma polymer films and self-assembled fluoroalkyl silane monolayers, *Nanotechnology*, **18**, p. 265303.

90. Chen, X. H., Xia, J. T., Peng, J. C., Li, W. Z., and Xie, S. S. (2000). Carbon-nanotube metal-matrix composites prepared by electroless plating, *Compos. Sci. Technol.*, **60**, pp. 301–306.

91. Dupuis, A. C. (2005). The catalyst in the CCVD of carbon nanotubes: a review, *Prog. Mater. Sci.*, **50**, pp. 929–961.

92. Li, Y., Liu, J., Wang, Y. Q., and Wang, Z. L. (2001). Preparation of monodispersed Fe–Mo nanoparticles as the catalyst for CVD synthesis of carbon nanotubes, *Chem. Mater.*, **13**, pp. 1008–1014.

93. Su, F. B., Zeng, J. H., Bao, X. Y., Yu, Y. S., Lee, J. Y., and Zhao, X. S. (2005). Preparation and characterization of highly ordered graphitic mesoporous carbon as a Pt catalyst support for direct methanol fuel cells, *Chem. Mater.*, **17**, pp. 3960–3967.

94. Puli, V. S., Martínez, R., Kumar, A., Scott, J. F., and Katiyar, R. S. (2011). A quaternary lead based perovskite structured materials with diffuse phase transition behavior, *Mater. Res. Bull.*, **46**, pp. 2527–2530.

95. Sreenivas Puli, V., Pradhan, D. K., Pérez, W., and Katiyar, R. S. (2013). Structure, dielectric tunability, thermal stability and diffuse phase transition behavior of lead free BZT–BCT ceramic capacitors, *J. Phys. Chem. Solids*, **74**, pp. 466–475.

96. Adireddy, S., Carbo, C. E., Rostamzadeh, T., Vargas, J. M., Spinu, L., and Wiley, J. B. (2014). Peapod-type nanocomposites through the in situ growth of gold nanoparticles within preformed hexaniobate nanoscrolls, *Angew. Chem. Int. Ed.*, **53**, pp. 4614–4617.

97. Adireddy, S., Carbo, C. E., Yao, Y., Vargas, J. M., Spinu, L., and Wiley, J. B. (2013). High-yield solvothermal synthesis of magnetic peapod nanocomposites via the capture of preformed nanoparticles in scrolled nanosheets, *Chem. Mater.*, **25**, pp. 3902–3909.

98. Adireddy, S., Lin, C., Cao, B., Zhou, W., and Caruntu, G. (2010). Solution-based growth of monodisperse cube-like BaTiO$_3$ colloidal nanocrystals, *Chem. Mater.*, **22**, pp. 1946–1948.

99. Adireddy, S., Lin, C., Palshin, V., Dong, Y., Cole, R., and Caruntu, G. (2009). Size-controlled synthesis of quasi-monodisperse transition-metal ferrite nanocrystals in fatty alcohol solutions, *J. Phys. Chem. C*, **113**, pp. 20800–20811.

100. Zheng, W. J., Liu, X. D., Yan, Z. Y., and Zhu, L. J. (2009). Ionic liquid-assisted synthesis of large-scale TiO$_2$ nanoparticles with controllable phase by hydrolysis of TiCl$_4$, *ACS Nano*, **3**, pp. 115–122.

101. Zhou, Y., Yu, S. H., Cui, X. P., Wang, C. Y., and Chen, Z. Y. (1999). Formation of silver nanowires by a novel solid–liquid phase arc discharge method, *Chem. Mater.*, **11**, pp. 545–+.

102. Yu, S. H. (2001). Hydrothermal/solvothermal processing of advanced ceramic materials, *J. Ceram. Soc. Jpn.*, **109**, pp. S65–S75.

103. Naik, S. R., Salker, A. V., Yusuf, S. M., and Meena, S. S. (2013). Influence of Co^{2+} distribution and spin–orbit coupling on the resultant magnetic

properties of spinel cobalt ferrite nanocrystals, *J. Alloys Compd.*, **566**, pp. 54–61.

104. Pietras, P., Przekop, R., and Maciejewski, H. (2013). New approach to preparation of gelatine/SiO$_2$ hybrid systems by the sol-gel process, *Ceram. Silik.*, **57**, pp. 58–65.

105. Serrano, E., Linares, N., Garcia-Martinez, J., and Berenguer, J. R. (2013). Sol–gel coordination chemistry: building catalysts from the bottom-up, *ChemCatChem*, **5**, pp. 844–860.

106. Valenzuela, F., Covarrubias, C., Martinez, C., Smith, P., Diaz-Dosque, M., and Yazdani-Pedram, M. (2012). Preparation and bioactive properties of novel bone-repair bionanocomposites based on hydroxyapatite and bioactive glass nanoparticles, *J. Biomed. Mater. Res. Part B-Appl. Biomater.*, **100B**, pp. 1672–1682.

107. Wiglusz, R. J., Grzyb, T., Bednarkiewicz, A., Lis, S., and Strek, W. (2012). Investigation of structure, morphology, and luminescence properties in blue-red emitter, europium-activated ZnAl$_2$O$_4$ nanospinels, *Eur. J. Inorg. Chem.*, pp. 3418–3426.

108. Amendola, V., and Meneghetti, M. (2009). Laser ablation synthesis in solution and size manipulation of noble metal nanoparticles, *Phys. Chem. Chem. Phys.*, **11**, pp. 3805–3821.

109. Asahi, T., Sugiyama, T., and Masuhara, H. (2008). Laser fabrication and spectroscopy of organic nanoparticles, *Acc. Chem. Res*, **41**, pp. 1790–1798.

110. Eliezer, S., Eliaz, N., Grossman, E., Fisher, D., Gouzman, I., Henis, Z., Pecker, S., Horovitz, Y., Fraenkel, M., Maman, S., and Lereah, Y. (2004). Synthesis of nanoparticles with femtosecond laser pulses, *Phys. Rev. B*, **69**, p. 144119.

111. Geohegan, D. B., Puretzky, A. A., Duscher, G., and Pennycook, S. J. (1998). Time-resolved imaging of gas phase nanoparticle synthesis by laser ablation, *Appl. Phys. Lett.*, **72**, pp. 2987–2989.

112. Sakamoto, M., Fujistuka, M., and Majima, T. (2009). Light as a construction tool of metal nanoparticles: synthesis and mechanism, *J. Photochem. Photobiol. C*, **10**, pp. 33–56.

113. Caro, J., Noack, M., Kolsch, P., and Schafer, R. (2000). Zeolite membranes: state of their development and perspective, *Microporous Mesoporous Mater.*, **38**, pp. 3–24.

114. Chichkov, B. N., Momma, C., Nolte, S., von Alvensleben, F., and Tunnermann, A. (1996). Femtosecond, picosecond and nanosecond laser ablation of solids, *Appl. Phys. A*, **63**, pp. 109–115.

115. Liu, X., Du, D., and Mourou, G. (1997). Laser ablation and micromachining with ultrashort laser pulses, *IEEE J. Quantum Electron.*, **33**, pp. 1706–1716.
116. Lowndes, D. H., Geohegan, D. B., Puretzky, A. A., Norton, D. P., and Rouleau, C. M. (1996). Synthesis of novel thin-film materials by pulsed laser deposition, *Science*, **273**, pp. 898–903.
117. Preuss, S., Demchuk, A., and Stuke, M. (1995). Sub-picosecond UV laser ablation of metals, *Appl. Phys. A*, **61**, pp. 33–37.
118. Willmott, P. R., and Huber, J. R. (2000). Pulsed laser vaporization and deposition, *Rev. Mod. Phys.*, **72**, pp. 315–328.
119. Yang, G. W. (2007). Laser ablation in liquids: applications in the synthesis of nanocrystals, *Prog. Mater. Sci.*, **52**, pp. 648–698.
120. Momma, C., Chichkov, B. N., Nolte, S., von Alvensleben, F., Tunnermann, A., Welling, H., and Wellegehausen, B. (1996). Short-pulse laser ablation of solid targets *Opt. Commun.*, **129**, pp. 134–142.
121. Russo, R. E., Mao, X. L., Liu, H. C., Gonzalez, J., and Mao, S. S. (2002). Laser ablation in analytical chemistry: a review, *Talanta*, **57**, pp. 425–451.
122. Lili, L., Xianwen, Z., and Jharna, C. (2014). Size control in the synthesis of 1–6 nm gold nanoparticles using folic acid-chitosan conjugate as a stabilizer, *Mater. Res. Express*, **1**, p. 035033.
123. Thanh, N. T. K., Maclean, N., and Mahiddine, S. (2014). Mechanisms of nucleation and growth of nanoparticles in solution, *Chem. Rev.*, **114**, pp. 7610–7630.
124. LaMer, V. K., and Dinegar, R. H. (1950). J. Theory, production and mechanism of formation of monodispersed hydrosols, *Am. Chem. Soc.*, **72**, 4847–4854.
125. Viswanatha, R., and Sarma, D. D. (2007). Growth of nanocrystals in solution, In *Nanomaterials Chemistry* (Wiley-VCH Verlag GmbH & Co. KGaA), pp. 139–170.
126. Ejaz, M., Puli, V. S., Elupula, R., Adireddy, S., Riggs, B. C., Chrisey, D. B., and Grayson, S. M. (2015). Core-shell structured poly(glycidyl methacrylate)/BaTiO$_3$ nanocomposites prepared by surface-initiated atom transfer radical polymerization: a novel material for high energy density dielectric storage, *J. Polym. Sci. Part A: Polym. Chem.*, **53**, pp. 719–728.
127. Xie, L., Huang, X., Yang, K., Li, S., and Jiang, P. (2014). "Grafting to" route to PVDF-HFP-GMA/BaTiO$_3$ nanocomposites with high dielectric

constant and high thermal conductivity for energy storage and thermal management applications, *J. Mater. Chem. A*, **2**, pp. 5244–5251.

128. Puli, V. S., Elupula, R., Riggs, B. C., Grayson, S. M., Katiyar, R. S., and Chrisey, D. B. (2014). Surface modified BaTiO$_3$-polystyrene nanocomposites for energy storage, *Int. J. Nanotechnol.*, **11**, pp. 910–920.

129. Zhang, W. L., Choi, H. J., and Seo, Y. (2013). Facile fabrication of chemically grafted graphene oxide–poly(glycidyl methacrylate) composite microspheres and their electrorheology, *Macromol. Chem. Phys.*, **214**, pp. 1415–1422.

130. Adireddy, S., Puli, V. S., Lou, T. J., Elupula, R., Sklare, S. C., Riggs, B. C., and Chrisey, D. B. (2014). Polymer-ceramic nanocomposites for high energy density applications, *J. Sol-Gel Sci. Technol.*, **73**, pp. 641–646.

Chapter 9

Nanoparticle Generation by Double-Pulse Laser Ablation

Emanuel Axente,[a] Tatiana E. Itina,[b] and Jörg Hermann[c]

[a]*National Institute for Lasers, Plasma and Radiation Physics, Lasers Department, "Laser-Surface-Plasma Interactions" Laboratory, PO Box MG-36, RO-77125, Bucharest-Magurele, Romania*
[b]*Laboratoire Hubert Curien, CNRS 5516, Université de Lyon, 42000 Saint-Etienne, France*
[c]*Laboratoire Lasers, Plasmas et Procédés Photoniques, LP3 CNRS – Aix-Marseille University, 13288 Marseille Cedex 9, France*
emanuel.axente@inflpr.ro, tatiana.itina@univ-st-etienne.fr, hermann@lp3.univ-mrs.fr

The purpose of this chapter is to give a critical review of the processes involved in the generation of nanoparticles by material ablation with two time-delayed laser pulses. Experimental investigations of the ablation characteristics during nanoparticle synthesis with short double pulses are presented. In particular, the composition and expansion dynamics of the ablated material are examined. The latest progress achieved in modeling laser–matter interactions in a double-pulse regime is discussed. The correlation between ablation efficiency and nanoparticle generation is assessed and compared

---

Pulsed Laser Ablation: Advances and Applications in Nanoparticles and Nanostructuring Thin Films
Edited by Ion N. Mihailescu and Anna Paola Caricato
Copyright © 2018 Pan Stanford Publishing Pte. Ltd.
ISBN 978-981-4774-23-9 (Hardcover), 978-1-315-18523-1 (eBook)
www.panstanford.com

with numerical simulations, the influence of interpulse delay on metals ablation being revealed.

9.1 Introduction

During the last decade, nanoparticles (NPs) have demonstrated huge potential for research and application in the fields of nanoscience and nanotechnology. Among them, noble metal NPs have proven to be in the forefront of developments for sustained advances in fuel cells, analytical sensors, nanobiotechnology, and nanomedicine due to their size-dependent electrical, optical, magnetic, and chemical properties [1–7]. Potential applications of noble metal NPs in biomedicine are related but not limited to targeted delivery of drugs and other substances [8, 9], chemical sensing and imaging applications [10, 11], detection and control of microorganisms [12], and cancer cell photothermolysis [13, 14]. This is first due to facile surface functionalization with specific biomolecules and second due to distinct optical properties related, for example, to localized surface plasmon resonance [3, 4]. Wet-chemical synthesis and functionalization of NPs are routinely used for obtaining a broad spectrum of nano-objects [4, 5], from the classical colloidal gold nanospheres, silver nanorods, and silica/gold nanoshells to "exotic architectures," such as nanocubes [15], nanorice [16], nanostars [17], and nanocages [18]. Detailed information about the toxicological hazard of NPs was reviewed by Murphy et al. [5] and De Jong et al. [3].

Laser ablation (LA) is a unique tool for the fabrication of NPs, exhibiting several advantages over the classical wet-chemical synthesis methods. On the one hand, the main advantage of the LA technique is the possibility to preserve material stoichiometry during the ablation process [19]. On the other hand, LA is a faster and cleaner procedure since toxicity is difficult to avoid in traditional chemical routes of synthesis [20]. Indeed, NPs are directly generated during laser irradiation of bulk targets in vacuum, gas, or liquid environments. The high versatility of the method that allows the independent variation of the irradiation parameters like laser wavelength, energy density, pulse duration, and repetition rate is worth mentioning. As a consequence, a certain degree of freedom for tailoring NP size and composition is possible while the fabrication of complicated nano-objects is a difficult task, most of the NPs being spherical. Recent advances in laser synthesis of NPs with both nanosecond and femtosecond pulses are reviewed in the book chapters published by Besner and Meunier [21], Voloshko and Itina [19], and Caricato et al. [20]. Moreover, the processes involved in short-pulse LA in vacuum and in a low-pressure background gas, the generation of NPs, and the deposition of NP-assembled films are discussed in Chapter 4 of this book.

Compared to nanosecond ablation, subpicosecond laser–matter interaction is significantly different, as illustrated in several experimental and theoretical studies [22–29]. It was shown that LA of metals with short laser pulses is an efficient tool to generate NPs having sizes in the range of a few nanometers [30, 31]. Contrarily, in the nanosecond regime, the particles ejected directly from the target surface have significantly larger sizes in the micrometer range [32], while smaller ones are formed by condensation during plume expansion. The difference in particle size is attributed to characteristic thermal regimes and thickness of the laser-heated layer [33, 34] for the two irradiation cases.

Despite the large number of experimental studies reported in literature, the physicochemical mechanisms involved in NP formation during LA are not yet fully understood. Two main approaches are currently applied for theoretical modeling of LA and NP formation. Both are based on the two-temperature model (TTM), considering that electrons and lattice have different temperatures under material excitation by short laser pulses [35]. The combination with a hydrodynamic model was employed in several numerical studies whereas the microscopic description using molecular dynamics (MD) was alternatively used in other numerical investigations. A large description of the theoretical models currently used in simulations, with their advantages and limitations, was reviewed (see Refs. [19, 20, 36, 37] and references therein). An improvement of the theoretical approaches requires, therefore, reliable experimental data for comparison.

A deeper understanding of short-pulse LA and NP generation from metal targets can be achieved from experiments using two time-delayed laser pulses. The so-called double-pulse (DP) technique became popular, in particular in material analysis by laser-induced breakdown spectroscopy (LIBS), where an enhanced analytical signal and a better signal-to-noise ratio are beneficial [38, 39]. Other experimental studies of DP-LA are related to investigations of plume dynamics [40], optimization of the NP size distribution in vacuum [41, 42] and liquids [43–47], probing electron–phonon coupling in metals [48], and modification of optical properties through ripple formation [49–52].

Modeling DP-LA with subpicosecond laser pulses is a difficult task compared to the description of the single-pulse regime. Several competing physical processes should be considered in the case of LA with time-delayed pulses. These are different laser absorption by the skin layer and in the subcritical plasma, electron thermodynamics, thermal conductivity, electron–phonon coupling, and the interaction of the pressure waves generated by time-delayed pulses [37]. Recent advances in the field of modeling short-pulse laser–matter interactions can be found in the studies reported by Povarnitsyn et al. [37, 53] and Roth et al. [54] and the references therein. Using a hybrid model that combines classical MD and an energy equation for free electrons [37], the authors evidenced an elevation of the electron temperature in the plume up to three times when increasing the interpulse delay from 0 to 200 ps. The effect was accompanied by a monotonic decrease of the ablation crater depth, in agreement with experimental studies on aluminum and copper [37, 53].

In this chapter, we give a critical review of short-pulse LA of metals and NP generation with two time-delayed femtosecond laser pulses. The next section is devoted to the description of typical experiments employed in DP laser–matter interactions. Both collinear and orthogonal irradiation geometries used for material processing and analyses are explained, and representative results are presented for both cases. In the following section, experimental results on the DP-LA of metals (gold and copper) and NP generation are presented and discussed with respect to the influence of the interpulse delay on plasma composition and crater depth. Examples from the literature, covering other materials and experimental parameters, are reviewed as well. The latest developments achieved in modeling laser–matter interactions in the DP regime are presented in the last section. Conclusions and perspectives of this fast-expanding research field are given at the end of the chapter.

9.2   Typical Experimental Design for Laser–Matter Interactions with Double Pulses

The latest technological developments offer a high degree of freedom in controlling the laser pulse characteristics, including the temporal shape, the spatial distribution, the spectral profile, and the polarization state [55]. There are several approaches for DP excitation and ablation experiments. Usually, in most of the studies and in particular in the nanosecond regime, material irradiation is performed using two laser sources synchronized by a delay generator in order to trigger the pulse-to-target energy delivery and to control the interpulse delay [56]. A second approach refers to a Michelson interferometric setup, mostly in the femtosecond regime, in which the laser pulses emitted by a sole source are splitted into two beams. A delay line is interposed on one arm of the interferometer, the DPs being generated after beam reconstruction [41, 48, 57]. Other recent concepts rely on programmable pulse-shaping techniques, emphasizing ultrafast pulse tailoring in the spatio-temporal domain. They are described in detail by Stoian et al. [55].

DP laser–matter interaction is generally performed in two distinct configurations with respect to the propagation direction of the pulses and their temporal sequence, the collinear geometry and the orthogonal geometry, as schematically depicted in Figs. 9.1 and 9.2. A high experimental versatility is available for material processing and analysis, several combinations being proposed in literature. Depending on the application, optimum energy delivery and subsequent material response can be achieved by the proper choice of irradiation parameters. Accordingly, different beam geometries, pulse widths, wavelengths, interpulse delays, and pulse energies [38, 58] are actively explored by the scientific community. The DP irradiation geometries are mostly used in material analyses by DP-LIBS, but they are also for the synthesis of thin films via pulsed laser deposition and for NP generation in vacuum, gas, or liquid environments.

9.2.1 Collinear Double-Pulse Interaction Geometry

The collinear configuration, in which two laser beams have the same propagation pathway, is the simplest but less versatile approach [58]. It is, however, the most used configuration, since it enables an easy alignment of the laser beams. For instance, it is mainly used in material analysis by DP-LIBS, combining different laser wavelengths (532/1064, 532/532, or 532/355 nm), as schematically depicted in Fig. 9.1a,b. Consequently, improved sensitivity and lower detection limits were achieved for lead detection in metal alloys when LIBS was performed with an additional laser pulse [59]. The authors attributed the enhanced sensitivity to the higher plasma temperature.

Experiments of collinear DP-LA in liquid were carried out for studying the formation mechanisms of silver NPs in water and revealed the fundamental role of the cavitation bubble dynamics in the synthesis of aqueous colloidal dispersions [45].

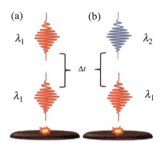

Figure 9.1 Scheme of double-pulse ablation experiments in collinear geometry using one (a) or two laser wavelengths (b).

9.2.2 Orthogonal Double-Pulse Interaction Geometry

In the orthogonal approach, one laser pulse ablates the sample (usually directed perpendicular to the surface) and the second pulse (propagating parallel to the sample surface) is applied either before, to form a preablation spark, or after, in order to reheat the plasma generated by the first pulse (Fig. 9.2a–c). Another possible arrangement is to irradiate the target at a specific angle (typically 45°), as schematically depicted in Fig. 9.2d. Although more complicated in terms of beam alignment, this experimental design is more versatile, and a broad range of pump-probe experiments are possible. Gautier et al. [60] demonstrated the intensity enhancement of emission lines, ranging from a factor of 2 to 100 during DP-LIBS analyses of aluminum alloys. The authors investigated the influence of interpulse delay and laser energy and the physical mechanisms responsible for signal optimization. A comprehensive review on material analyses by DP-LIBS, covering experimental studies and the fundamental mechanisms responsible for signal enhancement and applications, was published by Babushok and coworkers [38].

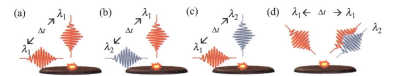

Figure 9.2 Orthogonal double-pulse interaction geometries using a unique laser (a), combining different wavelengths, energies, and pulse durations (b, c), and for prepulse/reheating mode or at specific angles (d).

9.2.3 Experiment for NP Generation with Delayed Short Laser Pulses

Although NP formation in vacuum, gas, or liquid environments is an intrinsic characteristic of short-pulse LA, studies demonstrated that the DP technique enables a better control of the NP synthesis. A proper choice of the interpulse delay leads to the reduction of NPs and to an increased atomization of the ablation plumes [41]. Both effects are suitable for material analyses by LIBS. Other possible applications are related to the deposition of ultrasmooth thin films by pulsed laser deposition [61] or for high-precision micromachining [62].

A schematic representation of the experimental setup used in the studies reviewed in this chapter is given in Fig. 9.3. The ablation experiments were performed with a Ti:sapphire laser source (λ = 800 nm, τ = 100 fs, and f_{rep} = 1 kHz). A square aperture of 2 × 2 mm^2 was used to select the central part of the Gaussian beam, which was imaged with an achromatic lens of 50 mm focal length onto the target surface. The two delayed laser pulses were obtained by turning the beam polarization, passing it through a half-wave plate, and splitting the beam with the aid of a polarized large band prism.

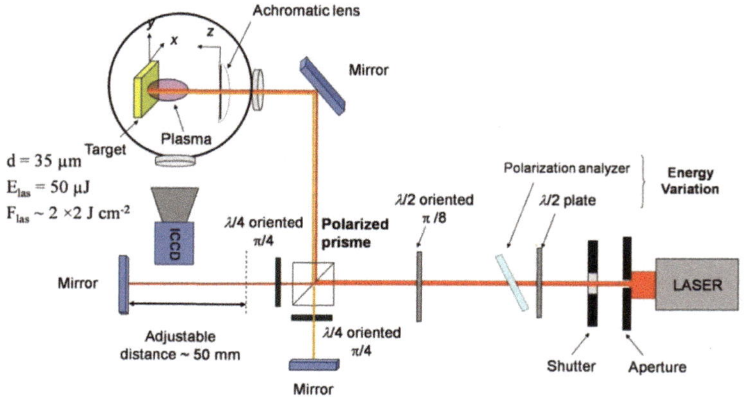

Figure 9.3 Schematic representation of the experimental setup.

Each beam crosses a quarter-wave plate before and after reflection on a mirror at 90° incidence. After recombination by the prism, two laser pulses of equal energy and orthogonal polarization were obtained. The interpulse delay was varied from 0 to 300 ps by adjusting the length of one beam path. According to the total laser pulse energy of 2 × 25 µJ incident onto the sample surface and a spot diameter of 35 µm, a maximum laser fluence of 2 × 2 J cm^{-2} was available. A mechanical shutter was used to control the number of applied DPs.

The metal targets were placed in a vacuum chamber of 10^{-4} Pa residual pressure. Inside the chamber (Fig. 9.4), the target holder and the focusing lens were mounted on motorized translation axes. A glass plate placed between target and lens prevented the latter to be coated by the ablated material. The glass plate was replaced regularly to minimize the laser beam extinction by the deposit. Fast plume imaging was performed with the aid of a focusing objective and an intensified charge-coupled device, orthogonal to the symmetry axis of plasma expansion (see Fig. 9.3). The delay between laser pulse and observation gate was set to 400 ns with the aid of a delayed pulse generator. During the experiments, 20 laser shots were applied at maximum to each irradiation site to avoid deep crater drilling and to keep the ablation process and plume dynamics reproducible.

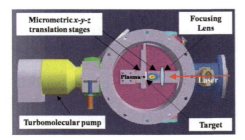

Figure 9.4 Up-side view of the reaction chamber used in the collinear ablation geometry.

For measurements that require data acquisition over more than 20 ablation events, several sites were irradiated by translating the target by 50 μm perpendicularly to the laser beam propagation. The laser-produced craters were investigated by optical microscopy after the ablation experiments. The ablated material was collected on mica substrates, placed parallel to the target at a separation distance of 20 mm from the target surface. Atomic force microscopy (AFM) operated in tapping mode was used to analyze the obtained NPs. Several areas of 2 × 2 μm^2 were scanned to characterize each sample in order to count a number of particles sufficiently large for statistical analysis of the particle size distribution.

9.3 Investigation of Nanoparticles Produced by Short Double-Pulse Laser Ablation of Metals

Several experimental and theoretical studies devoted to short-pulse LA evidenced that nanometer-sized particles represent a large fraction of the ablated material [25, 63–67]. This characteristic feature of short-pulse LA is attributed to the fast heating and energy relaxation. Two principal mechanisms of NP generation during the quasi-adiabatic plume expansion were proposed [68, 69]. One called "phase explosion" or "thermal decomposition" [70, 71] consists of the transformation of matter into a liquid-gas mixture that favors the NP generation. The other, "mechanical decomposition" of the metastable melt due to shock- and rarefaction waves, is considered another source of efficient NP generation [72]. The influence of pressure relaxation on LA physical processes was addressed by Chimier et al. [69] and Norman et al. [73]. Contrary to the nanosecond regime, NP formation by condensation is supposed to play a minor role in case of short-pulse LA under vacuum [22].

In the present section, we evaluate the influence of the interpulse delay on the composition of the laser-produced plasma and on the quantity of ablated copper and gold targets. To this purpose, a brief summary of the mechanisms involved in single-pulse metal ablation and NP generation during intense short-pulse laser irradiation is first presented in the next section.

9.3.1 Correlation between Ablation Efficiency and Nanoparticle Generation in the Single-Pulse Regime

Experimental investigations, correlating the properties of the plume with measurements of the laser-produced crater volumes, allowed us to evidence two distinct ablation regimes for copper and gold that were also observed for other metals [74–76]. The two regimes are related to the applied laser fluence (F_{las}) and strongly coupled with the plume composition [25].

In the low-fluence regime, a small increase in the ablation depth with F_{las} is observed and a large fraction of the ablated material is atomized. The high-fluence regime is characterized by efficient NP generation. As the ejection of clusters occurs when the absorbed energy density is smaller than the one required for complete metals atomization, the ablation efficiency increases with the amount of NPs within the plume [25].

Comparing the single-pulse LA of copper and gold, three main conclusions could be highlighted (see Fig. 9.5): (i) the transition fluence characteristic of each ablation regime is higher in the case of copper, (ii) both the ablation rate and efficiency are smaller for copper, and (iii) the relative fraction of NPs within the plume is smaller for copper.

These differences are attributed to the different heat regimes of the two metals. Compared to gold, copper has a much larger electron–lattice coupling and the energy transported toward the bulk via fast electron heat diffusion is lower and characterized by a smaller depth. Consequently, a steeper temperature depth profile and a thinner heat-affected zone are expected for copper. According to the larger energy density deposited in a thinner layer, the observed plume atomization is higher. Contrarily, the thicker melted zone in the case of gold favors NP generation. Since the thickness of the melted surface layer increases with F_{las} faster for gold, the onset of efficient NP generation occurs at a lower fluence for the precious metal.

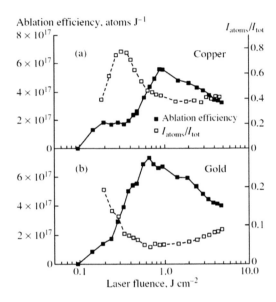

Figure 9.5 Ablation efficiency and atomization degree of copper (a) and gold (b) as functions of F_{las}. Reproduced from Ref. [25] with permission from Springer.

9.3.2 Influence of Interpulse Delay on Plume Composition

It was shown in the previous studies that the NP fraction generated during metal ablation only weakly depends on the laser fluence [64] but is strongly influenced by the strength of electron–lattice coupling [25]. This behavior was attributed to the competition between the fast electron heat transport and the energy transfer to the lattice that governs the metal's heat regime for irradiation with laser pulses shorter than the characteristic time of electron–lattice thermalization τ_{el}.

The results presented here were obtained with collinear DPs following the experimental procedure described in Section 9.2.3. Two plume images are shown in Fig. 9.6, one recorded during single-pulse LA of copper with F_{las} = 4 J cm^{-2} (Fig. 9.6a) and one with two time-delayed laser pulses of 2 J cm^{-2} delayed by 33 ps (Fig. 9.6b). The delay between observation gate and laser pulses was 400 ns. In both images, two distinct plume components are observed. A "slow" component, of high emission intensity, is located close to the target surface, whereas a "fast" component, of lower intensity, is observed at a larger distance. The splitting into two main velocity populations in the plasma expansion was observed for short-pulse LA of several metals [22, 24, 64]. Combined analysis by time-resolved imaging and optical emission spectroscopy revealed that the fast component contains neutral atoms mainly, whereas NPs dominate the slow one [25, 64, 77]. In the images below, the 10-level color palette is adjusted to the emission intensity of NPs.

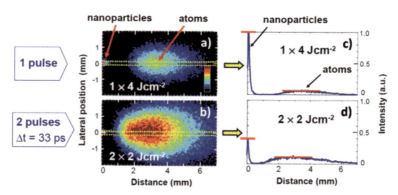

Figure 9.6 Plume images recorded during single-pulse LA of copper (a) and with two pulses delayed by 33 ps (b). The intensity profiles (c) and (d) were obtained from the images (a) and (b), respectively.

With respect to the single-pulse experiment (Fig. 9.6a), a large increase of the emission intensity of the atomized plume component is observed when two delayed pulses are applied (Fig. 9.6b). At the same time, the NP plume intensity decreases, as revealed by the intensity profiles presented in Fig. 9.6c,d. The profiles were obtained from the plume images (Fig. 9.6a,b) by averaging the signal intensity over several pixel rows along the plasma symmetry axis (see the yellow dashed lines in Fig. 9.6a,b). Moreover, it is shown that the maximum intensity position of the atomized plume component is shifted toward the target surface for the DP case. This behavior may be due to the interaction of the second pulse with the expanding plasma.

The emission intensities of atoms and NPs deduced from the plume images are presented in Fig. 9.7 as functions of interpulse delay for copper (Fig. 9.7a) and gold (Fig. 9.7b). Strong changes of the plume composition occur for both metals with increasing interpulse delay. Indeed, the emission intensity of atoms increases by a factor of four whereas the NP intensity decreases by the same amount. Moreover, the intensity changes occur at a shorter delay for the DP-LA of copper (Fig. 9.7a) with respect to the ablation of gold (Fig. 9.7b). The black lines in Fig. 9.7 present the intensity increase and decrease according to the formula in Eq. 9.1 below, suggesting thus different characteristic times for the observed plume composition changes.

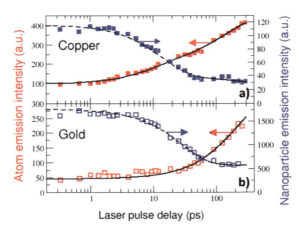

Figure 9.7 Emission intensities of atoms (red squares) and NPs (blue squares) deduced from plume images (see Fig. 9.6) versus the interpulse delay for two laser pulses of 2 J cm^{-2} [78].

The evolution of the emission intensity changes of both components versus interpulse delay Δt was described using the following biexponential function [41, 78]:

$$I(t) = I_0 + I_1(1 - e^{-\Delta t/\tau_1}) + I_2(1 - e^{-\Delta t/\tau_2}). \quad (9.1)$$

Here, I_0 is the intensity for an interpulse delay at $\Delta t = 0$, I_1 and I_2 are the amplitudes of intensity changes that occur on different timescales, characterized by τ_1 and τ_2, respectively. The continuous and dashed lines represent the emission intensities of atoms and NPs, respectively, calculated with the formula in Eq. 9.1 using the parameters listed in Table 9.1.

Table 9.1 Intensity values I_0, I_1, and I_2 and characteristic times τ_1 and τ_2 used to describe the emission intensity evolution of atoms and NPs as functions of interpulse delay using the biexponential function represented in Eq. 9.1

Metal	Plume component	I_0 (a.u.)	I_1 (a.u.)	I_2 (a.u.)	τ_1 (ps)	τ_2 (ps)
Cu	Atoms	95	105	240	10	140
Cu	NPs	117	−78	−12	13	200
Au	Atoms	45	0	240	–	200
Au	NPs	1740	−1170	0	30	–

The approximation by a biexponential function revealed two characteristic times for the observed plume compositional changes. The shorter time τ_1 was deduced to be about 10 ps for copper and 30 ps for gold. In good agreement with other studies [79, 80], these values are close to the characteristic times of electron-lattice relaxation τ_{el} = 1–10 ps and 30–100 ps for copper and gold, respectively. The second parameter τ_2 was found about 1 order of magnitude larger than τ_1. It was suggested that it characterizes a slower process, strongly influencing the atomic emission intensity increase, as revealed by the relatively large amplitude deduced (see I_2 value in Table 9.1). The slow process has a negligible influence on the NP plume component, as illustrated by the intensity ratio I_2/I_1 << 1.

The physical mechanisms involved in the modification of plume composition with increasing interpulse delay were explained by Noël and Hermann [41]. First, according to the TTM [35], during short-pulse LA the material properties depend on both electron and lattice temperatures, T_e and T_l, respectively. Thus, the electron heat conductivity k_e diminishes with the T_l, as predicted by the simplified expression $k_e \approx T_e/T_l$ proposed by Kanavin et al. [81] for a small T_e. Consequently, for delays $\Delta t \approx \tau_{el}$ the energy deposited by the second laser pulse is confined in a small volume since T_l is large and thus k_e is small at this stage and the energy transport toward the bulk reduced. The increased energy density promotes the growth of the atomization degree within the ablation plume. Second, for larger interpulse delays, the atomized plume component is heated up by inverse *Bremsstrahlung*, leading thus to an enhanced atomic emission intensity, as also reported by several authors [57, 82]. Indeed, complementary analyses by optical emission spectroscopy [78] evidenced the increase of plume temperature and ionization degree. The observation of plume temperature increase was also confirmed by modeling [37]. Moreover, an amplification of both mechanisms could be foreseen by the interaction of the second laser pulse with hot clusters and bubbles that start growing at times of the order of τ_{el} [41].

The decrease of the NP fraction with increasing interpulse delay was further confirmed by AFM analyses of the ablated material collected on mica substrates [41]. To demonstrate the strong decrease of NP number with Δt, we present in Fig. 9.8 AFM images of mica substrates after deposition with 500 laser DPs and different interpulse delays. With respect to ablation with $\Delta t = 0$ (Fig. 9.8a), the number of NPs is strongly reduced when increasing the interpulse delay to 6 ps (Fig. 9.8b). A further decrease was evidenced for longer Δt values, as illustrated in Fig. 9.8c, where only a few, smaller NPs are observed.

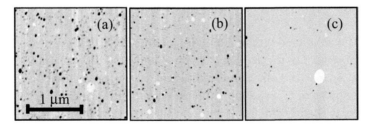

Figure 9.8 AFM images of copper NPs collected on mica substrates during ablation with two laser pulses of 2 J cm^{-2}, delayed by (a) 0, (b) 6, and (c) 90 ps [78].

9.3.3 Influence of Interpulse Delay on Ablation Depth and Crater Morphology

The ablation depth was measured in the following way: for each interpulse delay, a series of 10 craters was drilled, with an increasing number of applied DPs, as shown in Fig. 9.9. The depth of each crater was measured using optical microscopy by focusing on the sample's surface and the crater bottom [83]. According to the linear increase of crater depth z with the number of DPs n_{las}, the ablation depth was deduced from the slope $\Delta z/\Delta n_{las}$.

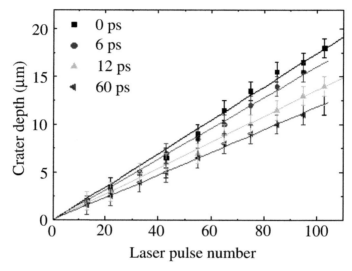

Figure 9.9 Crater depth versus the number of applied double pulses for ablation of copper with various interpulse delays. The ablation depth (per double pulse) was deduced from the slope [78].

The ablation depth as a function of interpulse delay is presented in Fig. 9.10 for the DP-LA of copper and gold. For an interpulse delay $\Delta t = 0$, the ablation depth equals the value obtained with a single pulse of $F_{las} = 4$ J cm^{-2}, whereas the depth measured for large delays is slightly smaller than the value observed for a single pulse of 2 J cm^{-2} fluence. The number of NPs deduced from the AFM analysis is presented in Fig. 9.10 as a function of interpulse delay and compared to the biexponential function (Eq. 9.1) that was previously used to characterize the NP emission intensity during copper ablation (see Fig. 9.7). Similar to the characterization of the plume emission intensities, we approximate the ablation depth by a biexponential function, replacing the intensity values in Eq. 9.1 by the appropriate depths (continuous and dashed lines for copper and gold, respectively). It is noted that the times τ_1 and τ_2 equal the characteristic times of the plume intensity evolution. The main contribution of the observed ablation depth decrease is attributed to the change of the sample's heat regime as we have $z_1 \gg z_2$, similar to the evolution of the NP emission intensity. This confirms that NPs present the major part of the ablated mass.

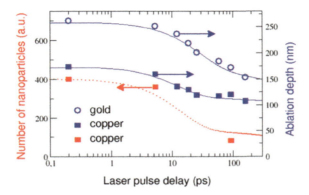

Figure 9.10 Number of Cu NPs (red color) deposited on mica substrates with double pulses of 2 × 2 J cm^{-2}. The decrease versus interpulse delay is similar to that of the NP emission intensity presented by the dotted line (taken from Fig. 9.7). The ablation depths of Cu (squares) and Au (circles) are also shown. The continuous and dashed lines were obtained using Eq. 9.1 for τ_1 = 12 and 30 ps, respectively, replacing the characteristic intensities by the appropriate depths [78].

This behavior is in agreement with the ablation depth measurements performed by Semerok and Dutouquet [57] after LA of aluminum and copper with short laser pulses of duration ranging from 50 fs to 2 ps. The authors reported that the craters obtained in the DP regime were almost two times deeper for $\Delta t < 1$ ps. For interpulse delays in the 1–10 ps range, the crater depth decreased with increasing interpulse delay. Finally, for $\Delta t \geq 10$ ps, the crater depth obtained with DPs equaled the depth drilled with a single pulse [57]. Moreover, the authors observed that, compared to single-pulse ablation, the use of DPs of equal total energy leads to a 10 times higher plasma emission intensity if interpulse delays in the range of 100 to 235 ps are applied.

The hybrid TTM-MD developed by Povarnitsyn et al. [37] confirms the measured crater depth evolution as a function of interpulse delay. It is concluded that the crater depth reduction is related to the formation of a high-pressure zone inside the plume. This suppresses the fragmentation in the rarefaction wave caused by the first pulse for $\Delta t \leq 20$ ps and merges the inner ablated layers of atoms with the target surface for delays ≥50 ps [37].

Although less investigated, the morphology of the craters drilled by DP in a material is also influenced by the interpulse delay. In the single-pulse regime, the best surface microprocessing results are usually obtained by applying low laser fluences, close to the ablation threshold. Le Harzic et al. [84] investigated DP laser-processing of metals (steel, Al, and Cu) in order to get insights into material microstructuring when a high-speed, high-fluence regime is required. The authors evidenced a significant quality improvement in metal microstructuring (less removed or recast matter) via DP-LA. For interpulse delays typically smaller than 1 ps, the structures formed in the crater bottoms are less pronounced when DPs were applied. They exhibit fewer spikes or pit craters, while the surface seems to be smoother.

Short DPs, of 180 fs duration and equal intensity, separated by an interpulse delay from a few hundred femtoseconds to 22 ps, were used by Spyridaki et al. [85] to irradiate silicon targets in vacuum. The authors found that for $\Delta t > 3$ ps, the second pulse couples efficiently to the liquid layer formed at the surface, leading to the total vaporization of the melt. As a result, a featureless structure without a residual cast is generated onto the silicon surface, in contrast to the single-pulse regime that produces significant thermal and hydrodynamic effects in the residual melt [85].

An improvement of the crater morphology was reported by Stoian et al. [86] in the case of short-pulse LA of dielectrics using temporally shaped pulse trains with subpicosecond separation. Cleaner structures with lower stress were thus obtained by the sequential energy delivery to the targets.

9.3.4 Overview of Other Investigations in the Field of Double-Pulse Laser–Matter Interactions

Experimental results obtained by DP-LA were reported in several studies, generally revealing similar observations. Semerok and Dutouquet [57] investigated the characteristics of LA and plasma reheating with ultrashort DPs, of 50 fs and 10 ps duration, time-delayed from 50 fs up to 250 ps. The authors studied the influence of the interpulse delay on the ablation features (efficiency, crater diameter, depth, volume, and shape) and on the plume properties (shape and dimensions, expansion velocity, intensity and lifetime, and pulse-to-pulse reproducibility) for aluminum and copper targets. They observed an increase of plume brightness and a decrease of crater depth for delays ranging from 1 to 10 ps. Both effects were attributed to partial plasma shielding, while a 100–200 ps delay was determined as optimum for plasma reheating by the second pulse. During DP-LA on the same metals, Le Harzic et al. [84] observed a reduced ablation depth and a smoother crater bottom when applying a second laser pulse with $\Delta t < 1$ ps. Scuderi et al. [82] performed plasma analyses via time-resolved optical emission spectroscopy during the DP-LA of titanium with pulses of 100 fs duration. The authors observed that the amount of NPs decreased with the interpulse delay, whereas the number of atoms and ions increased. The changes of plume composition were attributed to NP fragmentation. Moreover, they claimed the possibility of tailoring plume components over different characteristics, such as the kinetic energy of ions and neutrals and their relative fraction and over the NP production efficiency. Results on the DP-LA of nickel in vacuum using pulses of 250 fs duration delayed by 1–1000 ps were reported by Donnelly et al. [40]. The authors observed that the increase of the interpulse delay from 10 to 100 ps provoked a decrease of the crater volume by more than a factor of 2 (up to a value below the single-pulse regime), while the ion yield was strongly enhanced. They attributed this behavior to the interaction of the second laser pulse with the material ablated by the first pulse. The evolution of NPs ejected during the short-pulse LA (250 fs) of copper-based alloys and relative calibration plots of an fs-ns DP-LIBS orthogonal configuration were investigated by Guarnaccio et al. [87]. The authors stated that their particular DP-LIBS configuration can provide new perspectives for compositional analyses of materials. Recently, Garrelie et al. [88] investigated the effects of temporal laser pulse shaping during LA of aluminum and graphite targets in view of optimizing thin-film synthesis by femtosecond pulsed laser deposition. In situ optical diagnostic methods were used to monitor the expansion and excitation degree of atomic species and NPs.

The authors showed that the use of optimized pulse shapes leads to a decreased NP production by balancing the thermomechanical energy content. Short DP-LA (pulses of 100 fs separated by delays up to 90 ps) during drilling of aluminum and copper films was studied by Wang et al. [89]. The authors reported that the drilling efficiency in the DP regime is not higher compared to single-pulse drilling with the same total energy. The size of NPs redeposited around the drilled craters was found to depend on the metal (having different electron–phonon coupling parameters). In addition, the particle size was shown to decrease with an increasing interpulse delay. The authors concluded that the drilling process can be regulated by varying the interpulse delay and an even more efficient processing control is possible by using a shaped femtosecond laser pulse train [89].

Several studies were also devoted to the DP-LA of semiconductors and dielectrics [85, 90–93]. Analyses by time-of-flight mass spectrometry [85] and optical emission spectroscopy [92, 93] were performed during the DP-LA of silicon, demonstrating that the ion yield and the plume emission intensity significantly increase with the interpulse delay in the range of some tens of picoseconds. Moreover, the laser-produced crater seemed smoother [85]. These changes were attributed to alterations of the optical properties when the material reaches the melted state. With respect to solid silicon, the melt exhibits a decreased optical penetration depth, leading thus to increased laser energy coupling. The optical breakdown thresholds in silica and silicon were measured using a DP of 40 fs duration by Deng et al. [91]. By varying laser energy and interpulse delay, they found that the total energy required for breakdown decreases for silica and increases for silicon with the elevation of the first pulse energy. Chowdhury et al. [90] conducted pump-probe experiments on the femtosecond DP-LA of fused silica by investigating the plasma dynamics versus interpulse delays. They found that the total ablated volume decreased with increasing interpulse delay and attributed the lowering to the screening by the plasma generated by the first pulse [90]. Stoian et al. [86] demonstrated an improvement in the quality of femtosecond laser microstructuring of dielectrics by using temporally shaped pulse trains [55].

With respect to NP synthesis by conventional wet-chemical procedures, short-pulse LA of metals in liquids exhibits several promising advantages, such as environmental sustainability, a simple experimental setup, and long-term stability [44, 45, 94, 95]. Compared to NP generation in vacuum and/or low-pressure gas backgrounds (see Chapter 4 for details), LA in liquids evidenced the possibility for fabricating nanomaterials with special morphologies, microstructures, and phases and for designing various functionalized nanostructures in a single-step process, as reviewed by Zeng et al. [44]. The physical mechanisms of NP formation by the single-pulse LA of metals in a liquid environment were recently addressed by Povarnitsyn et al. [28].

Besner et al. [96] proposed a laser-based method to control the size characteristics of gold colloidal NPs by the generation of femtosecond laser-induced supercontinuum. Highly stable aluminum NPs generated via single-pulse LA in ethanol using either femtosecond (200 fs) or picosecond (30 and 150 ps) laser pulses were obtained by Stratakis et al. [97]. An exhaustive description of NP synthesis by single-pulse LA in liquid environments can be found in [98]. However, DP-LA attracted less interest and DP experiments in liquids were performed mostly in the nanosecond regime for analytical purposes.

Burakov et al. [99] reported that metal ablation by DPs in transparent liquids could lead to the fabrication of stable, size-selected NPs with an increased production efficiency. Silver NP colloidal solutions were obtained by Phuoc et al. [100] using DP-LA in the orthogonal configuration, with the goal of inducing fragmentation for an improved NP size control. Experiments performed in collinear geometry for the production of metal (Ag, Au, and Cu) NPs revealed an enhanced synthesis rate and an increased emission signal from the plasma atoms and ions due to more efficient material ablation [101]. Titanium DP irradiation under water has been investigated both theoretically and experimentally by Casavola et al. [102]. The authors evidenced that the dynamics of the plasma is strongly affected by chemical reactions between plume species and vaporized liquid within the laser-induced cavitation bubble. Later, the same group reported on collinear DP-LA in water for the synthesis of silver NPs [45]. Several optical and spectroscopic techniques were employed to study the dynamics of laser-induced plasma and cavitation bubbles and to monitor Ag NP generation as a function of interpulse delay. They concluded that (i) smaller NPs were produced at interpulse delays corresponding to the early expansion and late collapse stages of the cavitation bubble and (ii) higher NP concentrations were obtained with DP-LA when the interpulse delay matched the maximum volume of bubble expansion [45].

As mentioned before, DP laser irradiation was also applied for thin-film synthesis by pulsed laser deposition. However, in most of the studies, two targets are simultaneously ablated by two synchronized laser pulses in order to obtain doped thin films or multilayered structures. Typically, UV excimer lasers operated at either 193 or 248 nm are used in combination with Q-switched Nd:YAG lasers operated at 355 or 532 nm to grow ferromagnetic NiMnSb [103], Al-doped ZnO [104], Au-doped TiO_2 [105, 106], and Au-doped NiO thin films [107]. DPs were also used in reactive pulsed laser deposition where an improved quality of deposited carbon nitride films was obtained for an optimized interpulse delay [108].

Despite the success in obtaining stoichiometric thin films, pulsed laser deposition is often affected by the presence of micrometer-sized droplets that may lower the coating's quality in certain applications. György et al. [109] investigated the morphology of the Ta thin films obtained by DP laser irradiation. In their experiments, an UV laser beam (λ = 193 nm and $\tau_{FWHM} \approx$ 10 ns) was used for material ablation and a delayed IR laser (λ = 1.064 µm and $\tau_{FWHM} \approx$ 10 ns), propagating parallel to the target surface, was used for the interaction with the particulates present in the ablation plume. The authors analyzed the density of particulates on the surface of the films as a function of the delay between the UV and IR laser pulses and evidenced the possibility of obtaining particle-free Ta thin films.

9.4 Modeling of Double-Pulse Laser Ablation

9.4.1 Fundamentals of Laser–Matter Interactions

In the case of nanosecond LA, the most common and simplest model that describes laser–solid interaction is based on a thermal effect [110]. The problem of evaporation of a metal surface heated up to a certain temperature T_0 was considered by a number of authors, for example, by Anisimov et al. [111]. The equilibrium evaporation law may be then obtained by using the well-known Hertz-Knudsen equation. However, LA often occurs under nonequilibrium conditions. Therefore, several groups considered nonequilibrium surface processes [112]. In particular, the role of a so-called phase explosion mechanism was discussed [113–115]. In addition, photophysical ablation of organic polymer materials was examined [116].

Recently, femtosecond laser systems attracted particular attention [27, 71, 117–119]. Several theoretical investigations have already underlined the main physical processes involved in the ultrashort laser–matter interactions [34, 120–122]. The classical approach is based on the TTM [35]. In addition, several hydrodynamic simulations were carried out to describe the target material motion [123–125]. In particular, shock wave propagation was shown to play a crucial role in these calculations. MD simulations were, furthermore, performed to provide even more detailed insights into the LA mechanisms, such as phase explosion, fragmentation, evaporation, and mechanical spallation [80, 126, 127].

A typical hydrodynamic model is based on the solution of a system of either Lagrangian [128] or Eulerian hydrodynamics [129]. These equations were extended to the case of two-temperature hydrodynamics and supplemented by a laser energy absorption source, electron heat conductivity, and electron–phonon energy exchange terms [130]. To complete the model, a properly chosen equation of state (EOS) should be used. Previously, a semiempirical thermodynamically complete EOS with separated components of electrons and lattice (heavy particles) was used for metals, such as aluminum [131]. This EOS is developed to fulfil the following requirements: (i) to describe experimental results on compression and expansion for a wide range of densities and temperatures, including data on critical and triple points, (ii) to contain separate information about electron and ion/lattice subsystems, and (iii) to represent changes of thermodynamic parameters during phase transitions. Here, we switched between two different modifications of the EOS: (i) with metastable states and (ii) without metastable states.

To account for the kinetic processes, an estimation of a lifetime was introduced for the superheated liquid (metastable state). Thus, when the binodal line is crossed [27], a particular treatment was applied to each of the following two competitive effects: (i) for the spinodal decomposition, a criterion of the metastable liquid lifetime, based on the theory of homogeneous nucleation [132], is used and (ii) for the fragmentation, a mechanical failure algorithm of Grady [133] is applied. In the first case, we estimated the lifetime of the metastable liquid as $\tau = (CnV)^{-1} e^{(W/k_B T)}$, where $C = 10^{10}$ s^{-1} is the kinetic coefficient, n is the concentration, V is the volume, $W = 16\pi\sigma^3/3\Delta P^2$ is the work needed to cause the phase transition, ΔP is the difference between saturated vapor pressure at the same temperature and the pressure of substance, k_B is the Boltzmann constant, and T is the temperature of the subsystem of heavy particles. The temperature dependence of the surface tension is described as $\sigma = \sigma_0(1 - T/T_c)^{1.25}$, where T_c is the temperature in the critical point, and σ_0 is the surface tension in normal conditions. In this case, as soon as the lifetime τ in the volume V is expired, the phase state in this point is no more metastable. The EOS with metastable phase states is, therefore, no more relevant in this volume, so the corresponding thermodynamic properties should be rather calculated by using the stable EOS. To account for the second effect, a fragmentation criterion was used for the liquid phase with the spall strength $P_S = (6\rho^2 c^3 \sigma \dot{\varepsilon})^{1/3}$ and the time required for fragmentation $t_s = 1/c(6\sigma/\rho\dot{\varepsilon}^2)^{1/3}$, where ρ is the density, $\dot{\varepsilon}$ is the strain rate, and c is the sound speed. When this criterion was satisfied, we introduced vacuum into the cell and relaxed the pressure to zero. Both of these criteria are used simultaneously, and each of them can prevail in a given computational cell depending on the substance location on the phase diagram.

Laser energy absorption by the conduction band or free electrons is described by solving the following system of Helmholtz equations:

$$\frac{\partial^2 E}{\partial z^2} + k_0^2 \left[\varepsilon(z) - \sin^2 \theta_0 \right] E = 0 \tag{9.2}$$

and

$$\frac{\partial^2 B}{\partial z^2} + k_0^2 \left[\varepsilon(z) - \sin^2 \theta_0 \right] B - \frac{\partial \ln \varepsilon(z)}{\partial z} \frac{\partial B}{\partial z} = 0, \tag{9.3}$$

Where $E(z, t)$ and $B(z, t)$ are the slowly varying laser field amplitudes, $\varepsilon(z, t)$ is the dielectric function calculated for the given layer at a given time, and θ_0 is the angle of incidence. Equations 9.2 and 9.3 are solved by using a transfer-matrix method [134] for both s and p polarizations. The absorbed laser energy can be then calculated as follows:

$$Q_L(t,z) = \frac{\omega}{8\pi} Im\{\varepsilon(z,t)\} |E(t,z)|^2 \qquad (9.4)$$

MD simulations coupled to a TTM can be used as an alternative to such two-temperature hydrodynamic methods [37, 122]. In the TTM-MD method, the dynamics of the atomic subsystem is determined by an interatomic potential and boundary conditions. For metals, an embedded atom model [135] is often used. It accounts for both a pairwise interaction and a contribution of the electron charge density from nearest neighbors of an atom under consideration.

9.4.2 Numerical Simulations of Short Double-Pulse Interaction with Materials

DP experiments are first studied with the aid of the recently developed TTM-MD model [37]. Here, the target surface is placed at $z = 0$ nm and the laser beam propagates in the positive direction of z. The ablation dynamics is analyzed by using contour plots of the main thermodynamic parameters for different delays between the laser pulses. When $\Delta t = 0$, the dynamics of ablation corresponds to the case of a single pulse with the doubled incident fluence. At delays up to 20 ps, the target expansion sets in, resulting in the enlargement of the absorption region.

For longer delays, shielding of the target surface by the nascent plume enters into the play. At a delay $\Delta t \approx 50$ ps (Fig. 9.11, left) a rarefaction wave produced by the first laser pulse passes through the melted surface layer, leading to its fragmentation. The plasma temperature is as high as 50,000 K in the zone of absorption of the second laser pulse. In this case, the second shock wave, with an amplitude of about 2 GPa, starts its motion from the plume region toward the bulk. This wave is followed by the second rarefaction wave. Its power is, however, insufficient to cause spallation of the liquid layer [37].

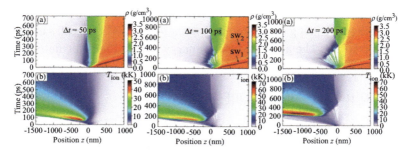

Figure 9.11 Contour plots of the density (a) and temperature (b) of ions for different interpulse delays: left 50 ps, middle 100 ps, and right 200 ps. Reprinted (figure) with permission from Ref. [37]. Copyright (2015) by the American Physical Society.

For the delay of 100 ps (Fig. 9.11, middle), the second pulse energy absorption is observed at $z \approx 500$ nm, while the plume temperature reaches about 70,000 K. The power of the second shock wave is sufficient to completely eliminate the fragmentation produced by the first pulse by erasing the created voids. As the absorption of the second pulse takes place far enough from the initial surface of the target (in the leading edge of the nascent ablation plume), the collapse of voids is finished by $\Delta t \approx 350$ ps. An additional melting of the target can also result from this event. Finally, for the delay of 200 ps (Fig. 9.11, right), the energy of the second pulse is absorbed at an even longer distance from the target. Again, the shock wave originates in this region and the motion of this wave is clearly seen in the figure. At 500 ps, a hot supercritical phase with a temperature higher than 50,000 K appears in the plume and plume merging with the target liquid layer takes place. In addition, melting is observed until ~1100 ps and the melted depth reaches ~450 nm.

Figure 9.12 shows the LA depth as a function of interpulse delay. The case of a 0 ps delay is identical to single-pulse ablation with the fluence $2 \times F_{las} = 4$ J cm^{-2}. Then, as the delay increases, the ablation depth starts to drop and, by the delay of about 30 ps, it reaches the single pulse ablation depth for F_{las}. For longer delays, the ablation depth monotonically drops below the depth of single-pulse ablation, as shown in the experiments [41, 57]. In fact, the plasma plume starts to expand and the second pulse reaches the target and raises the ablation crater depth. The experimentally observed linear growth of the crater depth with the number of pulses thus supports the final value of the extrapolated depth at $\Delta t \to \infty$.

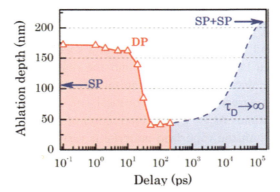

Figure 9.12 Ablation depth dependence on the delay between two succeeding pulses of 2 J cm^{-2} fluence each—empty (red) triangles. The extrapolation to long delays is shown by the dashed (blue) curve. The blue arrows show the ablation depth of a single pulse of 2 J cm^{-2} fluence and two succeeding pulses with the delay $\to \infty$. Reprinted (figure) with permission from Ref. [37]. Copyright (2015) by the American Physical Society.

On the basis of the above analysis, two main mechanisms could be identified to be responsible for the suppression of ablation in the DP ablation experiments [37]. The first one is associated with the suppression of the rarefaction wave, which leads to homogeneous nucleation in the liquid layer of the target under tensile stress. This mechanism dominates for the delays up to 20–50 ps and results in the monotonic decrease of the crater depth. For delays longer than 50 ps, mechanical fragmentation occurs after the first pulse but the second pulse generates a novel high-pressured plasma region ahead of the ablated liquid layers and pushes the large fraction of these ablated layers back to the target. In this case, the ablation depth can be even smaller than that in the case of a single-pulse regime.

9.5 Conclusions and Perspectives

The given summary of experimental and theoretical investigations of material ablation with time-delayed short laser pulses reveals strong changes of the plume composition when the interpulse delay reaches the characteristic time of electron–lattice thermalization. The changes of the relative emission intensities of atoms and NPs within the plume evidence two characteristic times, suggesting thus different mechanisms responsible for the plume composition modification. It is shown that for $\Delta t \leq \tau_{el}$, the energy deposited by the second laser pulse is confined in a smaller volume due to the increased lattice temperature and the reduced electron heat conductivity. Accordingly, the atomization degree of the plume increases, in agreement with theoretical and other experimental studies. For times $\Delta t \gg \tau_{el}$, the interaction of the second delayed laser pulse with the vaporized material is responsible for plasma reheating. These observations were confirmed by ablation depth measurements and AFM analysis of the NPs collected on mica substrates.

As a consequence of the efficient reduction of the NP fraction and the increase of the atomic emission intensity, the performances of material analysis via DP-LIBS could be considerably improved. Understanding the complex mechanisms of DP-LA leading to the change of the sample heat regime with the interpulse delay is mandatory for the optimization of several other applications, like ultraprecise micromachining, high-quality thin-film synthesis by pulsed laser deposition, and syntheses of new nanomaterials and nanostructures.

Recent theoretical studies of DP laser-metal ablation are in good agreement with experimental investigations. A further progress of laser-based applications will be possible only by a synergistic approach between numerical modeling and accurate experiments.

Indeed, developments could be foreseen in the field of laser interaction with dielectrics for applications in three-dimensional data storage and fabrication of micro- to nanofluidic devices for single-cell analyses. Although remarkable progress was achieved in the NP synthesis by DP-LA in liquid environments, further advances are awaited in the synthesis of new nano-objects and their direct functionalization for cancer diagnostics and treatment or for nanomedicine in general.

Finally, the simplified approach of using DPs could open the gate for more sophisticated experiments for material nanoprocessing and analyses using shaped pulses. Probing nanoscale phenomena, near-field and plasmon coupling [55] may stimulate new directions in nanoscience and nanotechnology.

Acknowledgments

E. Axente acknowledges the financial support of UEFISCDI under the PNII-RU-TE-2014-4-1790 contract.

References

1. Gao, J., Gu, H., and Xu, B. (2009). Multifunctional magnetic nanoparticles: design, synthesis, and biomedical applications, *Acc. Chem. Res.*, **42**, pp. 1097–1107.
2. Guo, S., and Wang, E. (2011). Noble metal nanomaterials: controllable synthesis and application in fuel cells and analytical sensors, *Nano Today*, **6**, pp. 240–264.
3. De Jong, W. H., and Borm, P. J. A. (2008). Drug delivery and nanoparticles: applications and hazards, *Int. J. Nanomed.*, **3**, pp. 133–149.
4. Khlebtsov, N. G., and Dykman, L. A. (2010). Optical properties and biomedical applications of plasmonic nanoparticles, *J. Quant. Spectrosc. Radiat. Transfer*, **111**, pp. 1–35.
5. Murphy, C. J., Gole, A. M., Stone, J. W., Sisco, P. N., Alkilany, A. M., Goldsmith, E. C., and Baxter, S. C. (2008). Gold nanoparticles in biology: beyond toxicity to cellular imaging, *Acc. Chem. Res.*, **41**, pp. 1721–1730.
6. Salata, O. V. (2004). Applications of nanoparticles in biology and medicine, *J. Nanobiotechnol.*, **2**, pp. 1–6.
7. Wang, H., Agarwal, P., Zhao, S., Yu, J., Lu, X., and He, X. (2016). A biomimetic hybrid nanoplatform for encapsulation and precisely controlled delivery of theranostic agents, *Nat. Commun.*, **6**, p. 10081.
8. Blanco, E., Shen, H., and Ferrari, M. (2015). Principles of nanoparticle design for overcoming biological barriers to drug delivery, *Nat. Biotechnol.*, **33**, pp. 941–951.
9. Han, G., Ghosh, P., and Rotello, V. M. (2007). Functionalized gold nanoparticles for drug delivery, *Nanomedicine*, **2**, pp. 113–123.
10. Jiang, J., Gu, H. W., Shao, H. L., Devlin, E., Papaefthymiou, G. C., and Ying, J. Y. (2008). Bifunctional Fe3O4-Ag heterodimer nanoparticles for two-photon fluorescence imaging and magnetic manipulation, *Adv. Mater.*, **20**, pp. 4403–4407.
11. Xu, C. J., Xie, J., Ho, D., Wang, C., Kohler, N., Walsh, E. G., Morgan, J. R., Chin, Y. E., Sun, S. H. (2008). Au-Fe3O4 dumbbell nanoparticles as dual-functional probes. *Angew. Chem. Int. Ed.*, **47**, pp. 173–176.

12. Luo, P. G., and Stutzenberger, F. J. (2008). Nanotechnology in the detection and control of microorganisms, *Adv. Appl. Microbiol.*, **63**, pp. 145–181.
13. Huang, X., Jain, P. K., El-Sayed, I. H., and El-Sayed, M. A. (2008). Plasmonic photo-thermal therapy (PPTT) using gold nanoparticles, *Lasers Med. Sci.*, **23**, pp. 217–228.
14. Lal, S., Clare, S. E., and Halas, N. J. (2008). Nanoshell-enabled photothermal cancer therapy: impending clinical impact, *Acc. Chem. Res.*, **41**, pp. 1842–1851.
15. Sun, Y., and Xia, Y. (2002). Shape-controlled synthesis of gold and silver nanoparticles, *Science*, **298**, pp. 2176–2179.
16. Wang, H., Brandl, D. W., Le, F., Nordlander, P., and Halas, N. J. (2006). Nanorice: a hybrid plasmonic nanostructure, *Nano Lett.*, **6**, pp. 827–832.
17. Nehl, C. L., Liao, H., and Hafner, J. H. (2006). Optical properties of star-shaped gold nanoparticles, *Nano Lett.*, **6**, pp. 683–688.
18. Sun. Y, and Xia, Y. (2003). Alloying and dealloying processes involved in the preparation of metal nanoshells through a galvanic replacement reaction, *Nano Lett.*, **3**, pp. 1569–1572.
19. Voloshko, A., and Itina T. E. (2015). Nanoparticle formation by laser ablation and by spark discharges—properties, mechanisms, and control possibilities, in *Nanoparticles Technology*, ed. Aliofkhazraei, M. (InTech), pp. 1–12.
20. Caricato, A. P., Luches, A., and Martino, M. (2016). Laser fabrication of nanoparticles, in *Handbook of Nanoparticles*, ed. Aliofkhazraei, M. (Springer International, Switzerland), pp. 407–428.
21. Besner, S., and Meunier M. (2010). Laser Synthesis of Nanomaterials, in *Laser Precision Microfabrication*, eds. Sugioka, K., Meunier, M., and Piqué, A., Springer Series in Materials Science (Springer-Verlag, Berlin, Heidelberg), Vol. 135, pp. 163–187.
22. Amoruso, S., Bruzzese, R., Spinelli, N., Velotta, R., Vitiello, M., Wang, X., Ausanio, G., Iannotti, V., and Lanotte, L. (2004). Generation of silicon nanoparticles via femtosecond laser ablation in vacuum, *Appl. Phys. Lett.*, **84**, pp. 4502–4504.
23. Amoruso, S., Ausanio, G., Bruzzese, R., Vitiello, M., and Wang, X. (2005). Femtosecond laser pulse irradiation of solid targets as a general route to nanoparticle formation in a vacuum, *Phys. Rev. B*, **71**, pp. 033406-1–033406-4.

24. Grojo, D., Hermann, J., and Perrone, A. (2005). Plasma analyses during femtosecond laser ablation of Ti, Zr, and Hf, *J. Appl. Phys.*, **97**, pp. 063306-1–063306-9.
25. Hermann, J., Noël, S., Itina, T. E., Axente, E., and Povarnitsyn, M. E. (2008). Correlation between ablation efficiency and nanoparticle generation during short-pulse laser ablation of metals, *Laser Phys.*, **18**, pp. 374–379.
26. Zhigilei, L. V. (2003). Dynamics of the plume formation and parameters of the ejected clusters in short-pulse laser ablation, *Appl. Phys. A*, **76**, pp. 339–350.
27. Povarnitsyn, M. E., Itina, T. E., Sentis, M., Khishchenko, K. V., and Levashov, P. R. (2007). Material decomposition mechanisms in femtosecond laser interactions with metals, *Phys. Rev. B*, **75**, pp. 235414-1–235414-5.
28. Povarnitsyn, M. E., Itina, T. E., Levashov, P. R., and Khishchenko, K. V. (2013). Mechanisms of nanoparticle formation by ultra-short laser ablation of metals in liquid environment, *Phys. Chem. Chem. Phys.*, **15**, pp. 3108–3114.
29. Delfour, L., and Itina, T. E. (2015). Mechanisms of ultrashort laser-induced fragmentation of metal nanoparticles in liquids: numerical insights, *J. Phys. Chem. C*, **119**, pp. 13893–13900.
30. Liu, B., Hu, Z., Che, Y., Chen, Y., and Pan, X. (2007). Nanoparticle generation in ultrafast pulsed laser ablation of nickel, *Appl. Phys. Lett.*, **90**, pp. 044103-1–044103-3.
31. Caricato, A. P., Luches, A., and Martino, M. (2016). Laser fabrication of nanoparticles, in *Handbook of Nanoparticles*, ed. Aliofkhazraei, M. (Springer International, Switzerland), pp. 407–428.
32. van de Riet, E., Nillesen, C. J. C. M., and Dieleman, J. (1993). Reduction of droplet emission and target roughening in laser ablation and deposition of metals, *J. Appl. Phys.*, **74**, pp. 2008–2012.
33. Bonn, M., Denzler, D. N., Funk, S., Wolf, M., Wellershoff, S. S., and Hohlfeld, J. (2000). Ultrafast electron dynamics at metal surfaces: competition between electron-phonon coupling and hot-electron transport, *Phys. Rev. B*, **61**, pp. 1101–1105.
34. Anisimov, S. I., and Luk'yanchuk, B. S. (2002). Selected problems of laser ablation theory, *Phys. Usp.*, **45**, pp. 293–324.
35. Anisimov, S. I., Kapeliovich, B. L., and Perel'man, T. L. (1974). Electron emission from metal surfaces exposed to ultrashort laser pulses, *Sov. Phys. JETP*, **39**, pp. 375–377.

36. Itina, T., and Gouriet, K. (2010). Mechanisms of nanoparticle formation by laser ablation, in *Laser Pulse Phenomena and Applications*, ed. Duarte, F. J. (InTech), pp. 309–322.

37. Povarnitsyn, M. E., Fokin, V. B., Levashov, P. R., and Itina, T. E. (2015). Molecular dynamics simulation of subpicosecond double-pulse laser ablation of metals, *Phys. Rev. B*, **92**, pp. 174104-1–174104-10.

38. Babushok, V. I., DeLucia, Jr. F. C., Gottfried, J. L., Munson, C. A., and Miziolek, A. W. (2006). Double pulse laser ablation and plasma: laser induced breakdown spectroscopy signal enhancement, *Spectrochim. Acta, Part B*, **61**, pp. 999–1014.

39. Beldjilali, S., Yip, W. L., Hermann, J., Baba-Hamed, T., and Belasri, A. (2011). Investigation of plasmas produced by laser ablation using single and double pulses for food analysis demonstrated by probing potato skins, *Anal. Bioanal. Chem.*, **400**, pp. 2173–2183.

40. Donnelly, T., Lunney, J. G., Amoruso, S., Bruzzese, R., Wang, X., and Ni, X. (2009). Double pulse ultrafast laser ablation of nickel in vacuum, *J. Appl. Phys.*, **106**, pp. 013304-1–013304-5.

41. Noël, S., and Hermann, J. (2009). Reducing nanoparticles in metal ablation plumes produced by two delayed short laser pulses, *Appl. Phys. Lett.*, **94**, pp. 053120-1–053120-3.

42. Loktionov, E., Ovchinnikov, A., Protasov, Y., Protasov, Y., and Sitnikov, D. (2014). Gas-plasma flows under femtosecond laser ablation for metals in vacuum, *High Temp.*, **52**, pp. 132–134.

43. Muto, H., Miyajima, K., and Mafuné, F. (2008). Mechanism of laser-induced size reduction of gold nanoparticles as studied by single and double laser pulse excitation, *J. Phys. Chem. C*, **112**, pp. 5810–5815.

44. Zeng, H., Du, X., Singh, S. C., Kulinich, S. A., Yang, S., He, J., and Cai, W. (2012). Nanomaterials via laser ablation/irradiation in liquid: a review, *Adv. Funct. Mater.*, **22**, pp. 1333–1353.

45. Dell'Aglio, M., Gaudiuso, R., El Rashedy, R., De Pascale, O., Palazzo, G., and De Giacomo, A. (2013). Collinear double pulse laser ablation in water for the production of silver nanoparticles, *Phys. Chem. Chem. Phys.*, **15**, pp. 20868–20875.

46. Karpukhin, V., Malikov, M., Borodina, T., Val'yano, G., Gololobova, O., and Strikanov, D. (2015). Formation of hollow micro- and nanostructures of zirconia by laser ablation of metal in liquid, *High Temp.*, **53**, pp. 93–98.

47. Li, X., Zhang, G., Jiang, L., Shi, X., Zhang, K., Rong, W., Duan, J., and Lu, Y. (2015). Production rate enhancement of size-tunable silicon

nanoparticles by temporally shaping femtosecond laser pulses in ethanol, *Opt. Express*, **23**, pp. 4226–4232.

48. Axente, E., Mihailescu, I. N., Hermann, J., and Itina, T. E. (2011). Probing electron-phonon coupling in metals via observations of ablation plumes produced by two delayed short laser pulses, *Appl. Phys. Lett.*, **99**, pp. 081502-1–081502-3.

49. Grunwald, R., Rohloff, M., Höhm, S., Bonse, J., Krüger, J., Das, S., and Rosenfeld, A. (2011). Formation of laser-induced periodic surface structures on fused silica upon multiple cross-polarized double-femtosecond-laser-pulse irradiation sequences, *J. Appl. Phys.*, **110**, pp. 014910-1–014910-4.

50. Barberoglou, M., Gray, D., Magoulakis, E., Fotakis, C., Loukakos, P. A., and Stratakis, E. (2013). Controlling ripples' periodicity using temporally delayed femtosecond laser double pulses, *Opt. Express*, **21**, pp. 18501–18508.

51. Höhm, S., Rosenfeld, A., Krüger, J., and Bonse, J. (2013). Area dependence of femtosecond laser-induced periodic surface structures for varying band gap materials after double pulse excitation, *Appl. Surf. Sci.*, **278**, pp. 7–12.

52. Derrien, T. J.-Y., Krüger, J., Itina, T. E., Höhm, S., Rosenfeld, A., and Bonse, J. (2014). Rippled area formed by surface plasmon polaritons upon femtosecond laser double-pulse irradiation of silicon: the role of carrier generation and relaxation processes, *Appl. Phys. A*, **117**, pp. 77–81.

53. Povarnitsyn, M. E., Itina, T. E., Khishchenko, K. V., and Levashov, P. R. (2009). Suppression of ablation in femtosecond double-pulse experiments, *Phys. Rev. Lett.*, **103**, pp. 195002-1–195002-4.

54. Roth, J., Krauß, A., Lotze, J., and Trebin, H.-R. (2014). Simulation of laser ablation in aluminum: the effectivity of double pulses, *Appl. Phys. A*, **117**, pp. 2207–2216.

55. Stoian, R., Wollenhaupt, M., Baumert, T., and Hertel, I. V. (2010). Temporal pulse tailoring in ultrafast laser manufacturing technologies, in *Laser Precision Microfabrication*, eds. Sugioka, K., Meunier, M., and Piqué, A., Springer Series in Materials Science (Springer-Verlag, Berlin, Heidelberg), Vol. 135, pp. 121–144.

56. Gautier, C., Fichet, P., Menut, D., Lacour, J.-L., L'Hermite, D., and Dubessy, J. (2004). Study of the double-pulse setup with an orthogonal beam geometry for laser-induced breakdown spectroscopy, *Spectrochim. Acta, Part B*, **59**, pp. 975–986.

57. Semerok, A., and Dutouquet, C. (2004). Ultrashort double pulse laser ablation of metals, *Thin Solid Films*, **453-454**, pp. 501-505.
58. Fortes, F. J., Moros, J., Lucena, P., Cabalin, L. M., and Laserna, J. J. (2013). Laser-induced breakdown spectroscopy, *Anal. Chem.*, **85**, pp. 640-669.
59. Piscitelli, V., Martínez, M. A., Fernandez, A. J., Gonzalez J. J., Mao X. L., and Russo, R. E. (2009). Double pulse laser induced breakdown spectroscopy: experimental study of lead emission intensity dependence on the wavelengths and sample matrix, *Spectrochim. Acta, Part B*, **64**, pp. 147-154.
60. Gautier, C., Fichet, P., Menut, D., Lacour, J.-L., L'Hermite, D., and Dubessy, J. (2005). Quantification of the intensity enhancements for the double-pulse laser-induced breakdown spectroscopy in the orthogonal beam geometry, *Spectrochim. Acta, Part B*, **60**, pp. 265-276.
61. Sasaki, A., Liu, J., Hara, W., Akiba, S., Saito, K., Yodo, T., and Yoshimoto, M. (2004). Room-temperature growth of ultrasmooth AlN epitaxial thin films on sapphire with NiO buffer layer, *J. Mater. Res.*, **19**, pp. 2725-2729.
62. Bruneau, S., Hermann, J., Dumitru, G., Sentis, M., and Axente, E. (2005). Ultra-fast laser ablation applied to deep-drilling of metals, *Appl. Surf. Sci.*, **248**, pp. 299-303.
63. Chimier, B., Tikhonchuk, V. T., and Hallo, L. (2007). Heating model for metals irradiated by a subpicosecond laser pulse, *Phys. Rev. B*, **75**, pp. 195124-1–195124-12.
64. Noël, S., Hermann, J., and Itina, T. E. (2007). Investigation of nanoparticle generation during femtosecond laser ablation of metals, *Appl. Surf. Sci.*, **253**, pp. 6310-6315.
65. Ganeev, R. A., Hutchison, C., Lopez-Quintas, I., McGrath, F., Lei, D. Y., Castillejo, M., and Marangos, J. P. (2013). Ablation of nanoparticles and efficient harmonic generation using a 1-kHz laser, *Phys. Rev. A*, **88**, pp. 033803-1–033803-10.
66. Starikov, S. V., and Pisarev, V. V. (2015). Atomistic simulation of laser-pulse surface modification: predictions of models with various length and time scales, *J. Appl. Phys.*, **117**, pp. 135901-1–135901-9.
67. Wu, C., Christensen, M. S., Savolainen, J.-M., Balling, P., and Zhigilei, L. V. (2015). Generation of subsurface voids and a nanocrystalline surface layer in femtosecond laser irradiation of a single-crystal Ag target, *Phys. Rev. B*, **91**, pp. 035413-1–035413-14.

68. Cheng, C., and Xu, X. (2005). Mechanisms of decomposition of metal during femtosecond laser ablation, *Phys. Rev. B*, **72**, pp. 165415-1–165415-15.
69. Chimier, B., Tikhonchuk, V. T., and Hallo, L. (2008). Effect of pressure relaxation during the laser heating and electron–ion relaxation stages, *Appl. Phys. A*, **92**, pp. 843–848.
70. Miotello, A., and Kelly, R. (1995). Critical assessment of thermal models for laser sputtering at high fluencies, *Appl. Phys. Lett.*, **67**, pp. 3535-1–3535-3.
71. Colombier, J. P., Combis, P., Bonneau, F., Le Harzic, R., and Audouard, E. (2005). Hydrodynamic simulations of metal ablation by femtosecond laser irradiation, *Phys. Rev. B*, **71**, pp. 165406-1–165406-6.
72. Perez, D., and Lewis, L. (2002). Ablation of solids under femtosecond laser pulses, *Phys. Rev. Lett.*, **89**, pp. 255504-1–255504-4.
73. Norman, G. E., Starikov, S. V., and Stegailov, V. V. (2012). Atomistic simulation of laser ablation of gold: effect of pressure relaxation, *Sov. Phys. JETP*, **114**, pp. 792–800.
74. Nolte, S., Momma, C., Jacobs, H., Tünnermann, A., Chichkov, B. N., Wellegehausen, B., and Welling, H. (1997). Ablation of metals by ultrashort laser pulses, *J. Opt. Soc. Am. B*, **14**, pp. 2716–2722.
75. Furusawa K., Takahashi K., Kumagai H., Midorikawa, K., and Obara M. (1999). Ablation characteristics of Au, Ag, and Cu metals using a femtosecond Ti:sapphire laser, *Appl. Phys. A*, **69**, pp. S359–S366.
76. Hirayama, Y., and Obara, M. (2002). Heat effects of metals ablated with femtosecond laser pulses, *Appl. Surf. Sci.*, **197–198**, pp. 741–745.
77. Axente, E., Noël, S., Hermann, J., Sentis, M., and Mihailescu, I. N. (2009). Subpicosecond laser ablation of copper and fused silica: initiation threshold and plasma expansion, *Appl. Surf. Sci.*, **255**, pp. 9734–9737.
78. Hermann, J., Mercadier, L., Axente, E., and Noël, S. (2012). Properties of plasmas produced by short double pulse laser ablation of metals, *J. Phys.: Conf. Ser.*, **399**, pp. 012006-1–012006-12.
79. Hohlfeld, J., Wellershoff, S. S., Grüdde, J., Conrad, U., Jänke, V., and Matthias, E. (2000). Electron and lattice dynamics following optical excitation of metals, *Chem. Phys.*, **251**, pp. 237–258.
80. Schäfer, C., Urbassek, H. M., and Zighilei, L. V. (2002). Metal ablation by picosecond laser pulses: a hybrid simulation, *Phys. Rev. B*, **66**, pp. 115404-1–115404-8.
81. Kanavin, A., Smetanin, I., Isakov, V., Afanasiev, Y., Chichkov, B., Wellegehausen, B., Nolte, S., Momma, C., and Tünnermann, A. (1998).

Heat transport in metals irradiated by ultrashort laser pulses, *Phys. Rev. B*, **57**, pp. 14698–14703.

82. Scuderi, D., Albert, O., Moreau, D., Pronko, P. P., and Etchepare, J. (2005). Interaction of a laser-produced plume with a second time delayed femtosecond pulse, *Appl. Phys. Lett.*, **86**, pp. 071502-1–071502-3.

83. Noël, S., and Hermann, J. (2007). Influence of irradiation conditions on plume expansion induced by femtosecond laser ablation of gold and copper, *Proc. SPIE*, **6785**, p. 67850F.

84. Le Harzic, R., Breitling, D., Sommer, S., Föhl, C., König, K., Dausinger, F., and Audouard, E. (2005). Processing of metals by double pulses with short laser pulses, *Appl. Phys. A*, **81**, pp. 1121–1125.

85. Spyridaki, M., Koudoumas, E., Tzanetakis, P., Fotakis, C., Stoian, R., Rosenfeld, A., and Hertel, I. V. (2003). Temporal pulse manipulation and ion generation in ultrafast laser ablation of silicon, *Appl. Phys. Lett.*, **83**, pp. 1474–1476.

86. Stoian, R., Boyle, M., Thoss, A., Rosenfeld, A., Korn, G., Hertel, I. V., and Campbell, E. E. B. (2002). Laser ablation of dielectrics with temporally shaped femtosecond pulses, *Appl. Phys. Lett.*, **80**, pp. 353–355.

87. Guarnaccio, A., Parisi, G.P., Mollica, D., De Bonis, A., Teghil, R., and Santagata, A. (2014). Fs–ns double-pulse laser induced breakdown spectroscopy of copper-based-alloys: generation and elemental analysis of nanoparticles, *Spectrochim. Acta, Part B*, **101**, pp. 261–268.

88. Garrelie, F., Bourquard, F., Loir, A.-S., Donnet, C., and Colombier J.-P. (2016). Control of femtosecond pulsed laser ablation and deposition by temporal pulse shaping, *Opt. Laser Technol.*, **78**, pp. 24–51.

89. Wang, Q., Luo, S., Chen, Z., Qi, H., Deng, J., and Hu, Z. (2016). Drilling of aluminum and copper films with femtosecond double-pulse laser, *Opt. Laser Technol.*, **80**, pp. 116–124.

90. Chowdhury, I. H., Xu, X., and Weiner, A. M. (2005). Ultrafast double-pulse ablation of fused silica, *Appl. Phys. Lett.*, **86**, pp. 151110-1–151110-3.

91. Deng, Y. P., Xie, X. H., Xiong, H., Leng, Y. X., Cheng, C. F., Lu, H. H., Li, R. X., and Xu, Z. Z. (2005). Optical breakdown for silica and silicon with double femtosecond laser pulses, *Opt. Express*, **13**, pp. 3096–3103.

92. Hu, Z., Singha, S., Liu, Y., and Gordon, R. J. (2007). Mechanism for the ablation of Si<111> with pairs of ultrashort laser pulses, *Appl. Phys. Lett.*, **90**, pp. 131910-1–131910-3.

93. Schiffern, J. T., Doerr, D. W., and Alexander, D. R. (2007). Optimization of collinear double-pulse femtosecond laser-induced breakdown spectroscopy of silicon, *Spectrochim. Acta, Part B*, **62**, pp. 1412–1418.
94. Amendola, V., and Meneghetti, M. (2013). What controls the composition and the structure of nanomaterials generated by laser ablation in liquid solution?, *Phys. Chem. Chem. Phys.*, **15**, pp. 3027–3046.
95. De Giacomo, A., Dell'Aglio, M., Santagata, A., Gaudiuso, R., De Pascale, O., Wagener, P., Messina, G. C., Compagnini, G., and Barcikowski, S. (2013). Cavitation dynamics of laser ablation of bulk and wire-shaped metals in water during nanoparticles production, *Phys. Chem. Chem. Phys.*, **15**, pp. 3083–3092.
96. Besner, S., Kabashin, A. V., and Meunier, M. (2006). Fragmentation of colloidal nanoparticles by femtosecond laser-induced supercontinuum generation, *Appl. Phys. Lett.*, **89**, pp. 233122-1–233122-3.
97. Stratakis, E., Barberoglou, M., Fotakis, C., Viau, G., Garcia, C., and Shafeev, G. A. (2009). Generation of Al nanoparticles via ablation of bulk Al in liquids with short laser pulses, *Opt. Express*, **17**, pp. 12650–12659.
98. Besner, S., and Meunier M. (2010). Laser synthesis of nanomaterials, in *Laser Precision Microfabrication*, eds. Sugioka, K., Meunier, M., and Piqué, A., Springer Series in Materials Science (Springer-Verlag, Berlin, Heidelberg), Vol. 135, pp. 163–187.
99. Burakov, V. S., Tarasenko, N. V., Butsen, A. V., Rotzantsev, V. A., and Nedel'ko, M. I. (2005). Formation of nanoparticles during double-pulse laser ablation of metals in liquids, *Eur. Phys. J.: Appl. Phys.*, **30**, pp. 107–112.
100. Phuoc, T. X., Soong, Y., and Chyu, M. K. (2007). Synthesis of Ag-deionized water nanofluids using multi-beam laser ablation in liquids, *Opt. Lasers Eng.*, **45**, pp. 1099–1106.
101. Burakov, V. S., Butsen, A. V., and Tarasenko, N. V. (2010). Laser-induced plasmas in liquids for nanoparticle synthesis, *J. Appl. Spectrosc.*, **77**, pp. 386–393.
102. Casavola, A., De Giacomo, A., Dell'Aglio, M., Taccogna, F., Colonna, G., De Pascale, O., and Longo, S. (2005). Experimental investigation and modelling of double pulse laser-induced plasma spectroscopy under water, *Spectrochim. Acta, Part B*, **60**, pp. 975–985.
103. Caminat, P., Valerio, E., Autric, M., Grigorescu, C., and Monnereau, O. (2004). Double beam pulse laser deposition of NiMnSb thin films at ambient temperature, *Thin Solid Films*, **453–454**, pp. 269–272.

104. György, E., Santiso, J., Giannoudakos, A., Kompitsas, M., Mihailescu, I. N., and Pantelica, D. (2005). Growth of Al doped ZnO thin films by a synchronized two laser system, *Appl. Surf. Sci.*, **248**, pp. 147–150.
105. György, E., Sauthier, G., Figueras, A., Giannoudakos, A., Kompitsas, M., and Mihailescu, I. N. (2006). Growth of Au–TiO2 nanocomposite thin films by a dual-laser, dual-target system, *J. Appl. Phys.*, **100**, pp. 114302-1–114302-5.
106. Kompitsas, M., Giannoudakos, A., György, E., Sauthier, G., Figueras, A., and Mihailescu, I. N. (2007). Growth of metal-oxide semiconductor nanocomposite thin films by a dual-laser, dual target deposition system, *Thin Solid Films*, **515**, pp. 8582–8585.
107. Fasaki, I., Kandyla, M., Tsoutsouva, M. G., and Kompitsas, M. (2013). Optimized hydrogen sensing properties of nanocomposite NiO:Au thin films grown by dual pulsed laser deposition, *Sens. Actuator B-Chem.*, **176**, pp. 103–109.
108. Kuzyakov, Yu. Ya., Varakin, V. N., Moskvitina, E. N., and Stolyarov, P. M. (2009). Double-pulse laser ablation applied to reactive PLD method for synthesis of carbon nitride film: a second laser shot delay, *Laser Phys.*, **19**, pp. 1159–1164.
109. György, E., Mihailescu, I. N., Kompitsas, M., and Giannoudakos, A. (2002). Particulates-free Ta thin films obtained by pulsed laser deposition: the role of a second laser in the laser-induced plasma heating, *Appl. Surf. Sci.*, **195**, pp. 270–276.
110. Gusarov, A. V., and Smurov, I. (2005). Thermal model of nanosecond pulsed laser ablation: analysis of energy and mass transfer, *J. Appl. Phys.*, **97**, pp. 014307-1–014307-13.
111. Anisimov, S. I., Bonch-Brouevich, A. M., El'yashevich, M. A., Imas, Ya. A., Pavlenko, N. A., and Romanov, G. S. (1967). Effect of powerful light fluxes on metals, *Sov. J. Tech. Phys.*, **11**, pp. 945–952.
112. Olstad, R. A., and Olander, D. R. (1975). Evaporation of solids by laser pulses. I. Iron, *J. Appl. Phys.*, **46**, pp. 1499–1508.
113. Garrison, B. J., Itina, T. E., and Zhigilei, L. V. (2003). Limit of overheating and the threshold behavior in laser ablation, *Phys. Rev. E*, **68**, pp. 041501-1–041501-4.
114. Martynyuk, M. M. (1974). Vaporization and boiling of liquid metal in an exploding wire, *Sov. Phys. Thech. Phys.*, **19**, pp. 793–797.
115. Martynyuk, M. M. (1983). Critical constants of metals, *Russ. J. Phys. Chem.*, **57**, pp. 494–501.

116. Luk'yanchuk B., Bityrin N., Anisimov S., Malyshev A., Arnold N., and Bäuerle, D. (1996). Photophysical ablation of organic polymers: the influence of stresses, *Appl. Surf. Sci.*, **106**, pp. 120–125.
117. Nolte, S., Chichkov, B. N., Welling, H., Shani, Y., Lieberman, K., and Terkel, H. (1999). Nanostructuring with spatially localized femtosecond laser pulses, *Opt. Lett.*, **24**, pp. 914–916.
118. Amoruso, S., Bruzzese, R., Vitiello, M., Nediakov, N. N., and Atanasov, P. A. (2005). Experimental and theoretical investigations of femtosecond laser ablation of aluminum in vacuum, *J. Appl. Phys.*, **98**, pp. 044907-1–044907-7.
119. Plech, A., Kotaidis, V., Lorenc, M., and Boneberg, J. (2006). Femtosecond laser near-field ablation from gold nanoparticles, *Nat. Phys.*, **2**, pp. 44–47.
120. Komashko, A. M., Feit, M. D., Rubenchik, A. M., Perry, M. D., and Banks, P. S. (1999). Simulation of material removal efficiency with ultrashort laser pulses, *Appl. Phys. A*, **69**, pp. S95–S98.
121. Rethfeld, B., Sokolowski-Tinten, K., von der Linde, D., and Anisimov, S. I. (2002). Ultrafast thermal melting of laser-excited solids by homogeneous nucleation, *Phys. Rev. B*, **65**, pp. 092103-1–092103-4.
122. Ivanov, D. S., and Zhigilei, L. V. (2003). Combined atomistic-continuum modeling of short-pulse laser melting and disintegration of metal films, *Phys. Rev. B*, **68**, pp. 064114-1–064114-22.
123. Eidmann, K., Meyer-ter-Vehn, J., Schlegel, T., and Hüller, S. (2000). Hydrodynamic simulation of subpicosecond laser interaction with solid-density matter, *Phys. Rev. E*, **62**, pp. 1202–1214.
124. Vidal, F., Johnston, T. W., Laville, S., Barthélemy, O., Chaker, M., Le Drogoff, B., Margot, J., and Sabasi, M. (2001). Critical-point phase separation in laser ablation of conductors, *Phys. Rev. Lett.*, **86**, pp. 2573–2576.
125. Glover, T. E. (2003). Hydrodynamics of particle formation following femtosecond laser ablation, *J. Opt. Soc. Am. B*, **20**, pp. 125–131.
126. Zhigilei, L. V., and Garrison, B. J. (2000). Microscopic mechanisms of laser ablation of organic solids in the thermal and stress confinement irradiation regimes, *J. Appl. Phys.*, **88**, pp. 1281–1298.
127. Perez, D., and Lewis, L. J. (2003). Molecular-dynamics study of ablation of solids under femtosecond laser pulses, *Phys. Rev. B*, **67**, pp. 184102-1–184102-15.
128. Itina, T. E., Vidal, F., Delaporte, P., and Sentis, M. (2004). Numerical study of ultra-short laser ablation of metals and of laser plume dynamics, *Appl. Phys. A*, **79**, pp. 1089–1092.

129. Miller, G. H., and Puckett, E. G. (1996). A high-order godunov method for multiple condensed phases, *J. Comput. Phys.*, **128**, pp. 134–164.
130. Povarnitsyn, M. E., Itina, T. E., Khishchenko, K. V., and Levashov, P. R. (2007). Multi-material two-temperature model for simulation of ultrashort laser ablation, *Appl. Surf. Sci.*, **253**, pp. 6343–6346.
131. Khishchenko, K. V. (2005). In *Physics of Extreme States of Matter—2005* (IPCP RAS, Chernogolovka), pp. 170–172.
132. Frenkel, J. (1946). *Kinetic Theory of Liquids* (Clarendon Press, Oxford).
133. Grady, D. E. (1988). The spall strength of condensed matter, *J. Mech. Phys. Solids*, **36**, pp. 353–384.
134. Born, M., and Wolf, E. (1964). *Principles of Optics: Electromagnetic Theory of Propagation, Interference and Diffraction of Light* (Pergamon Press, Oxford).
135. Zhakhovskii, V., Inogamov, N., Petrov, Y., Ashitkov, S., and Nishihara, K. (2009). Molecular dynamics simulation of femtosecond ablation and spallation with different interatomic potentials, *Appl. Surf. Sci.*, **255**, pp. 9592–9596.

Chapter 10

Ultrafast Laser-Induced Phenomena inside Transparent Materials

Felix Sima,[a,b] Jian Xu,[a] and Koji Sugioka[a]

[a]*RIKEN-SIOM Joint Research Unit, RIKEN Center for Advanced Photonics, 2-1 Hirosawa, Wako, Saitama 351-0198, Japan*
[b]*CETAL Department, National Institute for Lasers, Plasma and Radiation Physics, Atomistilor 409, Magurele, Ilfov 00175, Romania*
ksugioka@riken.jp

Ultrafast lasers, referred to as lasers with pulse durations shorter than a few picoseconds, are excellent tools for transparent material processing. Due to the high peak intensity, which induces multiphoton absorption, they are able to induce local modifications inside the transparent materials with high precision in micro- to nanoscale. Rapid prototyping technologies using ultrafast lasers for glass or polymer processing have been progressively developed in the last decade for photonic device and microfluidic biochip applications. We review herein the fundamentals and characteristics of interaction of ultrafast laser pulses (ULPs) with transparent materials, in particular glasses and photoresists. The phenomena of undeformative laser-induced modification, subtractive laser-

---

Pulsed Laser Ablation: Advances and Applications in Nanoparticles and Nanostructuring Thin Films
Edited by Ion N. Mihailescu and Anna Paola Caricato
Copyright © 2018 Pan Stanford Publishing Pte. Ltd.
ISBN 978-981-4774-23-9 (Hardcover), 978-1-315-18523-1 (eBook)
www.panstanford.com

assisted chemical etching of glasses, and additive processes such as two-photon polymerization with ULPs and their innovative hybrid approach will be described and their potential will be evaluated for the fabrication of highly functional true 3D biochips.

10.1 Introduction

Laser light is often used for the study of interaction with solid materials in which the deposition of pulsed beam energy produces a damage [1]. This damage is dependent on pulse energy, duration, and repetition rate, as well as sample translation speed and focusing optics. It is produced by thermal processes in the case of long pulses, while the physical aspect is predominant for ultrashort pulses (pulses shorter than a few picoseconds) [2]. The laser–matter interaction results in the heat diffusion process, which leads to a larger heat-affected zone (HAZ) in the case of, for example, nanosecond pulses. In contrast, HAZ is negligible, so-called HAZ-free, in the case of ultrashort pulses [3]. Additionally, a focused ultrashort pulsed laser beam with a very high peak intensity can deposit localized energy by nonlinear absorption in transparent materials such as glasses [4, 5]. The modifications are classified into voids, birefringent changes, and refractive index change and depend on the laser–material processing parameters [6]. Because the material is altered by nonlinear processes, the fabrication resolution can overcome optical diffraction limit [7]. The birefringent and refractive index change induces no visible change, so they can be categorized to an undeformative process, in other words, a zero process. Interestingly, the local modification can also enhance the glass ability in the chemical etch rate [8]. The confined modification inside glasses with controlled sample translation followed by chemical wet etching allows the fabrication with a high precision of various three-dimensionally embedded hollow structures with applications in fabrication of waveguides and photonic or lab-on-a-chip devices. This is a subtractive process since the irradiated areas are removed. Besides glasses, polymers are excellent materials for many applications in terms of transparency, chemical stability, and flexible design. Two-photon polymerization (TPP) by femtosecond lasers was applied in a 3D space for fabricating structures with feature sizes below the diffraction limit [9]. When applied to an epoxy negative-tone photoresist, this technology is categorized to an additive process. Flexible micro- and nanoprocessing with respect to structure, function, and scale are then possible by ultrafast lasers with accurate control on all 3D environments for both inorganic and organic materials by undeformative, subtractive, or additive technologies. Integration of technologies further enhances the performance of ultrafast laser 3D processing to achieve more complex configurations with higher functionalities.

In this chapter the phenomena induced by ultrafast laser pulses (ULPs) inside transparent materials such as glasses and photoresists are comprehensively reviewed with relevant applications to fabrication of some functional 3D microdevices by each process. The hybrid technique of ultrafast laser 3D processing is also introduced to offer rapid prototyping technologies for the fabrication of 3D multifunctional biochips.

10.2 Characteristics of Glass Material Processing by Ultrafast Laser Pulses

ULPs in interaction with transparent materials exhibit unique characteristics due to an ultrashort pulse width and an extremely high peak intensity and thus offer not only a lowered laser-induced breakdown threshold and an eliminated HAZ but also efficient energy deposition based on nonlinear processes [10]. The advantages of ULP can be then exploited in precision micromachining of various materials, from glasses to polymers, in the range of applications where micrometer and submicrometer feature sizes are required.

10.2.1 Interaction Mechanism of Ultrafast Laser Pulses with Glasses

10.2.1.1 Nonlinear multiphoton absorption

In the case of inorganic transparent materials, that is, glasses, the origin of ULP-induced modification could be densification by either pressure wave or fast heating-cooling processes [11–13]. The laser-induced optical breakdown threshold was investigated in fused silica made of Corning 7940 with laser pulse widths from 7 ns to 150 fs [14] and independently in fused silica and calcium fluoride for pulse duration ranging from 1 ns to 270 fs [15]. It was found that ULP-induced plasma remained localized at the threshold, with no essential collateral damage, providing precise control of the interaction region, with great potential for material processing. Femtosecond laser irradiation inside various glasses, that is, silica, borate, soda lime silicate, and fluorozirconate, induced the formation of defects, nonbridging oxygen hole centers (NBHOCs), and peroxy radicals, suggesting that multiphoton interactions occur, which is conducive to fabrication of 3D optical circuits in bulk glasses with a focused laser beam [16]. A real-time observation of refractive index change by a femtosecond laser pulse was performed inside a soda lime glass plate, and a mechanism of rapid temperature increase followed by the propagation of pressure wave was proposed [12]. Structural changes in the Corning 0211 glass were obtained by single- or multiple-pulse irradiation using tightly focused ULPs at low energies, and it was demonstrated that at high repetition rates a cumulative heating effect induced by successive laser pulses dominates the process [11]. Meanwhile, writing a 3D pattern inside silica glass by focused ULPs followed by etching of the optically damaged silica in a 5% aqueous solution of HF acid permitted microfabrication of 3D microfluidic channels in volume [17].

The breakdown threshold induced by laser pulses in optical materials depends on the pulse width. For longer pulses ($\tau > 10$ ps), the breakdown threshold laser fluence F_{Bth} follows the equation [18]

$$F_{Bth} \propto \sqrt{\tau}, \tag{10.1}$$

where τ is the laser pulse width. In this case, the bandgap energy of the material must be smaller than or equal to photon energy for the linear absorption according to Beer's law

$$I(z) = I(0)e^{-\alpha z}, \tag{10.2}$$

where α is the absorption coefficient depending on a material and z the propagation distance of the incident laser beam with an initial intensity $I(0)$.

Both linear and nonlinear absorption processes are possible in ULP interaction with matter. However, it was demonstrated that the breakdown threshold laser fluence does not follow the same scaling for pulses shorter than 10 ps [14]. For large-bandgap materials, such as fused silica and calcium fluoride, both damage mechanism and morphology are dramatically changed by ULPs shorter than 20 ps as compared with longer pulses; however, no evidence on the increase in the damage threshold with a decreasing pulse width was observed [15]. In addition, a transition from a thermally dominated regime for longer pulses to an ablative regime dominated by collisional and multiphoton ionization and plasma formation was evidenced [15]. Indeed, typical ultrafast lasers do not have sufficient photon energy to be absorbed linearly in glasses. Consequently, nonlinear multiphoton absorption (MPA) is responsible for exciting the electrons from the valence band to the conduction band (Fig. 10.1).

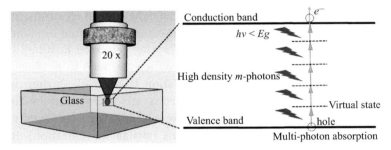

Figure 10.1 Schematic of ULP irradiation of glass and electron excitation by MPA.

A certain number of photons m is then required to satisfy the transition given by $mh\nu > E_g$, where ν is the light frequency and E_g the bandgap of the transparent material. Two-photon absorption (TPA) or MPA was theoretically predicted by Goeppert-Mayer and Born in 1931 [19, 20]. By definition, TPA is the simultaneous absorption of two photons via virtual states in the material. In the case of single-photon absorption, a single photon has sufficient photon energy to excite an electron in a molecule, while for TPA high-density photons with respect to both time and space domains are necessary to change the energy state. This requires a highly intense laser beam at the focal volume. Two possible mechanisms to generate free electrons that are dependent on the laser frequency and intensity are (i) multiphoton photoionization at low laser intensities and frequencies and (ii) tunneling photoionization at high laser intensities and low frequencies. The transition between the multiphoton ionization and tunneling ionization is given by the Keldysh parameter [21]

$$\gamma = \frac{\omega}{e}\sqrt{\frac{m_e c n \varepsilon_0 E_g}{I}}, \quad (10.3)$$

where ω represents the laser frequency, e the electron charge, m_e the effective electron mass, c the speed of light, n the refractive index, ε_0 the permittivity of free space, and I the laser intensity. It was then evidenced that for $\gamma \ll 1.5$ the tunneling effect is predominant while for $\gamma > 1.5$ the multiphoton ionization prevails [22]. There is an intermediate regime at $\gamma \approx 1.5$, where photoionization occurs by a combination of both tunneling and multiphoton ionization [22]. After MPA, the generated free electrons can be further excited by the latter part of identical ULP since electron excitation in glass typically occurs in the order of 100 fs. Specifically, the excited electrons can successively absorb several laser photons and be excited to higher energy states due to free carrier absorption (electron heating). In addition, when the laser intensity is sufficiently high to produce a very strong laser field, impact ionization is induced. During the impact ionization, the excited electrons are accelerated by the intense electric field of ULP and they collide with surrounding atoms, generating secondary electrons (avalanche ionization). By measurements and modeling of a damage threshold at the interaction of a single-shot ULP (20 fs) with fused silica, the roles of multiphoton ionization, tunnel ionization, and impact ionization in laser damage were evaluated [23]. It was then concluded that avalanche ionization predominates in the case of sub-100-fs pulses.

10.2.1.2 Heat accumulation effects

The reduction of HAZ is an important characteristic for high-precision laser material processing. In the case of longer pulses (larger than the material electron–phonon coupling time), the thermal diffusion length l_d is roughly given by the equation

$$l_d = \sqrt{k\tau}, \quad (10.4)$$

where k is the material thermal diffusivity and τ the laser pulse width. In the meanwhile, for ULP interaction with a metal, the thermal diffusion length l_d is the following [24]:

$$l_d = \left[\frac{128}{\pi}\right]^{1/8} \left[\frac{DC_i}{T_{im}G^2 C'_e}\right]^{1/4}, \qquad (10.5)$$

where D is the heat conductivity, C_i the lattice heat capacity, $C'_e = C_e/T_e$, C_e the electron heat capacity, T_e the electron temperature, and G the electron–phonon coupling constant. As a consequence, for example, the l_d for copper is estimated to be 1.5 µm for τ = 10 ns and 329 nm for τ = 130 fs [25]. Thus, the diffusion of heat to the surrounding laser-irradiated area is considered negligible for ULPs.

In the case of glasses, the heat effect is more noticeable when ULPs are focused above a critical intensity and the melting can be observed. Heat accumulation cannot be ignored when the repetition rate of ULPs exceeds hundreds of kilohertz, which is attractive for some applications, including rapid writing of low-loss optical waveguides in transparent materials and integrated micro-optical devices in novel 3D architectures [26]. ULPs with low repetition rates (1 to 200 kHz) and high pulse energy (~1 mJ) induce strong nonlinear absorption in glass with weak focusing optics, resulting in asymmetric refractive index profiles. Even if some studies with ULPs with a repetition rate of 166 kHz at a 100 µm/s scan speed demonstrated that the thermal diffusion in a tight focusing volume can produce symmetric waveguides [27], higher-repetition-rate ULPs (10 MHz) at a 1 mm/s scan speed can rapidly create a more symmetric cross section due to cumulative heating (interval between the successive laser pulses is less than the time required for sufficiently cooling the temperature at the focal volume) [28, 29]. A variable repetition rate (0.1 and 5 MHz) from IMRA's Fiber Chirped Pulse Amplification (FCPA) µJewel femtosecond fiber laser was used to study waveguide properties in fused silica and various borosilicate glasses and found that the repetition rate greatly influences the heat accumulation process [26]. The localized heat accumulation effect induced by a fiber laser at 1558 nm with a repetition rate of 500 kHz was able to efficiently weld nonalkali alumino silicate glass substrates [30].

10.2.2 Spatial Resolution in Ultrafast Laser Processing of Glass

Important parameters in the case of ULP-induced spatial resolution are (i) the laser repetition rate, (ii) energy and the numerical aperture (NA) of the objectives used to focus the beam inside the materials, and (iii) the material properties.

- The laser repetition rate divides the interaction mechanisms into two regimes in terms of heat accumulation, low repetition rate (below hundreds of kilohertz) and high repetition rate (above hundreds of kilohertz), discussed before [6, 31]. At the second regime, heat is accumulated so as to diffuse to the surrounding area and thus the modification volume increases with the number of pulses [26, 28, 32]. Therefore, the spatial resolution can be restricted within the laser focal volume for the first regime, while it becomes worse for the second regime.
- The deposited energy in glass materials is related to the NA of the objective lens used to focus the laser beam. The dependence of processed diameter $d(F)$ on the laser fluence F is correlated by

$$d(F) = \frac{d_0}{\sqrt{2}}\sqrt{\ln\left(\frac{F}{F_{\text{th}}}\right)}, \qquad (10.6)$$

where F_{th} is the processing threshold and d_0 is the diameter of the focused beam given by

$$d_0 = \frac{2\lambda M^2}{\pi \text{NA}} \approx \frac{\lambda}{\text{NA}}, \qquad (10.7)$$

in which λ is the laser beam wavelength and M^2 is the beam quality factor. Thus, the spatial resolution for the first regime should be the order of wavelength of ULPs. Decreasing the laser fluence can further improve the spatial resolution due to the threshold effect, since ULP typically has a Gaussian beam profile in space, as described below.
- The spatial resolution is enhanced by ULP-induced nonlinear MPA as compared with the single-photon absorption at the same wavelength. This means that m-photon absorptions with a different number of m give different spatial resolutions, assuming an ideal Gaussian spatial profile of the beam. For example, in the case of single-photon absorption, the spatial distribution of the laser energy absorbed by the material corresponds to the actual beam profile, as shown by the thick dashed line in Fig. 10.2. In the case of MPA, the distribution of the absorbed energy becomes narrower with an increase

in the number m, since the effective absorption coefficient for m-photon absorption is proportional to the mth power of the laser intensity. Then, the effective beam size ω for m-photon absorption can be expressed as

$$\omega = \frac{\omega_0}{\sqrt{m}}, \qquad (10.8)$$

where ω_0 is the actual spot size of the focused laser beam.

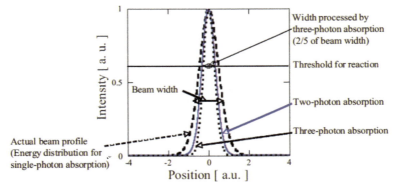

Figure 10.2 Actual ideal beam profile (thick dashed line) and spatial distributions of laser energy absorbed by two-photon (solid line) and three-photon (thin dashed line) absorption. The solid horizontal line indicates the reaction threshold.

Figure 10.2 shows the spatial distributions of the laser energy absorbed by single-photon absorption (thick dashed line), TPA (solid line), and three-photon absorption (thin dashed line), indicating that increasing the order in the MPA decreases the width. Furthermore, when there is a threshold laser intensity, above which a reaction occurs on absorption, the fabrication resolution can be further improved by adjusting the laser intensity, as shown by the straight solid line in Fig. 10.2. Thus, the synergetic contribution of negligible thermal diffusion, MPA, and a threshold effect combined with a high NA objective lens has enabled deep subwavelength fabrication resolutions that are below the diffraction limit. In principle, using the threshold effect offers no limitation to the improvement that can be achieved in the fabrication resolution by precise control of the laser intensity. However, the pulse-to-pulse fluctuation in ULP energy makes it extremely difficult to maintain the same fabrication resolution when the laser intensity is near the threshold intensity. Therefore, the reproducible fabrication resolution is limited to 100–200 nm.

10.3 Undeformative Processing: ULP-Induced Internal Modifications

Dependent on the laser deposited energy, the structural modifications are given by the concrete phenomena: at high energies, microexplosions produce a damage consisting of voids or cavities with diameters of 200–1000 nm [33, 34]. The moderate energies don't cause any shape change, which perform undeformative processing. At this regime, nanogratings of about 300 nm period stay at the origin of birefringent modifications [35–37]. Even at low energies fast localized melting and fast cooling cause an isotropic refractive index change [5, 16]. The less energy deposited in the material can still induce modification. By focusing a femtosecond laser operating at a nonresonant wavelength with different energies, four typical examples of an induced structural change are shown in Fig. 10.3. The energy increases from Fig. 10.3a to Fig. 10.3d: (a) coloration caused by the valence state change of active ions and the succeeding aggregation, (b) refractive index change as an effect of bond scission and local densification, (c) void formation by a microexplosion, and (d) microcrack formation due to destructive breakdown or other phenomena [33].

An array of voids obtained by ULPs in transparent materials was proposed to apply to devices for 3D optical data storage. Figure 10.4 shows an array of voids embedded in fused silica glass in which has been achieved read/write of digital data in volume layers. Such features inscribed inside fused silica showed a real potential for a driveless, multilayer read system for permanent digital storage with a recording density similar to a conventional compact disc [38].

Single-beam irradiation of linearly polarized ULPs can create undeformative nanogratings at the focal volume inside silica glass with defined periodicity consisting of stripelike regions of 20 nm width, which have been found to be the smallest embedded structures ever created by light (Fig. 10.5). Self-organized nanogratings with periodicities from 140 to 320 nm were then formed by tailoring the pulse energy and the number of pulses [35]. This phenomenon was interpreted in terms of interference of the incident light with the electric field of the bulk, which modulates the electron plasma concentration and the structural changes in glass. Interestingly, such nanogratings produce a permanent birefringent characteristic [39].

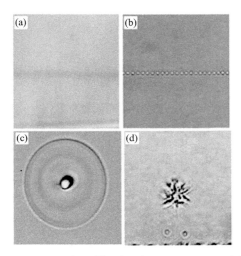

Figure 10.3 Various ULP-induced localized microstructures: (a) coloration, (b) refractive index change, (c) microvoid formation, and (d) microcrack formation. Reproduced with permission from Ref. [33]. Copyright 2005 The Japan Society of Applied Physics.

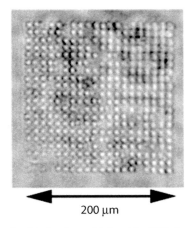

Figure 10.4 Array of voids in glass obtained by ULP. Reprinted from Ref. [6], Copyright (2016), with permission from Elsevier.

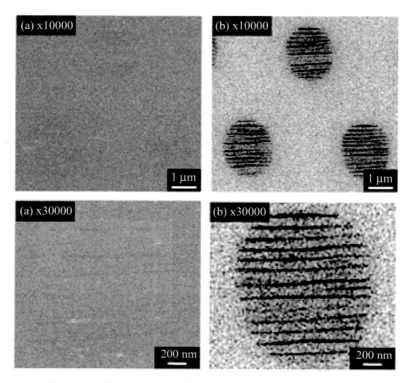

Figure 10.5 (a) SEM images of a silica glass surface polished close to the depth of the focal spot. (b) Light "fingerprints": Backscattering electron images of the same surface. The magnification of the upper and lower images is 10,000x and 30,000x, respectively. Reprinted (figure) with permission from Ref. [35]. Copyright (2003) by the American Physical Society.

Aiming at creating various optical devices, ULP-induced modifications were investigated for various glasses. A focused laser beam was found to induce visible, round-elliptical damage lines within the irradiated patterns inside silica, borate, soda lime silicate, fluoride, and chalcogenide glasses [40]. Mechanisms of ULP-induced refractive index change (Δn) were investigated for fused silica and a borosilicate glass. Two exposure situations were studied: (i) high repetition rate and low pulse energy and (ii) low repetition rate and high pulse energy. The results did not find a thermal origin of the induced Δn and confirmed that densification plays a small role in this change, while the defects contribute substantially to the index increase [41].

The refractive index change of the glass by ULPs was found to be of great interest for applications in waveguide writing. They were fabricated inside various glasses and crystals, generating 3D photonic devices, such as optical splitter in fused silica [42]. New optical properties of such waveguides open new possibilities for the fabrication of complex high-density integrated optical elements [43]. Furthermore, the refractive index change of the glass by ULPs was extensively applied to fabricate some other 3D photonic devices, including volume Bragg gratings [44], diffractive lenses [45], Mach–Zehnder interferometers (MZIs) [46], and waveguide lasers [47, 48].

The waveguides were classified into three types upon the material core modification, as shown in Fig. 10.6. One class of waveguides consists of a core region with a localized increase of material refractive index (Fig. 10.6a), a second one involves the inscription of two damage regions surrounding the waveguide core (Fig. 10.6b), and a third class consists of little change of core optical properties forming waveguide laser cavities resulting from an unmodified core region surrounded by localized areas with a refractive index change (Fig. 10.6c) [8].

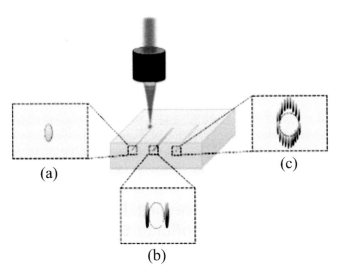

Figure 10.6 Laser inscription geometry with examples of waveguide cross sections for (a) type I, (b) type II, and (c) type III depressed cladding waveguide structures. The guiding regions for each type are indicated using red dashed lines. Reproduced from Ref. [8] with permission from Wiley.

The power of ULP writing of waveguides was demonstrated by writing both straight and curved patterns in a variety of silicate glasses for the fabrication of an optical interleaver [44]. Optical waveguides incorporating Bragg gratings were written in fused silica by a single process using ULP-induced direct-write without the need of lithography or ion-beam techniques [49]. In addition, integrated microfluidic channels and high-quality optical waveguides intersecting each other in a single glass substrate were obtained by ULP irradiation only and proposed for in situ sensing in lab-on-a-chip devices, as introduced below [50].

In the case of ULPs that are loosening the focus, a filamentation phenomenon appears with sufficient energy to produce a refractive index change by nonlinear absorption and modify the volume in a filamentary shape [51]. This phenomenon is effective for glass welding by focusing and depositing the energy that induces melting at the interface between two glasses [30] (Fig. 10.7).

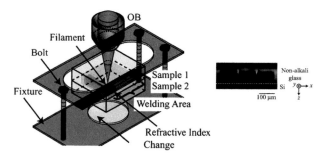

Figure 10.7 Welding two glass substrates by focusing femtosecond laser pulses at the interface between substrates. Optical image of the welding volume in the xz plane. Reproduced from Ref. [30] with permission from The Optical Society.

10.4 Subtractive Processing: Formation of 3D Micro- and Nanofluidic Structure

10.4.1 Ultrafast Laser-Induced Modification Followed by Selective Wet Etching

The regions in fused silica glass modified by focused ULPs at moderate energy can be selectively removed by etching in a 5%

aqueous solution of HF to fabricate 3D microfluidic channels in volume [17]. The etching selectivity originates from the weakened chemical bonds by the physical process of ULPs. Such etching selectivity is also available with photosensitive glass. However, the modification mechanism, which relies on a chemical process, is completely different from fused silica. Typical photosensitive glasses are silicate glasses that contain dopants, that is, metal ions such as gold, silver, or copper, capable of forming photographic images after the successive action of UV radiation and thermal treatment [52]. The treated areas can be chemically removed by acid solutions because they became more reactive than the nontreated surface. In the particular case of Ag-doped glasses, the nucleation and growth kinetics of Ag particles in glass irradiated by UV were studied to propose the following mechanism: in a first step of UV irradiation (two 125 W high-pressure Hg lamps for up to 250 h), some nuclei are formed by the generated photoelectrons that reduce some Ag^+ ions to Ag^0 atoms [53, 54],

$$Ag^+ + e^- = Ag^0, \qquad (10.9)$$

in which the photoelectrons e^- come from

$$Ag^+ + h\nu = Ag^{++} + e^- \qquad (10.10)$$

Alternatively, it was found that incorporation of small amounts of Ce ions sensitize the UV irradiation to more efficiently generate free electrons by acting as donors, according to Ref. [53],

$$Ce^{3+} + h\nu = Ce^{4+} + e^-, \qquad (10.11)$$

increasing the density of Ag atoms by the supplementary photoelectrons contributing to the conversion.

Foturan glass commercially available from Schott Glass Corporation is a photosensitive glass composed of lithium aluminosilicate doped with traces of Ag, Ce, and Sb ions. A pulsed UV Nd:YAG laser frequency tripled to 355 nm with pulse duration of 8 ns was used to directly write 3D patterned microstructures in Foturan glass [55–57]. The first step of the process involved irradiation to obtain an initial latent image in glass, followed by the second step—thermal treatment. By isotropic etching the material was preferentially removed from developed regions. The mechanism is similar to Kreibig's evaluation for Ag containing photosensitive glasses exposed to UV light [53]. Foturan photosensitive glasses were exposed to ULPs in a femtosecond laser system that generates a pulse width of 150 fs at a 775 nm wavelength. Remarkably, after exposure, latent images were written and etched away in HF after a postbaking treatment [58]. A mechanism from a model developed by Fuqua et al. [55] in which the laser fluence is related to a critical dose D_c for the threshold of photoreaction [58] was employed to discuss interaction of ULPs and Foturan photosensitive glasses. This dose is defined as the number of photons necessary to create a network of nuclei with the appropriate density for inducing chemical etching. Once this regime is achieved, the Foturan glass is modified within the irradiated regions. The evaluation of D_c allows determining the laser energy and number of pulses for the microfabrication of glass. There is a critical density Q of nuclei, which can follow the equation

$$Q = KF^m N, \qquad (10.12)$$

where K is a constant of proportionality, F is the laser fluence calculated for one pulse, m is the power dependence, and N is the number of pulses. If an arbitrary dose is described as $D = Q/K$ then the equation [12] can be written as

$$D = F^m N, \qquad (10.13)$$

As a consequence, one can deduce

$$D_c = F_c^m N \qquad (10.14)$$

The experimental investigation of the critical fluence dependence on the number of pulses allowed the calculation of m and D_c, respectively. It was found that the photoreaction is following a six-photon process ($m = 6$), while the dose $D_c = 1.3 \times 10^{-5}$ J^6/cm^{12} for the 775 nm wavelength femtosecond laser. A deeper study of the interaction mechanism of ULPs with Foturan photosensitive glass was carried out by exposing the sensitive material to (i) a femtosecond laser (150 fs, 775 nm, and 1 kHz) and nanosecond lasers of (ii) 266 nm, (iii) 355 nm, and (iv) 308 nm wavelengths [59, 60]. A significant increase in the absorption spectrum of the exposed samples to around 360 nm was found for femtosecond-irradiated glasses, corresponding to absorption from oxygen-deficient centers (ODC) [60]. This means that during irradiation, the excited electrons induce a Si-O bond scission, which results in ODC and NBOHCs [16]. Absorption of around 315 nm related to Ce ions was not observed, suggesting that Ce^{3+} ions don't contribute to electron generation for reduction to Ag atoms in the case of femtosecond-irradiated samples. Specifically, free electrons are generated by interband excitation based on the cascade process through an intermediate state. Each step is induced by three-photon absorption, resulting in the six-photon process for free electron generation. A similar photoreaction mechanism by successive interband excitation was evidenced for 266 nm laser-irradiated samples following a two-photon process (m = 2). On the other hand, in the case of 355 nm laser-irradiated samples free electrons are generated by Ce^{3+} (TPP) while in the case of 308 nm laser, both absorption by Ce^{3+} (single-photon absorption) and interband excitation (TPP) were found [60].

The multiphoton process using ULPs was found attractive for the microfabrication of embedded structures inside photosensitive glasses as it provides the capability of selective generation of complex 3D shapes [61]. The process was called femtosecond laser–assisted etching (FLAE). To obtain high-quality wall surfaces of channels, the fabrication process was continuously optimized [58, 62–67]. During FLAE of a photosensitive glass, the scanning speed, direction, and laser power are adjusted to achieve the desired shape. After laser irradiation and the first annealing treatment, in which Ag atoms cluster to act as nuclei for the growth of the crystalline phase of lithium metasilicate at the laser exposed regions, one can visibly observe the modified patterns inside photosensitive glass (Fig. 10.8A,D). The etching time is another critical parameter for developing uniform channels with similar widths over long lengths (Fig. 10.8B,E). A post-thermal treatment was proposed in order to decrease the roughness from several tens of nanometers of as-etched surface to the nanometer range (Fig. 10.8C,F). The additional annealing can increase the smoothening of glass and offers highly transparent samples for optical integration (Fig. 10.8F). Thus microfluidic platforms with the scale-down (μm) and scale-up (mm) characteristics for a functional microfluidic device with easy handling can be achieved.

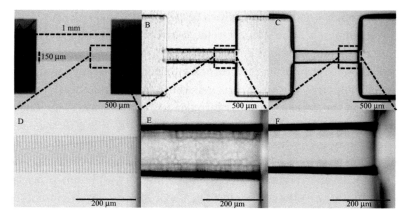

Figure 10.8 Optical images of photosensitive Foturan glasses after exposing to ULP laser irradiation and consequently to the first annealing treatment (A and D for details), 45 minutes of etching in a 10% HF etching (B and E for details), and successive postannealing treatment for the smoothening of etched surfaces (C and F for details).

10.4.2 Liquid-Assisted ULP Processing

Alternatively, 3D microfluidic structures inside glass are also available by liquid-assisted ultrafast laser drilling in which the rear surface of the glass sample is in contact with distilled water or other liquids [68–70]. Straight and bent through-channels were thus obtained by ULP drilling from the rear surface. The wetting fluid can penetrate the channels during their formation to clean them from ablated debris and thereby to drill deeper [71]. As a consequence, channels of tens of microns diameters with high aspect ratios and a good wall surface quality were fabricated. By water-assisted ablation with ULPs, true 3D microchannels consisting of longitudinal and transverse microholes were successfully achieved [72].

This is a rather simple single-step technique for 3D microfluidic structure fabrication in various transparent materials with very high aspect ratios [72–75]. Nanochannels with submicron diameters can be obtained, and 3D nanofluidic devices were proposed to be used for mixing femtoliter fluid volumes [76].

New strategies by using mesoporous glass were employed to fabricate microchannels of nearly unlimited lengths and arbitrary geometries by ULP write ablation in water immersion followed by thermal treatment. Long square-wave-shaped microchannels [77], large-volume microfluidic chambers [78], and integrated 3D microchannel systems consisting of a passive microfluidic mixer [79] (Fig. 10.9) can be thus fabricated.

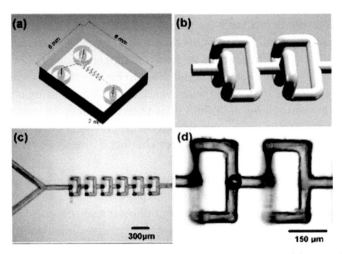

Figure 10.9 Schematics of the 3D passive microfluidic mixer: (a) general and (b) detailed images, (c) optical image of the microfluidic mixer, and (d) detailed optical image (top view) of two mixing units. Reproduced from Ref. [79] with permission from The Royal Society of Chemistry.

10.4.3 Pros and Cons of the ULP 3D Subtractive Process

For biochip applications a specific glass sample should offer robustness for easy manipulation and portability, transparency for optical interrogation, as well as chemical stability and good biocompatibility. The ULP-induced chemical modification followed by selective etching enables us to fabricate 3D microfluidic structures with flexible geometries inside glass. This technique can be extended to the fabrication of micro-optics, such as micromirrors and microlenses inside a glass substrate [62]. It can also be used to fabricate movable micromechanical components, such as microvalves and micropumps, which are free from the substrates inside the 3D microfluidic structures, because the ultrafast laser direct writing in this process simply inscribes the latent 3D images, which are then developed by wet etching [64, 80]. However, the resulting structures inevitably become wider than the laser-exposed regions and are tapered because of an etch selectivity ratio of approximately 50 between the laser-exposed and laser-unexposed regions. Specifically, it is only possible to achieve an embedded short microfluidic channel (approximately hundred μm length) 5 to 10 μm in width and height, even for submicron laser-focusing conditions used for irradiation. More complicated, longer channels are then impossible to fabricate at small dimensions; and, in general, for longer channels (>1 mm) widths and heights cannot be inhibited to be less than several tens of micrometers. In contrast, water-assisted ULP processing enables the formation of long microfluidic channels with complex structures. This technique relies on ablation; therefore, it can provide not only better uniformity but also narrower channel diameters over a long range.

Meanwhile, both techniques have poor resolution to fabricate 3D solid microstructures with complex shapes inside glass. Alternatively, ULP-induced photopolymerization of photoresists, described in the following section, provides the ability to decrease dimensions to the nanometric range.

10.5 Additive Processing: ULP-Induced Photopolymerization of Photoresists

10.5.1 Mechanisms and Limitations

Photoresists are transparent light-sensitive materials used in the microfabrication processes of various patterns onto substrates with specific applications. They are divided into two categories in terms of types of photoresists used, positive-tone photoresists and negative-tone photoresists. Negative-tone photoresists are the insoluble materials in developing solutions after light exposure, while the positive ones become soluble when they are exposed to optical radiation. The nonlinear absorption of ULPs is used for two-photon polymerization (TPP) of photoresists as a new method of maskless stereolithography.

In the case of TPP using the negative-tone resist, there is an **initiation** phase, in which the energy collected from multiple photons can excite photoinitiators present in the resin (photoresist). The photoinitiator is a molecule that can produce reactive species by absorption of light that catalyze successive chemical reactions, resulting in significant changes of solubility and physical properties of reaction products. The photoinitiators can rapidly form free radicals by a photolytic process able to polymerize monomers prone to free radical polymerization. The polymerization is followed by a **propagation** phase, in which the polymeric chain increases in length, and a **termination** phase, when chain radical molecules join to form a molecule with no radical activity. The mechanism of polymerization by TPA is described next [81]:

1. **Initiation**: $I + 2h\upsilon \rightarrow 2R \bullet \Rightarrow R \bullet + M \rightarrow RM \bullet$,
   where I is the initiator molecule, R the free radical, and M the monomer, and $\bullet$ represents an unpaired electron with a radical molecule.
2. **Propagation** : $R - (Mn)_x \bullet + M \rightarrow R - (Mn)_{x+1} \bullet$,
   where Mn is the polymer of n monomer units and $R - (Mn)_x$ is the chained radical of length x.
3. **Termination**: $-M \bullet + \bullet M - \rightarrow - M - M -$

ULPs can induce TPP at the focal point, solidifying a localized volume element called **voxel** by the mechanism described above. The rate of TPP is proportional to the square of light intensity and dependent on the photon density within the focal point [82], which contributes to a narrower spatial distribution as compared to linear absorption processes, as described in Section 10.2.2. The spatial resolution in TPP is also related to the focusing conditions, in particular to the NA of the employed objectives. High NA lenses can focalize a laser beam within a very small spot where the density of photons is large enough to induce photochemical reactions. The optical diffraction limit is given by the Rayleigh criterion:

$$R = k\frac{\lambda}{NA}, \qquad (10.15)$$

where k is a constant related to laser linewidth and projection geometry, λ the laser wavelength, and NA the numerical aperture. The resolution (the voxel) of TPP-fabricated features can go below the optical diffraction limit as the individual photochemical reactions induced in the material are dependent on the laser intensity [83].

This means that there is a material-dependent threshold of intensity given by different exposure times at specific pulse laser energies. Thus an important development of the TPP research is the discovery of novel two-photon initiators and sensitizers [84–87].

In the pioneering studies TPP demonstrated micronanofabrication with resolutions of subdiffraction limits. A lateral spatial resolution down to 120 nm was achieved by using high-NA (=1.4) oil immersion optics [83]. Later, lateral and axial diameters of the polymerized region smaller than 30 nm and 90 nm, respectively, (corresponding to aλ/25 resolution) were achieved by an appropriate selection of materials and spatial light intensity profile close-to-threshold conditions [88]. In this case, the axial cross section of the focal region of the beam was focused in the sample by an oil immersion objective lens with a high NA (=1.4) and a refractive index of 1.515, similar to that of the immersion oil. As predicted by Urey, the differences between lateral and axial resolution are related to the focal plane spot size s and the depth of focus Δz for a Gaussian beam–focused spot related to the axial irradiance drops to 80%–50% of the focal-plane axial irradiance [89]. Indeed, voxels are of an ellipsoidal shape, with the lengths of the axes given by 2a = 2b < 2c (2a, 2b, and 2c are the maximal dimensions along x, y, and z axes, respectively) [90]. It was interesting to observe that at intermediate irradiation levels obtained by a variable pinhole–tailored NA, a lower NA (0.88) gives a better lateral resolution as compared to a higher NA (1.12 or 1.4), although the smallest visible voxels were of 260 nm at 0.88 NA and 120 nm at 1.4 NA [91]. This phenomenon was interpreted by a threshold effect: at low NA focusing, the laser power is distributed in a larger volume, with a solidified front vertically expanded and laterally shrunken (slim voxel). The voxel size can be increased but limited by a laser power range, from the threshold necessary to induce the polymerization but below the laser-induced breakdown threshold.

The important laser parameter with the TPP process is related to the density and intensity of photons deposited on the photoresist material. This means that besides laser power (or energy), the laser repetition rate and the translation speed of the stage for developing the desired structure are critical parameters for creating the patterns. This is a so-called dynamic exposure time of interaction between the material and the appropriate laser beam. There are several regimes in which different laser systems were employed for TPP. These regimes in which ULPs are employed can be categorized as a function of the repetition rate, from low-repetition-rate lasers (1 kHz) to high-repetition-rate lasers (hundreds of kilohertz and megahertz). The most encouraging sources for TPP are the high-repetition-rate lasers as they can exceed the photoreaction threshold at lower pulse laser energies, providing a better resolution.

The surface roughness of the polymerized material by TPP (SU-8 from MicroChem, a negative photoresist) was evaluated as low as 2.5 nm [92]. The surface remodeling during development due to surface tension (self-smoothening effect) was found to improve the surface quality [93]. This high-resolution process clearly demonstrates the potential of TPP for the fabrication of large-scale nanostructures. There are, however, some issues related to the stability of photopolymerized structures, especially when they are fabricated with the highest resolution (finer features) for applications such as microphotonics, nanophotonics, and microfluidics. The stability problems are due to polymer shrinkages, deformations, or collapsing after developing. To increase stability, TPP was then applied to fabricate 3D microstructures in open-glass microfluidic [94]. The results suggested that TPP prototyping has the potential to functionalize and integrate general microfluidic chips.

10.5.2 Applications of Two-Photon Polymerization

TPP can tailor the shape and size of the solidified resin volume element (the voxel) that determines the resolution. Capability of controlling the feature sizes at the micro- and nanoscale and then approaching larger and functional devices has developed TPP as a powerful technology for fabricating microdevices using the appropriate design.

The first proposed devices were 3D photonic crystals that showed excellent photonic bandgaps [95]. These structures were later improved and, taking advantage of optimal laser parameters and CAD-CAM programs, almost perfect photonic woodpile lattices were fabricated from which a power rejection of 35% was obtained at a specific wavelength of 780 nm (80 fs pulse width and 82 MHz repetition rate) [96]. The ULP processing was further applied to fabricate photonic crystals with more complex, spiral architectures, showing also promising bandgap properties [97]. Such constructions are reproduced in Fig. 10.10.

Figure 10.10 SEM images of spiral structures obtained by TPP. Reproduced from Ref. [97] with permission from Wiley.

Complex micromachines, such as microrotors, were fabricated and successfully manipulated by laser tweezers, allowing the evaluation of rotation [98]. Microtweezers and microneedles were also achieved by TPP, and applications in manipulation of cells or other organisms in liquid media were proposed [99]. Three-dimensional microfluidic devices consisting of 3D filtering structures were grown by TPP on surfaces [100] or into open channels [94, 101]. Metamaterial architectures with submicron features have been used to fabricate core-shell elastostatic cloaks in which the high precision of the TPP process was needed to adjust mechanical properties of the cloaking environment [102]. Three-dimensional scaffolds imitating cellular media in organisms can be constructed with pore size control and desired geometry [103]. This particular application can be of great interest as the 3D environments provide surrounding support for cells, like in vivo, and could represent a real alternative to animal testing. Many studies with different biomimetic architectures tested with various cell lines provided specific cellular responses to 3D constructs [104–112]. A complex geometry of a biomimetic architecture with well-distributed pores is reproduced in Fig. 10.11.

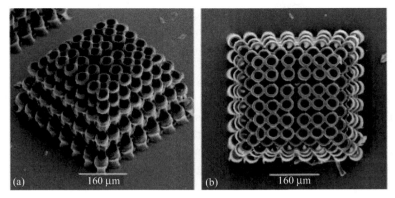

Figure 10.11 Three-dimensional scaffold fabricated by two-photon polymerization: (a) SEM image (perspective view) of produced structure and (b) SEM image (top view) of produced structure. Reprinted from Ref. [105], Copyright (2011), with permission from Elsevier.

10.6 Hybrid ULP 3D Processing

10.6.1 Combination of Subtractive and Undeformative Processing

Optofluidic circuit structures are of real interest in life sciences by approaching studies with individual cells for cell manipulation, isolation, or sorting. ULP 3D processing offers the possibility to combine advantages of both subtractive and undeformative processes to achieve the fabrication of functional, robust, and monolithic optofluidic chips. Hybrid structures consisting of microchannels integrated with optical waveguides can provide great advantages since there is no interconnection between the microfluidic device and a detector. An incident light beam is introduced and guided by optical waveguides through the device. These are connected with a fabricated microchannel and can probe local disturbance of the optical property at a junction, enabling single-cell detection by analyzing the respective transmission intensity change [113]. It was thus possible to detect a single red blood cell via two optical detection approaches by either recording the transmission intensity of a He-Ne laser beam or observing fluorescence after excitation by an Ar^+

laser beam. In another study, a Mach–Zehnder interferometer (MZI) was inscribed in fused silica by crossing orthogonally a microfluidic channel with the sensing arm and reference arm passing over it. The geometrical 3D capabilities of the device permitted a label-free sensing of glucose flowing in the microchannel, with a detection sensitivity down to 4 mM [114]. Optical waveguides and filters coupled with microfluidic channels allowed one to elucidate that the CO_2 secreted from the seedling root in the presence of light attracts *Phormidium*, a species of cyanobacteria. By determining the light intensity and its specific wavelength, a better understanding of the gliding mechanism of *Phormidium* was achieved, with consequences on improved vegetable production [115]. A typical scheme of the hybrid waveguide-microchannel device and the obtained signals in absorption measurements for determination of CO_2 concentration are presented in Fig. 10.12.

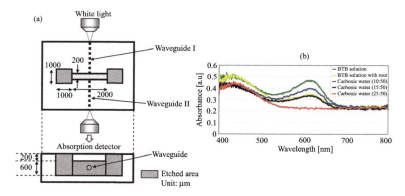

Figure 10.12 (a) Schematic illustration of the microchip integrated with optical waveguides. (b) Optical absorption spectra of water containing an aqueous bromothymol blue solution (green line), with a seedling root (yellow line), and with carbonic water with different CO_2 concentrations. Reproduced from Ref. [115] with permission from The Royal Society of Chemistry.

The illumination by laser light from a curved waveguide to water containing algae specimens in microchannels and the monitoring of intensity distribution of collected light by external photodetector allowed a size-sensitive detection and classification of five different algae species with 78% accuracy [116]. Furthermore, a complex optofluidic device capable of sorting specific cells by optical forces was introduced. The sorter is activated by detecting fluorescence from the specific cells flowing into a microchannel induced by laser illumination using the waveguide (fluorescence waveguide). An external laser source coupled to another waveguide (sorting waveguide) is automatically switched on with a proper delay time when the fluorescence is detected and triggers the impulse to push the target cell toward the lower stream in a directed flow for sorting [117].

10.6.2 Combination of Subtractive and Additive Processing

Another scheme of hybrid 3D processing is a combination of subtractive and additive processing, which enhances the advantages and compensates for the drawbacks of both techniques to create 3D micro- and nanostructures with more complicated shapes and increased functionalities. Hybrid additive and subtractive ULP processing with TPP followed by femtosecond laser multiphoton ablation was successfully demonstrated for the fabrication of submicrometer polymer fibers containing periodic holes with diameters of 500 nm and 3D microfluidic channels with diameters of 1 μm [118]. For the fabrication of polymer fibers containing periodic holes, TPP was first used to fabricate fiber structures with 2, 1, and 0.5 μm line widths. After TPP, subtractive femtosecond laser ablation formed periodic holes with diameters of approximately 500 nm in the polymer fibers, which acted as Bragg grating structures in the fibers. The diameter and the periodicity of the holes were adjustable through the control of the laser irradiation conditions. It should be noted that it is difficult to fabricate such fiber Bragg gratings by either TPP or femtosecond laser multiphoton ablation alone.

A "ship in a bottle" fabrication concept based on a hybrid subtractive and additive ULP processing was introduced in order to integrate 3D polymer structures inside (in volume) glass microchannels [119–122]. It consists of FLAE of photosensitive glass and TPP of a negative epoxy-resin inside the microchannels. The process allows lowering the size limit of 3D objects created inside channels to smaller details down to the dimensions of several μm or below and improving the structure stability at the same time as it offers the required robustness for assembling a concrete lab-on-a-chip device. The capability to directly fabricate 3D complex shapes of both glass channels and polymeric integrated patterns enables us to spatially design 3D biochips for specific applications.

The process starts with FLAE applied to a photosensitive glass in order to obtain embedded microchannels. The procedure is followed by TPP of a negative epoxy-resin inside the processed microchannels. The hybrid process is illustrated in Fig. 10.13. In the first step, a 20x magnification objective lens of NA 0.46 is used to focus the laser beam into glass to create predesigned micropatterns with the translation of the 3D stage (Fig. 10.13A). After laser irradiation, the sample is annealed and etched to fabricate the desired microchannel (Fig. 10.13C). A second annealing is needed for surface smoothening inside the glass microchannel. After fabrication, the channel is filled with a negative-tone photoresist (SU-8) after dilution in acetone. Prebaking is then carried out to evaporate the solvent from the SU-8 resin and, at the same time, improve adhesion to the surface. To integrate 3D polymer nano- and micropattern structures in the glass microchannel, the same laser system is employed for the TPP process. The laser beam of 100 µW power is focalized through a 100x magnification objective lens of NA 1.4 onto the sample through oil immersion (Fig. 10.13B). The translation speed of the sample is about 400 µm/s. The photoinitiator (triarylsulphonium/hexafluoroantimonate salt in propylene carbonate) generates an acid under the laser irradiation. The chemical chain reaction of the nano- and micropatterns by acid-assisted crosslinking was acquired by postbaking. Unpolymerized resin is removed by the SU-8 developer (1-methoxy-2-propylacetate) after sample immersion and polymeric patterns develop (Fig. 10.13D). To complete the procedure, an additional postdrying process is necessary.

Due to a closed environment the prebaking time is longer than the typical TPP because the evaporation of the solvent is slow. The laser power used for TPP in a microchannel is about two times larger than that on the surface since the optical loss is caused by reflection and scattering at both channel–polymer interfaces (Fig. 10.13E) and multiphoton absorption (MPA) of the glass and photoresist that fills the tens-of-micrometer channel height. In addition, the developing time is longer inside a microchannel, which could represent an advantage due to the slow process, which induces structure stability. Polymeric micro- and nanostructure pattern integration inside microfluidic systems covers the scale-down (nm) and scale-up (μm) aspects in a larger (mm) multifunctional microfluidic device (Fig. 10.14). It also allows nanomechanical manipulation on both 2D and 3D environments and optical visualization, while providing an increase in sensitivity and eventually in the performance of assembled devices.

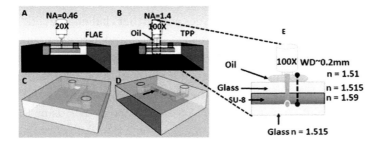

Figure 10.13 Schematic of a hybrid FLAE-TPP process for ship-in-a-bottle polymer integration: FLAE of Foturan glass (A), TPP of SU-8 photoresist inside a microchannel (B), a Z-shape microchannel fabricated by FLAE (C), polymeric patterns (indicated by black arrows) developed by TPP inside a microchannel (D), and a detailed view of the laser path inside three different media (immersion oil, glass, and photoresist) (E).

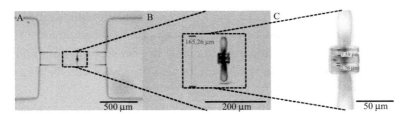

Figure 10.14 Optical images of microchannels (2.76 and 7.59 μm in width) by TPP inside a glass channel (A) during the developing procedure and (B) after developing and drying. (C) A detailed view and evaluation of microchannel dimensions.

10.7 Challenges and Perspectives

Ultrafast lasers have been rapidly developed during the last decade as their characteristics are more and more exploited for material processing. ULPs that can much reduce HAZ within the processed materials are now critical for the micro- and nanoscale fabrication of both inorganic and organic materials. The ULPs generating high peak power intensities that produce MPA in transparent materials can localize internal modifications with high precision and resolution below the diffraction limit.

Both fundamental research and applicative investigations are critical for understanding the issues and further challenging concrete applications in the photonic-related fields and bionanomedicine [123]. As the internal modification of glass, a brittle material, is possible only by ultrafast lasers, the fabrication of 3D optofluidic and lab-on-a-chip devices is then promising in applications of ULPs. Spatial resolution below 100 nm can be attained by TPP using ULPs, offering the potential of fabricating polymeric photonic crystals and micro- and nanosystems [95, 124–126].

The prospect of tailoring spatial and temporal profiles of ULPs is attractive for micro- and nanoscale fabrication by high-throughput and high-energy-efficiency laser processing [127]. Tailoring temporal profiles provides an advantage of inducing nanoscale ablation in transparent materials [128].

Hybrid glass-polymer devices are recently investigated as potential lab-on-a-chip 3D platforms for concrete applications [129]. A great perspective is the fabrication of a biochip with 3D architecture by laser only, able to mimic in vivo surroundings and provide the necessary environment for cells to grow in a media close to human conditions.

Still at its infancy, the distinct laser processing of transparent materials based on ULPs needs further development in the future in terms of reducing the laser production costs. With this view, novel high-power and high-repetition-rate sources are required for high-quality, inexpensive processing.

Acknowledgments

Felix Sima acknowledges the support of UEFISCDI under the contract PNII-RU-TE-2014-4-1273 (187/2015).

References

1. Allmen, M. V., and Blatter, A. (2013). *Laser-Beam Interactions with Materials: Physical Principles and Applications* (Springer Science & Business Media).
2. Chichkov, B. N., Momma, C., Nolte, S., Von Alvensleben, F., and Tünnermann, A. (1996). Femtosecond, picosecond and nanosecond laser ablation of solids, *Appl. Phys. A*, **63**, pp. 109–115.
3. Le Harzic, R., Huot, N., Audouard, E., Jonin, C., Laporte, P., Valette, S., Fraczkiewicz, A., and Fortunier, R. (2002). Comparison of heat-affected zones due to nanosecond and femtosecond laser pulses using transmission electronic microscopy, *Appl. Phys. Lett.*, **80**, pp. 3886–3888.
4. Itoh, K., Watanabe, W., Nolte, S., and Schaffer, C. B. (2006). Ultrafast processes for bulk modification of transparent materials, *MRS Bull.*, **31**, pp. 620–625.
5. Gattass, R. R., and Mazur, E. (2008). Femtosecond laser micromachining in transparent materials, *Nat. Photonics*, **2**, pp. 219–225.
6. Watanabe, W., Li, Y., and Itoh, K. (2016). [INVITED] Ultrafast laser micro-processing of transparent material, *Opt. Laser Technol.*, **78**, pp. 52–61.
7. Umran, F. A., Liao, Y., Elias, M. M., Sugioka, K., Stoian, R., Cheng, G., and Cheng, Y. (2013). Formation of nanogratings in a transparent material with tunable ionization property by femtosecond laser irradiation, *Opt. Express*, **21**, pp. 15259–15267.
8. Choudhury, D., Macdonald, J. R., and Kar, A. K. (2014). Ultrafast laser inscription: perspectives on future integrated applications, *Laser Photonics Rev.*, **8**, pp. 827–846.
9. Kawata, S., Sun, H.-B., Tanaka, T., and Takada, K. (2001). Finer features for functional microdevices, *Nature*, **412**, pp. 697–698.
10. Liu, X., Du, D., and Mourou, G. (1997). Laser ablation and micromachining with ultrashort laser pulses, *IEEE J. Quantum Electron.*, **33**, pp. 1706–1716.
11. Schaffer, C. B., Brodeur, A., García, J. F., and Mazur, E. (2001). Micromachining bulk glass by use of femtosecond laser pulses with nanojoule energy, *Opt. Lett.*, **26**, pp. 93–95.
12. Sakakura, M., and Terazima, M. (2004). Oscillation of the refractive index at the focal region of a femtosecond laser pulse inside a glass, *Opt. Lett.*, **29**, pp. 1548–1550.

13. Sakakura, M., Terazima, M., Shimotsuma, Y., Miura, K., and Hirao, K. (2007). Observation of pressure wave generated by focusing a femtosecond laser pulse inside a glass, *Opt. Express*, **15**, pp. 5674–5686.
14. Du, D., Liu, X., Korn, G., Squier, J., and Mourou, G. (1994). Laser-induced breakdown by impact ionization in SiO2 with pulse widths from 7 ns to 150 fs, *Appl. Phys. Lett.*, **64**, pp. 3071–3073.
15. Stuart, B., Feit, M., Rubenchik, A., Shore, B., and Perry, M. (1995). Laser-induced damage in dielectrics with nanosecond to subpicosecond pulses, *Phys. Rev. Lett.*, **74**, p. 2248.
16. Davis, K. M., Miura, K., Sugimoto, N., and Hirao, K. (1996). Writing waveguides in glass with a femtosecond laser, *Opt. Lett.*, **21**, pp. 1729–1731.
17. Marcinkevičius, A., Juodkazis, S., Watanabe, M., Miwa, M., Matsuo, S., Misawa, H., and Nishii, J. (2001). Femtosecond laser-assisted three-dimensional microfabrication in silica, *Opt. Lett.*, **26**, pp. 277–279.
18. Glass, A. J., and Guenther, A. H. (1975). Laser induced damage in optical materials: 6th ASTM symposium, *Appl. Opt.*, **14**, pp. 698–715.
19. Göppert-Mayer, M. (1931). Über elementarakte mit zwei quantensprüngen, *Ann. Phys.*, **401**, pp. 273–294.
20. Goeppert-Mayer, M., and Born, M. (1931). Dynamic lattice theory of crystals, *Handb. Phys.*, **24**, p. 623.
21. Keldysh, L. (1965). Ionization in the field of a strong electromagnetic wave, *Sov. Phys. JETP*, **20**, pp. 1307–1314.
22. Schaffer, C. B., Brodeur, A., and Mazur, E. (2001). Laser-induced breakdown and damage in bulk transparent materials induced by tightly focused femtosecond laser pulses, *Meas. Sci. Technol.*, **12**, p. 1784.
23. Tien, A.-C., Backus, S., Kapteyn, H., Murnane, M., and Mourou, G. (1999). Short-pulse laser damage in transparent materials as a function of pulse duration, *Phys. Rev. Lett.*, **82**, p. 3883.
24. Corkum, P., Brunel, F., Sherman, N., and Srinivasan-Rao, T. (1988). Thermal response of metals to ultrashort-pulse laser excitation, *Phys. Rev. Lett.*, **61**, p. 2886.
25. Borowiec, A., and Haugen, H. (2003). Subwavelength ripple formation on the surfaces of compound semiconductors irradiated with femtosecond laser pulses, *Appl. Phys. Lett.*, **82**, pp. 4462–4464.
26. Eaton, S., Zhang, H., Herman, P., Yoshino, F., Shah, L., Bovatsek, J., and Arai, A. (2005). Heat accumulation effects in femtosecond laser-

written waveguides with variable repetition rate, *Opt. Express*, **13**, pp. 4708–4716.

27. Osellame, R., Chiodo, N., Della Valle, G., Taccheo, S., Ramponi, R., Cerullo, G., Killi, A., Morgner, U., Lederer, M., and Kopf, D. (2004). Optical waveguide writing with a diode-pumped femtosecond oscillator, *Opt. Lett.*, **29**, pp. 1900–1902.

28. Schaffer, C. B., García, J. F., and Mazur, E. (2003). Bulk heating of transparent materials using a high-repetition-rate femtosecond laser, *Appl. Phys. A*, **76**, pp. 351–354.

29. Minoshima, K., Kowalevicz, A. M., Hartl, I., Ippen, E. P., and Fujimoto, J. G. (2001). Photonic device fabrication in glass by use of nonlinear materials processing with a femtosecond laser oscillator, *Opt. Lett.*, **26**, pp. 1516–1518.

30. Tamaki, T., Watanabe, W., and Itoh, K. (2006). Laser micro-welding of transparent materials by a localized heat accumulation effect using a femtosecond fiber laser at 1558 nm, *Opt. Express*, **14**, pp. 10460–10468.

31. Tan, D., Sharafudeen, K. N., Yue, Y., and Qiu, J. (2016). Femtosecond laser induced phenomena in transparent solid materials: fundamentals and applications, *Prog. Mater. Sci.*, **76**, pp. 154–228.

32. Shah, L., Arai, A., Eaton, S., and Herman, P. (2005). Waveguide writing in fused silica with a femtosecond fiber laser at 522 nm and 1 MHz repetition rate, *Opt. Express*, **13**, pp. 1999–2006.

33. Shimotsuma, Y., Hirao, K., Kazansky, P. G., and Qiu, J. (2005). Three-dimensional micro-and nano-fabrication in transparent materials by femtosecond laser, *Jpn. J. Appl. Phys.*, **44**, p. 4735.

34. Juodkazis, S., Nishimura, K., Tanaka, S., Misawa, H., Gamaly, E. G., Luther-Davies, B., Hallo, L., Nicolai, P., and Tikhonchuk, V. T. (2006). Laser-induced microexplosion confined in the bulk of a sapphire crystal: evidence of multimegabar pressures, *Phys. Rev. Lett.*, **96**, p. 166101.

35. Shimotsuma, Y., Kazansky, P. G., Qiu, J., and Hirao, K. (2003). Self-organized nanogratings in glass irradiated by ultrashort light pulses, *Phys. Rev. Lett.*, **91**, p. 247405.

36. Kazansky, P. G., Yang, W., Bricchi, E., Bovatsek, J., Arai, A., Shimotsuma, Y., Miura, K., and Hirao, K. (2007). "Quill" writing with ultrashort light pulses in transparent materials, *Appl. Phys. Lett.*, **90**, p. 151120.

37. Sudrie, L., Franco, M., Prade, B., and Mysyrowicz, A. (1999). Writing of permanent birefringent microlayers in bulk fused silica with femtosecond laser pulses, *Opt. Commun.*, **171**, pp. 279–284.

38. Watanabe, T., Shiozawa, M., Tatsu, E., Kimura, S., Umeda, M., Mine, T., Shimotsuma, Y., Sakakura, M., Nakabayashi, M., and Miura, K. (2013). A driveless read system for permanently recorded data in fused silica, *Jpn. J. Appl. Phys.*, **52**, p. 09LA02.

39. Bricchi, E., Klappauf, B. G., and Kazansky, P. G. (2004). Form birefringence and negative index change created by femtosecond direct writing in transparent materials, *Opt. Lett.*, **29**, pp. 119–121.

40. Hirao, K., and Miura, K. (1998). Writing waveguides and gratings in silica and related materials by a femtosecond laser, *J. Non-Cryst. Solids*, **239**, pp. 91–95.

41. Streltsov, A. M., and Borrelli, N. F. (2002). Study of femtosecond-laser-written waveguides in glasses, *J. Opt. Soc. Am. B*, **19**, pp. 2496–2504.

42. Nolte, S., Will, M., Burghoff, J., and Tuennermann, A. (2003). Femtosecond waveguide writing: a new avenue to three-dimensional integrated optics, *Appl. Phys. A*, **77**, pp. 109–111.

43. Will, M., Nolte, S., Chichkov, B. N., and Tünnermann, A. (2002). Optical properties of waveguides fabricated in fused silica by femtosecond laser pulses, *Appl. Opt.*, **41**, pp. 4360–4364.

44. Florea, C., and Winick, K. A. (2003). Fabrication and characterization of photonic devices directly written in glass using femtosecond laser pulses, *J. Lightwave Technol.*, **21**, pp. 246–253.

45. Bricchi, E., Mills, J. D., Kazansky, P. G., Klappauf, B. G., and Baumberg, J. J. (2002). Birefringent Fresnel zone plates in silica fabricated by femtosecond laser machining, *Opt. Lett.*, **27**, pp. 2200–2202.

46. Liao, Y., Xu, J., Cheng, Y., Zhou, Z., He, F., Sun, H., Song, J., Wang, X., Xu, Z., and Sugioka, K. (2008). Electro-optic integration of embedded electrodes and waveguides in LiNbO3 using a femtosecond laser, *Opt. Lett.*, **33**, pp. 2281–2283.

47. Della Valle, G., Taccheo, S., Osellame, R., Festa, A., Cerullo, G., and Laporta, P. (2007). 1.5 µm single longitudinal mode waveguide laser fabricated by femtosecond laser writing, *Opt. Express*, **15**, pp. 3190–3194.

48. Kawamura, K.-I., Hirano, M., Kurobori, T., Takamizu, D., Kamiya, T., and Hosono, H. (2004). Femtosecond-laser-encoded distributed-feedback color center laser in lithium fluoride single crystals, *Appl. Phys. Lett.*, **84**, pp. 311–313.

49. Marshall, G. D., Ams, M., and Withford, M. J. (2006). Direct laser written waveguide-Bragg gratings in bulk fused silica, *Opt. Lett.*, **31**, pp. 2690–2691.

50. Osellame, R., Maselli, V., Vazquez, R. M., Ramponi, R., and Cerullo, G. (2007). Integration of optical waveguides and microfluidic channels both fabricated by femtosecond laser irradiation, *Appl. Phys. Lett.*, **90**, p. 231118.
51. Yamada, K., Watanabe, W., Toma, T., Itoh, K., and Nishii, J. (2001). In situ observation of photoinduced refractive-index changes in filaments formed in glasses by femtosecond laser pulses, *Opt. Lett.*, **26**, pp. 19–21.
52. Stookey, S. (1953). Chemical machining of photosensitive glass, *Ind. Eng. Chem.*, **45**, pp. 115–118.
53. Kreibig, U. (1976). Small silver particles in photosensitive glass: their nucleation and growth, *Appl. Phys.*, **10**, pp. 255–264.
54. Stookey, S. D. (1949). Photosensitive glass, *Ind. Eng. Chem.*, **41**, pp. 856–861.
55. Fuqua, P. D., Janson, S. W., Hansen, W. W., and Helvajian, H. (1999). Fabrication of true 3D microstructures in glass/ceramic materials by pulsed UV laser volumetric exposure techniques, *Optoelectronics' 99-Integrated Optoelectronic Devices*, pp. 213–220.
56. Helvajian, H., Fuqua, P. D., and Hansen, W. W. (2005). Ultraviolet method of embedding structures in photocerams, Google Patents.
57. Livingston, F. E., Hansen, W. W., Huang, A., and Helvajian, H. (2002). Effect of laser parameters on the exposure and selective etch rate in photostructurable glass, *High-Power Lasers and Applications*, pp. 404–412.
58. Masuda, M., Sugioka, K., Cheng, Y., Aoki, N., Kawachi, M., Shihoyama, K., Toyoda, K., Helvajian, H., and Midorikawa, K. (2003). 3-D microstructuring inside photosensitive glass by femtosecond laser excitation, *Appl. Phys. A*, **76**, pp. 857–860.
59. Sugioka, K., Cheng, Y., and Midorikawa, K. (2005). Three-dimensional micromachining of glass using femtosecond laser for lab-on-a-chip device manufacture, *Appl. Phys. A*, **81**, pp. 1–10.
60. Hongo, T., Sugioka, K., Niino, H., Cheng, Y., Masuda, M., Miyamoto, I., Takai, H., and Midorikawa, K. (2005). Investigation of photoreaction mechanism of photosensitive glass by femtosecond laser, *J. Appl. Phys.*, **97**, p. 063517.
61. Sugioka, K., and Cheng, Y. (2014). Ultrafast lasers—reliable tools for advanced materials processing, *Light Sci. Appl.*, **3**, p. e149.
62. Cheng, Y., Sugioka, K., Midorikawa, K., Masuda, M., Toyoda, K., Kawachi, M., and Shihoyama, K. (2003). Three-dimensional micro-optical

components embedded in photosensitive glass by a femtosecond laser, *Opt. Lett.*, **28**, pp. 1144–1146.

63. Cheng, Y., Sugioka, K., Midorikawa, K., Masuda, M., Toyoda, K., Kawachi, M., and Shihoyama, K. (2003). Control of the cross-sectional shape of a hollow microchannel embedded in photostructurable glass by use of a femtosecond laser, *Opt. Lett.*, **28**, pp. 55–57.

64. Masuda, M., Sugioka, K., Cheng, Y., Hongo, T., Shihoyama, K., Takai, H., Miyamoto, I., and Midorikawa, K. (2004). Direct fabrication of freely movable microplate inside photosensitive glass by femtosecond laser for lab-on-chip application, *Appl. Phys. A*, **78**, pp. 1029–1032.

65. Cheng, Y., Tsai, H.-L., Sugioka, K., and Midorikawa, K. (2006). Fabrication of 3D microoptical lenses in photosensitive glass using femtosecond laser micromachining, *Appl. Phys. A*, **85**, pp. 11–14.

66. Wang, Z., Sugioka, K., and Midorikawa, K. (2007). Three-dimensional integration of microoptical components buried inside photosensitive glass by femtosecond laser direct writing, *Appl. Phys. A*, **89**, pp. 951–955.

67. Sugioka, K., Hanada, Y., and Midorikawa, K. (2010). Three-dimensional femtosecond laser micromachining of photosensitive glass for biomicrochips, *Laser Photonics Rev.*, **4**, pp. 386–400.

68. Sugioka, K., and Cheng, Y. (2012). Femtosecond laser processing for optofluidic fabrication, *Lab Chip*, **12**, pp. 3576–3589.

69. Xu, B.-B., Zhang, Y.-L., Xia, H., Dong, W.-F., Ding, H., and Sun, H.-B. (2013). Fabrication and multifunction integration of microfluidic chips by femtosecond laser direct writing, *Lab Chip*, **13**, pp. 1677–1690.

70. Sugioka, K., and Cheng, Y. (2014). Femtosecond laser three-dimensional micro-and nanofabrication, *Appl. Phys. Rev.*, **1**, p. 041303.

71. Hwang, D., Choi, T., and Grigoropoulos, C. (2004). Liquid-assisted femtosecond laser drilling of straight and three-dimensional microchannels in glass, *Appl. Phys. A*, **79**, pp. 605–612.

72. An, R., Li, Y., Dou, Y., Yang, H., and Gong, Q. (2005). Simultaneous multi-microhole drilling of soda-lime glass by water-assisted ablation with femtosecond laser pulses, *Opt. Express*, **13**, pp. 1855–1859.

73. Li, Y., Itoh, K., Watanabe, W., Yamada, K., Kuroda, D., Nishii, J., and Jiang, Y. (2001). Three-dimensional hole drilling of silica glass from the rear surface with femtosecond laser pulses, *Opt. Lett.*, **26**, pp. 1912–1914.

74. Kim, T. N., Campbell, K., Groisman, A., Kleinfeld, D., and Schaffer, C. B. (2005). Femtosecond laser-drilled capillary integrated into a microfluidic device, *Appl. Phys. Lett.*, **86**, p. 201106.

75. Li, Y., and Qu, S. (2011). Femtosecond laser-induced breakdown in distilled water for fabricating the helical microchannels array, *Opt. Lett.*, **36**, pp. 4236–4238.
76. Ke, K., Hasselbrink, E. F., and Hunt, A. J. (2005). Rapidly prototyped three-dimensional nanofluidic channel networks in glass substrates, *Anal. Chem.*, **77**, pp. 5083–5088.
77. Liao, Y., Ju, Y., Zhang, L., He, F., Zhang, Q., Shen, Y., Chen, D., Cheng, Y., Xu, Z., and Sugioka, K. (2010). Three-dimensional microfluidic channel with arbitrary length and configuration fabricated inside glass by femtosecond laser direct writing, *Opt. Lett.*, **35**, pp. 3225–3227.
78. Ju, Y., Liao, Y., Zhang, L., Sheng, Y., Zhang, Q., Chen, D., Cheng, Y., Xu, Z., Sugioka, K., and Midorikawa, K. (2011). Fabrication of large-volume microfluidic chamber embedded in glass using three-dimensional femtosecond laser micromachining, *Microfluid. Nanofluid.*, **11**, pp. 111–117.
79. Liao, Y., Song, J., Li, E., Luo, Y., Shen, Y., Chen, D., Cheng, Y., Xu, Z., Sugioka, K., and Midorikawa, K. (2012). Rapid prototyping of three-dimensional microfluidic mixers in glass by femtosecond laser direct writing, *Lab Chip*, **12**, pp. 746–749.
80. Sugioka, K., and Cheng, Y. (2014). Fabrication of 3D microfluidic structures inside glass by femtosecond laser micromachining, *Appl. Phys. A*, **114**, pp. 215–221.
81. Odian, G. (2004). Radical chain polymerization, in *Principles of Polymerization*, 4th ed. (Wiley), pp. 198–349.
82. Maruo, S., and Kawata, S. (1998). Two-photon-absorbed near-infrared photopolymerization for three-dimensional microfabrication, *J. Microelectromech. Syst.*, **7**, pp. 411–415.
83. Tanaka, T., Sun, H.-B., and Kawata, S. (2002). Rapid sub-diffraction-limit laser micro/nanoprocessing in a threshold material system, *Appl. Phys. Lett.*, **80**, pp. 312–314.
84. Cumpston, B. H., Ananthavel, S. P., Barlow, S., Dyer, D. L., Ehrlich, J. E., Erskine, L. L., Heikal, A. A., Kuebler, S. M., Lee, I.-Y. S., and McCord-Maughon, D. (1999). Two-photon polymerization initiators for three-dimensional optical data storage and microfabrication, *Nature*, **398**, pp. 51–54.
85. Kim, O.-K., Lee, K.-S., Woo, H.Y., Kim, K.-S., He, G. S., Swiatkiewicz, J., and Prasad, P. N. (2000). New class of two-photon-absorbing chromophores based on dithienothiophene, *Chem. Mater.*, **12**, pp. 284–286.

86. Kuebler, S. M., Rumi, M., Watanabe, T., Braun, K., Cumpston, B. H., Heikal, A. A., Erskine, L. L., Thayumanavan, S., Barlow, S., and Marder, S. R. (2001). Optimizing two-photon initiators and exposure conditions for three-dimensional lithographic microfabrication, *J. Photopolym. Sci. Technol.*, **14**, pp. 657–668.

87. Kuebler, S. M., Braun, K. L., Zhou, W., Cammack, J. K., Yu, T., Ober, C. K., Marder, S. R., and Perry, J. W. (2003). Design and application of high-sensitivity two-photon initiators for three-dimensional microfabrication, *J. Photochem. Photobiol. A*, **158**, pp. 163–170.

88. Juodkazis, S., Mizeikis, V., Seet, K. K., Miwa, M., and Misawa, H. (2005). Two-photon lithography of nanorods in SU-8 photoresist, *Nanotechnology*, **16**, p. 846.

89. Urey, H. (2004). Spot size, depth-of-focus, and diffraction ring intensity formulas for truncated Gaussian beams, *Appl. Opt.*, **43**, pp. 620–625.

90. Born, M., and Wolf, E. (2000) *Principles of Optics: Electromagnetic Theory of Propagation, Interference and Diffraction of Light* (CUP Archive).

91. Sun, H.-B., Maeda, M., Takada, K., Chon, J. W., Gu, M., and Kawata, S. (2003). Experimental investigation of single voxels for laser nanofabrication via two-photon photopolymerization, *Appl. Phys. Lett.*, **83**, pp. 819–821.

92. Wu, D., Chen, Q.-D., Niu, L.-G., Jiao, J., Xia, H., Song, J.-F., and Sun, H.-B. (2009). 100% fill-factor aspheric microlens arrays (AMLA) with sub-20-nm precision, *IEEE Photonics Technol. Lett.*, **21**, pp. 1535–1537.

93. Wu, D., Wu, S.-Z., Niu, L.-G., Chen, Q.-D., Wang, R., Song, J.-F., Fang, H.-H., and Sun, H.-B. (2010). High numerical aperture microlens arrays of close packing, *Appl. Phys. Lett.*, **97**, p. 031109.

94. Wang, J., He, Y., Xia, H., Niu, L.-G., Zhang, R., Chen, Q.-D., Zhang, Y.-L., Li, Y.-F., Zeng, S.-J., and Qin, J.-H. (2010). Embellishment of microfluidic devices via femtosecond laser micronanofabrication for chip functionalization, *Lab Chip*, **10**, pp. 1993–1996.

95. Sun, H.-B., Matsuo, S., and Misawa, H. (1999). Three-dimensional photonic crystal structures achieved with two-photon-absorption photopolymerization of resin, *Appl. Phys. Lett.*, **74**, pp. 786–788.

96. Sun, H.-B., Suwa, T., Takada, K., Zaccaria, R. P., Kim, M.-S., Lee, K.-S., and Kawata, S. (2004). Shape precompensation in two-photon laser nanowriting of photonic lattices, *Appl. Phys. Lett.*, **85**, pp. 3708–3710.

97. Seet, K. K., Mizeikis, V., Matsuo, S., Juodkazis, S., and Misawa, H. (2005). Three-dimensional spiral-architecture photonic crystals obtained by direct laser writing, *Adv. Mater.*, **17**, pp. 541–545.

98. Galajda, P., and Ormos, P. (2001). Complex micromachines produced and driven by light, *Appl. Phys. Lett.*, **78**, pp. 249–251.
99. Maruo, S., Ikuta, K., and Korogi, H. (2003). Submicron manipulation tools driven by light in a liquid, *Appl. Phys. Lett.*, **82**, pp. 133–135.
100. Wu, D., Chen, Q.-D., Niu, L.-G., Wang, J.-N., Wang, J., Wang, R., Xia, H., and Sun, H.-B. (2009). Femtosecond laser rapid prototyping of nanoshells and suspending components towards microfluidic devices, *Lab Chip*, **9**, pp. 2391–2394.
101. Lim, T. W., Son, Y., Jeong, Y. J., Yang, D.-Y., Kong, H.-J., Lee, K.-S., and Kim, D.-P. (2011). Three-dimensionally crossing manifold micro-mixer for fast mixing in a short channel length, *Lab Chip*, **11**, pp. 100–103.
102. Bückmann, T., Thiel, M., Kadic, M., Schittny, R., and Wegener, M. (2014). An elasto-mechanical unfeelability cloak made of pentamode metamaterials, *Nat. Commun.*, **5**, p. 4130.
103. Tayalia, P., Mendonca, C. R., Baldacchini, T., Mooney, D. J., and Mazur, E. (2008). 3D cell-migration studies using two-photon engineered polymer scaffolds, *Adv. Mater.*, **20**, pp. 4494–4498.
104. Ovsianikov, A., Schlie, S., Ngezahayo, A., Haverich, A., and Chichkov, B. N. (2007). Two-photon polymerization technique for microfabrication of CAD-designed 3D scaffolds from commercially available photosensitive materials, *J. Tissue Eng. Regener. Med.*, **1**, pp. 443–449.
105. Ovsianikov, A., Malinauskas, M., Schlie, S., Chichkov, B., Gittard, S., Narayan, R., Löbler, M., Sternberg, K., Schmitz, K.-P., and Haverich, A. (2011). Three-dimensional laser micro-and nano-structuring of acrylated poly (ethylene glycol) materials and evaluation of their cytoxicity for tissue engineering applications, *Acta Biomater.*, **7**, pp. 967–974.
106. Claeyssens, F., Hasan, E. A., Gaidukeviciute, A., Achilleos, D. S., Ranella, A., Reinhardt, C., Ovsianikov, A., Shizhou, X., Fotakis, C., and Vamvakaki, M. (2009). Three-dimensional biodegradable structures fabricated by two-photon polymerization, *Langmuir*, **25**, pp. 3219–3223.
107. Hidai, H., Jeon, H., Hwang, D. J., and Grigoropoulos, C. P. (2009). Self-standing aligned fiber scaffold fabrication by two photon photopolymerization, *Biomed. Microdevices*, **11**, pp. 643–652.
108. Sima, L., Buruiana, E., Buruiana, T., Matei, A., Epurescu, G., Zamfirescu, M., Moldovan, A., Petrescu, S., and Dinescu, M. (2013). Dermal cells distribution on laser-structured ormosils, *J. Tissue Eng. Regener. Med.*, **7**, pp. 129–138.

109. Raimondi, M. T., Eaton, S. M., Nava, M. M., Laganà, M., Cerullo, G., and Osellame, R. (2012). Two-photon laser polymerization: from fundamentals to biomedical application in tissue engineering and regenerative medicine, *J. Appl. Biomater. Funct. Mater*, **10**, pp. 55–65.
110. Klein, F., Richter, B., Striebel, T., Franz, C. M., Freymann, G. V., Wegener, M., and Bastmeyer, M. (2011). Two-component polymer scaffolds for controlled three-dimensional cell culture, *Adv. Mater.*, **23**, pp. 1341–1345.
111. Fadeeva, E., Deiwick, A., Chichkov, B., and Schlie-Wolter, S. (2014). Impact of laser-structured biomaterial interfaces on guided cell responses, *Interface Focus*, **4**, p. 20130048.
112. Marino, A., Filippeschi, C., Mattoli, V., Mazzolai, B., and Ciofani, G. (2015). Biomimicry at the nanoscale: current research and perspectives of two-photon polymerization, *Nanoscale*, **7**, pp. 2841–2850.
113. Kim, M., Hwang, D. J., Jeon, H., Hiromatsu, K., and Grigoropoulos, C. P. (2009). Single cell detection using a glass-based optofluidic device fabricated by femtosecond laser pulses, *Lab Chip*, **9**, pp. 311–318.
114. Crespi, A., Gu, Y., Ngamsom, B., Hoekstra, H. J., Dongre, C., Pollnau, M., Ramponi, R., van den Vlekkert, H. H., Watts, P., and Cerullo, G. (2010). Three-dimensional Mach-Zehnder interferometer in a microfluidic chip for spatially-resolved label-free detection, *Lab Chip*, **10**, pp. 1167–1173.
115. Hanada, Y., Sugioka, K., Shihira-Ishikawa, I., Kawano, H., Miyawaki, A., and Midorikawa, K. (2011). 3D microfluidic chips with integrated functional microelements fabricated by a femtosecond laser for studying the gliding mechanism of cyanobacteria, *Lab Chip*, **11**, pp. 2109–2115.
116. Schaap, A., Rohrlack, T., and Bellouard, Y. (2012). Optical classification of algae species with a glass lab-on-a-chip, *Lab Chip*, **12**, pp. 1527–1532.
117. Bragheri, F., Minzioni, P., Vazquez, R. M., Bellini, N., Paie, P., Mondello, C., Ramponi, R., Cristiani, I., and Osellame, R. (2012). Optofluidic integrated cell sorter fabricated by femtosecond lasers, *Lab Chip*, **12**, pp. 3779–3784.
118. Xiong, W., Zhou, Y. S., He, X. N., Gao, Y., Mahjouri-Samani, M., Jiang, L., Baldacchini, T., and Lu, Y. F. (2012). Simultaneous additive and subtractive three-dimensional nanofabrication using integrated two-photon polymerization and multiphoton ablation, *Light Sci. Appl.*, **1**, p. e6.

119. Wu, D., Wu, S. Z., Xu, J., Niu, L. G., Midorikawa, K., and Sugioka, K. (2014). Hybrid femtosecond laser microfabrication to achieve true 3D glass/polymer composite biochips with multiscale features and high performance: the concept of ship-in-a-bottle biochip, *Laser Photonics Rev.*, **8**, pp. 458–467.
120. Sugioka, K., Xu, J., Wu, D., Hanada, Y., Wang, Z., Cheng, Y., and Midorikawa, K. (2014). Femtosecond laser 3D micromachining: a powerful tool for the fabrication of microfluidic, optofluidic, and electrofluidic devices based on glass, *Lab Chip*, **14**, pp. 3447–3458.
121. Wu, D., Xu, J., Niu, L.-G., Wu, S.-Z., Midorikawa, K., and Sugioka, K. (2015). In-channel integration of designable microoptical devices using flat scaffold-supported femtosecond-laser microfabrication for coupling-free optofluidic cell counting, *Light Sci. Appl.*, **4**, p. e228.
122. Sima, F., Wu, D., Xu, J., Midorikawa, K., and Sugioka, K. (2015). Ship-in-a-bottle integration by hybrid femtosecond laser technology for fabrication of true 3D biochips, *SPIE LASE*, pp. 93500F-93500F-93508.
123. Lv, C., Xia, H., Guan, W., Sun, Y.-L., Tian, Z.-N., Jiang, T., Wang, Y.-S., Zhang, Y.-L., Chen, Q.-D., and Ariga, K. (2016). Integrated optofluidic-microfluidic twin channels: toward diverse application of lab-on-a-chip systems, *Sci. Rep.*, **6**, p. 19801.
124. Shoji, S., and Kawata, S. (2000). Photofabrication of three-dimensional photonic crystals by multibeam laser interference into a photopolymerizable resin, *Appl. Phys. Lett.*, **76**, pp. 2668–2670.
125. Serbin, J., Egbert, A., Ostendorf, A., Chichkov, B., Houbertz, R., Domann, G., Schulz, J., Cronauer, C., Fröhlich, L., and Popall, M. (2003). Femtosecond laser-induced two-photon polymerization of inorganic–organic hybrid materials for applications in photonics, *Opt. Lett.*, **28**, pp. 301–303.
126. Yu, Y.-H., Tian, Z.-N., Jiang, T., Niu, L.-G., and Gao, B.-R. (2016). Fabrication of large-scale multilevel phase-type Fresnel zone plate arrays by femtosecond laser direct writing, *Opt. Commun.*, **362**, pp. 69–72.
127. Bechtold, P., Zimmermann, M., Roth, S., Alexeev, I., and Schmidt, M. (2016). Beam guidance, focal position shifting and beam profile shaping in ultrashort pulsed laser materials processing, in *Ultrashort Pulse Laser Technology* (Springer), pp. 245–281.
128. Wollenhaupt, M., Bayer, T., and Baumert, T. (2016). Control of ultrafast electron dynamics with shaped femtosecond laser pulses: from atoms to solids, in *Ultrafast Dynamics Driven by Intense Light Pulses* (Springer), pp. 63–122.

129. Xu, B., Du, W.-Q., Li, J.-W., Hu, Y.-L., Yang, L., Zhang, C.-C., Li, G.-Q., Lao, Z.-X., Ni, J.-C., and Chu, J.-R. (2016). High efficiency integration of three-dimensional functional microdevices inside a microfluidic chip by using femtosecond laser multifoci parallel microfabrication, *Sci. Rep.*, **6**, p. 19989.

Chapter 11

Ultrafast Processes on Semiconductor Surfaces Initiated by Temporally Shaped Femtosecond Laser Pulses

P. A. Loukakos, G. D. Tsibidis, and E. Stratakis

Foundation for Research and Technology—Hellas, Institute of Electronic Structure and Laser, N. Plastira 100, PO Box 1385, Vassilika Vouton, 71110 Heraklion, Greece
loukakos@iesl.forth.gr

The application of temporally shaped femtosecond laser pulses in the micro-/nanostructuring of semiconductor surfaces is demonstrated. As an initial step toward full pulse shaping, sequences of double pulses with variable temporal spacing in the picosecond time domain with equal intensity have been used. Craters decorated with nanometer-size ripples are formed following the laser-surface interaction, depending on the irradiation conditions. The area, depth, and strikingly the ripple periodicity show a dependence on the temporal delay between the double pulses. Our analysis of and explanation for the dependence of the micro- and nanomorphological features on the pulse delay are based on a combination of mechanisms, including laser-triggered ultrafast excitation and relaxation on a semiconductor surface, such as carrier

---

Pulsed Laser Ablation: Advances and Applications in Nanoparticles and Nanostructuring Thin Films
Edited by Ion N. Mihailescu and Anna Paola Caricato
Copyright © 2018 Pan Stanford Publishing Pte. Ltd.
ISBN 978-981-4774-23-9 (Hardcover), 978-1-315-18523-1 (eBook)
www.panstanford.com

excitation, ultrafast carrier–lattice energy exchanges, and energy transport, along with the slower phenomenon of melting and the corresponding hydrodynamics and resolidification that follow until the final surface morphology is established. Our investigations on laser-irradiated Si and ZnO surfaces are discussed.

11.1 Introduction

Material processing with ultrashort-pulse lasers has received considerable attention over the past decades due to its important technological applications, in particular in industry and medicine [1–6]. Rapid energy delivery and reduction of the heat-affected areas are the most pronounced advantages compared to irradiation by longer pulses [7], which reflect the merit of the method as a tool for laser-assisted fabrication at the micro-/nanoscale. These abundant applications require thorough knowledge of the processes emerging by the interaction of the laser with the target material for enhanced controllability of the resulting modification of the target relief. Crater creation and development of periodic structures (ripples) are some of the observable results of surface processing after material irradiation with subpicosecond pulsed lasers, while other structures, such as grooves and spikes, are also possible. Therefore, there exists a great demand for continuing experimental and theoretical research aiming at controlling the generated structures. Researchers frequently use femtosecond laser pulses, which present a clean and precise method, to create nanostructures on surfaces using light.

The type of nanostructures produced depends on the level of irradiation. As Varlamova et al. have shown [8], following irradiation by only a few laser pulses subwavelength ripples are formed with their direction perpendicular to the polarization of the electric field. With increasing number of the incident pulses the ripples start to break up, and at greater levels of irradiation, the formation of craters decorated by characteristic microstructures with a conical shape is achieved. In a recent work of ours, experimental data have been additionally analyzed and suprawavelength periodic structures with a different polarization as well as the early stages of the formation of spike-like structures have been modeled [9]. Although that work focused on the exploration of surface morphological changes on semiconductors (i.e., Si) the investigation can be extended to other types of solids (i.e., metals, dielectrics, and polymers) [10]. Depending on the relation of the ripple wavelength with the laser wavelength it is possible to distinguish between low-spatial-frequency ripples, which have wavelengths near and below the laser wavelength, and high-spatial-frequency ripples, which have wavelengths that are much lower than the laser wavelength [11]. In the present chapter, the focus is on the first level of irradiation (i.e., a few laser pulses), which leads to the formation of ripples.

Ripples on semiconductor surfaces were first observed by Birnbaum [12], and their formation was attributed to the optical diffraction of the laser beam with itself at the focal position. Later on, other groups have given explanations on the ripple formation, such as Emmony et al. [13], where they attributed the ripple formation on Ge surfaces to the interference of the incident laser with surface scattered waves. Lastly the dominant explanation was presented by Sipe et al. [14], according to which the formation of ripples on Si surfaces is closely linked to the generation of surface plasmon waves and their interference with the incident laser wave. Other research groups also considered processes following laser excitation of the Si surface to explain ripple formation [11, 14–19]. Nevertheless, there is still ongoing experimental and theoretical work in order to explore the underlying fundamental mechanisms that govern the formation of nanostructures and to control the underlying processes and, therefore, engineer and tune the characteristics of the nanostructure formed in order to maximize the efficiency of the base material for the sake of a given application.

In this chapter we present a summary of our recent research results toward controlling and tuning of the ripples induced on semiconductor surfaces by irradiating the material with temporally shaped laser pulses in order to obtain materials with application-based optical properties. At the same time we pursue the fundamental understanding of the physical processes that are involved during the interaction of the laser beam with the surface and lead to the formation of ripples from the ultrafast to the ns temporal domain.

11.2 Experimental Details

Experiments were performed with a femtosecond Ti:sapphire laser system operating at a wavelength of 800 nm and a repetition rate of 1 kHz. The pulse duration was set to 80–430 fs and measured by cross-correlation techniques. The temporal pulse shaper consisted of a 4f optical configuration containing a spatial light modulator (SLM), which was used to filter the Fourier spectrum of the laser pulses and create double-pulse sequences with pulse separations varying from 0 to 7 ps. A primitive form of temporal pulse shaping was performed by splitting the initial pulse into two equal components and varying the temporal pulse separation to a time range of a few picoseconds, thus attempting to manipulate the ultrafast electron interaction and cooling processes that take place in this temporal range for many solids in general and in silicon in particular.

A Pockels cell controlled the repetition rate and the number of the pulses that irradiated the Si and ZnO surfaces. The beam was perpendicular to the semiconductor surface, giving a spot diameter of 15–30 µm located inside a vacuum chamber evacuated down to a residual pressure of 10^{-2} mbar. The number of laser pulses and fluences used ranged from 15 to 1000 and 0.55 to 4.5 J/cm², respectively. Field emission scanning electron microscopy (FESEM) was used for the morphological characterization of the irradiated areas. ZnO thin films with thicknesses up to 4 µm were deposited onto Corning glass (1737F) in an Alcatel D.C. magnetron system using a 99.999% pure metallic zinc target of a diameter of 15 cm. The base pressure of the ultrahigh vacuum (UHV) chamber was below 5×10^{-7} mbar, while during the deposition, the oxygen ambient pressure was 8×10^{-3} mbar and the substrate temperature was 27°C. All films were grown at a constant plasma current of 0.45 A. To avoid thickness-induced nonuniformities, the distance between the target and the substrate was set to 20 cm. The ZnO films thicknesses were measured using an AlphaStep profilometer.

More details on the experiments performed can be found in our previous works [20–23].

11.3 Theoretical Details

To explain the ripple formation mechanism both optical- and capillarity-driven effects have been considered assuming that the irradiation conditions are sufficient to induce minimal ablation. The proposed theoretical model comprises (i) an electromagnetic component that describes the surface plasmon interference with the incident beam that leads to a periodic energy deposition, (ii) a heat transfer component that accounts for carrier-lattice thermalization through particle dynamics and heat conduction, and (iii) a hydrodynamics component that describes the molten material dynamics and resolidification process [21].

Simulations were performed assuming a *p*-polarized laser pulse with a Gaussian (both spatially and temporally) distribution of fluence. A systematic analysis of the fundamental mechanism reveals that mass removal exists due to the fact that excessive temperatures are reached (larger than ~0.90 T_c; T_c = 3540 K for Si) [21, 24], leading to recoil pressure. Additionally, the inhomogeneous (i.e., periodic) deposition of the laser energy leads to surface tension variance and temperature gradients. More specifically, it is assumed that lattice points that obtain temperatures higher than T_c are ejected and therefore the associated material regions are removed (i.e., minimal ablation). It will be explained, later, that a minimal ablation is a prerequisite for surface plasmon excitation when subsequent irradiation follows. Returning to the dynamics of the part of the material that is in the liquid phase, it is noted that a Marangoni-driven flow and associated capillary waves eventually lead upon resolidification to the formation of a crater and a rippled profile. Furthermore, mass conservation and surface tension forces produce a protrusion near the periphery of the spot.

More details on the simulation, the proposed physical mechanism, the model, and the results produced can be found in our previous work [21–23].

11.4 Results and Discussion I: SI

Our experiments are parameterized by varying the level of irradiation, that is, the number of incident pulses and the interpulse delay (the temporal delay between the double fs laser pulses). FESEM pictures of the irradiated spots are shown in Fig. 11.1. These illustrate the dependence of the morphology of the irradiated spots on both the number of incident pulses as well as the interpulse delay. To examine the evolution of the irradiated spot morphology with an increasing irradiation level, first the result of the irradiation of the Si surface by just one laser shot is examined. This is shown in Fig. 11.2. Formation of a shallow crater is observed, which is confirmed by the intensity cross section obtained from the SEM images. Additionally, as observed by atomic force microscopy (AFM), the shallow crater is surrounded by a rim, which is formed at the periphery of the crater.

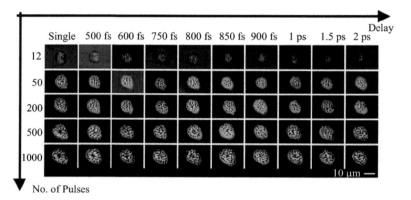

Figure 11.1 FESEM images of the irradiated Si surface with varying interpulse delays and varying incident numbers of double pulses. Reproduced from Ref. [23] with permission from Springer.

To understand and explain the morphology of the crater that is experimentally observed, the following basic considerations in the following basic steps are made:

1. Energy absorption and dissipation are considered, which include electron transport and electron cooling by electron–phonon interaction. These processes are described theoretically by the so-called two-temperature model by Anisimov et al. [25, 26]. Due to the length of the pulse duration (i.e., much larger than the time required for carrier thermalization), the two-temperature model provides an adequate description of the carrier excitation and carrier–phonon relaxation processes.

2. Lattice heating, followed by melting and evaporation of some part of the irradiated surface, which exceeds the critical temperature T_c for lattice evaporation, is considered.
3. Hydrodynamics are introduced in order to describe particle flow and capillary wave formation. The fluid is considered to be an incompressible Newtonian fluid, and a Navier-Stokes equation to describe the fluid movement is incorporated in the model.
4. Lattice cooling that leads to resolidification takes place, leaving a surface that is modified and has micro- and nanostructures. Note that the resolidification process is determined by the movement of the isothermal line corresponding to the melting temperature rather than a typical equation of state (EOS).

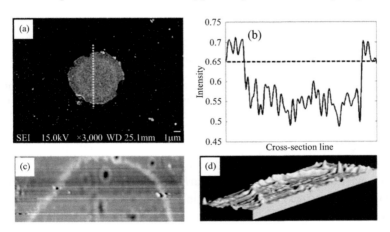

Figure 11.2 (a) FESEM image of an irradiated Si surface by one laser shot. (b) Intensity cross section of (a). (c) AFM image of the irradiated Si surface by just one laser shot. (d) AFM intensity cross section at a viewing angle. Reprinted (figure) with permission from Ref. [21]. Copyright (2012) by the American Physical Society.

Past considerations of the processes following laser excitation of the Si surface [11, 14–19] have overlooked the contribution of hydrodynamics. Thus, the distinct and novel contribution here is the introduction and consideration of the particle flow and the hydrodynamics that follow the laser irradiation and the ultrafast processes involved, as described in detail in Ref. [21]. The simulated crater profile is shown in Fig. 11.3 and is in very close qualitative and quantitative agreement with the crater profile as measured by the SEM and AFM pictures. This demonstrates that the proposed theoretical model describes adequately the experimental data concerning the crater formation on the Si surface. Therefore, the next step will be to introduce the irradiation by multiple and single (i.e., nontemporally shaped) laser pulses.

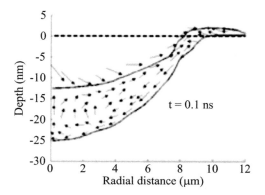

Figure 11.3 Surface profile and flow pattern of the irradiated Si surface at 0.1 ns after laser excitation. Reprinted (figure) with permission from Ref. [21]. Copyright (2012) by the American Physical Society.

A characteristic SEM image of the Si surface irradiated by four laser pulses is shown in Fig. 11.4. Ripples are clearly observed, with the orientation vertical to the polarization of the electric field of the laser pulses. To explain the formation of this periodic nanostructure, the mechanism that was first proposed by Sipe et al. [14], that is, the generation of surface plasmons and the interference of the incident laser field with the surface plasmon is introduced as a first step in our theoretical model. This interference leads to spatially periodic distributions of energy deposition, surface temperature, carrier densities, and hydrodynamic quantities.

The excitation of surface plasmons on Si (i.e., a semiconductor) is possible due to the fact that (i) the produced carrier density number is sufficient to allow validity of the condition that is necessary for plasmon excitation (i.e., real part of the dielectric constant smaller than −1) and (ii) the surface profile possesses a roughness (i.e., a crater after irradiation with one pulse and a periodic grating for subsequent pulses) so the laser dispersion curve will meet the surface plasmon dispersion curve [27].

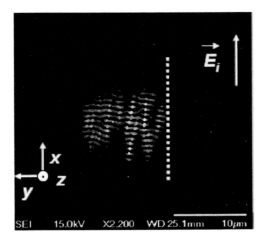

Figure 11.4 SEM image of the irradiated Si surface by four consecutive laser shots. The electric field denoting the polarization of the incident pulses is shown. Reprinted (figure) with permission from Ref. [21]. Copyright (2012) by the American Physical Society.

After the introduction of wave interference, the calculated crater profile is modified and is shown in Fig. 11.5. The calculated profile is in very good qualitative agreement with the rippled nanostructures observed by SEM and AFM pictures. Moreover, the predicted, by the theoretical simulations, periodicity of the ripples is very close to the experimental value, that is, 738 nm.

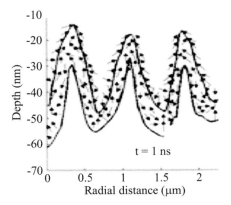

Figure 11.5 Calculated crater profile and flow pattern for the Si surface irradiated by four pulses at 1 ns delay time. Reprinted (figure) with permission from Ref. [21]. Copyright (2012) by the American Physical Society.

While the ripple periodicity size is well described by the surface plasmon model, the orientation of the periodic structures can be explained by the deposition of the inhomogeneous energy and the efficacy factor that constitutes a measure of the energy deposition. To correlate the excited electron density with surface structures, the inhomogeneous energy deposition into the irradiated material is computed through the calculation of the product $\eta(\mathbf{k}_L,\mathbf{k}_i) \times |b(\mathbf{k}_L)|$, as described in the Sipe–Drude model [28]. In the above expression, η describes the efficacy with which the surface roughness at the wave vector $\mathbf{k}_L$ (i.e., normalized wave vector $|\mathbf{k}_L| = \lambda_L/\Lambda$) induces inhomogeneous radiation absorption, $\mathbf{k}_i$ is the component of the wave vector of the incident beam on the material's surface plane, b represents a measure of the amplitude of the surface roughness at $\mathbf{k}_L$, λ_L is the laser wavelength, and Λ is the grating periodicity [29]. Movement of the fluid is shown with arrows at time $t = 1$ ns in Fig. 11.6a, while the pseudocolor represents the temperature at the same time. The periodicity of the ripples as a function of the number of pulses is illustrated in Fig. 11.6b.

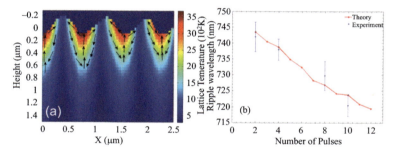

Figure 11.6 (a) Spatial distribution of lattice temperature at $t = 1$ ns for NP = 10 (arrows indicate the flow movement). The laser field polarization is horizontal. (b) Ripple periodicity versus the number of pulses. (E_d = 0.7 J/cm^2, τ_p = 430 fs, and R_0 = 15 μm). Reprinted (figure) with permission from Ref. [21]. Copyright (2012) by the American Physical Society.

To demonstrate qualitatively and quantitatively the agreement of theoretical simulations with experimental observations, the surface profile is illustrated in Fig. 11.7a (a quadrant) for 10 pulses, while the ripple height along the dashed white line in Fig. 11.7a is shown in Fig. 11.7b.

Having demonstrated the development of a working theoretical model for the prediction of the crater characteristics and the ripple periodicity, temporal pulse shaping can now be introduced in its most basic form, that is, sequences of double fs laser pulses where we can control the interpulse delay. Irradiation of Si surfaces with laser pulses and the resulting morphological profile are evaluated by SEM and AFM images, as shown in Fig. 11.8. It is evident, by the instances of zero interpulse delay time (i.e., a single pulse with the same total energy as the sum of the energies of the double pulses) and 0.5 ps and 2 ps interpulse delay time, the crater size is reduced when the interpulse delay time is increased.

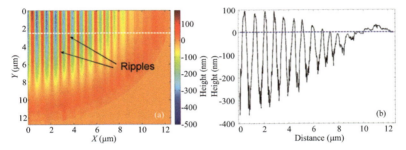

Figure 11.7 Theoretical results for (a) upper view of the ripple pattern in a quadrant for NP = 10 (the double-ended arrow indicates beam polarization), (b) the height of ripples (side view) for NP = 10 at Y = 2.5 µm, along the dashed line in (a). (E_d = 0.7 J/cm^2, τ_p = 430 fs, and R_0 = 15 µm). Reprinted (figure) with permission from Ref. [21]. Copyright (2012) by the American Physical Society.

This is also confirmed by the cross sections obtained by the corresponding AFM images. The results are summarized for all the irradiation levels used (i.e., the number of incident laser pulse sequences) in Fig. 11.10. Thereby, we deduce that the crater spot area as measured by the SEM images is monotonically reduced with an increasing interpulse delay time for values from 0 to 2 ps and remains stable thereafter (not shown).

To evaluate the ripple periodicity intensity cross sections are obtained from the SEM images. The results for the ripple periodicity versus the interpulse delay time shown in Fig. 11.9 indicate that there is a sharp decrease in the ripple periodicity within the first few ps of the interpulse delay time, followed by a smaller slope of decrease with further increasing interpulse delay time. The decrease in the ripple periodicity within the first few ps is roughly 20 nm, that is, from 742 nm for 0 ps to about 723 nm for 14 ps, which is the largest value of interpulse delay that could be achieved in our experiments.

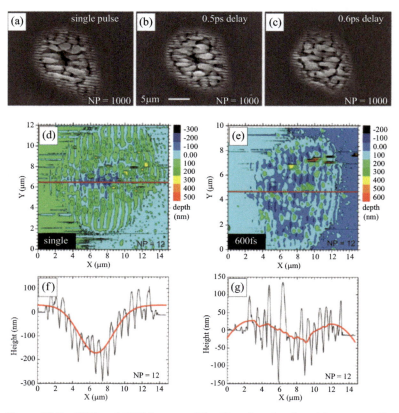

Figure 11.8 SEM and AFM images of the Si surface irradiated by temporally shaped (double) fs laser pulses. (a,b,c) SEM images for 0, 0.5, and 2 ps, respectively. (d,e) AFM images of the crater profile for 0 and 2 ps, respectively. (g,h) The intensity profile of the craters for the same interpulse delays as in (d) and (e). Reproduced from Ref. [23] with permission from Springer.

To correlate the theoretical simulations and experimental findings, both the actual real-time temporal profile of the simulated laser pulse (shown in the inset of Fig. 11.11) and the electron temperature profile as a result of such an irradiation (shown in the main part of Fig. 11.11) are employed. It is interesting to examine the maximum carrier and lattice temperatures that can be achieved using our theoretical model as a function of the interpulse delay time. This is shown in Fig. 11.13. It is evident that the maximum carrier temperature falls monotonously, while the maximum lattice temperature exhibits a nonmonotonous behavior: it firstly increases and it reaches its maximum value at an interpulse delay of about 2.2 ps, which is roughly 4- to 5-pulse duration apart, while it subsequently drops monotonically with increasing interpulse delay time.

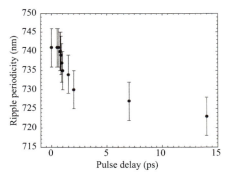

Figure 11.9 Ripple wavelength as measured by the SEM images versus interpulse delay time. Reproduced from Ref. [23] with permission from Springer.

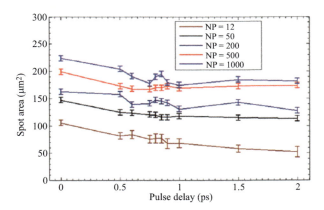

Figure 11.10 Measured crater area on the Si surfaces as irradiated by 12, 50, 200, 500, and 1000 double pulses. Reproduced from Ref. [23] with permission from Springer.

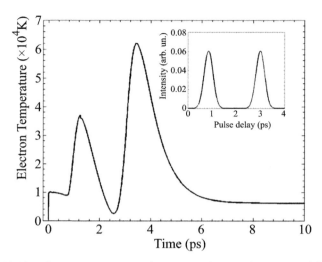

Figure 11.11 Electron temperature dynamics at the Si surface vicinity following excitation by double laser pulses. (Inset) The calculated temporal profile of the double laser pulses that are assumed in the calculations. Reproduced from Ref. [23] with permission from Springer.

The monotonous drop of the maximum electronic temperature is consistent with the decreasing value of heat conductivity $k(T)$. On the other hand, the maximum electron temperature is achieved when the two pulses overlap in time. When a delay between the two pulse components is introduced the maximum electron temperature on the surface that is achieved drops, as expected.

On the other hand, the optimum lattice temperature at a nonzero interpulse delay results from the competing mechanisms of electron transport versus electron–phonon scattering. Due to efficient electron transport at zero interpulse delay time (due to the highest values of heat conductivity) the electrons move very efficiently away from the surface region and, therefore, they do not interact a lot with the surface. Thus the energy is dissipated by transport into the bulk. Energy is transferred to the lattice by electron–phonon coupling, which tends to take over only after some time where the heat conductivity drops due to the now-decreased electron maximum temperature and, therefore, the decrease of the electron heat conductivity. So the electrons stay longer in the vicinity of the surface. Therefore, there is more time available for the transfer of the energy toward the lattice, that is, greater values of the maximum lattice temperature that is finally reached. This optimum could be potentially used as feedback to applications related to genetic algorithms and feedback loops in pulse-shaping optimization applications.

By looking at the experimental results of the crater area and the crater depth versus the interpulse delay time in Fig. 11.12 a monotonous decay for both graphs is observed and is in very good agreement with the predictions of our theoretical model shown with the solid line. Moreover, the evolution of the crater area and the crater depth is qualitatively similar to the evolution of the maximum electron temperature dynamics with interpulse delay time, as shown in Fig. 11.13. This observation confirms that the electron dynamics are strongly correlated and related to the ablation-like mechanisms, as is also previously reported and analyzed by previous experimental and theoretical works [30].

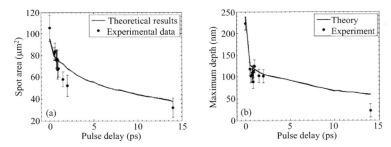

Figure 11.12 Crater area (a) and crater depth (b) versus interpulse delay time. Points represent experimental data, while the solid line represents the calculated values. Reproduced from Ref. [23] with permission from Springer.

In Fig. 11.14 the evolution of the ripple periodicity with interpulse delay time is illustrated, comparing experimental results with the predictions of the theoretical model. Again, the agreement is very good, and this helps us to conclude that we may assign the decreasing ripple period with interpulse delay time to the decreasing maximum electron temperature with interpulse delay time and to the wavelength of the surface plasmon that follows the dynamics of the maximum electron temperature and the number of excited carriers. Therefore, a decreasing number of carriers and a decreasing maximum electron temperature with interpulse delay time lead to a decreasing surface plasmon wavelength and, therefore, to a decreasing ripple period, since the ripple emergence has been attributed to the interference of the laser wave with the surface plasmon wave. The theoretical values come even closer to the experimental ones if in our model the recoil pressure is included that is induced by the ejected species during the ablation process. This is described in detail in our previous work [23].

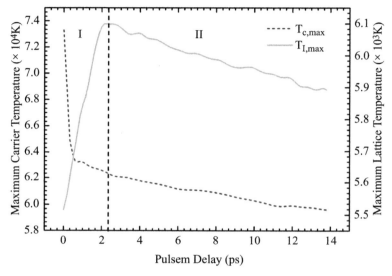

Figure 11.13 Maximum electron (left y axis, dashed line) and maximum lattice (right y axis, solid line) temperatures that are achieved at the surface vicinity versus the interpulse delay time. Reproduced from Ref. [23] with permission from Springer.

Note here that although the ripple wavelength could be tuned and, in particular, could be decreased by increasing the level of irradiation of the surface, that is, by increasing the number of incident laser pulses, as was shown in previous experimental work [21] (see also Fig. 11.15), this may prove to be undesirable in many applications since an increased number of incident pulses lead inevitably to an increased level of damage effects, which is of course undesirable for applications that demand fine interaction of the laser beams with optical materials and the creation of fine nanostructures on these materials. Therefore, we demonstrate a unique way of creating these nanostructures by exploiting the temporal regulation of the energy of the laser beam on a material's surface.

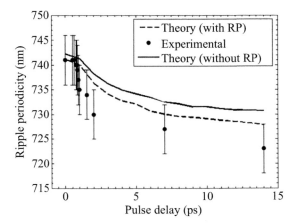

Figure 11.14 Ripple wavelength versus interpulse delay time. Points are experimental results, and solid and dashed lines are theoretical calculations based on our model, either including (dashed) or excluding (solid) the influence of the recoil pressure by the ablated species. Reproduced from Ref. [23] with permission from Springer.

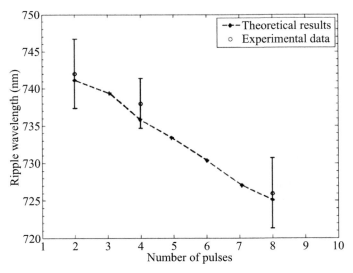

Figure 11.15 Ripple wavelength versus the incident number of pulses. Theoretical calculations are represented by the dashed line. Reprinted (figure) with permission from Ref. [21]. Copyright (2012) by the American Physical Society.

11.5 Results and Discussion II: ZnO

Next, the laser-induced formation of nanostructures on ZnO is explored. ZnO was selected as a representative material of semiconductors with high bandgaps that belong to a class of photonic materials important for use in light-related applications. For the case of ZnO, similar to Si, the experiments are parametrized with respect to the incident number of laser pulses and increasing interpulse delay time. As shown in Fig. 11.16 we obtain various SEM images for varying the previously mentioned parameters as well as the fluence of the incident pulses. By careful examination of a SEM image at a higher magnification, as in Fig. 11.17a,b, obtained for fluences close to (0.55 J/cm^2) and above (0.92 J/cm^2) the damage threshold, one observes three distinct classes of ripples with different periodicities. Firstly, at the center of the laser-irradiated spot the low-frequency ripples with a period of 640 nm are evident. When zooming closer to the periphery of the irradiated spots, two more ripple periodicities can be distinguished after careful inspection: one with a period of about 260 nm and one with an even higher periodicity, of about 170 nm. The ripple periodicities are summarized in Fig. 11.17c as a function of the incident laser fluence.

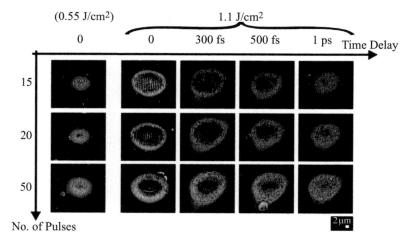

Figure 11.16 FESEM images of craters formed on ZnO irradiated by double fs laser pulses as a function of the incident number of pulses and interpulse delay [20].

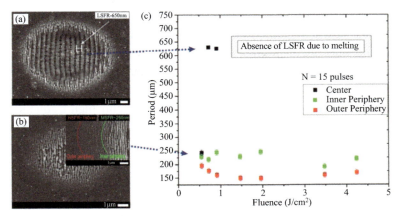

Figure 11.17 High-magnification SEM images of a typical crater on ZnO for fluences 0.55 J/cm² (a) and 0.92 J/cm² (b), focusing on the regions with apparently three different ripple classes. (c) Ripple period versus fluence for all three ripple classes [20].

The formation of the lowest-periodicity ripples on the ZnO surface can be attributed to the generation of a surface plasmon wave and the interaction with the incident laser radiation, as in the case of the Si surface. As for the medium- and high-periodicity ripples they are probably created by the generation of a surface wave at the second harmonic frequency. The second harmonic wave, in turn, excites a surface plasmon wave and the medium-periodicity ripples are the result of the interference of these two waves, as was previously analyzed and explained by the work of Dufft et al. [31].

By inspection of Fig. 11.17c it is found that the low-frequency ripples disappear when the incident laser fluence is increased beyond 1 J/cm² and this observation is attributed to melting effects at the surface, which are considered responsible for washing out the low-frequency ripples. More experimental investigations are required to obtain further insights into this particular effect.

When turning the focus toward the dependence of the ripple periodicities on the temporal shape of the laser pulses, that is, the interpulse delay time, a remarkable effect is observed, as shown in Fig. 11.18: for interpulse delay time from 0 to about 0.5 ps the three different ripple periodicities are found to coexist at the ZnO film surface. However, when the interpulse delay time is increased beyond 0.5 ps, the low-spatial-frequency ripples immediately disappear, leaving only the medium- and high-frequency ripples at the surface. Even at the central part of the crater, where the low-frequency ripples existed, now for larger than 0.5 ps interpulse delays, the crater surface is dominated by the medium- and high-frequency ripples. This is also corroborated in Fig. 11.20, where the area coverage of each of the ripple classes is examined for various values of interpulse delay time. Again, it is found that all three different ripple periodicities coexist at the crater surface for low interpulse delay time but when the interpulse delay time is increased, then the coverage by the low-frequency ripples rapidly decreases and drops to practically zero coverage for interpulse delays above 0.5 ps. On the other hand, the coverage by the medium- and high-frequency ripples increases for delays up to 0.5 ps at the expense of the coverage by the low-frequency ripples. For even larger interpulse delay time the coverage decreases overall as the total area of the crater decreases, similar to the case of the Si surfaces analyzed previously in this book.

It is worthwhile to emphasize at this point the significance of the aforementioned observations for applications in the nanostructuring of photonic materials. This is done by demonstrating the printing of gratings at a larger scale on ZnO surfaces, shown in Fig. 11.19. The surface of the sample below the laser beam is scanned with a constant velocity, and the interaction of the laser beam with the surface prints large surfaces whose area can in principle be infinitely large. In Fig. 11.19a the SEM image of a printed surface is shown with zero interpulse delay time and the result is a grating with a periodicity of about 650 nm, as was observed by the SEM images of Fig. 11.19a. On the other hand, in Fig. 11.19b the temporal shape of the laser pulses has been altered and interpulse delay time of 1 ps has been introduced. Remarkably, the grating has switched now to a significantly smaller periodicity, measured to be ~220 nm, which is in accordance with the results presented in Fig. 11.18. Thus, the ability to drastically control the morphology of a nanostructure imprinted on the surface of a photonic material is demonstrated by changing nothing else in the laser pulse but the shape of its temporal profile. Thus the flow of laser energy onto the material and the rate of its absorption by the material have been regulated, thereby controlling the excitation of its internal degrees of freedom. Thus, the interaction of the laser beam with the material, and as a result its macroscopic properties, has been demonstrated to be controlled at the nanoscale utilizing nonstandard laser beam parameter control, that is, temporal pulse shaping.

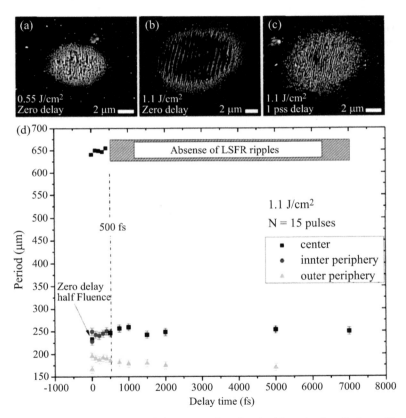

Figure 11.18 SEM images of the ZnO surface for double-pulse irradiation with (a) $F = 0.55$ J/cm^2, zero delay, (b) $F = 1.1$ J/cm^2, zero delay, and (c) $F = 1.1$ J/cm^2, 1 ps delay. (d) Summary of the results for the ripple periodicity for all three different ripple classes versus interpulse delay time [20].

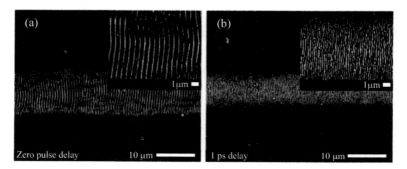

Figure 11.19 Formed gratings on the ZnO surface by scanning it under the incident laser beam with the interpulse delay at (a) 0 and (b) 1 ps [20].

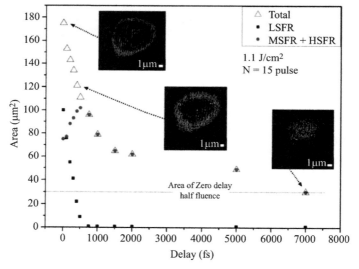

Figure 11.20 Measured area coverage by the different ripple classes on the ZnO surface versus interpulse delay time [20].

11.6 Conclusions

To summarize, an experimental approach has been utilized employing temporal pulse shaping and a unified theoretical model has been developed in order to describe the crater formation and ripple formation on semiconductor surfaces following interaction with sequences of double fs laser pulses with varying interpulse delays. It has been shown that it is possible to control and fine-tune the morphological characteristics of the craters as well as the morphology of the nanostructures formed. The underlying mechanism for formation of the latter structures was associated to the interference of the incident laser pulses with a surface scattered wave, the surface plasmon. On the other hand, the interpulse delay results in the control of the maximum achieved electron temperature and carrier density at the surface, which in turn produces fine tuning of the plasmon wavelength and thus the nanostructure periodicity. This effect was also observed on ZnO surfaces, where additionally the striking change in the morphology with a vast change of the periodicity of the printed gratings has been demonstrated. Next investigations involve the extension of the current studies to more complex nanostructures, such as grooves and microcones, and the effect of combining temporal shape and polarization. Also, there are plans to investigate different materials, such as metals, ceramics, and polymers important for solar applications. Lastly, there are plans to develop and employ more complex pulse shapes and feedback and genetic algorithms to allow the nanogratings to self-evolve and self-optimize their optical properties.

References

1. Baeuerle, D. (2011). *Laser Processing and Chemistry*, 4th ed. (Springer, Berlin).
2. Diels, J.-C., and Rudolph, W. (2006). Ultrashort laser pulse phenomena fundamentals, techniques, and applications on a femtosecond time scale, in *Optics and Photonics* (Elsevier/Academic Press, Amsterdam, London), pp. 1 online resource (xxi, 652 p.).
3. Stratakis, E., Ranella, A., Farsari, M., and Fotakis, C. (2009). Laser-based micro/nanoengineering for biological applications, *Prog. Quant. Electron.*, **33**, pp. 127–163.
4. Magoulakis, E., Papadopoulou, E. L., Stratakis, E., Fotakis, C., and Loukakos, P. A. (2010). Ultrafast electron dynamics in ZnO/Si microcones, *Appl. Phys. A*, **98**, pp. 701–705.

5. Ranella, A., Barberoglou, M., Bakogianni, S., Fotakis, C., and Stratakis, E. (2010). Tuning cell adhesion by controlling the roughness and wettability of 3D micro/nano silicon structures, *Acta Biomater.*, **6**, pp. 2711–2720.

6. Vorobyev, A. Y., and Guo, C. L. (2010). Laser turns silicon superwicking, *Opt. Express*, **18**, pp. 6455–6460.

7. Herrmann, R. F. W., Gerlach, J., and Campbell, E. E. B. (1998). Ultrashort pulse laser ablation of silicon: an MD simulation study, *Appl. Phys. A*, **66**, pp. 35–42.

8. Varlamova, O., Bounhalli, M., and Reif, J. (2013). Influence of irradiation dose on laser-induced surface nanostructures on silicon, *Appl. Surf. Sci.*, **278**, pp. 62–66.

9. Tsibidis, G. D., Fotakis, C., and Stratakis, E. (2015). From ripples to spikes: a hydrodynamical mechanism to interpret femtosecond laser-induced self-assembled structures, *Phys. Rev. B*, **92**, p. 041405(R).

10. Vorobyev, A. Y., and Guo, C. L. (2013). Direct femtosecond laser surface nano/microstructuring and its applications, *Laser Photonics Rev.*, **7**, pp. 385–407.

11. Bonse, J., Munz, M., and Sturm, H. (2005). Structure formation on the surface of indium phosphide irradiated by femtosecond laser pulses, *J. Appl. Phys.*, **97**, p. 013538.

12. Birnbaum, M. (1965). Semiconductor surface damage produced by ruby lasers, *J. Appl. Phys.*, **36**, pp. 3688–3689.

13. Emmony, D. C., Howson, R. P., and Willis, L. J. (1973). Laser mirror damage in germanium at 10.6 Mu, *Appl. Phys. Lett.*, **23**, pp. 598–600.

14. Sipe, J. E., Young, J. F., Preston, J. S., and Vandriel, H. M. (1983). Laser-induced periodic surface-structure. 1. Theory, *Phys. Rev. B*, **27**, pp. 1141–1154.

15. Zhou, G. S., Fauchet, P. M., and Siegman, A. E. (1982). Growth of spontaneous periodic surface-structures on solids during laser illumination, *Phys. Rev. B*, **26**, pp. 5366–5381.

16. Tan, B., and Venkatakrishnan, K. (2006). A femtosecond laser-induced periodical surface structure on crystalline silicon, *J. Micromech. Microeng.*, **16**, pp. 1080–1085.

17. Costache F., Eckert, S., and Reif, J. (2008). Near-damage threshold femtosecond laser irradiation of dielectric surfaces: desorbed ion kinetics and defect dynamics, *Appl. Phys. A*, **92**, pp. 897–902.

18. Huang, M., Zhao, F. L., Cheng, Y., Xu, N. S., and Xu, Z. Z. (2009). Origin of laser-induced near-subwavelength ripples: interference between surface plasmons and incident laser, *ACS Nano*, **3**, pp. 4062–4070.
19. Han, Y. H., and Qu, S. L. (2010). The ripples and nanoparticles on silicon irradiated by femtosecond laser, *Chem. Phys. Lett.*, **495**, pp. 241–244.
20. Barberoglou, M., Gray, D., Magoulakis, E., Fotakis, C., Loukakos, P. A., and Stratakis, E. (2013). Controlling ripples' periodicity using temporally delayed femtosecond laser double pulses, *Opt. Express*, **21**, pp. 18501–18508.
21. Tsibidis, G. D., Barberoglou, M., Loukakos, P. A., Stratakis, E., and Fotakis, C. (2012). Dynamics of ripple formation on silicon surfaces by ultrashort laser pulses in subablation conditions, *Phys. Rev. B*, **86**, p. 115316.
22. Tsibidis, G. D., Stratakis, E., Loukakos, P. A., and Fotakis, C. (2014). Controlled ultrashort-pulse laser-induced ripple formation on semiconductors, *Appl. Phys. A*, **114**, pp. 57–68.
23. Barberoglou, M., Tsibidis, G. D., Gray, D., Magoulakis, E., Fotakis, C., Stratakis, E., and Loukakos, P. A. (2013). The influence of ultra-fast temporal energy regulation on the morphology of Si surfaces through femtosecond double pulse laser irradiation, *Appl. Phys. A*, **113**, pp. 273–283.
24. Kelly, R., and Miotello, A. (1996). Comments on explosive mechanisms of laser sputtering, *Appl. Surf. Sci.*, **96–98**, pp. 205–215.
25. Anisimov, S. I., and Luk'yanchuk, B. S. (2002). Selected problems of laser ablation theory, *Phys. Usp.*, **45**, pp. 293–324.
26. Anisimov, S. I., Kapeliov, B., and Perelman, T. L. (1967). Electron-emission from surface of metals induced by ultrashort laser pulses, *Zhurnal Eksperimentalnoi Teor. Fiz.*, **66**, (1974) [*Sov. Phys. Tech. Phys.*, **11**, p. 945], pp. 776–781.
27. Raether, H. (1998). *Surface Plasmons on Smooth and Rough Surfaces and on Gratings* (Springer-Verlag, Berlin, New York).
28. Sipe, J. E., Young, J. F., Preston, J. S., and van Driel, H. M. (1983). Laser-induced periodic surface structure. I. Theory, *Phys. Rev. B*, **27**, pp. 1141–1154.
29. Bonse, J., Munz, M., and Sturm, H. (2005). Structure formation on the surface of indium phosphide irradiated by femtosecond laser pulses, *J. Appl. Phys.*, **97**, p. 013538.

30. Bulgakova, N. M., Stoian, R., Rosenfeld, A., Hertel, I. V., and Campbell, E. E. B. (2004). Electronic transport and consequences for material removal in ultrafast pulsed laser ablation of materials, *Phys. Rev. B*, **69**, p. 054102.
31. Dufft, D., Rosenfeld, A., Das, S. K., Grunwald, R., and Bonse, J. (2009). Femtosecond laser-induced periodic surface structures revisited: a comparative study on ZnO, *J. Appl. Phys.*, **105**, p. 034908.

Chapter 12

Atomistic Simulations of the Generation of Nanoparticles in Short-Pulse Laser Ablation of Metals: Effect of Background Gas and Liquid Environments

Cheng-Yu Shih,[a] Chengping Wu,[a] Han Wu,[a,b,c] Maxim V. Shugaev,[a] and Leonid V. Zhigilei[a]

[a]*Department of Materials Science and Engineering, University of Virginia, 395 McCormick Road, Charlottesville, VA 22904-4745, USA*
[b]*Institute of Modern Optics, Nankai University, 94 Weijin Road, Tianjin 300071, China*
[c]*School of Mechanical Science and Engineering, Huazhong University of Science and Technology, 1037 Luoyu Road, Wuhan, 430074, China*
lz2n@virginia.edu

Atomistic simulations are playing an increasingly important role in the investigation of the fundamental mechanisms of laser-material interactions. The advancements in the computational methodology and fast growth of available computing resources are rapidly expanding the range of problems amenable to atomistic modeling. This chapter provides an overview of the results obtained in recent simulations of laser ablation of metal targets in vacuum, a background gas, and a liquid environment. A comparison of the

---

Pulsed Laser Ablation: Advances and Applications in Nanoparticles and Nanostructuring Thin Films
Edited by Ion N. Mihailescu and Anna Paola Caricato
Copyright © 2018 Pan Stanford Publishing Pte. Ltd.
ISBN 978-981-4774-23-9 (Hardcover), 978-1-315-18523-1 (eBook)
www.panstanford.com

results of the simulations of laser ablation of Al targets in vacuum and in a 1 atm Ar background gas reveals a surprisingly strong effect of the gas environment on the initial plume dynamics and the cluster size distribution, with almost complete suppression of the generation of small atomic clusters in the front part of the ablation plume. A stronger spatial confinement of laser ablation by a liquid environment, investigated for Ag targets in water, is found to suppress the material ejection and produce large frozen subsurface voids at low laser fluences and a dense layer of superheated liquid metal at the front of the ejected ablation plume at higher fluences. The implications of the computational predictions for interpretation of experimental data on the effect of background gas and liquid environments on the generation of nanoparticles in laser ablation are discussed.

12.1 Introduction

"Laser ablation" is a term used to describe material removal from a target irradiated by a laser pulse. A wide range of practical applications of laser ablation includes generation of chemically clean and environmentally friendly nanoparticles [1]. The production of nanoparticles through direct laser ablation of an irradiated target eliminates the need for chemical precursors and presents a number of important advantages over conventional multistep chemical synthesis methods that introduce contamination from intermediate reactants and/or produce agglomerated structures with degraded functionality. The size, shape, structure, and composition of nanoparticles generated by laser ablation can be controlled by changing the target structure and composition [2], varying the background gas environment [3, 4], or mixing ablation plumes generated by double-pulse irradiation [5, 6]. Short-pulse (fs-ps) laser sources are especially suitable for nanoparticle production due to more localized and intense laser heating compared to nanosecond laser pulses [7–10] that increase the fraction of nanoparticles in the ablation plume [11–14].

Laser ablation in a liquid environment has recently emerged as a particularly promising approach to the generation of colloidal solutions of contamination-free nanoparticles [1, 15–18]. The characteristics of nanoparticles in this case are affected by the choice of the liquid medium [19–21], its temperature [22], and the presence of surfactants [23–27]. The highly nonequilibrium conditions created by the interaction of the ablation plume with a liquid environment can result in the formation of nanoparticles with unusual structures, shapes, and composition, such as nanocubes [28], hollow spheroids [29, 30], patch-joint football-like AgGe microspheres [31], and diamond nanocrystallites produced by the laser ablation of graphite [32].

The main obstacles to broadening the range of practical applications benefiting from nanoparticle synthesis by laser ablation are the relatively low productivity [33] and wide (and often bimodal) nanoparticle size distributions [34–36]. The latter can be related to the variability of the nanoparticle formation mechanisms in different parts of the ablation plume [37–40], as well as the generation of large micron-size droplets by hydrodynamic sputtering of the melted pool or rupturing of liquid layers separated/spalled from the target in the course of the relaxation of laser-induced pressure [38, 41, 42]. Further progress in the optimization of experimental parameters for the efficient generation of nanoparticles with a narrow size distribution, required for advanced sensing, catalysis, and biomedical applications, can be facilitated by improved physical understanding of the involved processes.

While the general mechanisms of laser melting, spallation, and ablation in vacuum have been extensively studied experimentally, theoretically, and computationally, the effect of spatial confinement by a high-pressure (1 atm or higher) background gas, a liquid environment, or a solid overlayer on the laser-induced processes still remains largely unexplored. Introduction of spatial confinement and the interaction between the ejected plume and the surrounding medium adds another layer of complexity to the description of short-pulse laser ablation, which by itself is a complex and highly nonequilibrium phenomenon. As a result, the theoretical analysis of the nanoparticle formation by laser ablation in liquids [43] is largely based on semiquantitative models that adopt the concepts developed for the plume expansion in a background gas to the much stronger confinement by a liquid environment, and describes the nanoparticle formation as a process of coalescence of clusters in a supersaturated solution formed by the mixing of the ablation plume and the liquid. The analysis in this case relies on the assumptions of the initial cluster size distribution in the solution, the temperature evolution in the plume–liquid mixing region, the thickness of the mixing region, and other parameters. While the continuum-level modeling can provide additional insights into the ablation dynamics [44, 45], some of the key processes, such as the mixing of the ablation plume with a liquid environment and the generation of nanoparticles in the mixed region, cannot be easily included in the continuum models.

Under conditions when the analytical and continuum-level numerical descriptions of spatially confined laser ablation are hindered by the complexity and highly nonequilibrium nature of laser-induced processes, the molecular dynamics (MD) computer simulation technique can serve as a useful alternative approach, capable of providing atomic-level insights into the laser-induced processes. The main advantage of the MD technique is that no assumptions are made on the processes or mechanisms under study. The only input in the MD model is the interatomic interaction potential that defines the equilibrium structure and thermodynamic properties of the material. The interatomic potentials are typically designed via ab initio calculations and fitted to reproduce basic material properties of interests. Once the interatomic potential is chosen and initial conditions are defined, the MD trajectories (positions and velocities) are obtained through numerical solution of the equations of motion for all atoms in the system without any further assumptions. This advantage makes MD an ideal technique for exploring nonequilibrium processes and revealing new physical phenomena.

Indeed, over the last 20 years MD simulations have been actively used in investigations of laser-induced generation of crystal defects, melting, and resolidification [46–57], as well as photomechanical spallation [38, 42, 55, 58–61] and ablation of various material systems [35, 37, 38, 55, 59, 60, 62–82]. Some of the results of MD simulation of laser–material interaction have been reviewed in Refs. [83–85]. Most of the MD simulations of laser ablation, however, have been performed for vacuum conditions, with the exception of a series of simulations of shock wave formation in laser ablation of an argon target in a background gas [77–80] and a two-dimensional simulation of laser ablation of a target covered by a thin (400 monolayers thick) wetting layer [86]. The focus of the simulations reported in Refs. [77–80] was on the characteristics of the shock waves produced by the interaction of the ablation plume with the surrounding gas, and the effect of the background gas on the generation of clusters and nanoparticles was not analyzed. The qualitative conclusion of Ref. [86] on the decrease of the size of the ejected clusters in the presence of a wetting layer has to be verified and quantified in more realistic three-dimensional simulations.

In this chapter, we provide an overview of the results obtained in recent MD simulations of laser ablation of metal targets in vacuum, a high-pressure background gas, and a liquid environment. The computational methodology developed for MD simulations of laser interactions with metal targets is briefly discussed in Section 12.2, with a particular focus on a computationally efficient description of the ablation plume interaction with the background gas and liquid environment. A brief overview of the relatively well-established mechanisms of laser ablation in vacuum is provided and illustrated by the results of recent large-scale MD simulations in Section 12.3. The effect of the background gas on the dynamics of the ablation plume and nanoparticle size distribution is discussed in Section 12.4. The results of the first MD simulations of laser ablation in water are described in Section 12.5. The overall conclusions on the effect of the background gas and liquid environment on the mechanisms of nanoparticle generation are provided in Section 12.6.

12.2 Computational Setup for the Simulation of Laser Interactions with Metals in a Background Gas or Liquid Environment

The simulations discussed in this chapter are performed with a hybrid computational model combining the classical MD method with a continuum-level description of the laser excitation and subsequent relaxation of the conduction band electrons. The basic concepts of the model as well as the new computational developments enabling the simulations of laser ablation in a background gas and in liquids are briefly described below.

12.2.1 Representation of Laser Interaction with Metals

Although the MD method is capable of providing detailed information on the microscopic mechanisms of laser ablation, several modifications have to be made in order to apply the classical MD for simulations of laser interactions with metals. In particular, a realistic description of the laser coupling to the target material, the kinetics of thermalization of the absorbed laser energy, and the fast electron heat conduction should be incorporated into the MD technique. These processes can be accounted for by incorporating the MD method into the general framework of the two-temperature model (TTM) [87, 88] commonly used in the simulations of short-pulse laser interactions with metals. The idea of the combined TTM-MD model [38, 48, 55] is schematically illustrated in Fig. 12.1 and is briefly explained below.

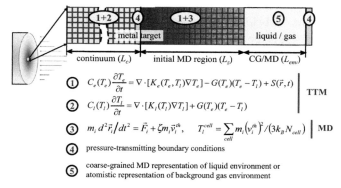

Figure 12.1 Schematic representation of the combined continuum–coarse-grained–atomistic model for the simulation of laser interactions with metals in a background gas or liquid environment. The metal target is represented by the TTM-MD model [38, 48], the liquid environment is simulated with a coarse-grained model [105], described in Section 12.2.2, and the simulations in a background gas are performed in the presence of a large gas-phase region equilibrated at the desired temperature and pressure. The spatial discretization in the continuum part of the model and the dimensions of the atomistic and continuum regions are not drawn to scale.

In the original TTM, the time evolution of the lattice and electron temperatures, T_l and T_e, is described by two coupled differential equations (Eqs. 12.1 and 12.2 in Fig. 12.1) that account for the electron heat conduction in the metal target and the energy exchange between the electrons and atomic vibrations. In the combined TTM-MD method, MD substitutes the TTM equation for the lattice temperature in the surface region of the target, where laser-induced structural and phase transformations take place. The diffusion equation for the electron temperature, T_e, is solved by a finite difference method simultaneously with MD integration of the equations of motion of atoms. The cells in the finite difference discretization are related to the corresponding volumes of the MD system, and the local lattice temperature, T_l^{cell}, is defined for each cell from the average kinetic energy of the thermal motion of atoms. The electron temperature enters a coupling term, $\xi m_i \vec{v}_i^{th}$, that is added to the MD equations of motion to account for the energy exchange between the electrons and atomic vibrations. In this coupling term, ξ is a coefficient that depends on the instantaneous difference between the local lattice and electron temperatures as well as the strength of the electron–phonon coupling [48], m_i is the mass of an atom i, $\vec{v}_i^{th}$ is the thermal velocity of the atom defined as $\vec{v}_i^{th} = \vec{v}_i - \vec{v}^c$, where $\vec{v}_i$ is the actual velocity of atom i and $\vec{v}^c$ is the velocity of the center of mass of a cell to which atom i belongs. The expansion, density variation, and at higher fluences, disintegration of the irradiated target predicted in the MD part of the model are accounted for through the corresponding changes of the parameters of the TTM equation for electron temperature. The three-dimensional solution of the diffusion equation for T_e is used in simulations of laser spallation and ablation [38, 56, 61], where the dynamic material decomposition may result in lateral density and temperature variations.

In the continuum part of the model, beyond the surface region represented by the MD method, the electron heat conduction and the energy exchange between the electrons and the lattice are described by the conventional TTM equations, with L_c = 2–6 µm chosen to ensure negligible temperature changes at the bottom of the computational domain during the simulation time. A dynamic pressure-transmitting boundary condition [89–92] is applied at the bottom of the MD part of the system (marked as ④ in Fig. 12.1) to ensure nonreflecting propagation of the laser-induced stress wave from the MD region of the computational system to the bulk of the target. The energy carried away by the stress wave is monitored, allowing for control over the total energy conservation in the combined model [91].

In the simulations of laser interactions with thin films deposited on a substrate [35], the atomistic TTM-MD representation is extended to the whole metal film, with a thickness L_z, and the boundary condition marked as ④ in the left part of Fig. 12.1 is modified to reproduce the elastic response of the substrate to the laser-induced pressure waves and the adhesion between the metal film and the substrate [35, 92]. The acoustic impedance matching boundary condition at the substrate–film interface is parametrized in this case to ensure the partial reflection of the laser-induced pressure wave from the interface, as well as the work of adhesion between the silica substrate and metal film. In an alternative method for the description of the metal–substrate interface, a part of the substrate is represented with atomic resolution [92], thus accounting not only for the elastic response of the substrate but also for the possibility of plastic deformation, melting, and atomic mixing in the region of the substrate adjacent to the metal film.

The choice of the interatomic potential in MD simulations defines all the thermal and elastic properties of the target material, such as the lattice heat capacity, elastic moduli, coefficient of thermal expansion, melting temperature, volume, and enthalpy of melting and vaporization. In the simulations discussed in this chapter, the interatomic interactions are described by the embedded atom method (EAM) potentials [93] that provide a computationally efficient but still realistic description of bonding in metals. In particular, an EAM potential for Al developed by Mishin et al. [94, 95] is used in the simulations discussed in Sections 12.3 and 12.4, and an EAM potential for Ag designed by Foiles, Baskes, and Daw [96] is used in Section 12.5.

The electron temperature dependences of the thermophysical material properties included in the TTM equation for the electron temperature (electron–phonon coupling factor G, the electron heat capacity C_e, and the heat conductivity K_e; see Fig. 12.1) are highly sensitive to details of the electronic structure of the material and can exhibit large deviations (up to an order of magnitude) from the commonly used approximations of a linear temperature dependence of the electron heat capacity and a constant electron–phonon coupling [97–99]. These deviations have important implications for quantitative computational analysis of ultrafast processes associated with fs laser interaction with metals [52–54, 97, 98] and are accounted for in the TTM-MD model.

12.2.2 Representation of Background Gas and Liquid Environments

The background gas environment is introduced into the model in a rather straightforward manner, by simply adding a region with gas-phase molecules/atoms equilibrated at the desired temperature and pressure above the surface of the target, as shown in Fig. 12.1. The size of the region, L_{env} in Fig. 12.1, is chosen to be sufficiently large to ensure that the shock wave generated in the background gas by the ejection of the ablation plume would not reach the upper end of the region during the time of the simulation. A simple rigid plane is used in this case as the upper boundary of the background gas region instead of the pressure-transmitting boundary shown in Fig. 12.1. This brute force approach is possible because the computational overhead added by the treatment of the background gas is relatively small. For example, in the simulations of laser ablation of Al targets discussed in Section 12.4, an Ar gas region with L_{env} = 4 μm contributes only 0.5 to 5% of the total number of atoms for the values of the background gas pressure ranging from 1 to 10 atm.

The interatomic interactions between Ar and Al atoms are described by the Lennard–Jones (LJ) potential, with the parameters for Ar-Ar adopted from Ref. [100] and the ones for Ar-Al fitted to the results of ab initio calculations of the adsorption energy of an Ar on Al (111) surface [101]. At short distances (high interaction energies), the LJ potential for Ar–Al interactions is substituted by the Ziegler–Biersack–Littmark (ZBL) potential [102], which provides a more realistic description of the energetic collisions between the Al and Ar atoms at the initial stage of the plume expansion. The LJ and ZBL potentials are smoothly connected with each other by a second-order polynomial applied in the range of distances that correspond to the energy of the repulsive interaction ranging from –0.026 to 0.220 eV.

The computational description of the liquid environment presents a bigger computational challenge as compared to that of the background gas. The direct application of the conventional all-atom MD representation of liquids in large-scale simulations of laser processing or ablation is not feasible due to the high computational cost. Thus, a coarse-grained representation of the liquid environment [59, 70], where each particle represents several molecules, is adapted in the simulations described in Section 12.5. The coarse-graining reduces the number of degrees of freedom treated in the MD simulations and, as a result, significantly increases the time and length scales accessible for the simulations. At the same time, however, the smaller number of the dynamic degrees of freedom results in a severe underestimation of the heat capacity of the liquid. To resolve this problem, the degrees of freedom missing in the coarse-grained model are accounted for through a heat bath approach that associates an internal energy variable with each coarse-grained particle [103–106]. The energy exchange between the internal (implicit) and dynamic (explicit) degrees of freedom is controlled by the dynamic coupling between the translational degrees of freedom and the vibrational (breathing) mode associated with each particle (the particles are allowed to change their radii, or to "breathe" [70, 105]). The energy exchange is implemented through the addition of a damping or viscosity force to the breathing mode, which connects the breathing mode to the energy bath with a capacity chosen to reproduce the real heat capacity of the group of atoms represented by each coarse-grained particle [105, 106]. In effect, the breathing mode serves as a "gate" for accessing the energy stored in the molecular heat bath.

The first implementation of the coarse-grained model with the heat bath approach was recently developed for water and applied in simulations of laser interactions with the water-lysozyme system [105] and ablation of thin Ag films in a water environment [106]. While one cannot expect the coarse-grained model to provide an accurate representation of all the structural and thermodynamic properties of water, the key physical properties predicted by the model, such as density, speed of sound, bulk modulus, viscosity, surface energy, melting temperature, critical temperature, and critical density, are found to not deviate from the experimental values by more than 25% [105, 106]. In the simulations discussed in Section 12.5, the potential describing the interactions between Ag atoms and the coarse-grained particles is fitted to match the diffusion of metal atoms and small clusters in water, predicted in atomistic simulations performed at different temperatures.

12.3   Large-Scale MD Simulations of Laser Ablation in Vacuum

The MD simulations are based on the solution of the equations of motion for all atoms included in the computational system and, as a result, are generally limited to the treatment of submicron regions of the irradiated targets containing fewer than a billion (10^9) atoms. With a typical laser spot of tens to hundreds of microns and an ablation depth of tens to hundreds of nanometers, the number of atoms in a computational system that would be needed for the direct MD simulation of processes occurring on the scale of the whole laser spot can easily exceed trillions and is clearly beyond the current capabilities of the MD technique. Under these circumstances, the MD computational cell is typically assumed to represent a local volume within the laser spot, and the material response to local laser energy deposition is investigated, as schematically shown in the left part of Fig. 12.1. The periodic boundary conditions in the lateral directions, parallel to the surface of the target, are used in this case to reproduce the interaction of atoms in the MD computational cell with the surrounding material.

The processes occurring at the scale of the whole laser spot can still be investigated by combining results of simulations performed at different laser fluences and mapping them to the different locations within the laser spot. This mosaic approach is illustrated in Fig. 12.2 for an Al target irradiated by a 100 fs pulse with a Gaussian spatial profile of the laser beam. The snapshots shown in the figure are from five TTM-MD simulations performed at different laser fluences. The computational systems used in these simulations have lateral dimensions of 94 × 94 nm^2 and the depth of the atomistic part of the computational cell (L_z in Fig. 12.1) ranging from 150 nm for the lowest fluence to 300 nm for the highest one (the corresponding number of atoms is between 77 to 159 millions). The snapshots from individual TTM-MD simulations are shown for the same time of 150 ps after the laser pulse and are aligned with locations within the laser spot that correspond to the values of local fluence used in the simulations, as shown by the vertical and horizontal dashed lines in the lower part of Fig. 12.2.

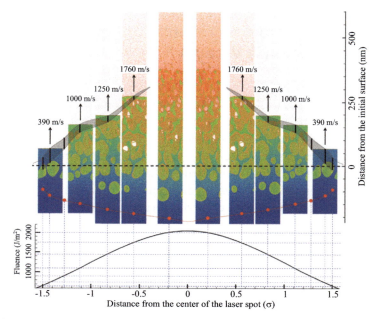

Figure 12.2 An illustration of the mosaic approach to the representation of the laser-induced processes at the scale of the whole laser spot, based on the results of a series of atomistic simulations performed for different laser fluences. The integral visual picture of melting, generation of subsurface voids, and material ejection is shown for an Al target irradiated by a 100 fs laser pulse. The laser beam has a Gaussian spatial profile with the standard deviation σ and a peak-absorbed laser fluence of 2050 J/m², as shown in the lower part of the figure. All snapshots are taken at the same time (150 ps) after the laser pulse and are aligned with locations within the laser spot that correspond to the values of the local fluence used in the simulations. The atoms in the snapshots are colored by their potential energy, from blue for low-energy atoms in the bulk of the target to red for the vapor-phase atoms. The red dots connected by the red line mark the location of the liquid–crystal interface. Reproduced from Ref. [38] with permission of Springer.

A mere visual inspection of Fig. 12.2 suggests coexistence of two distinct regimes of the material response to the laser energy deposition within the laser spot. In the periphery of the laser spot, a well-defined liquid layer can be seen at the top of the region of an expanding foamy structure of interconnected liquid regions. In the central part of the laser spot, the top surface layer of the target undergoes an explosive decomposition into vapor and small liquid droplets. The transition between the two regimes is manifested by the disappearance of the top liquid layer and a sharp increase in the density of the vapor emitted from the irradiated surface.

The two distinct visual pictures of the ablation process can be related to the differences in the physical mechanisms driving the material ejection. In the periphery of the laser spot, the energy density is not sufficient enough to cause an explosive release of the vapor (so-called phase explosion or explosive boiling [37, 60, 107–109]), and the material ejection (commonly called "spallation" [41, 42]) is driven by the relaxation of laser-induced stresses. The photomechanical nature of the spallation is evident from Fig. 12.3a, where the evolution of pressure and temperature is shown for a sequence of consecutive 5 nm thick layers located at different depths in the initial target irradiated at an absorbed laser fluence of 1100 J/m^2. The initial increase in temperature and pressure in different layers (Fig. 12.3a) is related to the laser excitation and electron–phonon energy transfer taking place under conditions of stress confinement [55, 42], when the heating rate exceeds the rate of the mechanical equilibration (expansion) of the material. The degree of stress confinement is different in different layers, with the top layers starting to expand before the electron–phonon equilibration is completed, thus reducing the maximum pressure reached in these layers. The relaxation of the compressive pressure in the presence of the free surface results in the development of an unloading (tensile) component of the stress wave propagating from the irradiated surface to the bulk of the target. The strength of the unloading component of the wave increases with depth down to the layer located between 110 and 115 nm. The temperature–pressure trajectories for layers located between 15 and 80 nm cross the red line, marking the limit of the stability of the metastable liquid against the onset of the cavitation (determined in a series of constant-pressure MD simulations [38]) (Fig. 12.3b). The observation that the void formation is only observed in a region that corresponds to the layers that cross the red line supports the notion of the photomechanical nature of the driving forces responsible for the nucleation, growth, and coalescence of subsurface voids, which eventually lead to the ejection of liquid droplets in the spallation regime.

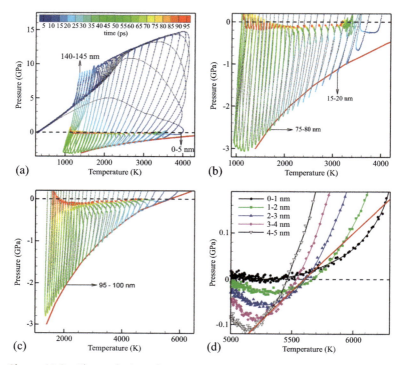

Figure 12.3 The evolution of temperature and pressure averaged over 5 nm thick (a–c) and 1 nm thick (d) consecutive layers located at different depths in the initial Al targets irradiated by 100 fs laser pulses at absorbed fluences of 1100 J/m² (a, b) and 2000 J/m² (c, d). The red line shows the temperature and pressure conditions for the onset of cavitation or phase explosion in the metastable liquid, calculated for Al represented by the EAM potential. Reproduced from Ref. [38] with permission of Springer.

In the central part of the laser spot, the temperature–pressure trajectories are similar to the ones discussed above, except that the higher energy density deposited in the top layers of the target makes the trajectories for the top layers cross the red line at higher temperatures and lower magnitudes of negative pressure, and even at positive pressure, as is the case for the three top 1 nm thick layers in Fig. 12.3c. The material decomposition in this case proceeds through the rapid (explosive) decomposition of the superheated liquid into vapor and small liquid droplets, signifying the transition to the phase explosion regime of laser ablation [55, 110]. Deeper into the target, the propagation of the tensile component of the stress wave leads to the additional cavitation in the superheated liquid, with the density of vapor in the pores decreasing with depth, as can be seen from the series of snapshots shown in Fig. 12.4 for a simulation performed at a fluence of 2000 J/m^2. The pressure–temperature trajectories for the material initially located down to ~100 nm below the surface cross the red line that defines the limit of the stability of the metastable liquid against the onset of the cavitation at large negative stresses, suggesting that the relaxation of the tensile stresses rather than the release of vapor-phase atoms is providing the main driving force for the nucleation and growth of voids in this region. Hence, the dominant driving force responsible for the material decomposition changes with depth and contributes to the effect of spatial segregation of droplets that is commonly observed in the ablation plume.

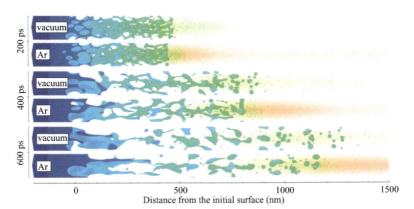

Figure 12.4 Snapshots of atomic configurations predicted in the TTM-MD simulation of laser ablation of a bulk Al target irradiated by a 100 fs laser pulse at an absorbed fluence of 2000 J/m^2 in vacuum and in the presence of a 1 atm Ar gas environment. The irradiation regime in these simulations corresponds to the phase explosion in the top part of the target. Only parts of the computational system are shown in the snapshots. The Al atoms are colored according to their potential energies.

The expansion of the fine "Swiss cheese"-like cellular structure generated in an ~100 nm wide surface region of the target during the first 50 ps after the laser pulse [38] leads to the coarsening of the liquid regions and eventual disintegration of the evolving foamy structure into individual liquid droplets (see snapshots for 200, 400, and 600 ps in Fig. 12.4). We can conclude that, under conditions of stress confinement realized in fs laser ablation of metals, the clusters and droplets generated in the phase explosion of the top part of the target and the ones emerging from photomechanical cavitation and disintegration of the deeper melted material have different characteristics and contribute to different parts of the ablation plume. The results of the cluster analysis shown Fig. 12.5 indicate that the front part of the expanding plume consists of vapor-phase atoms and small droplets, the medium part of the plume consists of medium-size droplets, and the rear part of the plume consists of large liquid droplets. This cluster segregation effect can be related to the results of plume imaging experiments [7, 8, 39, 40, 111–115], where the plume splitting into a fast component with an optical emission characteristic for neutral atoms and a slow component with blackbody-like emission of hot clusters is observed.

The occurrence of both spallation and phase explosion processes within the same laser spot (Fig. 12.2) can be related to the results of pump-probe experiments [116–118], where the observation of optical interference patterns (Newton rings) can be explained by the spallation of a thin liquid layer from the irradiated target, and the disappearance of the interference fringes in the central part of the laser spot [116–118] can be attributed to the transition to the phase explosion regime. Moreover, the sharp increase in the fraction of the vapor-phase atoms in the ablation plume upon the transition from spallation to phase explosion [38, 55, 110] can be related to the results of plume imaging experiments [111], where the maximum ejection of nanoparticles in laser ablation of Ni targets is observed at low fluences, whereas the degree of plume atomization increases at higher fluences. Note that the ejection of liquid layers observed in the simulations performed in the spallation regime can be related to the experimental observation of micro- and nanoparticles since the thin liquid layers can be expected to become unstable and decompose into individual droplets, as discussed in Ref. [38] on the basis of thin-film instability theory [119].

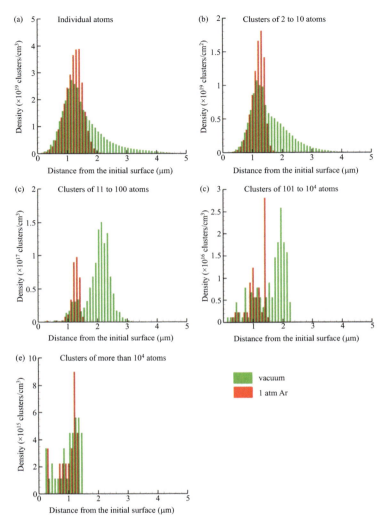

Figure 12.5 Number density of individual atoms (a) and clusters of different sizes (b–e) as a function of the distance from the initial surface in simulations of laser ablation of a bulk Al target irradiated by a 100 fs laser pulse at an absorbed fluence of 2000 J/m^2 with and without the presence of a 1 atm Ar background gas. The distributions are plotted for 600 ps after the laser pulse, which corresponds to the last pair of snapshots shown in Fig. 12.4. The analysis is performed for groups of clusters of similar sizes to obtain statistically adequate representations of the spatial distribution of clusters of different sizes in the ablation plume.

12.4 Ablation in a Background Gas

The presence of a background gas is recognized as a key factor enabling an effective control over nanoparticle structure and size distribution in short-pulse laser ablation, for example, Refs. [3, 4, 120–122]. Experimental probing of the ablation plume dynamics with various time- and spatially resolved optical and spectroscopic imaging techniques [3, 4, 8, 121–128] has provided a wealth of information on the evolution of different plume components and revealed a number of interesting phenomena, such as the plume splitting into fast and slow components, shock wave formation, onset of oscillatory behavior and turbulence, plume stagnation, and redeposition of ablated material.

A number of distinct regimes of ablation plume interaction with the background gas have been identified [122, 128, 129] and related to suitable theoretical models. At low pressure, up to about 1 Pa, the effect of scattering of the ablated species by the background gas is weak and the vacuum-like plume dynamics can be described as a self-similar adiabatic expansion [130]. The confining effect of the background gas becomes noticeable as the pressure approaches 10 Pa, when the plume shape starts to change [128], the angular distribution of the ablated species broadens [129], and the plume slows down. In this regime, the collective motion of the background gas due to the collisions with ablated species can still be neglected and the random scattering of the ablated species in the background gas can be simulated with the Monte Carlo computational technique [131, 132]. For intermediate pressure, between 10 and 100 Pa, a combination of scattering and collective motion leads to a complex interplay of compression and interpenetration of the plume and the background gas, thus presenting a challenge for a theoretical description. The empirical drag model [133], a model combining scattering and gas dynamic concepts [134], a mixed-propagation (diffusion and drag) model incorporating cluster growth kinetics [122, 135], and a combined Monte Carlo–hydrodynamic model [136] are among the approaches developed for the description of the ablation plume dynamic in this regime. Finally, for the background pressure of hundreds of pascals and higher, the shock wave formation and propagation dominate the plume dynamics and can be described by a "point explosion" analytical model [137–140] or gas dynamics computational models that neglect the diffusional mixing [127, 141–143].

A complete picture of the ablation plume interaction with the background gas, including the formation of the plume–ambient gas mixing zone and the shock front characteristics, can be provided by MD simulations of the ablation process. While, in principle, MD simulations can be performed for any magnitude of the background pressure, the high computational cost of the simulations makes it practical to only consider the high pressures, when the active processes of the shock wave and mixing zone formation take place during the first nanoseconds after the laser pulse. Indeed, the first MD simulations of laser ablation in a background gas were performed for 0.03 to 8.6 atm [77–80] and provided important insights into the initial ablation plume dynamics, shock wave formation, and characteristics of the plume–ambient gas interaction zone.

An important advantage of the MD technique is the ability to account for the direct ejection of atomic clusters and liquid droplets from the targets irradiated by short laser pulses. It has been demonstrated in experiments [8, 14, 108, 144, 145] and predicted in MD simulations performed for metals and molecular systems [37, 38, 55, 59, 60, 76, 110] that the clusters/droplets are unavoidable products of the explosive material disintegration in short-pulse laser ablation and constitute a major fraction of the total mass of ejected material. Most of the theoretical and computational approaches developed for the description of laser ablation in a background gas and briefly discussed above, however, treat the initial state of the ejected material as a hot dense vapor and only consider the formation of clusters and nanoparticles through the collision-induced condensation in the dense regions of the ejected plume. A realistic representation of the material ejection in MD simulations, therefore, makes this technique uniquely suitable for the analysis of the implications of the multiphase composition of the ablation plume. The ability of MD simulations to reproduce the evolution of the cluster size distribution in the ablation plume interacting with a background gas is illustrated below by the results of a simulation of laser ablation of an Al target in a 1 atm Ar gas environment.

To facilitate a comparison with laser ablation in vacuum, the simulation of laser ablation in Ar gas is performed for the same absorbed fluence of 2000 J/m^2 and pulse duration of 100 fs as in one of the simulations discussed above, in Section 12.3. The size of the Al sample is also the same, with a 4 µm thick Ar gas region added above the surface of the sample, as shown in Fig. 12.1. As can be seen from the snapshots shown in Fig. 12.4, the overall picture of the ablation process is similar in vacuum and in the 1 atm Ar gas. In both cases, the top surface regions of Al targets undergo explosive decomposition into mixtures of vapor and liquid droplets, while deeper regions undergo cavitation caused by the propagation of the tensile (unloading) components of the laser-induced pressure waves (see Section 12.3 for a discussion of the ablation mechanisms). However, one can still clearly observe the differences in the dynamics of the plume expansion. Even at the very early stage of the ablation process, at 200 ps after the laser pulse, there is a noticeable effect of the background gas on the expansion of the small liquid droplets, atomic clusters, and vapor-phase metal atoms generated in the phase explosion of the top surface layer of the target. The suppression of the expansion of the front part of the ablation plume becomes more pronounced with time and results in the formation of a well-defined layer of liquid droplets generated by the coalescence of small clusters and droplets decelerated by the interaction with the background gas. On the side of the background gas, the strong push from the ablation plume results in the shock wave formation in the Ar gas, with the temperature of the shocked Ar increasing up to more than 8000 K and the pressure behind the shock front reaching 80 atm.

The interaction of the plume with the compressed background gas has a dramatic effect on the cluster composition of the expanding plume as well as the spatial distribution of clusters in the plume, as can be seen from Fig. 12.5, where the distributions of vapor-phase atoms and clusters of different sizes are shown for a time of 600 ps after the laser pulse. The collisions of the metal atoms and small atomic clusters with the background gas atoms slow down the fastest plume species, reduce their total populations, and produce more narrow spatial distributions in Fig. 12.5a,b. The interaction of the ablation plume with the background gas results in the formation of a dense front layer of the plume, where collisions and coalescence of small droplets and clusters increase the population of intermediate and large droplets. The effect of the background gas is particularly strong on atomic clusters consisting of 11 to 100 atoms

(Fig. 12.5c) and small droplets consisting of up to 10,000 atoms (Fig. 12.5d), which are prominently present in the front part of the plume generated in the laser ablation in vacuum but almost completely eliminated by the plume interaction with the background gas. The larger droplets, with more than 10,000 atoms, are less affected by the background gas pressure, as most of them originate from the spallation process in the deeper part of the plume and do not directly interact with the Ar atoms during the time of the simulation.

While the strong effect of the background gas on the cluster composition of the ablation plume and the dynamics of different plume components is generally recognized and supported by experimental evidence [3, 4, 8, 120–122], the computational prediction of the short timescale of the drastic changes in the plume composition, occurring within the first nanosecond of the plume expansion, is rather startling and unexpected. Additional simulations performed at higher levels of background pressure (up to 10 atm, to be reported elsewhere) reveal the formation of an increasingly thick and continuous liquid layer at the interface with the shock-compressed background gas. The formation of the dense liquid layer at the front of the plume due to the spatial confinement of the plume by the background gas can be related to a similar phenomenon observed in simulations of laser ablation under much stronger confinement by a liquid environment, as discussed in the next section.

12.5 Ablation in Liquids

Pulse laser ablation in liquids (PLAL) has flourished since the early 2000s, when several groups successfully demonstrated the synthesis and size control of noble metal nanoparticles through PLAL [34, 36, 146, 147]. While the experimental setup in PLAL is simple, the liquid environment induces complicated plume–liquid interactions that not only define the nanoparticle size distribution but also facilitate the formation of metastable nanoparticles with unusual structures and composition [28–32, 148]. The size of the nanoparticles can be controlled by using different types of liquids [19–21], as well as by adding organic ligands [23–26] or inorganic salts [27] into the liquid. Moreover, a number of techniques based on post-irradiation of colloidal solutions, such as laser melting in liquids and laser fragmentation in liquids, have been demonstrated to be effective in further modifying the size, shape, and composition of the nanoparticles [149–153]. The ability of PLAL to produce stable contamination-free colloidal solutions of nanoparticles makes this technique very attractive for various fields of application, including biomedicine [154] and chemical catalysis [155, 156]. While there are comprehensive reviews of the experimental progress in the development of PLAL [15–18, 157], the fundamental mechanisms of nanoparticle formation by laser ablation in liquids are still not fully understood.

The general picture of laser ablation in liquids is very different from the one in vacuum. The ablation plume does not expand freely but is confined by the liquid environment. The liquid in contact with the plume is quickly heated and vaporized to form a thin layer of vapor surrounding the plume. This thin vapor layer can be directly observed via shadowgraphy [158] or X-ray radiography [159] as a dark zone surrounding the plume. The supply of heat from the plume creates and maintains high-temperature and high-pressure conditions in the vapor layer and drives the expansion of the layer. The expansion of the vapor layer away from the target leads to the formation of a cavitation bubble, while the pressure it exerts in the opposite direction pushes against the ablation plume and suppresses its expansion. It is often observed that the cavitation bubble undergoes a series of expansion, contraction, and collapse cycles [159, 160]. The interpretation of experimental results on the evolution of the cavitation bubble is typically based on the Rayleigh–Plesset equation used in combination with the van der Waals equation of state [158–162]. The dynamics of the cavitation bubble expansion and collapse are expected to play a major role in the generation of nanoparticles in PLAL, although the exact mechanisms of nanoparticle formation have not been established yet.

Currently, an accepted view is that the ablated material is likely to be confined and trapped inside the cavitation bubble, where favorable pressure and temperature conditions for nanoparticle nucleation, growth, coalescence, and solidification are realized [163, 164]. This general scenario, however, cannot be directly verified and detailed with conventional optical methods, since an optically dense interface between the liquid and the cavitation bubble blocks the view of the processes occurring inside the bubble. Recently, the novel use of small-angle X-ray scattering (SAXS) is demonstrated to be capable of relating the cavitation bubble dynamics to nanoparticle growth [159, 165]. In situ SAXS can be used to scan different positions in the bubble with different time delays, thus enabling mapping of the particle size distribution with respect to the time and position inside the bubble [159, 165, 166].

Signals yielded by SAXS have revealed two distinct nanoparticle size populations in the cavitation bubble: the "primary particles," with a size distribution centered around 8–10 nm, and the "secondary particles," with sizes around 50 nm. The primary particles are detected at the early stage of the first bubble expansion. A higher density of primary particles is detected near the bottom of the bubble, and their abundance decays toward the top of the bubble. The signal from the secondary particles is weak during the first cycle of the bubble expansion and collapse but becomes stronger after the rebound. The secondary particles are speculated to form due to the agglomeration of primary particles during the bubble collapse, when the primary particles are forced to collide with each other due to the sudden volume contraction. The sizes of the secondary particles are found to be highly variable, with rapid changes observed in the course of the bubble dynamics, suggesting that these "particles" may not be compact objects but some forms of loose networks of molten material or nanoparticle agglomerates. While the interpretation of SAXS data provides firsthand quantitative understanding of nanoparticle generation in PLAL, other approaches are needed to fully capture the nanoparticle generation mechanisms. In particular, atomic clusters and nanoparticles smaller than 5 nm cannot be detected with SAXS but may be responsible for the nanoparticle nucleation and growth in the ablation process, as well as the slower growth and coarsening of nanoparticles in the colloidal solutions generated by PLAL. Moreover, the primary particles detected at the early stage of the first bubble expansion are likely to be directly ejected from the irradiated target or formed during the first nanoseconds of the plume expansion, which is currently beyond the temporal resolution of SAXS. One needs to push beyond the current SAXS limits on particle sizes and temporal/spatial resolution to observe the ablation phenomena in more detail. There is also a need to resolve the thermodynamic states of the plume confined in the cavitation bubble to enable a reliable theoretical description of PLAL.

Computational modeling can play an important role in the interpretation of experimental observations and exploration of mechanisms of the nanoparticle generation in PLAL. Only the first attempts to address the nanoparticle generation mechanisms in one-dimensional continuum-level hydrodynamic modeling, however, have been reported so far [44, 45]. The liquid in these simulations is assumed to be a transparent, thermally insulating, and nonmixable overlayer with mechanical properties described by a single-phase equation of state parametrized for water. The highly simplified representation of the water environment prevents the realistic description of some of the key processes, such as the formation of a layer of water vapor driving the expansion of the cavitation bubble, water–metal mixing, and rapid cooling of the metal species in the mixing region. Moreover, the one-dimensional nature of the model does not allow for the direct simulation of nanoparticle generation. Nevertheless, the results of the hydrodynamic simulations have provided important insights into the initial dynamics of laser ablation under conditions of spatial confinement by a liquid overlayer and demonstrated that the confinement alone can facilitate generation of nanoparticles with a bimodal size distribution. The predictions of these simulations are illustrated in Fig. 12.6, where the contour plots show the phase transformations in an Au target irradiated by a 200 fs laser pulse at two different laser fluences. In the low-fluence regime, the ejection of multiple liquid layers is observed (Fig. 12.6a) and attributed to the photomechanical spallation. The spalled layers merge into one thick layer due to the deceleration of the front layer, in contact with water. One can speculate that breakdown of this liquid layer in the course of the spallation process or due to the thin-film instability may produce nanoparticles comparable to the thickness of the spalled layers, of the order of tens of nanometers. At a higher laser fluence, the top layer of the irradiated target undergoes an explosive decomposition into a mixture of liquid droplets and vapor, which pushes the water overlayer away from the target and creates a low-density region where small nanoparticles are expected to grow through condensation from the vapor phase.

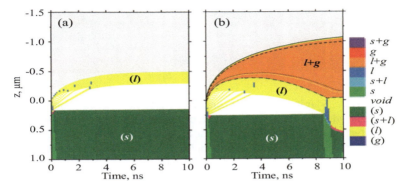

Figure 12.6 Time–space diagrams showing the phase composition of different regions of gold targets irradiated in a water environment by 200 fs laser pulses at absorbed fluences of 2600 J/m² (a) and 5500 J/m² (b), as predicted in continuum-level hydrodynamic simulations reported in Ref. [45].

As has been demonstrated above, in Sections 12.3 and 12.4, large-scale MD simulations are capable of not only predicting the thermodynamic conditions leading to the material ejection in laser ablation but also providing detailed atomic-level information on the mechanisms of nanoparticle formation. With the development of a computationally efficient coarse-grained representation of the liquid environment, discussed in Section 12.2.2, it is now possible to apply the MD technique to the simulation of PLAL, as has been demonstrated in recent MD simulations of thin Ag film ablation in water [106]. The extension of this work to a simulation of laser ablation of a bulk silver target in water is presented below and illustrated in Figs. 12.7–12.9. The schematic of the computational setup is provided in Fig. 12.1, with the depth of the atomistic TTM-MD part of the target, L_z, equal to 200 nm in the lower-fluence simulation (Fig. 12.7) and 400 nm at a higher fluence (Fig. 12.8); the size of the region where the coarse-grained representation of water overlayer is used, L_{env}, is 300 nm; the lateral dimensions of the computational cell are 100 nm × 100 nm; and the laser pulse duration is 100 fs.

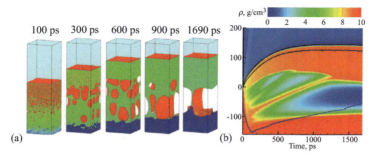

Figure 12.7 Snapshots of atomic configurations (a) and a density contour plot (b) predicted in the atomistic simulation of laser ablation of a bulk silver target irradiated in water by a 100 fs laser pulse at an absorbed fluence of 1500 J/m². Only parts of the computational system, from −160 to 200 nm with respect to the initial surface of the silver target, are shown in the snapshots. The atoms are colored according to their potential energies: from blue for the crystalline part of the target to green for liquid Ag and red for internal surfaces of the voids and the water–Ag interface. The molecules representing the water environment are blanked, and the presence of water is illustrated schematically as a bright-blue region above the Ag target. In the contour plot, the blue line shows the location of the melting and solidification fronts, while the two black lines outline the water–Ag mixing region, defined as a region where both water molecules and Ag atoms are present.

The absorbed fluence of 1500 J/m², used in the first simulation illustrated in Fig. 12.7, is about 50% above the spallation threshold for irradiation in vacuum [92], where the ejection of multiple liquid layers/droplets is observed (see Section 12.3 for the discussion of the spallation mechanism). The initial response of the Ag target to the laser excitation in the simulation of PLAL is similar to that in vacuum, with multiple voids nucleating, growing, and coalescing in the subsurface region of the target (Fig. 12.7a) in response to the generation of tensile stresses (Fig. 12.9a). The resistance of the water environment to the outward motion of the top layer, however, prevents the complete separation of the melted layer from the target, slows down the layer, and, at about 1.35 ns after the laser pulse, reverses the direction of its motion. The conductive cooling through the remaining liquid bridge connecting the top layer to the bulk of the target, combined with an additional cooling due to the interaction with water environment, brings the average temperature of the liquid layer down to the equilibrium melting temperature of the EAM Ag, T_m = 1139 K [52] by 740 ps after the laser pulse and undercools the layer down to 0.85 T_m by the end of the simulation at 1690 ps. While the simulation was not continued beyond 1690 ps, it can be estimated from the rate of cooling and the velocity of the layer in the direction toward the surface that the connecting bridge and the entire layer will solidify well before the time that would be needed for redeposition of the liquid layer to the target. The solidification is expected to proceed by a combination of epitaxial regrowth of the single-crystal target through the bridge into the top layer (the front of the epitaxial regrowth is shown by the blue line in the contour plot in Fig. 12.7b) and a massive nucleation of new crystallites at ~4 ns, when the temperature of the layer T_m is projected to decrease to ~0.7 [56].

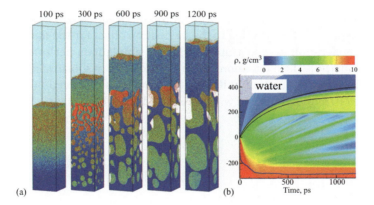

Figure 12.8 Snapshots of atomic configurations (a) and the density contour plot (b) predicted in atomistic simulation of laser ablation of a bulk silver target irradiated in water by a 100 fs laser pulse at an absorbed fluence of 4000 J/m². Only parts of the computational system, from −220 to 450 nm with respect to the initial surface of the silver target, are shown in the snapshots. The atoms are colored according to their potential energies, from blue for liquid Ag to red for the vapor-phase Ag atoms. The molecules representing the water environment are blanked, and the presence of water is illustrated schematically as a bright-blue region above the Ag target. In the contour plot, the blue line shows the location of the melting and solidification fronts, while the two black lines outline the water–Ag mixing region, defined as a region where both water molecules and Ag atoms are present.

The computational prediction of the formation of large subsurface voids, stabilized by the rapid cooling and solidification of the surface region, has important implications for both generation of nanoparticles in a multipulse irradiation regime and surface modification by laser processing in a liquid environment. The subsurface voids generated by laser spallation confined by water are many times larger than the ones observed close to the spallation threshold in vacuum [56]. The final structure expected to form after complete solidification in the simulation illustrated in Fig. 12.7a is essentially a thin metal layer loosely connected to the bulk of the target by thin walls and bridges. The irradiation of such a target by a subsequent laser pulse would result in the partial confinement of the deposited laser energy in the surface layer, leading to a substantial reduction of the threshold fluence for the onset of the phase explosion. Thus, while the generation of nanoparticles is not observed at this fluence upon the first pulse laser irradiation, the appearance of large subsurface voids can help to produce nanoparticles through the decomposition of the upper layer of the target upon irradiation by subsequent pulses, leading to the incubation effect for the nanoparticle generation demonstrated earlier in simulations performed in vacuum [56]. The incubation due to the presence of subsurface voids in pulsed laser ablation in liquids has also been observed in large-scale MD simulations and will be reported elsewhere.

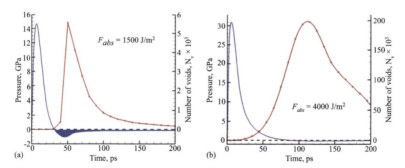

Figure 12.9 The evolution of pressure (blue curves) and the total number of voids (red curves) in atomistic simulations of 100 fs laser irradiation of Ag targets at absorbed fluences of 1500 J/m² (a) and 4000 J/m² (b). For both plots, the pressure is averaged over a region between 10 and 60 nm under the initial surface of the Ag target. The region of negative (tensile) pressure is colored blue in (a). The decrease in the number of voids is related to the void coalescence and coarsening, with the total volume of the voids rapidly increasing in both simulations during the time shown in the plots.

The stabilization of large subsurface voids and frozen surface structures by the interaction of the laser-generated transient melted structures with a liquid environment, predicted in the simulations, is also consistent with experimental observations of distinct surface nanomorphologies generated in laser processing in liquids [121, 167–172]. Further improvement in the computational efficiency of the atomistic simulations and the development of new multiscale approaches are needed to enable an efficient computational exploration of surface nanostructuring in liquids at experimentally relevant length scales and timescales.

The second simulation, illustrated in Figs. 12.8 and 12.9b, is performed at a higher absorbed laser fluence of 4000 J/m^2, which is about two times the threshold fluence for the transition from spallation to phase explosion regimes of laser ablation in vacuum [92]. The superheated liquid that in vacuum undergoes an explosive decomposition into small liquid droplets and vapor is now confined by water and is collected into a dense hot layer that pushes the water away from the target. The layer grows as the porous subsurface region of the Ag target expands, and more melted and vapor-phase Ag join the layer. Note that despite the visual similarity of the subsurface void evolution in Figs. 12.7 and 12.8, the main driving forces behind the void generation in the two simulations are different, as can be clearly seen from Fig. 12.9. In the low-fluence regime, at the absorbed fluence of 1500 J/m^2, the sharp increase of the number of subsurface voids coincides with the time when tensile stresses, highlighted by blue color in Fig. 12.9a, are generated due to the relaxation of the initial compressive pressure. At the higher fluence of 4000 J/m^2, the superheated top layer confined by the water environment keeps the positive pressure on the underlying melted part of the target for a longer time (Fig. 12.9b). The generation of voids in this case is mainly driven by the release of vapor and can be described as homogeneous boiling.

At a later time, beyond the timescale of the simulation, the top Ag layer is expected to completely separate from the target and slowly cool down due to the interaction with water. The water–silver mixing region, outlined by two black lines in Fig. 12.8b, is expected to grow and evolve into a low-density vapor region expanding under the action of water vapor pressure. The condensation of Ag vapor in the mixing region is expected to result in the formation of small Ag clusters, as have been observed in simulations of thin-film ablation [106] as well as ablation of bulk targets that will be reported elsewhere. At the same time, the top liquid layer is likely to rupture into larger liquid droplets due to the inherent instability of thin liquid films [38, 119] and the dynamic interaction of the liquid layer with the expanding and collapsing vapor bubble. The coexistence of the two distinct mechanisms of nanoparticle formation, the nucleation and growth in the mixing region and the decomposition of the thin liquid layer [106], may be related to the common observation of bimodal nanoparticle size distributions in PLAL [34, 36].

12.6 Concluding Remarks

The advancement of computational methods for the simulation of laser interactions with materials and the fast growth of available computing resources are expanding the range of research problems that can be addressed in large-scale MD simulations. Recent developments of computationally efficient approaches to the description of laser ablation in the presence of a background gas or liquid environment have enabled an exploration of the effect of various degrees of spatial confinement on the material response to short-pulse laser irradiation and generation of nanoparticles in laser ablation.

The comparison of the results of large-scale atomistic simulations of laser ablation of Al targets in vacuum and in a 1 atm Ar gas environment has revealed a surprisingly strong effect of the background gas on the initial plume expansion and evolution of the cluster size distribution. The formation of a strong shock wave in Ar, the rapid coalescence of smaller droplets and clusters in the dense front part of the ablation plume, and the suppression of the generation of small and medium-size Al clusters are among the effects predicted in the simulations.

The effect of a much stronger confinement of the ablation process by a liquid environment is explored in MD simulations of laser spallation and ablation of Ag targets in water. In the spallation regime, the confinement by a water environment is found to prevent the complete separation of the spalled layers from the target, leading to the stabilization of large subsurface voids frozen in a rapid solidification process and suggesting the incubation effect for the generation of nanoparticles. The computational prediction of the formation of a porous surface region with a complex morphology of frozen subsurface structures can also be related to experimental observations of distinct surface nanomorphologies generated in laser processing in liquids. At a higher laser fluence, which results in the phase explosion in vacuum, the confinement by water suppresses the phase decomposition of the superheated top layer of the target and collects the ejected material into a dense hot layer that pushes the water away from the target.

While the simulations reported in this chapter demonstrate the ability of the MD simulations to provide important insights into the mechanisms of laser interactions with metals in different environments, further advancements in the computational efficiency of MD simulations and the development of new multiscale approaches are needed to enable effective computational exploration of the longer-term processes of mixing of the ablation plume with a liquid environment, cavitation bubble formation, and surface nanostructuring in liquids at experimentally relevant length- and timescales.

Acknowledgments

Financial support for this work was provided by the National Science Foundation (NSF) through grants CMMI-1301298, CMMI-1436775, and DMR-1610936, as well as the Austrian Science Fund (FWF) through the Lise Meitner Programme (project M 1984). Computational support was provided by the Oak Ridge Leadership Computing Facility (INCITE project MAT130) and NSF through the Extreme Science and Engineering Discovery Environment (Project TG-DMR110090).

References

1. Barcikowski, S., Hahn, A., Kabashin, A. V., and Chickhov, B. N. (2007). Properties of nanoparticles generated during femtosecond laser machining in air and water, *Appl. Phys. A*, **87**, pp. 47–55.
2. Ruffino, F., Pugliara, A., Carria,E., Romano, L., Bongiorno, C., Spinella, C., and Grimaldi, M. G. (2012). Novel approach to the fabrication of Au/silica core–shell nanostructures based on nanosecond laser irradiation of thin Au films on Si, *Nanotechnology*, **23**, p. 045601.
3. Geohegan, D. B., Puretzky, A. A., Duscher, G., and Pennycook, S. J. (1998). Time-resolved imaging of gas phase nanoparticle synthesis by laser ablation, *Appl. Phys. Lett.*, **72**, pp. 2987–2989.

4. Umezu, I., Sugimura, A., Inada, M., Makino, T., Matsumoto, K., and Takata, M. (2007). Formation of nanoscale fine-structured silicon by pulsed laser ablation in hydrogen background gas, *Phys. Rev. B*, **76**, p. 045328.
5. Jo, Y. K., and Wen, S. B. (2011). Direct generation of core/shell nanoparticles from double-pulse laser ablation in a background gas, *J. Phys. D: Appl. Phys.*, **44**, p. 305301.
6. Umezu, I., Sakamoto, N., Fukuoka, H., Yokoyama, Y., Nobuzawa, K., and Sugimura, A. (2013). Effects of collision between two plumes on plume expansion dynamics during pulsed laser ablation in background gas, *Appl. Phys. A*, **110**, pp. 629–632.
7. Amoruso, S., Bruzzese, R., Wang, X., Nedialkov, N. N., and Atanasov, P. A. (2007). Femtosecond laser ablation of nickel in vacuum, *J. Phys. D: Appl. Phys.*, **40**, pp. 331–340.
8. Amoruso, S., Bruzzese, R., Wang, X., and Xia, J. (2008). Propagation of a femtosecond pulsed laser ablation plume into a background atmosphere, *Appl. Phys. Lett.*, **92**, p. 041503.
9. Chichkov, B. N., Momma, C., Nolte, S., Alvensleben, F. V., and Tunnermann, A. (1996). Femtosecond, picosecond and nanosecond laser ablation of solids, *Appl. Phys. A*, **63**, pp. 109–115.
10. Rethfeld, B., Sokolowski-Tinten, K., Von Der Linde, D., and Anisimov, S. I. (2004). Timescale in the response of materials to femtosecond laser excitation, *Appl. Phys. A*, **79**, pp. 767–769.
11. Amoruso, S., Nedyalkov, N. N., Wang, X., Ausanio, G., Bruzzese, R., and Atanasov, P. A. (2011). Ultrafast laser ablation of gold thin film targets, *J. Appl. Phys.*, **110**, p. 124303.
12. Oguri, K., Okano, Y., Nishikawa, T., and Nakano, H. (2009). Dynamics of femtosecond laser ablation studied with time-resolved x-ray absorption fine structure imaging, *Phys. Rev. B*, **79**, p. 144106.
13. Haustrup, N., and O'Connor, G. M. (2012). Impact of wavelength dependent thermo-elastic laser ablation mechanism on the generation of nanoparticles from thin gold films, *Appl. Phys. Lett.*, **101**, p. 263107.
14. Donnelly, T., Lunney, J. G., Amoruso, S., Bruzzese, R., Wang, X., and Ni, X. (2010). Dynamics of the plumes produced by ultrafast laser ablation of metals, *J. Appl. Phys.*, **108**, p. 043309.
15. Asahi, T., Mafune, F., Rehbock, C., and Barcikowski, S. (2015). Strategies to harvest the unique properties of laser-generated nanomaterials in biomedical and energy applications, *Appl. Surf. Sci.*, **348**, pp. 1–3.

16. Rehbock, C., Jakobi, J., Gamrad, L., van der Meer, S., Tiedemann, D., Taylor, U., Kues, W., Rath, D., and Barcikowski, S. (2014). Current state of laser synthesis of metal and alloy nanoparticles as ligand-free reference materials for nano-toxicological assays, *Beilstein J. Nanotechnol.*, **5**, pp. 1523–1541.

17. Yang, G. W. (2007). Laser ablation in liquids: application in the synthesis of nanocrystals, *Prog. Mater. Sci.*, **52**, pp. 648–698.

18. Tarasenko, N. V., and Butsen, A. V. (2010). Laser synthesis and modification of composite nanoparticles in liquids, *Quantum Electron.*, **40**, pp. 986–1003.

19. Tilaki, R. M., Iraji Zad, A., and Mahdavi, S. M. (2006). Stability, size and optical properties of silver nanoparticles prepared by laser ablation in different carrier media, *App. Phys. A*, **84**, pp. 215–219.

20. Bärsch, N., Jakobi, J., Weiler, S., and Barcikowski, S. (2009). Pure colloidal metal and ceramic nanoparticles from high-power picosecond laser ablation in water and acetone, *Nanotechnology*, **20**, p. 445603.

21. Gökce, B., VantZand, D. D., Menendez-Manjon, A., and Barcikowski, S. (2015). Ripening kinetic of laser-generated plasmonic nanoparticle in different solvents, *Chem. Phys. Lett.*, **626**, pp. 96–101.

22. Menendez-Manjon, A., Chichkov, B., and Barcikowski, S. (2010). Influence of water temperature on the hydrodynamic diameter of gold nanoparticles from laser ablation, *J. Phys. Chem. C*, **114**, pp. 2499–2504.

23. Mafuné, F., Kohno, J.-Y., Takeda, Y., Kondow, T., and Sawabe, H. (2001). Formation of gold nanoparticles by laser ablation in aqueous solution of surfactant, *J. Phys. Chem. B*, **105**, pp. 5114–5120.

24. Compagnini, G., Scalisi, A. A., Puglisi, O., and Spinella, C. (2004). Synthesis of gold colloids by laser ablation in thiol-alkane solutions, *J. Mater. Res.*, **19**, pp. 2795–2798.

25. Sylvestre, J. P., Kabashin, A. V., Sacher, E., Meunier, M., and Luong, J. H. T. (2004). Stabilization and size control of gold nanoparticles during laser ablation in aqueouscyclodextrins, *J. Am. Chem. Soc.*, **126**, pp. 7176–7177.

26. Tsuji, T., Thang, D.-H., Okazaki, Y., Nakanishi, M., Tsuboi, Y., and Tsuji, M. (2008). Preparation of silver nanoparticle by laser ablation in polyvinylpyrrolidone solutions, *Appl. Surf. Sci.*, **254**, pp. 5224–5230.

27. Rehbock, C., Merk, V., Gamrad, L., Streubel, R., and Barcikowski, S. (2013). Size control of laser-fabricated surfactant-free gold nanoparticles with highly diluted electrolytes and their subsequent bioconjugation, *Phys. Chem. Chem. Phys.*, **15**, pp. 3057–3067.

28. Yan, Z., Compagnini, G., and Chrisey, D. B. (2011). Generation of AgCl cubes by excimer ablation of bulk Ag in aqueous NaCl solutions, *J. Phys. Chem. C*, **115**, pp. 5058–5062.

29. Yan, Z., Bao, R., Huang, Y., and Chrisey, D. B. (2010). Hollow particles formed on laser-induced bubbles by excimer laser ablation in liquid, *J. Phys. Chem. C*, **114**, pp. 11370–11374.

30. Yan, Z., Bao, R., and Chrisey, D. B. (2011). Hollow nanoparticle generation on laser-induced cavitation bubbles via bubble interface pinning, *Appl. Phys. Lett.*, **97**, p. 124106.

31. Zhang, D., Gökce, B., Notthoff, C., and Barcikowski, S. (2015). Layered seed-growth of AgGe football-like mircospheres via precursor-free picosecond laser synthesis in water, *Sci. Rep.*, **5**, p. 13661.

32. Pearce, S. R. J., Henley, S. J., Claeyssens, F., May, P. W., Hallam, K. R., Smith, J. A., and Rosser, K. N. (2004). Production of nanocrystalline diamond by laser ablation at the solid/liquid interface, *Diamond Relat. Mater.*, **13**, pp. 661–665.

33. Streubel, R., Barcikowski, S., and Gökce, B. (2016). Continuous multigram nanoparticle synthesis by high-power, high-repetition-rate ultrafast laser ablation in liquids, *Opt. Lett.*, **41**, pp. 1486–1489.

34. Kabashin, A. V., and Meunier, M. (2003). Synthesis of colloidal nanoparticle during femtosecond laser ablation of gold in water, *J. Appl. Phys.*, **94**, pp. 7941–7943.

35. Rouleau, C. M., Shih, C.-Y., Wu, C., Zhigilei, L. V., Puretzky, A. A., and Geohegan, D. B. (2014). Nanoparticle generation and transport resulting from femtosecond laser ablation of ultrathin metal films: time-resolved measurements and molecular dynamics simulations, *Appl. Phys. Lett.*, **104**, p. 193106.

36. Sylvestre, J.-P., Kabashin, A. V., Sacher, E., and Meunier, M. (2005). Femtosecond laser ablation of gold in water: influence of the laser-produced plasma on the nanoparticle size distribution, *Appl. Phys. A*, **80**, pp. 753–758.

37. Zhigilei, L. V. (2003). Dynamics of the plume formation and parameters of the ejected clusters in short-pulse laser ablation, *Appl. Phys. A*, **76**, pp. 339–350.

38. Wu, C., and Zhigilei, L. V. (2014). Microscopic mechanisms of laser spallation and ablation of metal targets from large-scale molecular dynamics simulations, *Appl. Phys. A*, **114**, pp. 11–32.

39. Noel, S., Hermann, J., and Itina, T. (2007). Investigation of nanoparticle generation during femtosecond laser ablation of metals, *Appl. Surf. Sci.*, **253**, pp. 6310–6315.

40. Itina, T. E., Gouriet, K., Zhigilei, L. V., Noël, S., Hermann, J., and Sentis, M. (2007). Mechanisms of small clusters production by short and ultra-short pulse laser ablation, *Appl. Surf. Sci.*, **253**, pp. 7656–7661.
41. Paltauf, G., and Dyer, P. E. (2003). Photomechanical processes and effects in ablation, *Chem. Rev.*, **103**, pp. 487–518.
42. Leveugle, E., Ivanov, D. S., and Zhigilei, L. V. (2004). Photomechanical spallation of molecular and metal targets: molecular dynamics study, *Appl. Phys. A*, **79**, pp. 1643–1655.
43. Itina, T. E. (2011). On nanoparticle formation by laser ablation in liquids, *J. Phys. Chem. C*, **115**, pp. 5044–5048.
44. Povarnitsyn, M. E., and Itina, T. E. (2014). Hydrodynamic modeling of femtosecond laser ablation of metals in vacuum and in liquid, *Appl. Phys. A*, **117**, pp. 175–178.
45. Povarnitsyn, M. E., Itina, T. E., Levashov, P. R., and Khishchenko, K. V. (2013). Mechanisms of nanoparticle formation by ultra-short laser ablation of metals in liquid environment, *Phys. Chem. Chem. Phys.*, **15**, pp. 3108–3114.
46. Richardson, C. F., and Clancy, P. (1991). Picosecond laser processing of copper and gold: a computer simulation study, *Mol. Sim.*, **7**, pp. 335–355.
47. Hakkinen, H., and Landman, U. (1993). Superheating, melting, and annealing of copper surfaces, *Phys. Rev. Lett.*, **71**, pp. 1023–1026.
48. Ivanov, D. S., and Zhigilei, L. V. (2003). Combined atomistic-continuum modeling of short pulse laser melting and disintegration of metal films, *Phys. Rev. B*, **68**, p. 064114.
49. Ivanov, D. S., and Zhigilei, L. V. (2003). The effect of pressure relaxation on the mechanisms of short pulse laser melting, *Phys. Rev. Lett.*, **91**, p. 105701.
50. Lin, Z., and Zhigilei, L. V. (2006). Time-resolved diffraction profiles and atomic dynamics in short pulse laser induced structural transformations: molecular dynamics study, *Phys. Rev. B*, **73**, p. 184113.
51. Lin, Z., Johnson, R. A., and Zhigilei, L. V. (2008). Computational study of the generation of crystal defects in a bcc metal target irradiated by short laser pulses, *Phys. Rev. B*, **77**, p. 214108.
52. Wu, C., Thomas, D. A., Lin, Z., and Zhigilei, L. V. (2011). Runaway lattice-mismatched interface in an atomistic simulation of femtosecond laser irradiation of Ag film - Cu substrate system, *Appl. Phys. A*, **104**, pp. 781–792.

53. Thomas, D. A., Lin, Z., Zhigilei, L. V., Gurevich, E. L., Kittel, S., and Hergenröder, R. (2009). Atomistic modeling of femtosecond laser-induced melting and atomic mixing in Au film - Cu substrate system, *Appl. Surf. Sci.*, **255**, pp. 9605–9612.

54. Lin, Z., Bringa, E. M., Leveugle, E., and Zhigilei, L. V. (2010). Molecular dynamics simulation of laser melting of nanocrystalline Au, *J. Phys. Chem. C*, **114**, pp. 5686–5699.

55. Zhigilei, L. V., Lin, Z., and Ivanov, D. S. (2009). Atomistic modeling of short pulse laser ablation of metals: connections between melting, spallation, and phase explosion, *J. Phys. Chem. C*, **113**, pp. 11892–11906.

56. Wu, C., Christensen, M. S., Savolainen, J.-M., Balling, P., and Zhigilei, L. V. (2015). Generation of sub-surface voids and a nanocrystalline surface layer in femtosecond laser irradiation of a single crystal Ag target, *Phys. Rev. B*, **91**, p. 035413.

57. Sedao, X., Shugaev, M. V., Wu, C., Douillard, T., Esnouf, C., Maurice, C., Reynaud, S., Pigeon, F., Garrelie, F., Zhigilei, L. V., and Colombier, J.-P. (2016). Growth twinning and generation of high-frequency surface nanostructures in ultrafast laser-induced transient melting and resolidification, *ACS Nano*, **10**, pp. 6995–7007.

58. Anisimov, S. I., Zhakhovskii, V. V., Inogamov, N. A., Nishihara, K., Oparin, A. M., and Petrov, Yu. V. (2003). Destruction of a solid film under the action of ultrashort laser pulse, *Pis'ma Zh. Eksp. Teor. Fiz.*, **77**, pp. 731. [*JETP Lett.*, **77**, pp. 606–610 (2003)].

59. Zhigilei, L. V., Leveugle, E., Garrison, B. J., Yingling, Y. G., and Zeifman, M. I. (2003). Computer simulations of laser ablation of molecular substrates, *Chem. Rev.*, **103**, pp. 321–348.

60. Zhigilei, L. V., and Garrison, B. J. (2000). Microscopic mechanisms of laser ablation of organic solids in the thermal and stress confinement irradiation regimes, *J. Appl. Phys.*, **88**, pp. 1281–1298.

61. Wu, C., and Zhigilei, L. V. (2016). Nanocrystalline and polyicosahedral structure of a nanospike generated on metal surface irradiated by a single femtosecond laser pulse, *J. Phys. Chem. C*, **120**, pp. 4438–4447.

62. Ohmura, E., Fukumoto, I., and Miyamoto, I. (1998). Molecular dynamics simulation of laser ablation of metal and silicon, *Int. J. Jpn Soc. Precis. Eng.*, **32**, pp. 248–253.

63. Herrmann, R. F. W., Gerlach, J., and Campbell, E. E. B. (1998). Ultrashort pulse laser ablation of silicon: an MD simulation study, *Appl. Phys. A*, **66**, pp. 35–42.

64. Wu, X., Sadeghi, M., and Vertes, A. (1998). Molecular dynamics of matrix assisted laser desorption of leucine enkephalin guest molecules from nicotinic acid host crystal, *J. Phys. Chem. B*, **102**, pp. 4770–4778.
65. Cheng, C., and Xu, X. (2005). Mechanisms of decomposition of metal during femtosecond laser ablation, *Phys. Rev. B*, **72**, p. 165415.
66. Wang, X. W., and Xu, X. F. (2002). Molecular dynamics simulation of heat transfer and phase change during laser material interaction, *J. Heat Transfer*, **124**, pp. 265–274.
67. Lorazo, P., Lewis, L. J., and Meunier, M. (2006). Thermodynamic pathways to melting, ablation, and solidification in absorbing solids under pulsed laser irradiation, *Phys. Rev. B*, **73**, p. 134108.
68. Nedialkov, N. N., Atanasov, P. A., Imamova, S. E., Ruf, A., Berger, P., and Dausinger, F. (2004). Dynamics of the ejected material in ultra-short laser ablation of metals, *Appl. Phys. A*, **79**, pp. 1121–1125.
69. Anisimov, S. I., Zhakhovskii, V. V., Inogamov, N. A., Nishihara, K., Petrov, Y. V., Khokhlov, V. A. (2006). Ablated matter expansion and crater formation under the action of ultrashort laser pulse, *J. Exp. Theor. Phys.*, **103**, pp. 183–197.
70. Zhigilei, L. V., Kodali, P. B. S., and Garrison, B. J. (1997). Molecular dynamics model for laser ablation of organic solids, *J. Phys. Chem. B*, **101**, pp. 2028–2037.
71. Zhigilei, L. V., Kodali, P. B. S., and Garrison, B. J. (1997). On the threshold behavior in the laser ablation of organic solids, *Chem. Phys. Lett.*, **276**, pp. 269–273.
72. Zhigilei, L. V., and Garrison, B. J. (1999). Mechanisms of laser ablation from molecular dynamics simulations: dependence on the initial temperature and pulse duration, *Appl. Phys. A*, **69**, pp. S75–S80.
73. Yingling, Y. G., Zhigilei, L. V., and Garrison, B. J. (2001). The role of photochemical fragmentation in laser ablation: a molecular dynamics study, *J. Photochem. Photobiol. A*, **145**, pp. 173–181.
74. Schäfer, C., Urbassek, H. M., and Zhigilei, L. V. (2002). Metal ablation by picosecond laser pulses: a hybrid simulation, *Phys. Rev. B*, **66**, p. 115404.
75. Zhigilei, L. V., Yingling, Y. G., Itina, T. E., Schoolcraft, T. A., and Garrison, B. J. (2003). Molecular dynamics simulations of matrix assisted laser desorption - connections to experiment, *Int. J. Mass Spectrom.*, **226**, pp. 85–106.

76. Leveugle, E., and Zhigilei, L. V. (2007). Molecular dynamics simulation study of the ejection and transport of polymer molecules in matrix-assisted pulsed laser evaporation, *J. Appl. Phys.*, **102**, p. 074914.
77. Gacek, S., and Wang, X. (2008). Secondary shock wave in laser-material interaction, *J. Appl. Phys.*, **104**, p. 126101.
78. Gacek, S., and Wang, X. (2009). Dynamics evolution of shock waves in laser–material interaction, *Appl. Phys. A*, **94**, pp. 675–690.
79. Guo, L. Y., and Wang, X. W. (2009). Effect of molecular weight and density of ambient gas on shock wave in laser-induced surface nanostructuring, *J. Phys. D: Appl. Phys.*, **42**, p. 015307.
80. Li, C., Zhang, J. C., and Wang, X. W. (2013). Phase change and stress wave in picosecond laser-material interaction with shock wave formation, *Appl. Phys. A*, **112**, pp. 677–687.
81. Demaske, B. J., Zhakhovsky, V. V., Inogamov, N. A., and Oleynik, I. I. (2010). Ablation and spallation of gold films irradiated by ultrashort laser pulses, *Phys. Rev. B*, **82**, p. 064113.
82. Gill-Comeau, M., and Lewis, L. J. (2011). Ultrashort-pulse laser ablation of nanocrystalline aluminum, *Phys. Rev. B*, **84**, p. 224110.
83. Zhigilei, L. V., Lin, Z., Ivanov, D. S., Leveugle, E., Duff, W. H., Thomas, D., Sevilla, C., and Guy, S. J. (2010). Atomic/molecular-level simulations of laser-materials interactions, in *Laser-Surface Interactions for New Materials Production: Tailoring Structure and Properties*, eds. Miotello, A., and Ossi, P. M., Springer Series in Materials Science (Springer-Verlag, New York), Vol. 130, pp. 43–79.
84. Wu, C., Karim, E. T., Volkov, A. N., and Zhigilei, L. V. (2014). Atomic movies of laser-induced structural and phase transformations from molecular dynamics simulations, in *Lasers in Materials Science*, eds. Castillejo, M., Ossi, P. M., and Zhigilei, L. V., Springer Series in Materials Science (Springer International, Switzerland), Vol. 191, pp. 67–100.
85. Karim, E. T., Wu, C., and Zhigilei, L. V. (2014). Molecular dynamics simulations of laser-materials interactions: general and material-specific mechanisms of material removal and generation of crystal defects, in *Fundamentals of Laser-Assisted Micro- and Nanotechnologies*, eds. Veiko, V. P., and Konov, V. I., Springer Series in Materials Science (Springer International, Switzerland), Vol. 195, pp. 27–49.
86. Perez, D., Béland, L. K., Deryng, D., Lewis, L. J., and Meunier, M. (2008). Numerical study of the thermal ablation of wet solids by ultrashort laser pulses, *Phys. Rev. B*, **77**, p. 014108.

87. Kaganov, M. I., Lifshitz, I. M., and Tanatarov, L. V. (1957). Relaxation between electrons and crystalline lattices, *Sov. Phys. JETP*, **4**, pp. 173–178.
88. Anisimov, S. I., Kapeliovich, B. L., and Perel'man, T. L. (1974). Electron emission from metal surfaces exposed to ultrashort laser pulses, *Sov. Phys. JETP*, **39**, pp. 375–377.
89. Zhigilei, L. V., and Garrison, B. J. (1999). Pressure waves in microscopic simulations of laser ablation, *Mat. Res. Soc. Symp. Proc.*, **538**, pp. 491–496.
90. Schäfer, C., Urbassek, H. M., Zhigilei, L. V., and Garrison, B. J. (2002). Pressure-transmitting boundary conditions for molecular dynamics simulations, *Comp. Mater. Sci.*, **24**, pp. 421–429.
91. Zhigilei, L. V., and Ivanov, D. S. (2005). Channels of energy redistribution in short-pulse laser interactions with metal targets, *Appl. Surf. Sci.*, **248**, pp. 433–439.
92. Karim, E. T., Shugaev, M., Wu, C., Lin, Z., Hainsey, R. F., and Zhigilei, L. V. (2014). Atomistic simulation study of short pulse laser interactions with a metal target under conditions of spatial confinement by a transparent overlayer, *J. Appl. Phys.*, **115**, p. 183501.
93. Foiles, S. M. (1996). Embedded-atom and related methods for modeling metallic systems, *MRS Bull.*, **21**, pp. 24–28.
94. Mishin, Y., Farkas, D., Mehl, M. J., and Papaconstantopoulos, D. A. (1999). Interatomic potentials for monoatomic metals from experimental data and ab initio calculations, *Phys. Rev. B*, **59**, pp. 3393–3407.
95. Purja Pun, G. P., and Mishin, Y. (2009). Development of an interatomic potential for the Ni-Al system, *Phil. Mag.*, **89**, pp. 3245–3267.
96. Foiles, S. M., Baskes, M. I., and Daw, M. S. (1986). Embedded-atom-method functions for the fcc metals Cu, Ag, Au, Ni, Pd, Pt, and their alloys, *Phys. Rev. B*, **33**, pp. 7983–7991.
97. Lin, Z., and Zhigilei, L. V. (2006). Thermal excitation of d band electrons in Au: implications for laser-induced phase transformations, *Proc. SPIE*, **6261**, p. 62610U.
98. Lin, Z., and Zhigilei, L. V. (2007). Temperature dependences of the electron-phonon coupling, electron heat capacity and thermal conductivity in Ni under femtosecond laser irradiation, *Appl. Surf. Sci.*, **253**, pp. 6295–6300.
99. Lin, Z., Zhigilei, L. V., and Celli, V. (2008). Electron-phonon coupling and electron heat capacity of metals under conditions of strong electron-phonon nonequilibrium, *Phys. Rev. B*, **77**, p. 075133.

100. Matyushov, D. V., and Schmid, R. (1996). Calculation of Lennard-Jones energies of molecular fluids, *J. Chem. Phys.*, **104**, pp. 8627–8638.

101. Niu, W. X., and Zhang, H. (2012). Ar adsorptions on Al (111) and Ir (111) surfaces: a first-principles study, *Chin. Phys. B*, **21**, p. 026802.

102. Ziegler, J. F., Biersack, J. P., and Littmark, U. (1985). *The Stopping and Range of Ions in Solids* (Pergamon Press, New York).

103. Phares, D. J., and Srinivasa, A. R. (2004). Molecular dynamics with molecular temperature, *J. Phys. Chem. A*, **108**, pp. 6100–6108.

104. Jacobs, W. M., Nicholson, D. A., Zemer, H., Volkov, A. N., and Zhigilei, L. V. (2012). Acoustic energy dissipation and thermalization in carbon nanotubes: atomistic modeling and mesoscopic description, *Phys. Rev. B*, **86**, p. 165414.

105. Tabetah, M., Matei, A., Constantinescu, C., Mortensen, N. P., Dinescu, M., Schou, J., and Zhigilei, L. V. (2014). The minimum amount of "matrix" needed for matrix-assisted pulsed laser deposition of biomolecules, *J. Phys. Chem. B*, **118**, pp. 13290–13299.

106. Shih, C.-Y., Wu, C., Shugaev, M. V., and Zhigilei, L. V. (2016). Atomistic modeling of nanoparticle generation in short pulse laser ablation of thin metal films in water, *J. Colloid Interface Sci.*, **489**, pp. 3–17.

107. Miotello, A., and Kelly, R. (1999). Laser-induced phase explosion: new physical problems when a condensed phase approaches the thermodynamic critical temperature, *Appl. Phys. A*, **69**, pp. S67–S73.

108. Bulgakova, N. M., and Bulgakov, A. V. (2001). Pulsed laser ablation of solids: transition from normal vaporization to phase explosion, *Appl. Phys. A*, **73**, pp. 199–208.

109. Garrison, B. J., Itina, T. E., and Zhigilei, L. V. (2003). The limit of overheating and the threshold behavior in laser ablation, *Phys. Rev. E*, **68**, p. 041501.

110. Karim, E. T., Lin, Z., and Zhigilei, L. V. (2012). Molecular dynamics study of femtosecond laser interactions with Cr targets, *AIP Conf. Proc.*, **1464**, pp. 280–293.

111. Amoruso, S., Bruzzese, R., Pagano, C., and Wang, X. (2007). Features of plasma plume evolution and material removal efficiency during femtosecond laser ablation of nickel in high vacuum, *Appl. Phys. A*, **89**, pp. 1017–1024.

112. Albert, O., Roger, S., Glinec, Y., Loulergue, J. C., Etchepare, J., Boulmer-Leborgne, C., Perriere, J., and Millon, E. (2003). Time-resolved spectroscopy measurements of a titanium plasma induced by nanosecond and femtosecond lasers, *Appl. Phys. A*, **76**, pp. 319–323.

113. Jegenyes, N., Etchepare, J., Reynier, B., Scuderi, D., Dos-Santos, A., and Tóth, Z. (2008). Time-resolved dynamics analysis of nanoparticles applying dual femtosecond laser pulses, *Appl. Phys. A*, **91**, pp. 385–392.

114. Okano, Y., Oguri, K., Nishikawa, T., and Nakano, H. (2006). Observation of femtosecond-laser-induced ablation plumes of aluminum using space- and time-resolved soft X-ray absorption spectroscopy, *Appl. Phys. Lett.*, **89**, p. 221502.

115. Nakano, H., Oguri, K., Okano, Y., and Nishikawa, T. (2010). Dynamics of femtosecond-laser-ablated liquid-aluminum nanoparticles probed by means of spatiotemporally resolved X-ray absorption spectroscopy, *Appl. Phys. A*, **101**, pp. 523–531.

116. Agranat, M. B., Anisimov, S. I., Ashitkov, S. I., Zhakhovskii, V. V., Inogamov, N. A., Nishihara, K., Petrov, Yu. V., Fortov, V. E., and Khokhlov, V. A. (2007). Dynamics of plume and crater formation after action of femtosecond laser pulse, *Appl. Surf. Sci.*, **253**, pp. 6276–6282.

117. Sokolowski-Tinten, K., Bialkowski, J., Cavalleri, A., von der Linde, D., Oparin, A., Meyer-ter-Vehn, J., and Anisimov, S. I. (1998). Transient states of matter during short pulse laser ablation, *Phys. Rev. Lett.*, **81**, pp. 224–227.

118. Ionin, A. A., Kudryashov, S. I., Seleznev, L. V., and Sinitsyn, D. V. (2011). Dynamics of the spallative ablation of a GaAs surface irradiated by femtosecond laser pulses, *JETP Lett.*, **94**, pp. 753–758.

119. Vrij, A. (1966). Possible mechanism for the spontaneous rupture of thin, free liquid films, *Discuss. Faraday Soc.*, **42**, pp. 23–33.

120. Amoruso, S., Tuzi, S., Pallotti, D. K., Aruta, C., Bruzzese, R., Chiarella, F., Fittipaldi, R., Lettieri, S., Maddalena, P., Sambri, A., Vecchione, A., and Wang, X. (2013). Structural characterization of nanoparticles-assembled titanium dioxide films produced by ultrafast laser ablation and deposition in background oxygen, *Appl. Surf. Sci.*, **270**, pp. 307–311.

121. De Giacomo, A., Dell'Aglio, M., Gaudiuso, R., Amoruso, S., and De Pascale, O. (2012). Effects of the background environment on formation, evolution and emission spectra of laser-induced plasmas, *Spectrochim. Acta B*, **78**, pp. 1–19.

122. Ossi, P. M., Agarwal, N. R., Fazio, E., Neri, F., and Trusso, S. (2014). Laser-mediated nanoparticle synthesis and self-assembling, in *Lasers in Materials Science*, eds. Castillejo, M., Ossi, P. M., and Zhigilei, L. V., Springer Series in Materials Science (Springer International, Switzerland), Vol. 191, pp. 175–212.

123. Geohegan, D. B., and Puretzky, A. A. (1995). Dynamics of laser ablation plume penetration through low pressure background gases, *Appl. Phys. Lett.*, **67**, pp. 197–199.
124. Harilal, S. S., Bindhu, C. V., Tillack, M. S., Najmabadi, F., and Gaeris, A. C. (2003). Internal structure and expansion dynamics of laser ablation plumes into ambient gases, *J. Appl. Phys.*, **93**, pp. 2380–2388.
125. Freeman, J. R., Harilal, S. S., Diwakar, P. K., Verhoff, B., and Hassanein, A. (2013). Comparison of optical emission from nanosecond and femtosecond laser produced plasma in atmosphere and vacuum conditions, *Spectrochim. Acta B*, **87**, pp. 43–50.
126. Porneala, C., and Willis, D. A. (2006). Observation of nanosecond laser-induced phase explosion in aluminum, *Appl. Phys. Lett.*, **89**, p. 211121.
127. Miloshevsky, A., Harilal, S. S., Miloshevsky, G., and Hassanein, A. (2014). Dynamics of plasma expansion and shockwave formation in femtosecond laser-ablated aluminum plumes in argon gas at atmospheric pressures, *Phys. Plasmas*, **21**, p. 043111.
128. Diwakar, P. K., Harilal, S. S., Phillips, M. C., and Hassanein, A. (2015). Characterization of ultrafast laser-ablation plasma plumes at various Ar ambient pressures, *J. Appl. Phys.*, **118**, p. 043305.
129. Amoruso, S., Toftmann, B., and Schou, J. (2004). Thermalization of a UV laser ablation plume in a background gas: from a directed to a diffusionlike flow, *Phys. Rev. E*, **69**, p. 056403.
130. Anisimov, S. I., Bäuerle, D., and Luk'yanchuk, B. S. (1993). Gas dynamics and film profiles in pulsed-laser deposition of materials, *Phys. Rev. B*, **48**, pp. 12076–12081.
131. Itina, T. E., Marine, W., and Autric, M. (1997). Monte Carlo simulation of pulsed laser ablation from two-component target into diluted ambient gas, *J. Appl. Phys.*, **92**, pp. 3536–3542.
132. Itina, T. E., Katassonov, A. A., Marine, W., and Autric, M. (1998). Numerical study of the role of a background gas and system geometry in pulsed laser deposition, *J. Appl. Phys.*, **83**, pp. 6050–6054.
133. Geohegan, D. B. (1992). Fast intensified-CCD photography of $YBa_2Cu_3O_{7-x}$ laser ablation in vacuum and ambient oxygen, *Appl. Phys. Lett.*, **60**, pp. 2732–2734.
134. Wood, R. F., Chen, K. R., Leboeuf, J. N., Puretzky, A. A., and Geohegan, D. B. (1997). Dynamics of plume propagation and splitting during pulsed-laser ablation, *Phys. Rev. Lett.*, **79**, pp. 1571–1574.
135. Bailini, A., and Ossi, P. M. (2007). Expansion of an ablation plume in a buffer gas and cluster growth, *Europhys. Lett.*, **79**, p. 35002.

136. Itina, T. E., Hermann, J., Delaporte, P., and Sentis, M. (2002). Laser-generated plasma plume expansion: combined continuous-microscopic modeling, *Phys. Rev. E*, **66**, p. 066406.
137. Sedov, L. I. (1959). *Similarity and Dimensional Methods in Mechanics* (Academic Press, London).
138. Zel'dovich, Y. B. (2002). *Physics of Shock Waves and High-Temperature Hydrodynamic Phenomena* (Courier Corporation).
139. Arnold, N., Gruber, J., and Heitz, J. (1999). Spherical expansion of the vapor plume into ambient gas: an analytical model, *Appl. Phys. A*, **69**, pp. S87–S93.
140. Amoruso, S., Schou, J., and Lunney, J. G. (2010). Energy balance of a laser ablation plume expanding in a background gas, *Appl. Phys. A*, **101**, pp. 209–214.
141. Morel, V., Bultel, A., Annaloro, J., Chambrelan, C., Edouard, G., and Grisolia, C. (2015). Dynamics of a femtosecond/picosecond laser-induced aluminum plasma out of thermodynamic equilibrium in a nitrogen background gas, *Spectrochim. Acta B*, **103–104**, pp. 112–123.
142. Harilal, S. S., Miloshevsky, G. V., Diwakar, P. K., LaHaye, N. L., and Hassanein, A. (2012). Experimental and computational study of complex shockwave dynamics in laser ablation plumes in argon atmosphere, *Phys. Plasmas*, **19**, p. 083504.
143. Chen, Z. Y., and Bogaerts, A. (2005). Laser ablation of Cu and plume expansion into 1 atm ambient gas, *J. Appl. Phys.*, **97**, p. 063305.
144. Song, K. H., and Xu, X. (1998). Explosive phase transformation in excimer laser ablation, *Appl. Surf. Sci.*, **127–129**, pp. 111–116.
145. Yoo, J. H., Jeong, S. H., Greif, R., and Russo, R. E. (2000). Explosive change in crater properties during high power nanosecond laser ablation of silicon, *J. Appl. Phys.*, **88**, pp. 1638–1649.
146. Mafune, F., Kohno, J.-Y., Takeda, Y., and Kondow, T. (2000). Formation of gold nanoparticle by laser ablation in aqueous solution of surfactant, *J. Phys. Chem. B*, **105**, pp. 5114–5120.
147. Compagnini, G., Scalisi, A. A., and Puglisi, O. (2002). Ablation of noble metals in liquids: a method to obtain nanoparticles in a thin polymeric film, *Phys. Chem. Chem. Phys.*, **4**, pp. 2787–2791.
148. Yan, Z., and Chrisey, D. B. (2012). Pulsed laser ablation in liquid for micro-/nanostructure generation, *J. Photochem. Photobiol. C*, **13**, pp. 204–223.

149. Amendola, V., and Meneghetti, M. (2007). Controlled size manipulation of free goldnanoparticles by laser irradiation and their facile bioconjugation, *J. Mater. Chem.*, **17**, pp. 4705–4710.

150. Hashimoto, S., Werner, D., and Uwada, T. (2012). Studies on the interaction of pulsed laserswith plasmonic gold nanoparticles toward light manipulation, heat management, and nanofabrication, *J. Photochem. Photobiol. C*, **13**, pp. 28–54.

151. Besner, S., Kabashin, A. V., Winnik, F. M., and Meunier, M. (2009). Synthesis of size-tunable polymer-protected gold nanoparticle by femtosecond laser-based ablation and seed growth, *J. Phys. Chem. C*, **113**, pp. 9526–9531.

152. Pyatenko, A., Wang, H. Q., Koshizaki, N., and Tsuji, T. (2013). Mechanism of pulse laser interaction with colloidal nanoparticles, *Laser Photonics Rev.*, **7**, pp. 596–604.

153. Wang, H., Pyatenko, A., Kawaguchi, K., Li, X., Swiatkowska-Warkocka, Z., and Koshizaki, N. (2010). Selective pulsed heating for the synthesis of semiconductor and metal submicrometer spheres, *Angew. Chem. Int. Ed.*, **49**, pp. 6361–6364.

154. Hess, C., Schwenke, A., Wagener, P., Franzka, S., Sajti, C. L., Pflaum, M., Wiegmann, B., Haverich, A., and Barcikowski, S. (2014). Dose-dependent surface endothelializationand biocompatibility of polyurethane noble metal nanocomposites, *J. Biomed. Mater. Res. A*, **102**, pp. 1909–1920.

155. Kumar, V. L., Siddhardha, R. S. S., Kaniyoor, A., Podila, R., Molli, M., Kumar, S. M. V., Venkataramaniah, K., Ramaprabhu, S., Rao, A. M., and Ramamurthy, S. S. (2014). Gold decoratedgraphene by laser ablation for efficient electrocatalytic oxidation of methanoland ethanol, *Electroanalysis*, **26**, pp. 1850–1857.

156. Zhang, J., Chen, G., Guay, D., Chaker, M., and Ma, D. (2014). Highly active PtAu alloy nanoparticle catalysts for the reduction of 4-nitrophenol, *Nanoscale*, **6**, pp. 2125–2130.

157. Amendola, V., and Meneghetti, M. (2009). Laser ablation synthesis in solution and size manipulation of noble metal nanoparticles, *Phys. Chem. Chem. Phys.*, **11**, pp. 3805–3821.

158. Tamura, A., Sakka, T., Fukami, K., and Ogata, Y. H. (2013). Dynamics of cavitation bubbles generated by multi-pulse laser irradiation of a solid target in water, *Appl. Phys. A*, **112**, pp. 209–213.

159. Ibrahimkutty, S., Wagener, P., Dos Santos Rolo, T., Karpov, D., Menzel, A., Baumbach, T., Barcikowski, S., and Plech, A. (2015). A hierarchical view

on material formation during pulsed-laser synthesis of nanoparticles in liquid, *Sci. Rep.*, **5**, p. 16313.

160. Lam, J., Lombard, J., Dujardin, C., Leodoux, G., Merabia, S., and Amans, D. (2016). Dynamical study of bubble expansion following laser ablation in liquids, *Appl. Phys. Lett.*, **108**, p. 074104.

161. Soliman, W., Nakano, T., Takada, N., and Sasaki, K. (2010). Modification of Rayleigh–Plesset theory for reproducing dynamics of cavitation bubbles in liquid-phase laser ablation, *Jpn. J. Appl. Phys.*, **49**, p. 116202.

162. Hilgenfeldt, S., Grossman, S., and Lohse, D. (1999). A simple explanation of light emission in sonoluminescence, *Nature*, **398**, pp. 402–405.

163. Amendola, V., and Meneghetti, M. (2013). What controls the composition and the structure of nanomaterials generated by laser ablation in liquid solution? *Phys. Chem. Chem. Phys.*, **15**, pp. 3027–3046.

164. Dell'Aglio, M., Gaudiuso, R., De Pascale, O., and De Giacomo, A. (2015). Mechanisms and processes of pulsed laser ablation in liquids during nanoparticle production, *Appl. Surf. Sci.*, **348**, pp. 4–9.

165. Ibrahimkutty, S., Wagener, P., Menzel, A., Plech, A., and Barcikowski, S. (2012). Nanoparticle formation in a cavitation bubble after pulsed laser ablation in liquid studied with high time resolution small angle x-ray scattering, *Appl. Phys. Lett.*, **101**, p. 103104.

166. Wagener, P., Ibrahimkutty, S., Menzel, A., Plech, A., and Barcikowski, S. (2013). Dynamics of silver nanoparticle formation and agglomeration inside the cavitation bubble after pulsed laser ablation in liquid, *Phys. Chem. Chem. Phys.*, **15**, pp. 3068–3074.

167. Zavedeev, E. V., Petrovskaya, A. V., Simakin, A. V., and Shafeev, G. A. (2006). Formation of nanostructures upon laser ablation of silver in liquids, *Quantum Electron.*, **36**, pp. 978–980.

168. Stratakis, E., Zorba, V., Barberoglou, M., Fotakis, C., and Shafeev, G. A. (2009). Femtosecond laser writing of nanostructure on bulk Al via its ablation in air and liquids, *Appl. Surf. Sci.*, **255**, pp. 5346–5350.

169. Stratakis, E., Zorba, V., Barberoglou, M., Fotakis, C., and Shafeev, G. A. (2009). Laser writing of nanostructures on bulk Al via its ablation in liquids, *Nanotechnology*, **20**, p. 105303.

170. Barmina, E. B., Stratakia, E., Fotakis,C., and Shafeev, G. A. (2010). Generation of nanostructures on metals by laser ablation in liquids: new results, *Quantum Electron.*, **40**, pp. 1012–1020.

171. Podagtlapalli, G. K., Hamad, S., Sreedhar, S., Tewari, S. P., and Rao, S. V. (2012). Fabrication and characterization of aluminum nanostructures

and nanoparticles obtained using femtosecond ablation technique, *Chem. Phys. Lett.*, **530**, pp. 93–97.

172. Fan, P., Zhong, M., Li, L., Schmitz, P., Lin, C., Long, J., and Zhang, H. (2014). Angle-independent colorization of copper surfaces by simultaneous generation of picosecond-laser-induced nanostructures and redeposited nanoparticles, *J. Appl. Phys.*, **115**, p. 124302.

Chapter 13

Laser Nanostructuring of Polymers

Esther Rebollar,[a] Tiberio A. Ezquerra,[b] and Marta Castillejo[a]

[a]*Instituto de Química Física Rocasolano, IQFR-CSIC, Serrano 119, Madrid 28006, Spain*
[b]*Instituto de Estructura de la Materia, IEM-CSIC, Serrano 121, Madrid 28006, Spain*
e.rebollar@csic.es

The generation of nanostructured polymer films has been a challenge during the last decades. Laser-based methods enable high-spatial-resolution patterning and afford versatility and reliability without the need of stringent ambient conditions. Different laser techniques have been applied for nanostructuring of polymer films, among them, laser-induced periodic surface structures (LIPSS) using a polarized laser source. Structures with different sizes and shapes can be obtained by varying laser wavelength, polarization, angle of incidence, fluence, and duration and number of pulses. Potential applications of LIPSS in polymer films are illustrated in this chapter. In particular, the use of rippled polymer films as substrates for cell culture, surface-enhanced Raman scattering sensors, and nonvolatile organic memory devices will be shown.

Pulsed Laser Ablation: Advances and Applications in Nanoparticles and Nanostructuring Thin Films
Edited by Ion N. Mihailescu and Anna Paola Caricato
Copyright © 2018 Pan Stanford Publishing Pte. Ltd.
ISBN 978-981-4774-23-9 (Hardcover), 978-1-315-18523-1 (eBook)
www.panstanford.com

13.1 Introduction

Polymers and their composites are an essential part of the technological revolution of the information age through their multiple applications, ranging from lithographic masks to electronic connections and packaging [1–3]. The surface properties of polymers are of crucial importance in order to provide certain chosen functionalities. Nanostructuring has a significant effect on the physical and chemical properties of polymeric substrates, which underlies many advanced nanotechnologies.

Common techniques for generating polymer structures at the nanoscale are mainly based in soft lithography methods, like microcontact printing and nanoimprint lithography (NIL) [4–6]. These methods aim at reproducibility and low cost and generally provide versatile processing strategies. A drawback of these methods is the need of multiple steps involving clean room facilities and high-vacuum or complex mask fabrication procedures. Advanced lithographic methods are attracting a lot of interest as a complement to standard lithography, aiming to avoid the necessity of the mentioned demanding experimental conditions [7, 8]. In this respect e-beam technologies [9]; electrical [10], chemical [11], and mechanical [12] methods; block copolymer self-assembly [13]; use of templates [14]; and laser-induced patterning [15] of polymer surfaces are some versatile strategies for obtaining functional polymer materials.

Laser nanoablation and nanoswelling can be used for nanostructuring of polymer substrates. Laser ablation, a technique based on material removal by focusing the laser beam on the surface of the substrate, has been extensively studied in the past decades. The main trends in the development of this technique were reviewed by Bityurin [16], and models for the laser ablation of strongly absorbing polymers by nano- and femtosecond laser pulses have been proposed [17]. Other laser-based techniques are laser-induced periodic surface structures (LIPSS) [18], laser foaming [19], laser interference lithography (LIL) [20], and techniques based on transfer and deposition, such as laser-induced forward transfer (LIFT) [21], pulsed laser deposition (PLD) [22], and matrix-assisted pulsed laser evaporation (MAPLE) [23]. More recently, other specific laser processing techniques, including those taking advantage of field enhancement effects beneath an atomic force microscope tip [24], near-field optical microscope probes, and various near-field masks [25], have been developed and applied to the nanopatterning of soft polymer materials. A promising area of research is laser particle nanolithography, in which fabrication of submicron- and nanometer-scale structures is performed via laser radiation field localization using transparent micro- and nano-objects placed on the surface of the material. In particular, colloidal particle lens arrays have proved to be efficient near-field focusing devices for laser nanoprocessing of materials [26].

When using lasers with pulses in the nanosecond range, the material thermal properties determine the outcome of the laser fabrication process. However, for lasers with femtosecond pulses the laser–material interaction is the key factor governing the surface modification mechanism [27–29]. In this case, the possibility of temporally shaping femtosecond pulses offers new avenues for controlling and tailoring the features of the created structures [30, 31]. Additionally, laser nanostructuring within the material bulk can be achieved by a tightly focused laser beam taking advantage of the nonlinear light absorption of the material [32]. In particular, 3D nanopolymerization is a rapidly developing technology in the past few years [33].

Periodic nanostructures present a particular interest since they can be used as substrates in many fields, such as in organic photonics, microelectronics, and biomedicine [34–36]. Two laser techniques to produce periodic structures on polymers are LIL and LIPSS. In the case of LIL, interference of several beams is needed, while one single laser beam is used for LIPSS formation. Figure 13.1 shows an example of nanostructures obtained by both techniques in poly(bisphenol A carbonate) (PBAC).

In the present chapter we will focus on LIPSS formation on polymer substrates. LIPSS were first observed by Birnbaum [37] after ruby-laser irradiation of several semiconductor materials, and since then, these types of structures have been generated on a wide variety of materials. LIPSS originate from the interference of the incident and reflected or refracted laser light with the scattered light near the interface. The interference between the different waves leads to an inhomogeneous energy input, which, together with positive feedback mechanisms, can cause surface instabilities

[38–40]. LIPSS have been reported on the surface of different materials, that is, metals, semiconductors, and dielectrics with lasers of different pulse duration, from nanosecond to femtosecond, and different wavelengths, from the ultraviolet (UV) to the infrared (IR) [41–45]. In the case of polymers, LIPSS were first reported more than 20 years ago [18, 39], and several studies have shown that irradiation by a polarized laser beam induces self-organized ripple structure formation within a narrow fluence range, well below the ablation threshold [38, 43, 44, 46, 47]. In general, LIPSS can be prepared in both spin-coated [48–52] and freestanding polymer films [53–55].

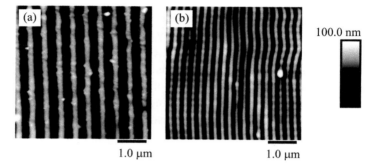

Figure 13.1 AFM height images of (PBAC) (a) nanostructured by LIL with two beams at 266 nm, a fluence of 80 mJ/cm², and an angle of 13° between them and (b) with LIPSS upon irradiation at 266 nm with 1200 pulses at 7 mJ/cm² at normal incidence.

According to the classical interference model the period of the ripples (L) depends on the laser wavelength (λ), the angle of incidence (θ) of the radiation, and the effective refractive index (n) of the material. Their direction is related to the direction of the laser beam polarization. The spacing of the structures can be described by the expression [56]

$$L = \frac{\lambda}{n - \sin(\theta)} \tag{13.1}$$

A large processed surface area and good-quality samples can be obtained by this method, and substrates nanostructured in this way can be used to tailor a great variety of surface properties, such as adhesion and friction [57], induced cell alignment [54], liquid crystal alignment [58], and color generated by superficial gratings [59].

13.2 LIPSS Formation Using Nanosecond Pulses

For polymers irradiated with nanosecond laser pulses different processes, such as thermal and nonthermal scissoring of polymer chains, amorphization of crystalline domains, local surface melting, ablation, photolytic shrinkage, photo-oxidation, and material transport and rearrangement, have been proposed to be involved in ripple formation [18, 38–40, 47, 57].

More recently it has been reported [48, 49] that in order to obtain LIPSS in amorphous polymers, a minimum fluence value is necessary to ensure that the surface is heated above the glass transition temperature (T_g), therefore allowing polymer segmental and chain dynamics. In the case of semicrystalline polymers, the thermal properties are governed not only by the glass transition temperature but also by the melting temperature (T_m). Thus, the temperature reached at the sample surface upon laser irradiation must overcome T_m so that superficial crystallites melt, providing enough polymer dynamics [60].

As a main rule, a high absorption coefficient at the laser wavelength is needed in order to obtain LIPSS upon irradiation with nanosecond pulses. Additionally, it is important that the original samples present a smooth surface with mean roughness in the range of a few nanometers. In principle, LIPSS can be formed on the surface of a great variety of polymers, such as polystyrene (PS), poly(ethylene terephthalate) (PET), poly(trimethylene terephthalate) (PTT), and PBAC [48, 49, 54]. Also naturally derived polymers such as chitosan and its 50/50 wt% blend with the synthetic polymer polyvinyl pyrrolidone (PVP) [55] and the polyconjugated polymer poly(3-hexylthiophene) (P3HT) have been reported to form LIPSS [60]. LIPSS are generated upon irradiation at an appropriate wavelength and fluence range, well below the ablation threshold, and after tens or hundreds of pulses, depending on the material and the irradiation conditions. The ripple axis, in the case of polymers, is parallel to the polarization vector of the laser beam. The nanostructured area is restricted to the footprint of the laser beam and typically, areas of several tens of square millimeters can be structured while keeping the sample at a fixed position upon irradiation. Alternatively, sample scanning may be used in case larger nanostructured areas are needed.

As a typical example of LIPSS formation, Fig. 13.2a–c displays AFM height images of areas of PTT irradiated with 600 pulses at different laser fluences. For a certain value of fluence (Fig. 13.2a), surface modification becomes apparent, as revealed by an increase of roughness. For higher fluences the periodic structures begin to be formed (Fig. 13.2b) as parallel and well-defined ripples in a narrow range of fluences (Fig. 13.2c). When the fluence is further increased, LIPSS become distorted. This trend is general for most polymers, although the specific fluence values at which LIPSS are generated depend on the particular polymer properties.

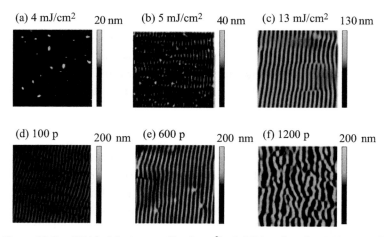

Figure 13.2 AFM height images (5 × 5 μm^2) of PTT irradiated at 266 nm with 600 pulses (a–c) as a function of fluence and at a fluence of 7 mJ/cm^2 (d–f) as a function of the number of pulses. Reprinted with permission from Ref. [49]. Copyright (2012) American Chemical Society.

The value of the period depends on the laser fluence, as shown in Fig. 13.3a for PET, PTT, and PBAC. In general for polymers, the period of the LIPSS increases with increasing fluence until it reaches a plateau for a value of the period similar to the irradiating laser wavelength [48, 49]. This effect is exemplified in Fig. 13.3a.

For a fixed value of fluence, LIPSS formation depends on the number of pulses, as shown in Fig. 13.2d–f. A certain number of pulses, in this case $N > 100$, are required to obtain well-defined LIPSS in thin films of PTT. Concerning the dependence of the LIPSS period on the number of pulses, a trend similar to the one shown for the dependence on the fluence is observed (Fig. 13.3b).

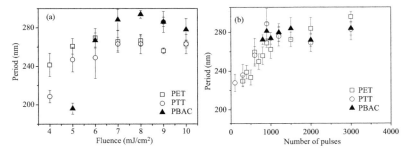

Figure 13.3 Period of LIPSS formed in PET, PTT, and PBAC upon irradiation at 266 nm (a) as a function of fluence for a fixed number of pulses N = 600 and (b) as a function of the number of pulses at 7 mJ/cm^2. Reprinted from Ref. [61], Copyright (2015), with permission from Elsevier.

Regarding the mechanism of LIPSS formation upon irradiation with nanosecond pulses, as mentioned above, it is commonly accepted that interference between the incident laser light and the surface scattered wave plays a crucial role [38, 56]. The irradiated surface scatters the incident light, which interferes with the surface wave, resulting in a modulated distribution of the energy on the surface. Laser irradiation causes the heating of the upper layer of the polymeric film, and the corresponding temperature increase can be estimated as a function of time at different depths from the surface, by solving the one-dimensional heat conduction equation [38, 48]. The results of these calculations have shown that to generate LIPSS, a minimum value of the laser fluence is necessary to ensure that the surface temperature overcomes T_g or T_m for amorphous or semicrystalline polymers, respectively. This allows polymer segmental and chain dynamics [48, 49]. The heating above T_g and T_m is expected to induce an increase of surface roughness caused by capillary waves [62], thus enhancing surface inhomogeneities and facilitating the feedback mechanism involved in LIPSS formation. When comparing polymers with different T_g values, like PET, PTT, and PBAC, it is observed that the one with a higher T_g, PBAC, requires higher fluences and number of pulses for LIPSS formation [48].

The feedback mechanism is confirmed by in situ monitoring of LIPSS formation using grazing incidence small-angle X-ray scattering (GISAXS) [63] with synchrotron radiation. This technique is appropriate to deal with dynamic processes even in the millisecond time range [64] and allows the analysis of larger areas in comparison to conventional microscopy techniques. In situ monitoring of LIPSS formation has shown that the dynamics of nanostructure formation also depends on the laser repetition rate. For irradiation at 10 Hz, a much lower number of pulses, around 30, is needed for the appearance of LIPSS in comparison to the more than 100 pulses that are required for irradiation at lower repetition rates. These results indicate that the process is more efficient for a shorter time separation between successive pulses.

13.3 LIPSS Formation Using Femtosecond Pulses

Despite the extensive studies on LIPSS, only a few references report about LIPSS on polymers upon irradiation in the femtosecond regime [50, 51, 65–68].

The great variety of experimental work on LIPSS using femtosecond pulses was subsumed into various theoretical approaches [69–71]. In the most common type of surface topography, a periodicity of about the wavelength of the laser radiation is observed, similar to the nanosecond case and also attributed to the interference between the incident and scattered light or excited surface waves [56, 70]. However, periodic surface structures induced by femtosecond laser irradiation do not fully comply with previous models on LIPSS formation developed for nanosecond laser irradiation. Thus, the interference model first suggested by Emmony et al. [72] was later improved by several other authors [69, 70].

Particularly, the work of Sipe et al. [70] represents a first-principle theory, which takes into consideration the interaction of an electromagnetic wave with a microscopically rough surface and also includes the possible excitation of surface plasmon polaritons (SPPs). Also according to Sipe at al. [70], efficient surface rippling can be enhanced due to the redistribution of the energy from subsequent laser pulses through a feedback effect. In later publications, Young and coworkers as well as Clark and Emmony applied this first-principle theory to explain the formation of LIPSS in semiconductors and metals [73, 74]. Some other theoretical attempts have been made to describe LIPSS formed upon irradiation with femtosecond pulses on the basis of well-known models from hydrodynamics of thin films [75] and on the Sipe–Drude model [42]. This model considers the laser-induced changes in the complex refractive index due to transient generation of quasi-free electrons in the conduction band of the solid. It has also been shown experimentally that the laser polarization plays a very important role in determining the orientation of the ripples [76]. Other authors [76] have proposed a different type of mechanism, in which the laser irradiation first creates a highly electrostatically unstable surface. This is followed by emission of high-momentum surface ions, which in turn induce an appreciable recoil pressure on the substrate. In this case, ripples would arise from relaxation rather than from wave interference effects.

In the case of irradiation with femtosecond pulses, to determine the temperatures involved, a two-temperature model has been proposed [77], in which electrons and lattice subsystems are described separately by their own temperature distribution. The electronic temperature is estimated assuming that the laser energy is first transferred to the electrons, and then the lattice temperature is determined by assuming energy transfer from the electrons to the lattice. This model has been mainly used to simulate the interaction of short pulses with metals but has also been applied to other materials, such as silicon [77]. However, the strong nonequilibrium conditions associated with ultrashort-pulse laser interaction with solids make a detailed theoretical analysis difficult. In the case of soft materials, such as polymers, numerical simulations of the evolution of temperature distribution on the surface and within the bulk have also been reported, considering again the generation of free electrons [78].

As mentioned before, significant absorption at the laser wavelength used for irradiation is required in order to obtain LIPSS with nanosecond pulses. However, LIPSS can be fabricated upon irradiation with femtosecond pulses at a wavelength at which the material has a low linear absorption coefficient. Figure 13.4a–c shows as an example AFM height images of PET irradiated at 795 nm with 50,000 pulses and different fluences [50, 51].

LIPSS are clearly observed at the fluences of 35 and 36 mJ/cm^2. However, above 37 mJ/cm^2, the polymer thin-film structure appears distorted and higher fluences induce ablation of the film surface. The period and height of ripples as a function of fluence in this case are shown in Fig. 13.5a. Periods of the ripples are again of the order of the laser wavelength.

The dependence of ripple parameters on the number of pulses was also studied. Figure 13.4d–f shows, as an example, AFM height images of LIPSS formed upon irradiation with different number of pulses at a constant irradiation fluence of 37 mJ/cm^2, the corresponding period and height being shown in Fig. 13.5b.

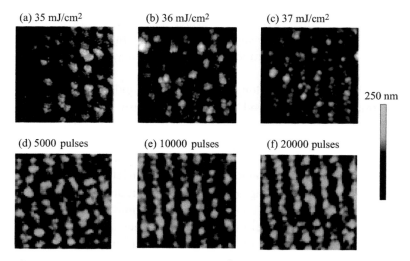

Figure 13.4 AFM height images (5 × 5 μm^2) of PET irradiated at 795 nm with 50,000 pulses (a–c) as a function of fluence and at a fluence of 37 mJ/cm^2 (d–f) as a function of the number of pulses. Reproduced from Ref. [51] with permission from the PCCP Owner Societies.

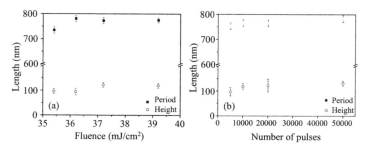

Figure 13.5 Period and height of LIPSS formed in PET upon irradiation at 795 nm (a) as a function of fluence for a fixed amount of pulses N = 50,000 and (b) as a function of the number of pulses at 37 mJ/cm^2.

The formation of LIPSS at this long wavelength, at which the absorbance of the polymers is very small, is possible due to the fact that, upon irradiation with femtosecond pulses, large laser intensities are involved and thus multiphoton absorption and ionization processes mediate in the coupling of laser light with the outer layer of the polymer film.

As the periodicities of the ripples reported here nearly coincide with the laser irradiation wavelengths, the phenomenon can be attributed to scattering, diffraction, and subsequent local-field enhancement effects [70, 74, 79, 80]. It is important to point out that these periodic structures were not created at the bottom of ablation craters but on the original film surface, differently to results previously reported [65, 68]. The absence of ablation in the nanostructuring of the thin polymer films upon irradiation in the present conditions accounts for a self-organization mediated process that does not require any surface material removal. Additionally, a feedback process is clear, as repetitive irradiation is needed in order to observe LIPSS formation. The fact that LIPSS are parallel to the polarization of the laser beam is supported by Sipe theory [70]. The orientation of the ripples with respect to the polarization of the incident laser light depends on the value of the dielectric permittivity of the substrate, $\varepsilon = \varepsilon' - \iota\varepsilon''$, in such a way that it is parallel to the polarization when $|\varepsilon'| < 1$, as is the case of dielectrics [81] and, in particular, of polymers.

13.4 Formation of LIPSS on Nonabsorbing Polymers

As mentioned, when irradiating with nanosecond pulses, absorption at the irradiation wavelength is an important condition in order to observe LIPSS formation. One possibility in order to nanostructure nonabsorbing polymers is the use of femtosecond lasers, as explained in the previous section. A different approach consists of the use of a dopant to increase the absorption of the polymer system at the irradiation wavelength. This has been proposed by Kalachyova et al. [82] for LIPSS formation in poly(methyl methacrylate) (PMMA) by irradiation with a KrF laser at 248 nm. At this wavelength PMMA absorbs weakly, but by using Fast Red ITR as a dopant it is possible to form LIPSS after irradiation with only 15 pulses at a fluence of 12 mJ/cm^2.

A different alternative is the use of a bilayer approach [83]. In this case, an example was recently introduced in which LIPSS were prepared on a ferroelectric random copolymer poly(vinylidene fluoride-tri fluoroethylene) (PVDF-TrFE). This copolymer (76:24 VDF:TrFE content) does not absorb in the UV-visible range and does not form LIPSS even if irradiation is performed at fluences higher than 100 mJ/cm^2 and with a large number of pulses [48]. Nanostructuring is possible by using a bilayer approach in which an absorbing polymer like P3HT is used as the bottom layer. Surface structuring takes place upon irradiation at 532 nm, and LIPSS similar to those formed on P3HT monolayers are observed at the surface of (PVDF-TrFE). The continuity of the nanostructured upper polymer layer is corroborated by the cross-sectional scanning electron microscopy (SEM) image shown in Fig. 13.6.

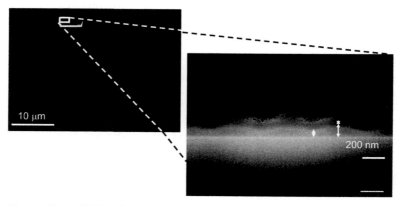

Figure 13.6 SEM and cross-sectional SEM images after focused ion beam milling of a (PVDF-TrFE)/P3HT thin bilayer film after irradiation at 532 nm with 3600 pulses at a laser fluence of 27 mJ/cm^2.

This approach is thought to be used for a range of nonabsorbing polymers, as is the case of (PVDF-TrFE)., making LIPSS more versatile and thus paving the way for a wider range of applications.

13.5 Formation of Alternative Periodic Structures

The period of LIPSS can be modified by changing the laser wavelength but also by varying the angle of incidence of the laser with respect to the surface according to Eq. 13.1 [56]. On increasing the angle of incidence the period of the structures increases. Figure 13.7a–c shows LIPSS formed on PET with periods of around 250, 320, and 490 nm measured for angles of incidence of 0°, 22°, and 45°, respectively, upon irradiation at a wavelength of 266 nm [52]. It is important to mention that the height of the structures obtained also increases with the angle of incidence, from 50 to 120 nm.

Control over the morphology of the superficial structures can also be achieved by changing the polarization of the laser beam from linear to circular. While Fig. 13.7a–c shows structures obtained using a linearly polarized laser beam, Fig. 13.7d shows the surface pattern formed upon irradiation with circular polarization. The sample is covered by uniformly distributed nanodots with a diameter of around 260 nm, similar to the laser wavelength. This type of structure has also been observed for nonpolymeric materials [84]. The change of morphology is related to the fact that the amplitude of the electric vector for circularly polarized light is constant and rotates around the propagation direction so that the irradiated material aligns to the successive polarization directions and gives rise to a pattern that is uniformly distributed.

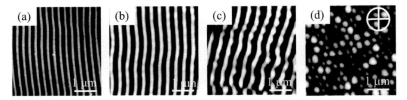

Figure 13.7 AFM height images (4 × 4 µm^2) of a PET film irradiated with a linearly polarized beam at 266 nm using 1200 pulses and 7 mJ/cm^2 (a) at normal incidence, (b) at 22°, (c) at 45°, and (d) with circular polarization.

13.6 Applications of Polymer LIPSS

Once it has been shown that it is possible to generate LIPSS with different sizes and shapes, we will describe how these substrates can be appropriate for different applications. In particular, in the following sections examples of the use of polymer LIPSS for cell culture, surface-enhanced Raman scattering (SERS) substrates, and nonvolatile organic memory devices will be shown.

13.6.1 Polymer LIPSS for Cell Culture

LIPSS generated on PET films with different sizes and shapes have been used for mesenchymal cell culture, as illustrated in Fig. 13.8. By comparing the culture on nonirradiated samples (Fig. 13.8a) with that on a polymer with LIPSS, it can be inferred that nanostructures promote in all cases cell adhesion and growth [52]. Also, nanostructuring seems to favor differentiation and communication among cells. For samples irradiated at 266 nm with circularly polarized light and at 193 nm, enhanced adhesion, proliferation, and differentiation have been observed. For these substrates a larger basic character and a higher value of surface free energy were determined by contact angle measurements, indicating that these characteristics are relevant for cell culture [52].

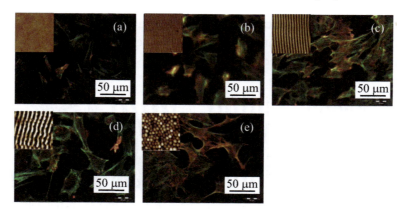

Figure 13.8 Microscopy images of mesenchymal cell–cultured PET films: (a) nonirradiated, (b) irradiated at 193 nm, (c) irradiated at 266 nm, linearly polarized laser at normal incidence, (d) irradiated at 266 nm, linearly polarized, incidence at 45°, and (e) irradiated at 266 nm, circularly polarized. (Insets) AFM height images of the nanostructured samples.

In addition to the enhanced adhesion and proliferation, it has been reported that in LIPSS generated on PS films [54], human embryonic kidney (HEK-293) cells, Chinese hamster ovary (CHO-K1) cells, and skeletal myoblasts align along the direction of the ripples. This alignment is cell-type dependent and occurs only when the period of the nanostructures is above a critical value. Thus, it is found that HEK-293 and CHO-K1 cells align along the direction of the ripples for periods equal to or above 340 nm, rat skeletal myoblasts align with periodicities equal to or above 430 nm, and human myoblasts only align for periodicities equal to or above 270 nm.

13.6.2 Polymer LIPSS for SERS Substrates

Polymer substrates with LIPSS can be used as SERS sensors [85, 86]. To that purpose LIPSS-nanostructured polymer films were coated with a gold layer by PLD. Gold layers of up to 45 nm thickness were deposited, and for all the gold thicknesses the morphology of the gold-coated polymer samples retained the initial periodic relief after PLD coating. The morphology of the gold deposits was characterized by AFM, and the corresponding height images are shown in Fig. 13.9. Figure 13.9a shows both the nonirradiated and irradiated polymer films upon different irradiation conditions in terms of laser wavelength, angle of incidence, and polarization, while Fig. 13.9b shows the same samples after gold coating.

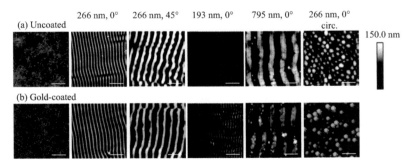

Figure 13.9 AFM height images of PTT nanostructured upon irradiation at different conditions. Substrates (a) before and (b) after a gold coating. The scale bar is 1 µm. Adapted from Ref. [86] with permission from Wiley.

Raman spectra were acquired on drops of aqueous solutions of the model analyte benzenethiol (BT) poured onto the gold-coated polymer substrates and dried in air. The enhancement factor of the Raman signal for the gold-coated substrates is of 8 orders of magnitude as a result of the coupling of localized surface plasmon on the gold nanoparticles. While concentrations of BT of the order 10 M on a silicon substrate give a negligible Raman signal, the use of gold-coated polymer films as substrates allows detection of down to a 10^{-8} M solution of this analyte [85]. An additional enhancement is achieved by the presence of the periodic structures. Figure 13.10 compares the spectra obtained for a BT concentration of 10^{-4} M on the different nanostructured substrates coated with gold by PLD with 12,000 laser pulses of 213 nm and 2 J/cm². The bands observed are readily assigned to the analyte. The largest Raman response is achieved for the substrates prepared by irradiation at 266 nm at an angle of incidence of 45° and by irradiation at 795 nm, where the enhancement factor is 33 and 30. The presence of circular structures increases the signal by a factor of 19, while irradiation at 193 nm only yields an improvement of the Raman signal by a factor of 2.

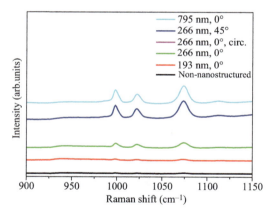

Figure 13.10 SERS spectra of BT at a concentration of 10^{-4} M on gold-coated nonnanostructured and nanostructured PTT substrates prepared using different irradiation conditions. Adapted from Ref. [86] with permission from Wiley.

13.6.3 Polymer LIPSS for Nonvolatile Organic Memory Devices

As described above, (PVDF-TrFE) can be nanostructured using a bilayer approach in which the bottom layer absorbs efficiently at the wavelength used for irradiation. It is also important that the ferroelectric material keep their properties. The ferroelectric response, as evaluated by piezoresponse AFM (PFM), of the LIPSS bilayer film is shown in Fig. 13.11a. The presence of hysteresis in the PFM phase indicates that the films with LIPSS display a ferroelectric response. Evidence of ferroelectricity in the nanostructured layer further supports that the top layer follows the topography of the underlying P3HT film. In addition, since the ferroelectricity of P(VDF-TrFE) depends strongly on its crystalline structure [87], results suggest that laser irradiation did not inhibit the recrystallization of the PVDF-TrFE copolymer.

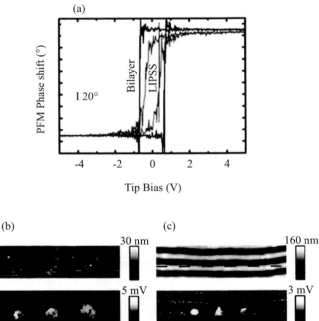

Figure 13.11 (a) PFM phase shift as a function of applied bias for the nonirradiated bilayer and LIPSS on the bilayer film. Topography and PFM amplitude of the bilayer film after poling at three points for 5 minutes at +12 V in the (b) nonirradiated sample and (c) irradiated film. Bottom panels are the amplitudes across single lines indicated in the height images.

PFM has been used to store nanoscale information on ferroelectric organic films [88–90], using the tip as a stylus to write the information by applying a bias above the coercive field. Figure 13.11b shows 5 × 1 µm² PFM images for the nonirradiated sample and Fig. 13.11c the sample with LIPSS. The ferroelectric state of the nonirradiated bilayer film, before and after poling for 5 minutes at a +12 V bias, is observed. PFM amplitude shows the development of ferroelectric contrast in the poled areas. For the nonirradiated samples the dipole orientation caused by the applied tip bias extends from 400 to 800 nm. Thus, considering the spreading of the ferroelectric information in the bilayer film, information storage density can be estimated as around 2 ± 1 Gbit/inch². In contrast, in the sample with LIPSS, the electrical poling generates circular domains of about 100 to 200 nm in diameter, which correspond to an information density of 35 ± 5 Gbit/inch².

13.7 Conclusions

LIPSS with periods close to the laser wavelength used for irradiation, and parallel to the laser polarization direction, are observed in films of strongly absorbing polymers in a narrow range of fluences well below the ablation threshold. LIPSS formation takes place by heating the film surface either above the characteristic glass transition temperature of amorphous polymers or above the melting temperature of semicrystalline ones. A feedback process is involved in LIPSS formation as supported by in situ monitoring of nanostructuring formation using X-ray scattering techniques.

Femtosecond lasers may be used for LIPSS formation, and in this case wavelengths that are weakly absorbed by the polymer also induce nanostructure formation due to the large laser intensities involved and the multiphoton absorption and ionization processes that take place. This allows tuning and controlling of the morphology and size of periodic structures on the surface of polymer films.

Laser-induced periodic surface structures can be prepared on polymer thin films, in spite of the absence of light absorption at the wavelength of the irradiating laser, by using a bilayer approach.

Different types of structures can be obtained by changing the laser parameters. Thus, while the period of the ripples is close to the laser wavelength for irradiation at normal incidence, it increases for irradiation at different angles. Additionally, the use of circularly polarized light gives rise to circular structures with diameters close to the laser wavelength.

Several applications of polymer substrates with LIPSS are proposed and demonstrated. The first application refers to cell culture, which is enhanced for laser nanostructured substrates in comparison to nonirradiated ones. A second application concerns the use of polymer LIPSS as SERS substrates. Coating with gold by PLD has been shown to preserve the relief of the nanostructured polymer films and to ensure good adhesion. Gold coating of the films induces an enhancement of 8 orders of magnitude, while an additional enhancement arises by the presence of LIPSS. The increase of SERS signal depends on the nanostructure size and morphology. The highest enhancement factor caused by nanostructuring of the underlying polymer film is ≥30 for polymer nanostructures obtained by laser irradiation with the highest values of period, depth, and roughness, which is attributed to the larger effective area of interaction of the substrate with the analyte. This approach allows a reduction in the quantity of precious metal needed to prepare SERS substrates. LIPSS-nanostructured ferroelectric copolymers keep their ferroelectric behavior, as evidenced by the hysteresis cycles recorded by PFM. The laser-fabricated nanogratings showed an increase of the storage information density of about 1 order of magnitude in comparison to the nonstructured bilayer.

The LIPSS technique has demonstrated to be advantageous due to its simplicity and versatility. Structures in the range 170–800 nm have been reported. It is important to mention that this laser treatment can easily be extended to large surfaces. The characteristic surface structures formed in this way are being used to tailor a great variety of surface properties and, as shown, display a high potential for applications in different fields.

Acknowledgments

The authors gratefully acknowledge financial support from the MINECO, Spain (CTQ 2013-43086-P, MAT2015-66443-C2-1-R, and CTQ2016-75880-P). E.R. also thanks MINECO for a Ramón y Cajal contract (RYC-2011-08069).

References

1. Aissou, K., Kogelschatz, M., Baron, T., and Gentile, P. (2007). Self-assembled block polymer templates as high resolution lithographic masks, *Surf. Sci.*, **601**, pp. 2611–2614.
2. Liu, G., Xun, S., Vukmirovic, N., Song, X., Olalde-Velasco, P., Zheng, H., Battaglia, V. S., Wang, L., and Yang, W. (2011). Polymers with tailored electronic structure for high capacity lithium battery electrodes, *Adv. Mater.*, **23**, pp. 4679–4683.
3. Hanle, J. E., Merz, E. H., and Mesrobian, R. B. (1966). Polymers in packaging, *J. Polym. Sci. Polym. Symp.*, **12**, pp. 185–195.
4. Schift, H. (2008). Nanoimprint lithography: an old story in modern times? A review, *J. Vac. Sci. Technol. B*, **26**, pp. 458–480.
5. Qin, D., Xia, Y. N., and Whitesides, G. M. (2010). Soft lithography for micro- and nanoscale patterning, *Nat. Protoc.*, **5**, pp. 491–502.
6. Ho, C. C., Chen, P. Y., Lin, K. H., Juan, W. T., and Lee, W. L. (2011). Fabrication of monolayer of polymer/nanospheres hybrid at a water-air interface, *ACS Appl. Mater. Interfaces*, **3**, pp. 204–208.
7. Schift, H. (2012). *NaPANIL-Library of Processes* (NaPANIL consortium, Berlin).
8. Zhang, P., Huang, H.-Y., Chen, Y., Yu, S., Krywka, C., Vayalil, S., Roth, S., and He, T.-B. (2014). Preparation of long-range ordered nanostructures in semicrystalline diblock copolymer thin films using micromolding, *Chin. J. Polym. Sci.*, **32**, pp. 1188–1198.
9. del Campo, A., and Arzt, E. (2008). Fabrication approaches for generating complex micro- and nanopatterns on polymeric surfaces, *Chem. Rev.*, **108**, pp. 911–945.
10. Schaffer, E., Thurn-Albrecht, T., Russell, T. P., and Steiner, U. (2000). Electrically induced structure formation and pattern transfer, *Nature*, **403**, pp. 874–877.
11. Wohlfart, E., Fernandez-Blazquez, J. P., Knoche, E., Bello, A., Perez, E., Arzt, E., and del Campo, A. (2010). Nanofibrillar patterns by plasma etching: the influence of polymer crystallinity and orientation in surface morphology, *Macromolecules*, **43**, pp. 9908–9917.
12. Gong, J. L., Lipomi, D. J., Deng, J. D., Nie, Z. H., Chen, X., Randall, N. X., Nair, R., and Whitesides, G. M. (2010). Micro- and nanopatterning of inorganic and polymeric substrates by indentation lithography, *Nano Lett.*, **10**, pp. 2702–2708.

13. Ruiz, R., Kang, H. M., Detcheverry, F. A., Dobisz, E., Kercher, D. S., Albrecht, T. R., de Pablo, J. J., and Nealey, P. F. (2008). Density multiplication and improved lithography by directed block copolymer assembly, *Science*, **321**, pp. 936–939.

14. Garcia-Gutierrez, M. C., Linares, A., Hernandez, J. J., Rueda, D. R., Ezquerra, T. A., Poza, P., and Davies, R. J. (2010). Confinement-induced one-dimensional ferroelectric polymer arrays, *Nano Lett.*, **10**, pp. 1472–1476.

15. Balan, L., Turck, C., Soppera, O., Vidal, L., and Lougnot, D. J. (2009). Holographic recording with polymer nanocomposites containing silver nanoparticles photogenerated in situ by the interference pattern, *Chem. Mater.*, **21**, pp. 5711–5718.

16. Bityurin, N. (2005). Studies on laser ablation of polymers, *Annu. Rep. Prog. Chem., Sect. C: Phys. Chem.*, **101**, pp. 216–247.

17. Bityurin, N., Luk'yanchuk, B. S., Hong, M. H., and Chong, T. C. (2003). Models for laser ablation of polymers, *Chem. Rev.*, **103**, pp. 519–552.

18. Bolle, M., Lazare, S., Le Blanc, M., and Wilmes, A. (1992). Submicron periodic structures produced on polymer surfaces with polarized excimer laser ultraviolet radiation, *Appl. Phys. Lett.*, **60**, pp. 674–676.

19. Castillejo, M., Rebollar, E., Oujja, M., Sanz, M., Selimis, A., Sigletou, M., Psycharakis, S., Ranella, A., and Fotakis, C. (2012). Fabrication of porous biopolymer substrates for cell growth by UV laser: the role of pulse duration, *Appl. Surf. Sci.*, **258**, p. 9.

20. Yu, F., Li, P., Shen, H., Mathur, S., Lehr, C.-M., Bakowsky, U., and Mücklich, F. (2005). Laser interference lithography as a new and efficient technique for micropatterning of biopolymer surface, *Biomaterials*, **26**, pp. 2307–2312.

21. Thomas, B., Alloncle, A. P., Delaporte, P., Sentis, M., Sanaur, S., Barret, M., and Collot, P. (2007). Experimental investigations of laser-induced forward transfer process of organic thin films, *Appl. Surf. Sci.*, **254**, pp. 1206–1210.

22. Rebollar, E., Gaspard, S., Oujja, M., Villavieja, M. M., Corrales, T., Bosch, P., Georgiou, S., and Castillejo, M. (2006). Pulsed laser deposition of polymers doped with fluorescent molecular sensors, *Appl. Phys. A*, **84**, pp. 171–180.

23. Pique, A., McGill, R. C. R. A., Chrisey, D. B., Callahan, J., and Mlsna, T. E. (1998). Matrix assisted pulsed laser evaporation (MAPLE) of polymeric materials: methodology and mechanistic studies, *Mater. Res. Soc. Symp. Proc.*, **526**, pp. 375–383.

24. Yong-Feng, L., Bing, H., Zhi-Hong, M., Wei-Jie, W., Wai-Kin, C., and Tow-Chong, C. (2001). Laser-scanning probe microscope based nanoprocessing of electronics materials, *Jpn. J. Appl. Phys.*, **40**, p. 4395.

25. Chong, T. C., Hong, M. H., and Shi, L. P. (2010). Laser precision engineering: from microfabrication to nanoprocessing, *Laser Photonics Rev.*, **4**, pp. 123–143.

26. Khan, A., Wang, Z., Sheikh, M. A., Whitehead, D. J., and Li, L. (2011). Laser micro/nano patterning of hydrophobic surface by contact particle lens array, *Appl. Surf. Sci.*, **258**, pp. 774–779.

27. Rebollar, E., Cordero, D., Martins, A., Chiussi, S., Reis, R. L., Neves, N. M., and León, B. (2011). Improvement of electrospun polymer fiber meshes pore size by femtosecond laser irradiation, *Appl. Surf. Sci.*, **257**, pp. 4091–4095.

28. Chichkov, B. N., Momma, C., Nolte, S., von Alvensleben, F., and Tünnermann, A. (1996). Femtosecond, picosecond and nanosecond laser ablation of solids, *Appl. Phys. A*, **63**, pp. 109–115.

29. Liu, X., Du, D., and Mourou, G. (1997). Laser ablation and micromachining with ultrashort laser pulses, *IEEE J. Quantum Electron.*, **33**, pp. 1706–1716.

30. Englert, L., Wollenhaupt, M., Haag, L., Sarpe-Tudoran, C., Rethfeld, B., and Baumert, T. (2008). Material processing of dielectrics with temporally asymmetric shaped femtosecond laser pulses on the nanometer scale, *Appl. Phys. A*, **92**, pp. 749–753.

31. Rebollar, E., Mildner, J., Götte, N., Otto, D., Sarpe, C., Köhler, J., Wollenhaupt, M., Baumert, T., and Castillejo, M. (2014). Microstructuring of soft organic matter by temporally shaped femtosecond laser pulses, *Appl. Surf. Sci.*, **302**, pp. 231–235.

32. Oujja, M., Pérez, S., Fadeeva, E., Koch, J., Chichkov, B. N., and Castillejo, M. (2009). Three dimensional microstructuring of biopolymers by femtosecond laser irradiation, *Appl. Phys. Lett.*, **95**, p. 263703.

33. Farsari, M., and Chichkov, B. N. (2009). Materials processing: two-photon fabrication, *Nat. Photonics*, **3**, pp. 450–452.

34. Paquet, C., and Kumacheva, E. (2008). Nanostructured polymers for photonics, *Mater. Today*, **11**, pp. 48–56.

35. Li, Y., Duan, G., Liu, G., and Cai, W. (2013). Physical processes-aided periodic micro/nanostructured arrays by colloidal template technique: fabrication and applications, *Chem. Soc. Rev.*, **42**, pp. 3614–3627.

36. Luong-Van, E., Rodriguez, I., Low, H. Y., Elmouelhi, N., Lowenhaupt, B., Natarajan, S., Lim, C. T., Prajapati, R., Vyakarnam, M., and Cooper, K.

(2013). Review: micro- and nanostructured surface engineering for biomedical applications, *J. Mater. Res.*, **28**, pp. 165–174.

37. Birnbaum, M. (1965). Semiconductor surface damage produced by ruby lasers, *J. Appl. Phys.*, **36**, pp. 3688–3689.

38. Csete, M., and Bor, Z. (1998). Laser-induced periodic surface structure formation on polyethylene-terephthalate, *Appl. Surf. Sci.*, **133**, pp. 5–16.

39. Bolle, M., and Lazare, S. (1993). Characterization of submicrometer periodic structures produced on polymer surfaces with low-fluence ultraviolet laser radiation, *J. Appl. Phys.*, **73**, pp. 3516–3524.

40. Prendergast, U., Kudzma, S., Sherlock, R., O'Connell, C., and Glynn, T. (2007). TEM investigation of laser-induced periodic surface structures on polymer surfaces, *Proc. SPIE*, **6458**, p. 64581V.

41. Reif, J., Varlamova, O., Ratzke, M., Schade, M., Leipner, H., and Arguirov, T. (2010). Multipulse feedback in self-organized ripples formation upon femtosecond laser ablation from silicon, *Appl. Phys. A*, **101**, pp. 361–365.

42. Bonse, J., and Kruger, J. (2010). Pulse number dependence of laser-induced periodic surface structures for femtosecond laser irradiation of silicon, *J. Appl. Phys.*, **108**, pp. 034903–034905.

43. Csete, M., Hild, S., Plettl, A., Ziemann, P., Bor, Z., and Marti, O. (2004). The role of original surface roughness in laser-induced periodic surface structure formation process on poly-carbonate films, *Thin Solid Films*, **453–454**, pp. 114–120.

44. Li, M., Lu, Q. H., Yin, J., Sui, Y., Li, G., Qian, Y., and Wang, Z. G. (2002). Periodic microstructure induced by 532 nm polarized laser illumination on poly(urethane-imide) film: orientation of the azobenzene chromophore, *Appl. Surf. Sci.*, **193**, pp. 46–51.

45. Sanz, M., Rebollar, E., Ganeev, R. A., and Castillejo, M. (2013). Nanosecond laser-induced periodic surface structures on wide band-gap semiconductors, *Appl. Surf. Sci.*, **278**, pp. 325–329.

46. Lazare, S., and Benet, P. (1993). Surface amorphization of Mylar[sup(R)] films with the excimer laser radiation above and below ablation threshold: ellipsometric measurements, *J. Appl. Phys.*, **74**, pp. 4953–4957.

47. Csete, M., Eberle, R., Pietralla, M., Marti, O., and Bor, Z. (2003). Attenuated total reflection measurements on poly-carbonate surfaces structured by laser illumination, *Appl. Surf. Sci.*, **208–209**, pp. 474–480.

48. Rebollar, E., Pérez, S., Hernández, J. J., Martín-Fabiani, I., Rueda, D. R., Ezquerra, T. A., and Castillejo, M. (2011). Assessment and formation mechanism of laser-induced periodic surface structures on polymer spin-coated films in real and reciprocal space, *Langmuir*, **27**, pp. 5596–5606.
49. Martín-Fabiani, I., Rebollar, E., Pérez, S., Rueda, D. R., García-Gutiérrez, M. C., Szymczyk, A., Roslaniec, Z., Castillejo, M., and Ezquerra, T. A. (2012). Laser-induced periodic surface structures nanofabricated on poly(trimethylene terephthalate) spin-coated films, *Langmuir*, **28**, pp. 7938–7945.
50. Rebollar, E., R. Vazquez de Aldana, J., Perez-Hernandez, J. A., Ezquerra, T. A., Moreno, P., and Castillejo, M. (2012). Ultraviolet and infrared femtosecond laser induced periodic surface structures on thin polymer films, *Appl. Phys. Lett.*, **100**, p. 041106.
51. Rebollar, E., Vazquez de Aldana, J. R., Martin-Fabiani, I., Hernandez, M., Rueda, D. R., Ezquerra, T. A., Domingo, C., Moreno, P., and Castillejo, M. (2013). Assessment of femtosecond laser induced periodic surface structures on polymer films, *Phys. Chem. Chem. Phys.*, **15**, pp. 11287–11298.
52. Rebollar, E., Perez, S., Hernandez, M., Domingo, C., Martin, M., Ezquerra, T. A., Garcia-Ruiz, J. P., and Castillejo, M. (2014). Physicochemical modifications accompanying UV laser induced surface structures on poly(ethylene terephthalate) and their effect on adhesion of mesenchymal cells, *Phys. Chem. Chem. Phys.*, **16**, pp. 17551–17559.
53. Slepicka, P., Rebollar, E., Heitz, J., and Svorcik, V. (2008). Gold coatings on polyethyleneterephthalate nano-patterned by F-2 laser irradiation, *Appl. Surf. Sci.*, **254**, pp. 3585–3590.
54. Rebollar, E., Frischauf, I., Olbrich, M., Peterbauer, T., Hering, S., Preiner, J., Hinterdorfer, P., Romanin, C., and Heitz, J. (2008). Proliferation of aligned mammalian cells on laser-nanostructured polystyrene, *Biomaterials*, **29**, pp. 1796–1806.
55. Pérez, S., Rebollar, E., Oujja, M., Martín, M., and Castillejo, M. (2013). Laser-induced periodic surface structuring of biopolymers, *Appl. Phys. A*, **110**, pp. 683–690.
56. Bäuerle, D. (2000). *Laser Processing and Chemistry* (Springer-Verlag, Berlin).
57. Bolle, M., and Lazare, S. (1993). Submicron periodic structures produced on polymer surfaces with polarized excimer laser ultraviolet radiation, *Appl. Surf. Sci.*, **65–66**, pp. 349–354.

58. Niino, H., Kawabata, Y., and Yabe, A. (1989). Application of excimer laser polymer ablation to alignment of liquid crystals: periodic microstructure on polyethersulfone, *Jpn. J. Appl. Phys.*, **28**, pp. L2225–L2227.

59. Lochbihler, H. (2009). Colored images generated by metallic subwavelength gratings, *Opt. Express*, **17**, pp. 12189–12196.

60. Rodríguez-Rodríguez, Á., Rebollar, E., Soccio, M., Ezquerra, T. A., Rueda, D. R., Garcia-Ramos, J. V., Castillejo, M., and Garcia-Gutierrez, M.-C. (2015). Laser-induced periodic surface structures on conjugated polymers: poly(3-hexylthiophene), *Macromolecules*, **48**, pp. 4024–4031.

61. Rebollar, E., Castillejo, M., and Ezquerra, T. A. (2015). Laser induced periodic surface structures on polymer films: from fundamentals to applications, *Eur. Polym. J.*, **73**, pp. 162–174.

62. Mate, C. M., Toney, M. F., and Leach, K. A. (2001). Roughness of thin perfluoropolyether lubricant films: influence on disk drive technology, *IEEE Trans. Magn.*, **37**, pp. 1821–1823.

63. Rebollar, E., Rueda, D. R., Martín-Fabiani, I., Rodríguez-Rodríguez, Á., García-Gutiérrez, M.-C., Portale, G., Castillejo, M., and Ezquerra, T. A. (2015). In situ monitoring of laser-induced periodic surface structures formation on polymer films by grazing incidence small-angle x-ray scattering, *Langmuir*, **31**, pp. 3973–3981.

64. Schwartzkopf, M., Buffet, A., Korstgens, V., Metwalli, E., Schlage, K., Benecke, G., Perlich, J., Rawolle, M., Rothkirch, A., Heidmann, B., Herzog, G., Muller-Buschbaum, P., Rohlsberger, R., Gehrke, R., Stribeck, N., and Roth, S. V. (2013). From atoms to layers: in situ gold cluster growth kinetics during sputter deposition, *Nanoscale*, **5**, pp. 5053–5062.

65. Baudach, S., Bonse, J., and Kautek, W. (1999). Ablation experiments on polyimide with femtosecond laser pulses, *Appl. Phys. A: Mater. Sci. Process.*, **69**, pp. S395–S398.

66. Forster, M., Kautek, W., Faure, N., Audouard, E., and Stoian, R. (2011). Periodic nanoscale structures on polyimide surfaces generated by temporally tailored femtosecond laser pulses, *Phys. Chem. Chem. Phys.*, **13**, pp. 4155–4158.

67. Krüger, J., and Kautek, W. (2004). Ultrashort Pulse Laser Interaction with Dielectrics and Polymers, in *Polymers and Light*, ed. Lippert, T. K. (Springer Berlin / Heidelberg), pp. 247–290.

68. Baudach, S., Krüger, J., and Kautek, W. (2001). Femtosecond laser processing of soft materials, *Rev. Laser Eng.*, **29**, pp. 705–709.

69. Guosheng, Z., Fauchet, P. M., and Siegman, A. E. (1982). Growth of spontaneous periodic surface structures on solids during laser illumination, *Phys. Rev. B*, **26**, pp. 5366–5381.
70. Sipe, J. E., Young, J. F., Preston, J. S., and van Driel, H. M. (1983). Laser-induced periodic surface structure. I. Theory, *Phys. Rev. B*, **27**, pp. 1141.
71. Barborica, A., Mihailescu, I. N., and Teodorescu, V. S. (1994). Dynamical evolution of the surface microrelief under multiple-pulse-laser irradiation: an analysis based on surface-scattered waves, *Phys. Rev. B*, **49**, pp. 8385–8395.
72. Emmony, D. C., Howson, R. P., and Willis, L. J. (1973). Laser mirror damage in germanium at 10.6 μm, *Appl. Phys. Lett.*, **23**, pp. 598–600.
73. Clark, S. E., and Emmony, D. C. (1989). Ultraviolet-laser-induced periodic surface structures, *Phys. Rev. B*, **40**, pp. 2031–2041.
74. Young, J. F., Preston, J. S., van Driel, H. M., and Sipe, J. E. (1983). Laser-induced periodic surface structure. II. Experiments on Ge, Si, Al, and brass, *Phys. Rev. B*, **27**, p. 1155.
75. Bestehorn, M., Pototsky, A., and Thiele, U. (2003). 3D large scale Marangoni convection in liquid films, *Eur. Phys. J. B*, **33**, pp. 457–467.
76. Costache, F., Henyk, M., and Reif, J. (2002). Modification of dielectric surfaces with ultra-short laser pulses, *Appl. Surf. Sci.*, **186**, pp. 352–357.
77. Derrien, T. J., Sarnet, T., Sentis, M., and Itina, T. E. (2010). Application of a two-temperature model for the investigation of the periodic structure formation on Si surface in femtosecond laser interactions, *J. Optoelectron. Adv. Mater.*, **12**, pp. 610–615.
78. Vogel, A., Noack, J., Hüttman, G., and Paltauf, G. (2005). Mechanisms of femtosecond laser nanosurgery of cells and tissues, *Appl. Phys. B: Lasers Opt.*, **81**, pp. 1015–1047.
79. Bonse, J., Munz, M., and Sturm, H. (2005). Structure formation on the surface of indium phosphide irradiated by femtosecond laser pulses, *J. Appl. Phys.*, **97**, pp. 013538–013539.
80. Young, J. F., Sipe, J. E., and van Driel, H. M. (1984). Laser-induced periodic surface structure. III. Fluence regimes, the role of feedback, and details of the induced topography in germanium, *Phys. Rev. B*, **30**, p. 2001.
81. Ashkenasi, D., Rosenfeld, A., Varel, H., Wähmer, M., and Campbell, E. E. B. (1997). Laser processing of sapphire with picosecond and sub-picosecond pulses, *Appl. Surf. Sci.*, **120**, pp. 65–80.

82. Kalachyova, Y., Lyutakov, O., Slepicka, P., Elashnikov, R., and Svorcik, V. (2014). Preparation of periodic surface structures on doped poly(methyl metacrylate) films by irradiation with KrF excimer laser, *Nanoscale Res. Lett.*, **9**, pp. 1–10.

83. Martínez-Tong, D. E., Rodríguez-Rodríguez, Á., Nogales, A., García-Gutiérrez, M.-C., Pérez-Murano, F., Llobet, J., Ezquerra, T. A., and Rebollar, E. (2015). Laser fabrication of polymer ferroelectric nanostructures for nonvolatile organic memory devices, *ACS Appl. Mater. Interfaces*, **7**, pp. 19611–19618.

84. Varlamova, O., Reif, J., Varlamov, S., and Bestehorn, M. (2011). The laser polarization as control parameter in the formation of laser-induced periodic surface structures: comparison of numerical and experimental results, *Appl. Surf. Sci.*, **257**, pp. 5465–5469.

85. Rebollar, E., Sanz, M., Perez, S., Hernandez, M., Martin-Fabiani, I., Rueda, D. R., Ezquerra, T. A., Domingo, C., and Castillejo, M. (2012). Gold coatings on polymer laser induced periodic surface structures: assessment as substrates for surface-enhanced Raman scattering, *Phys. Chem. Chem. Phys.*, **14**, pp. 15699–15705.

86. Rebollar, E., Hernández, M., Sanz, M., Pérez, S., Ezquerra, T. A., and Castillejo, M. (2015). Laser-induced surface structures on gold-coated polymers: influence of morphology on surface-enhanced Raman scattering enhancement, *J. Appl. Polym. Sci.*, **132**, p. 42770.

87. Naber, R. C. G., Asadi, K., Blom, P. W. M., de Leeuw, D. M., and de Boer, B. (2010). Organic nonvolatile memory devices based on ferroelectricity, *Adv. Mater.*, **22**, pp. 933–945.

88. Oh, S., Kim, Y., Choi, Y.-Y., Kim, D., Choi, H., and No, K. (2012). Fabrication of vertically well-aligned P(VDF-TrFE) nanorod arrays, *Adv. Mater.*, pp. 5708–5712.

89. Zhu, L., and Wang, Q. (2012). Novel ferroelectric polymers for high energy density and low loss dielectrics, *Macromolecules*, **45**, pp. 2937–2954.

90. Martínez-Tong, D. E., Soccio, M., García-Gutiérrez, M. C., Nogales, A., Rueda, D. R., Alayo, N., Pérez-Murano, F., and Ezquerra, T. A. (2013). Improving information density in ferroelectric polymer films by using nanoimprinted gratings, *Appl. Phys. Lett.*, **102**, p. 191601.

Chapter 14

Laser Materials Processing for Energy Storage Applications

Heungsoo Kim,[a] Peter Smyrek,[b,c] Yijing Zheng,[b,c] Wilhelm Pfleging,[b,c] and Alberto Piqué[a]

[a]*Materials Science & Technology Division, Naval Research Laboratory, Washington, DC 20375, USA*
[b]*Karlsruhe Institute of Technology, IAM-AWP, PO Box 3640, 76021 Karlsruhe, Germany*
[c]*Karlsruhe Nano Micro Facility, Hermann-von-Helmholtz-Platz 1, 76344 Egg.-Leopoldshafen, Germany*
Heungsoo.Kim@nrl.navy.mil; Wilhelm.Pfleging@kit.edu

This chapter will review the use of laser-based material processing techniques, such as pulsed laser deposition (PLD), laser-induced forward transfer (LIFT), and material processing via 3D laser structuring (LS) and laser annealing (LA) techniques for energy storage applications. PLD is a powerful tool for fabricating high-quality layers of materials for cathodes, anodes, and solid electrolytes for thin-film microbatteries. LIFT is a versatile technique for printing complex materials with highly porous structures for the fabrication of micropower sources, such as ultracapacitors and thick-film batteries. LS is a recently developed technique for modifying the

---

Pulsed Laser Ablation: Advances and Applications in Nanoparticles and Nanostructuring Thin Films
Edited by Ion N. Mihailescu and Anna Paola Caricato
Copyright © 2018 Pan Stanford Publishing Pte. Ltd.
ISBN 978-981-4774-23-9 (Hardcover), 978-1-315-18523-1 (eBook)
www.panstanford.com

active material by forming advanced 3D electrode architectures and increasing the overall active surface area. LA is a rapid technology for adjusting the crystalline battery phase and for controlling the grain size on the micro- and nanoscale. This chapter will review recent work using these laser processing techniques for the fabrication of micropower sources and lessons learned from the characterization of their electrochemical properties.

14.1 Introduction

In the past two decades, there has been significant progress toward developing smaller-scale and portable electronic devices, such as autonomous microelectronic sensors, microelectromechanical systems (MEMS), and portable personal electronics. However, the miniaturization of the power sources required to operate these microscale devices has shown little progress due to their volumetric 3D characteristics, limiting the ability of these devices to function autonomously. Therefore, the development of micropower systems and small-scale energy storage systems is essential to capitalize on the advantages of miniaturized devices. A lot of techniques have been applied to fabricate microbatteries, such as thin-film, stretchable, and 3D batteries [1–4]. Among the many techniques investigated to date, laser-based processes, such as pulsed laser deposition (PLD), laser-induced forward transfer (LIFT), and laser structuring (LS), are useful approaches to integrating various types of micropower sources, such as thin-film microbatteries, ultracapacitors, and solar cells for various microdevices. For the purposes of this review, micropower sources are defined as miniaturized electrochemical cells, such as Li ion batteries (LIBs) and ultracapacitors, where the total footprint area of the devices is in the mm^2 to cm^2 range, the thickness of the active layers is less than 10 μm (for thin-film batteries) to 10–100 μm (for thick-film batteries), and the total mass of active material does not exceed 10 g.

The power demand for micropower sources is largely dependent on their anticipated applications. For example, a few microwatts of power is required for applications, such as MEMS, microelectronic sensors, and some small personal electronics [5], while 1–100 μW of power is consumed by radio frequency identification (RFID) tags

[6]. To meet the power demand for certain types of microdevices, a combination of different types of power sources is required [7]. For example, wireless sensors consume very little power (<1 mW) in stand-by mode, while they consume many millwatts of power in data collection mode and several hundreds of millwatts are required to transmit the data to a remote system. In this example, a Li ion microbattery can deliver the constant low power (a few millwatts), an ultracapacitor can deliver the short burst of high power, and the power sources can be recharged by energy-harvesting modules during stand-by mode. As this chapter will show, laser processing techniques are ideally suited to prototype and fabricate these types of electrochemical energy storage components for hybrid micropower systems.

Recently, microbatteries with 3D architectures have been suggested to enhance the power density, while maintaining the high energy density of these systems [8-10]. These proposed 3D configurations can now be realized using laser-based processes to modify the active materials found in 3D Li ion microbatteries. Furthermore, laser processing techniques can also directly embed the electrochemical components into the device package, leading to a reduction in size and weight of the entire system [11]. Thus, laser-based processing techniques are promising approaches for developing micropower sources for microelectronic devices. This chapter will present a brief review of PLD techniques for thin-film microbatteries, LIFT techniques for ultracapacitors and thick-film microbatteries, LA for controlling the grain size and crystalline phase of thin-film electrodes, and LS processes for modifying active electrode materials for Li ion microbatteries. Finally, we will conclude with a discussion of challenges and future directions offered by the use of lasers to develop the next generation of microbatteries.

14.2 Background and Overview of Materials for Energy Storage

The two main electrochemical energy storage systems used in microscale devices are ultracapacitors and microbatteries. Ultracapacitors, also known as supercapacitors, are electrochemical energy storage devices with high power densities that can be fully charged and discharged in very short periods of time [12]. Although the energy density of an ultracapacitor is lower than that of a microbattery, it can deliver the stored energy in a very short time and can be charged/discharged millions of times without losing its energy storage capacity. Ultracapacitors can be classified into two possible classes on the basis of their charge storage mechanisms or the type of electrode materials. The first category is electrochemical double-layer capacitors, which can store charge electrostatically on the electrode–electrolyte interface. Carbon materials are commonly used in these types of ultracapacitors because they provide high surface areas and accordingly large amounts of charge storage in the system. The second category is pseudocapacitors or redox capacitors, which store charge through surface redox reactions. Transition metal oxides are typical examples of pseudocapacitors [12]. Both types of ultracapacitors show typical capacitor behavior: at constant current, the voltage across the device will decrease linearly with time.

Rechargeable LIBs have been considered the most promising energy storage system due to their highest energy per unit weight within the known energy storage systems. The commonly used cathode materials in LIBs are transition metal oxides or phosphates ($LiCoO_2$ [lithium cobalt oxide, LCO], $LiMn_2O_4$ [lithium manganese oxide, LMO], $LiNi_{1/3}Mn_{1/3}Co_{1/3}O_2$ [NMC], $LiNi_{1/3}Co_{1/3}Al_{1/3}O_2$, and $LiFePO_4$ [LFP]), while graphite is the most commonly used anode material. Both electrodes are separated by a polypropylene membrane filled with electrolyte that contains lithium salts, such as $LiPF_6$. The separator blocks the electrical contact between the electrodes, while it allows the diffusion of Li ions during the charging/discharging processes. Figure 14.1 provides a graphical comparison of the performance of various active electrode materials used in LIBs [13]. In general, the higher the specific capacity, the better, while higher potentials allow for higher power operation.

Although graphite is the most commonly used anode material in LIBs due to its low working potential, good cycle life, and low cost, it allows only one Li ion with six carbon atoms for intercalation. Thus, its low reversible capacity (372 mAh g^{-1}) cannot meet the high capacity demand of many electronic devices. There have been many efforts searching for new anode materials for LIBs. Among them, silicon has been considered the most attractive anode material due to its high theoretical capacity ($Li_{22}Si_5$, 4000 mAh/g) and its abundance [14]. However, its application suffers from the significant inhomogeneous volume expansion in silicon during insertion and extraction of Li ions. This volume expansion quickly destroys the electrical contacts in the Si anode after a few initial cycles. Therefore, it is necessary to develop ways to use Si while maintaining its conductivity. Some efforts to overcome these limitations using laser-based processes will be discussed in this chapter.

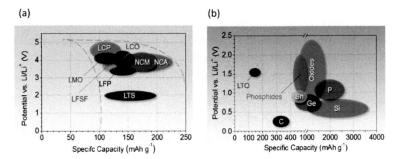

Figure 14.1 Discharge potentials versus specific capacity of some of the most commonly used (a) cathode and (b) anode active materials. The acronyms represent the following: LCO is for $LiCoO_2$, LMO for $LiMn_2O_4$, NCM for $LiNi_{1/3}Co_{1/3}Mn_{1/3}O_2$, NCA for $LiNi_{1/3}Co_{1/3}Al_{1/3}O_2$, LFP for $LiFePO_4$, and LTS for $Li_4Ti_5S_{12}$. Reprinted from Ref. [13], Copyright (2015), with permission from Elsevier.

14.3 Growth of Energy Storage Materials by Pulsed Laser Deposition

Figure 14.2 shows a schematic illustration of the basic components for a PLD setup. A UV excimer laser, such as ArF (193 nm), KrF (248 nm), or XeCl (308 nm), is utilized with a pulse width of tens of nanoseconds or shorter. Other types of lasers, such as the various harmonics of a Nd:YAG laser, have also been employed for PLD. The PLD process is very simple. A pulsed laser beam is used to evaporate material from a target, forming a thin film on a substrate, retaining the target composition. The high-power laser beam is focused inside a vacuum chamber onto the target typically at an incident angle of 45°. The target is typically rotated in order to avoid its fast deterioration and keep its surface under approximately constant conditions. The ablation of the target produces a visible plasma plume that expands from the target to the substrate. The target-to-substrate distance is in the range of 4–10 cm. This process can be performed in any kind of inert or reactive atmosphere and also performed at a dynamic range of pressures from ultrahigh vacuum (UHV) to high pressures (typically up to 1 torr). An important feature of the PLD technique is its capability to deposit multilayer thin films using a multitarget carousel system. The laser beam can be alternatively focused over two or more targets, inducing a sequential ablation process. This configuration is well suited for manufacturing complex stacks, such as typical layered battery structures.

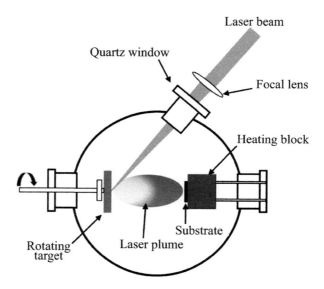

Figure 14.2 A schematic illustration of the pulsed laser deposition setup.

Another advantage of PLD is that the deposited films have compositions close to the targets. This is a result of the extremely high heating rates absorbed by the target surface due to its strong interaction with the laser irradiation. At low laser fluences, the laser pulse simply heats the target and evaporates the target species. In this case, the evaporative flux from a multicomponent target can be determined by the vapor pressures of its constituents. As the laser fluence is raised above the ablation threshold where the absorbed laser energy becomes higher than that needed for evaporation, the flux of vaporized species is independent of the vapor pressures of its constituents. This nonequilibrium process induces the congruent evaporation of the target. This stoichiometry transfer between target and substrate is difficult to achieve with other physical vapor deposition techniques, such as sputtering.

14.3.1 PLD of Cathodes

The PLD technique has been used to grow various cathode materials, such as LCO and LMO for Li ion microbatteries. An ideal thin-film cathode in a Li ion microbattery is a layered structure of lithiated oxides with a well-crystallized structure. The film crystallinity can be improved by postdeposition annealing, which is often required at relatively high temperatures. However, this high-temperature annealing process may produce microcracks in the cathode films, resulting in a nonuniform surface morphology of the solid electrolyte in the sequential deposition step and consequently a shorting problem in the thin-film microbatteries. Therefore, it would be great if a crystallized LCO thin-film cathode could be deposited without a high-temperature annealing step. One of the important features of PLD is that the kinetic energy (10 to 100 eV) of the ejected species in the laser-induced plasma plume is much higher than the kinetic energy of those from other physical vapor deposition techniques, such as sputtering (5 to 10 eV) and evaporation (<1 eV) [15–17]. Thus, LCO cathode films grown by the PLD technique can be crystallized at relatively low temperatures (300°C–600°C) without any postdeposition annealing, compared to other physical vapor deposition techniques (600°C–800°C) [18–21]. It was reported that PLD-grown LCO thin films are single-phase and well-crystallized structures with a high density and smooth surfaces, without cracks, essential for high performance in thin-film batteries [22].

The crystallographic orientation of the cathode thin film is another key parameter that affects the electrochemical activity. In PLD, the high supersaturation of the ablated flux creates 2D island nucleation on the film surface, which is suitable for layer-by-layer growth mode to form highly oriented thin films [16]. In PLD of the LCO cathode thin films, the grains with a (003) preferred orientation tend to reduce the Li ion diffusivity, while the grains with (101) and (104) orientations increase the Li ion diffusion [23]. The substrate is also an important parameter to determine the microstructure of the cathode films. LiNi$_{0.5}$Mn$_{0.5}$O$_2$ cathode thin films were grown by PLD on two different substrates (stainless steel and Au) [24]. The films grown on stainless-steel substrates formed a spinel structure, whereas the films grown on Au substrates showed a layered structure. The reversible capacity of the Li/LiNi$_{0.5}$Mn$_{0.5}$O$_2$ battery on the Au substrate showed ~150 mAh/g (between 2.5 and 4.3 V), which is close to the theoretical capacity (~160 mAh/g) and is much higher than that of the stainless-steel substrate (~40 mAh/g between 2.5 and 4.5 V).

Although one of advantages of PLD is that the stoichiometry of the films is normally close to that of the targets, it is not the case when the targets contain volatile light elements, such as lithium [25]. In the case of lithium-containing targets, the lithium content in the deposited films tends to be lower than that in the targets because these light elements are easily scattered by background gas molecules or other species in the plasma. This Li-content deficiency in the deposited film leads to structural changes and impurity formation, which will eventually degrade the battery performance. This problem can be resolved by using targets containing excess Li$_2$O [18, 23, 25]. Simmen et al. reported that LMO cathode thin films with a desired lithium content can be achieved using a composite target of Li$_{1.03}$Mn$_2$O$_4$ and 7.5 mol% Li$_2$O [25] with a discharge capacity of ~42 µAh/cm^2 µm. It was also reported that textured LCO cathode thin films were successfully prepared by a LCO target containing 15% excess Li$_2$O and their batteries showed reasonably good electrochemical performance (50–60 µAh/µm cm^2) [23].

As discussed before, crystallinity, smooth surface morphology, composition, and preferred orientation of the cathode films are very important conditions to produce high-quality cathode thin films. Table 14.1 shows electrochemical properties of thin-film cathodes prepared by PLD reported in the literature.

Table 14.1 Electrochemical properties of thin-film cathodes and anodes prepared by PLD

Material	Substrate	Voltage/Current	Capacity	References
Cathode				
$LiCoO_2$	SiO_2/Si	3.0–4.2 V/ 15 μA/cm²	64 μAh/μm cm²	[20]
$LiCoO_2$	$SnO_2/glass$	3.8–4.2 V/ 5 μA	89 mAh/g	[26]
$LiCo_{0.5}Al_{0.5}O_2$	SnO_2-glass	3.8–4.2 V/ 5 μA	30 mAh/g	[26]
$LiNi_{0.8}Co_{0.2}O_2$	-	3.0–4.3 V/ 10 μA/cm²	125 mAh/g	[27]
$LiNi_{0.8}Co_{0.15}Al_{0.05}O_2$	Si, Ni	2.5–4.2 V/ ~5 μA/cm²	98 μAh/μm cm²	[28]
$LiFePO_4$-C	Pt/Si	2.0–4.0 V/ ~8 μA/cm²	20 μAh/μm cm²	[29]
$Li_{1.13}Mn_2O_{3.73}$	SS*	3.5–4.4 V/ 2C rate	30 μAh/μm cm²	[25]
$LiNi_{0.5}Mn_{0.5}O_2$	Au	2.5–4.3 V/ 2 μA/cm²	150 mAh/g	[24]
$LiNi_{0.5}Mn_{0.5}O_2$	SS	2.5–4.5 V/ 2 μA/cm²	40 mAh/g	[24]
$Li_{1.2}Mn_{0.54}Ni_{0.13}Co_{0.13}O_2$	Au	2–4.8 V/ 2 μA/cm²	70 μAh/μm cm²	[30]
Anode				
Si-C	Cu	0.05–1.5V/ 54 μA/cm²	70 μAh/cm²	[31]
Si-graphene	Ni	0.05–1V/ C/5 rate	2400 mAh/g	[32]
a-Si	Si, SS	0.01–1.5V/ 100 μA/cm²	64 μAh/cm²	[23]
a-SnO	-	1.5–2.7V/ 100 μA/cm²	4–10 μAh/cm²	[33]

*SS: Stainless steel

14.3.2 PLD of Anodes

Although graphitized carbon is the most commonly used anode material in LIBs, its low theoretical capacity (i.e., LiC_6, 372 mAh/g) cannot meet the capacity demand of batteries in many electronic devices. There have been many research efforts on searching for new anode materials for LIBs. Among them, silicon has been considered as the most attractive anode material due to its high theoretical capacity ($Li_{22}Si_5$, 4000 mAh/g) and its abundance [34]. However, its application suffers from the significant inhomogeneous volume expansion in silicon during insertion and extraction of Li ions, leading to fast capacity fading. Amorphous silicon (a-Si) material has been studied as a promising anode for LIBs due to its homogeneous volume expansion during lithium insertion and accordingly improved cycling performance [35]. The a-Si anode thin films were successfully grown on stainless-steel substrates at room temperature by PLD (KrF excimer laser, 248 nm) [23]. The 120 nm thick a-Si anode thin films showed a reversible capacity of 64 μAh/cm^2 between 0.01 and 1.5 V, with a stable cycling behavior for 50 cycles, with a small fade rate in the capacity of 0.2% per cycle. The a-Si/LCO cell exhibited a stable discharge capacity of ~20 μAh/cm^2 between 1 and 4 V for 20 cycles [23]. Table 14.1 shows some of the electrochemical properties of a-Si anode thin films. The film thickness also affects the cycle life of the a-Si anode thin film. As the film thickness increases, capacity fading occurs faster due to an increased Li ion diffusion length.

PLD was also utilized to deposit thin Si layers on a multilayer graphene (MLG)-coated Ni foam substrate to build a Si-MLG composite anode for LIBs [32]. The MLG layer, first grown by chemical vapor deposition (CVD), serves as a conducting platform, thereby preventing contact loss from volume expansion during charging/discharging processes for the Si anode. The Ni foam substrate serves as a current collector and provides a larger surface area compared to thin metal foils. The cells based on the Si-MLG anodes displayed a stable capacity of ~2400 mAh/g during the cycling test, whereas the cells on the pure Si anodes showed a higher capacity during the first cycle (~2800 mAh/g) but their capacity faded rapidly during subsequent cycles. Thus, this combination of silicon and graphene offers an alternative route for reducing volume expansion issues on

Si anodes. Recently, Biserni et al. fabricated composite Si-C anodes by depositing nanostructured porous a-Si films by PLD at room temperature, followed by CVD of a thin carbon coating [31]. The mesoporosity of a nanostructured Si film helps to reduce the volume expansion issues, while the thin CVD-grown carbon layer helps to promote the formation of a stable solid electrolyte interface (SEI) layer and protect the Si from direct contact with the electrolyte. The cells based on these composite Si-C anodes exhibited a capacity of ~75 μAh/cm² after 1000 cycles at a current rate of 540 μA/cm², indicating their excellent capacity retention and cycle performances (see Fig. 14.3).

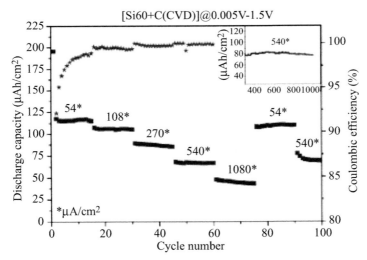

Figure 14.3 Curves of discharge capacity (red squares) and coulombic efficiency (green stars) of Si-C anodes. Integer numbers within the graphs represent the constant current rates (μA cm⁻²) applied during test. (Inset) Cycle performance of the Si-C anode at a constant current rate of 540 μA cm⁻² for 1000 cycles. Reprinted from Ref. [31], Copyright (2015), with permission from Elsevier.

In addition, PLD can be used to generate more complex Si anodes comprising multilayer stacks, sandwiched structures, or layers with a controlled gradient in composition. Such nonuniform geometries might help minimize cracking and delamination through improved handling of the large volume expansion of the silicon, thus resulting in improved cycling performance of the silicon anodes. Table 14.1

lists the electrochemical properties of thin-film anodes prepared by PLD. Although significant improvement has been reported on the Si-based anode thin films, their properties, such as cycle life, high rate, capacity retention, and film thickness, must be improved before they are used as a reliable anode material.

14.3.3   PLD of Solid-State Electrolytes

A solid-state electrolyte is one of the key materials for fabricating thin-film LIBs since there is a safety issue on traditional LIBs that use highly flammable organic liquid electrolytes. There have been considerable studies on all solid-state thin-film-type LIBs using nonflammable solid electrolytes due to their excellent potential for improving their safety and reliability. The PLD technique has been utilized in growing various inorganic solid-state electrolytes, such as lithium phosphorous oxynitride (LiPON), $Li_{2.2}V_{0.54}Si_{0.46}O_{3.4}$ lithium vanadium silicon oxide [LVSO], -Li_3PO_4, $Li_{3.25}Ge_{0.25}P_{0.75}S_4$ (thio-LISICON), and Li_2S-P_2O_5 films [33, 36–41]. Ideally, the solid-state electrolyte should have high ionic conductivity and low electronic conductivity for thin-film batteries. Table 14.2 shows data from various solid-state electrolyte thin films prepared by PLD. LiPON was successfully deposited by PLD and showed that the ionic conductivity can be adjusted with the concentration of nitrogen in the films by changing the N_2 background gas pressure during deposition [36]. The Li ion conductivity of the PLD-grown electrolyte films ranges from 10^{-4} to 10^{-7} S/cm, while the electric conductivity ranges from 10^{-7} to 10^{-13} S/cm. Due to their high Li ion conductivity and low electron conductivity, these PLD-grown solid-state thin-film electrolytes are promising for developing all solid-state Li ion microbatteries.

Another advantage of PLD in fabricating thin-film batteries is that all solid-state thin-film layers can be fabricated by a sequential PLD process. For example, all solid-state thin-film microbatteries, consisting of LCO cathode, LVSO solid electrolyte, and amorphous SnO anode, were fabricated by a sequential PLD process [33]. A schematic cross-sectional view of this thin-film LIB in Fig. 14.4a–c shows SEM micrographs of cross-sectional views of the batteries before and after testing for 100 cycles. The interface between the LCO cathode, LVSO electrolyte, and the SnO anode remained smooth before and after the cycling test. This thin-film microbattery showed an initial discharge capacity of 9.5 µAh/cm² and 45% of the initial discharge capacity was retained after 100 cycles at a constant current rate of 44 µA/cm² between 0.01 and 3 V (Fig. 14.4).

Table 14.2 Electrochemical properties of electrolyte thin films prepared by PLD

Material	Substrate	Li ion conductivity (S/cm)	References
LiPON	Si, Au/Si	1.6×10^{-6}	[36]
$Li_{2.2}V_{0.54}Si_{0.46}O_{3.4}$ (LVSO)	Quartz	2.5×10^{-7}	[33]
$Li_{6.16}V_{0.61}Si_{0.39}O_{5.36}$ (LVSO)[a]	Si, Al/glass	3.98×10^{-7}	[37]
$Li_{3.4}V_{0.6}Si_{0.4}O_4$ (LVSO)	Si, fused silica	10^{-7}	[38]
$Li_{3.25}Ge_{0.25}P_{0.75}S_4$ (thio-LISCON)	Quartz	1.7×10^{-4}	[39]
$80Li_2S$-$20P_2S_5$	Si	2.8×10^{-4} [b]	[40]
Li_4SiO_4-Li_3PO_4	Quartz	1.6×10^{-6}	[41]

[a]The target composition
[b]Heat-treated at 200°C for 1 h

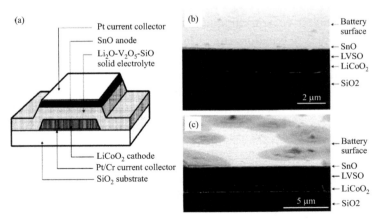

Figure 14.4 (a) Schematic illustration of a thin-film Li ion battery. SEM micrographs showing cross-sectional views of thin-film Li ion batteries (b) before and (c) after 100 cycles at a current density of 44 µAh cm². Reprinted from Ref. [33], Copyright (2004), with permission from Elsevier.

In summary, PLD is a powerful technique for fabricating high-quality layers of materials for cathodes, anodes, and solid electrolytes for thin-film microbatteries. Due to the relatively low crystallization temperature, highly textured cathode films can be prepared by PLD. PLD has been successfully employed to deposit a-Si, silicon-graphene, and Si-C composite anodes, while various solid-state thin-film electrolytes have been fabricated by PLD with high Li ion conductivity and low electronic conductivity. Finally, all solid-state thin-film microbatteries have been demonstrated by applying sequential PLD growth. In addition, from an applications point of view, fabricating thin-film microbatteries on flexible plastic substrates is highly desirable. However, most of the cathode and anode thin films are required to be grown at elevated temperatures (500°C–600°C). This is one of the limitations of thin-film batteries grown by physical vapor deposition techniques such as e-beam and sputtering. However, since films grown by PLD can be crystallized at relatively low temperatures, it might be possible to grow active materials by PLD at relatively low temperatures (<300°C) on flexible substrates such as polyimide (Kapton).

14.4 Printing of Energy Storage Materials by LIFT

LIFT is a direct-write technique that allows high-resolution printing from a variety of functional materials. Figure 14.5 shows a schematic illustration of the basic elements for the LIFT process. The concept of the LIFT process is simple. It uses a pulsed laser beam to induce the transfer of material from a donor substrate onto a receiver substrate. The donor substrate is prepared by coating the material of interest onto a laser-transparent quartz wafer and is also referred to as the ribbon. The receiver substrate is placed on the *x-y* stage, facing the donor substrate, at a distance of tens of microns. Laser pulses are focused onto the donor film by a microscope objective. When the incident laser pulse is higher than the threshold energy, material is ejected from the ribbon and transferred to the receiver substrate. Typical laser fluences for LIFT of battery materials, depending on the donor film thickness, range from 10 to 100 mJ/cm^2. In general, the laser beam travels by galvanometric scanning mirrors for the fast pattern generation during the LIFT process and a computer-controlled X-Y motion control system translates the ribbon and receiving substrates. A CCD camera can provide a real-time plan view of the transfer process. Unlike other film deposition and patterning processes, LIFT processes can be performed at ambient conditions without using vacuum or clean room environments.

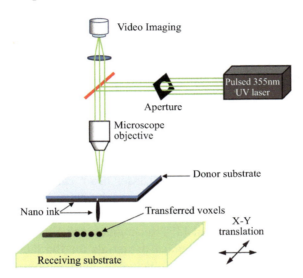

Figure 14.5 Schematic illustration (not to scale) of the laser-induced forward transfer (LIFT) setup.

One of the advantages of LIFT is that the deposited materials typically show porous structures with high surface areas. These porous structures are very important features for electrochemical devices due to an increased contact area between the electrodes and the electrolyte, leading to improved charge transfer and accordingly a more complete utilization of the electrode materials. Thus, the LIFT technique is ideally suited for printing the active electrodes in most energy storage systems, such as batteries, capacitors, and solar cells [42–49]. Another important feature of the LIFT process is that unlike inkjet printing, the LIFT technique is suitable for printing high-viscosity nanoinks due to its nozzle-free nature [50–52]. Although inkjet printing is a simple technique, it is limited by the transfer of only low-viscous nanoparticle suspensions in order to avoid clogging of the dispensing nozzles [53]. Thus, printing of precise patterns by inkjet is very difficult due to the variable behavior of fluids on different types of surfaces and their unstable wetting effects [54, 55]. Recently, LIFT has been applied to print high-viscosity pastes by laser-decal transfer (LDT) [56–63]. The LDT process is a new type of direct-writing technique in which voxel shape and size become controllable parameters, allowing the creation of thin-film-like structures for a wide range of applications, such as 3D interconnects, freestanding structures, metamaterials, membranes, and circuit repair. Furthermore, the complex shapes can be printed by a single laser pulse in one step, reducing the processing time and avoiding problems related to the merging of multiple voxels. Thus, in the case of small-scale energy storage, LIFT is beneficial as it can directly print the active materials into the substrate housing the circuit, that is, embedding the entire micropower source, eliminating excess weight due to packing.

In addition, by combining LIFT with a digital micromirror device, the size and shape of an incident laser beam can be dynamically controlled in real time, resulting in laser-printed functional materials with geometries identical to those of the projected beam [57, 62]. In this section, the use of LIFT techniques will be demonstrated to process active electrochemical materials for the fabrication of various micropower systems, including ruthenia-based planar ultracapacitors, rechargeable thick-film Li ion microbatteries, and solid-state electrolyte membranes.

14.4.1   LIFT of Ultracapacitors

The LIFT technique has been successfully employed to fabricate ultracapacitors [46, 64]. An ultracapacitor is another type of electrochemical energy storage and power generation system that displays electronic properties similar to both batteries and capacitors. Like a battery, it has the ability to store a large amount of energy during charging state, and like a capacitor, an ultracapacitor has the ability to discharge its energy very rapidly with a high power density. Thus, the ultracapacitor, like a battery capable of high discharge rates, is typically used for load leveling and applications where a short burst of power is needed. Hydrous ruthenium oxide is an excellent electrode material for microultracapacitors due to its high specific capacitance. A large pseudocapacitance effect can be achieved by rapid insertion and release of electrons and protons through this material due to its high specific surface area [65]. This effect can be enhanced by the creation of structural water in the lattice, providing more percolation pathways for proton conduction into the material [66]. The chemistry of this system involves two identical electrodes composed of a hydrous metal oxide whose electrochemical performance is sensitive to the processing temperature [67]. One of the important features of LIFT is that it can print ink composed of the electrode material and relevant electrolyte. Arnold et al. [46] demonstrated a planar type of hydrous ruthenium oxide ultracapacitor using the LIFT process with significantly improved discharge behavior. The active ink was composed of hydrous ruthenium oxide powder and sulfuric acid electrolyte. As shown in Fig. 14.6a,b, a 1 cm² gold-coated glass substrate laser

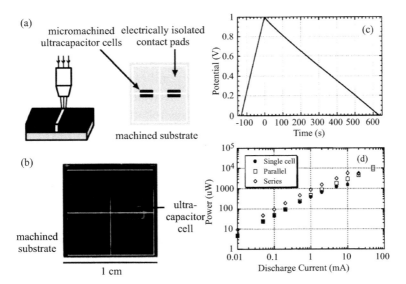

Figure 14.6 (a) Micromachining to produce a planar RuO$_2$-0.5H$_2$O microultracapacitor. (b) Picture of laser-machined planar microultracapacitors on a gold-coated quartz substrate. (c) Plot showing the charging/discharging behavior of a hydrous ruthenium oxide ultracapacitor printed by LIFT. The cell is charged at 50 µA and discharged at 10 µA. Starting lines are indicative of ideal capacitor behavior. The total electrode mass is 100 µg, with a footprint of 2 mm². (d) Power as a function of discharge current for a single ultracapacitor cell as well as parallel and series combinations. The power is calculated over 0–1 V for the single-cell and parallel combination and 0–2 V for the series combination. From Ref. [64]. Reproduced by permission of The Electrochemical Society.

micromachined into four electrically isolated regions serves as the electrodes for the cells. Figure 14.6c shows a typical charge/discharge plot for these ultracapacitor cells. The cell is charged at 50 µA until the voltage across the cell reaches 1 V. A constant current was applied during discharging until the voltage returned to zero. The cells exhibited the linear charge/discharge behavior at constant current, indicating ideal capacitor behavior. Figure 14.6d shows that the ultracapacitors added in series and parallel can be discharged at currents above 50 mA without damaging the cells [64].

14.4.2 LIFT of Li Ion Microbatteries

The LIFT technique has been successfully employed to print thick-film electrodes for microbatteries with much higher capacities per electrode area than those made by sputter-deposited thin-film electrodes [42–47]. This enhanced performance is related to the previously mentioned porous structure of the layers typically printed by LIFT. This porosity is what allows improved diffusion of the Li ions across the electrodes without a significant internal resistance loss. Kim et al. fabricated thick-film electrodes (LCO cathode and carbon anode) by LIFT for Li ion microbatteries to study their electrochemical properties [42]. The laser-printed electrodes were separated by a laser-cut porous membrane that is soaked with a gel polymer electrolyte to build mm-size solid-state packaged Li ion microbatteries. Figure 14.7a,b shows a cross-sectional schematic diagram and SEM micrograph of a typical Li ion microbattery fabricated by LIFT.

One of the advantages of LIFT is that the thickness of the printed films can be easily adjusted by the number of LIFT printing passes. For example, the thickness and mass of the LCO cathode thick films increase linearly with the number of LIFT passes (see Fig. 14.7c). Accordingly, the discharge capacities can also be easily controlled by the number of LIFT passes (see Fig. 14.7d). The viscosity of the ink also influences the film thickness, that is, for high-viscosity inks, thicker films are deposited than with lower-viscosity inks. With increasing ink viscosity, the laser power required for their printing increases as well, which can affect the printing resolution, as well as alter the film components, thus requiring optimization of the transfer conditions for each material.

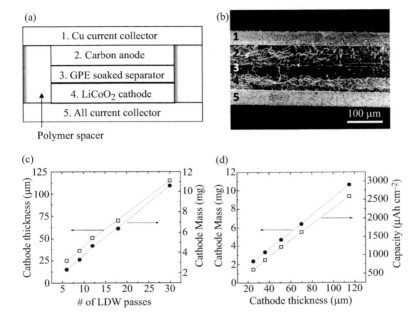

Figure 14.7 (a) Cross-sectional schematic diagram of a typical Li ion microbattery. (b) Cross-sectional SEM micrograph of a packaged thick-film Li ion microbattery. (c) LiCoO$_2$ cathode thickness versus cathode mass as a function of the number of LIFT passes. (d) LiCoO$_2$ cathode mass versus discharge capacity as a function of the cathode thickness. Reprinted from Ref. [42], Copyright (2007), with permission from Elsevier.

The electrochemical properties of Li ion microbatteries based on laser-printed LCO cathode thick films are highly dependent on the cathode thicknesses. Figure 14.8a shows charge/discharge curves at the 5th cycle for Li ion microbatteries with laser-printed LCO cathodes of different thicknesses (35–115 µm). Microcarbon microbead (MCMB) graphite is used as an anode layer, and a polyolefin-based microporous membrane soaked with a gel polymer electrolyte is used as a separator for these microbatteries. The microbatteries are charged and discharged at a constant current of 100 µA/cm^2. As shown in Fig. 14.8a, the discharge capacity per active electrode area is proportional to the cathode film thickness, suggesting that the discharge capacity increases with the cathode thickness (up to 115 μm thick). This increased capacity is related to their high-surface-area porous structure, allowing better ionic and electronic transport through the thick electrodes (~115 μm) without any significant internal cell resistance. Figure 14.8b shows the cycle performance of the Li ion microbattery with a 35 μm thick LCO cathode. The microbattery shows excellent cycling performance of over 200 cycles with a slow fade rate. The battery retains about 80% of its initial capacity after 200 cycles. The specific capacity based on the cathode mass is ~100 mAh/g after 200 cycles (see Fig. 14.8b).

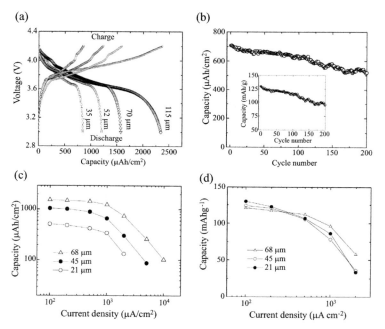

Figure 14.8 (a) Voltage versus capacity per unit area for a Li ion microbattery with different thicknesses of LiCoO$_2$ cathodes. (b) Cycle performance of the Li ion microbattery with a 35 μm thick LiCoO$_2$ cathode. Discharge capacity per (c) active electrode area and (d) cathode mass for the packaged thick-film microbatteries with three cathode thicknesses (21, 45, and 68 μm). The active electrode area was 0.49 cm^2. Microbatteries were charged at 0.1 mA/cm^2 between 4.2 and 3 V. Reprinted from Ref. [42], Copyright (2007), with permission from Elsevier.

The discharge rate is also one of the important properties of the electrodes, especially for high-power-density applications. Figure 14.8c shows the discharge capacity per active area for LCO thick-film microbatteries with three different thicknesses (21, 45, and 68 mm) as a function of discharge current rate. The slopes of the curves for all the cells are approximately parallel for current densities between 100 and 2000 µAh/cm², indicating that the discharge rate is not limited by the cathode thickness for Li ion transport during discharge. This suggests that the energy per unit area can be increased simply by increasing the thickness of the electrodes, without any significant internal resistance loss. The cell with the 68 mm thick cathode produces a maximum power density of ~38 mW/cm² (~10² mAh/cm²) at a discharge current of 10 mA/cm². For the entire current density range, almost the same fraction of active material in all the cells is accessed for charge/discharge activities (see Fig. 14.8d). These results confirm that the cathode thickness is not a rate-limiting factor for the thick-film microbatteries prepared by LIFT due to the porous structure of the laser-printed electrodes.

It is worth comparing the properties of the laser-printed thick-film electrodes with those of sputter-deposited thin-film electrodes. For example, when we consider the discharge capacity per active electrode area, sputter-deposited 2.5 µm thick LCO cathodes (active area = 1 cm²) exhibited a capacity of 160 µAh/cm² (or ~64 µAh/µm cm²) at a current density of 100 µA/cm² [4], while the laser-printed 115 µm thick LCO cathodes (active area = 0.49 cm²) demonstrated a capacity of 2586 µAh/cm² (or ~22.5 µAh/µm cm²) at the same current rate. Despite a three times smaller volumetric capacity than that of the sputter-deposited thin-film cells, the laser-printed thick-film microbatteries can produce orders of magnitude higher capacities per unit area. Specifically, to achieve the same discharge capacity (2586 µAh/cm²) as of the 115 µm thick laser-printed LCO cathodes employing 0.5 cm², the sputter-deposited 2.5 µm thick LCO cathodes would have to occupy an area of ~8 cm². On the basis of this comparison, it is clear that the laser-printed thick-film microbatteries are well suited for applications where limited space is available for the power source, such as wireless network sensors and autonomous microelectronic devices.

14.4.3 LIFT of Solid-State Electrolytes

A liquid electrolyte is mostly used in commercial LIB systems due to its high ionic conductivity. However, the use of the liquid electrolyte is limited by leakage and safety issues. To resolve these limitations it is important to replace the liquid electrolyte with a solid-state polymer electrolyte since the volatility of the liquid electrolyte is no longer a concern. This solid-state polymer electrolyte would allow for easier packaging with less material. Ollinger et al. reported on the printing by LIFT of an ionic liquid-based nanocomposite solid-state electrolyte for the fabrication of Li and Li ion microbatteries [43, 44]. The key feature of this solid-state electrolyte is that the laser-printed solid-state membranes exhibited the proper electrochemical behavior for ionic liquids with a high ionic conductivity of 1–3 mS cm^{-1}, while maintaining the strength and flexibility of the poly(vinylidene fluoride-co-hexafluoropropylene) (PVD-HFP) copolymer matrix. Accordingly, they can serve as both electrolyte and separator [43]. Figure 14.9a shows the optical micrographs of laser-printed solid-state polymer membranes on a glass substrate. These laser-printed membranes were dried at 75°C for 1 h, forming a continuous flexible pinhole-free membrane. It is clear from these micrographs that even 5 μm thick membranes are strong enough and flexible enough to be lifted off using tweezers without damaging the membranes (see Fig. 14.9b). Figure 14.9c shows the first four cycles for a Li ion microbattery fabricated with the laser-printed solid-state membrane (~20 μm thick), the 30 μm thick LCO cathode, and the Li-metal anode. The LIFT process was also employed to print sequential layers of cathode (LCO), solid-state electrolyte, and anode (MCMB graphite) into a laser-micromachined pocket on a thin polyimide substrate (Kapton) to build an embedded all solid-state Li ion microbattery [11]. This microbattery was charged and discharged at C/3 rate (~110 μA/cm^2) and exhibited an energy density of 1.32 mWh/cm^2 (or 0.41 mWh/cm^3), corresponding to a specific energy of 330 mWh/g (~100 mAh/g). The C-rate is a measure of the rate at which a battery is charged/discharged relative to its maximum capacity. A "1C" rate means that the discharge current will discharge the entire battery in 1 h.

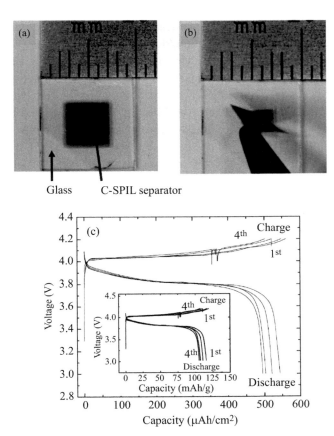

Figure 14.9 Optical micrographs of (a) the laser-printed c-SPIL membranes on a glass substrate and (b) membrane partially lifted off a glass slide by tweezers. (c) Charge–discharge performance of a packaged Li microbattery (LiCoO$_2$/c-SPIL separator/Li) cycled between 4.2 V and 3 V at current density of 40 µA/cm^2. The battery was tested in air at 25°C. The active electrode area is 1 cm^2. Reprinted from Ref. [43], Copyright (2006), with permission from Elsevier.

14.5 3D Processing of Energy Storage Materials by LS and LA

The development of 3D architectures for electrodes in LIBs is a promising approach to overcome problems such as 1D Li ion diffusion, inhomogeneous current densities, power losses, high interelectrode ohmic resistances, and mechanical stresses due to high-volume changes resulting from lithiation/de-lithiation during charging and discharging processes. By applying 3D battery architectures, one can achieve large areal energy capacities while maintaining high power densities at the same time. This feature is important, for example, for thin-film batteries where the Li ion diffusion is limited by the thickness of the compact film. A common approach for realization of 3D architectures in electrodes is the structuring of the current collector. An increased active surface achieved by 3D electrode architectures can induce large areal energy densities. Unfortunately, this approach is in a very early stage of development and, in general, it is not feasible for state-of-the-art electrodes.

With the direct structuring of thick-film and thin-film electrodes, a new process for generating 3D batteries was developed at the Kalsruhe Institute of Technology (KIT): patterning of thin-film and thick-film electrodes by direct laser modification and ablation. KIT introduced a new battery design concept that was successfully assigned to the 3D concept (3D battery) for achieving large areal energy capacities and power densities [8, 68–70]. This concept is not limited to microbatteries only since it can be applied to the other technical approaches listed in Ref. [8]. That means that batteries that combine high power densities and high energy densities at the same time can be realized through the application of laser annealing (LA) and laser structuring (LS) processes.

14.5.1 LA and LS of Thin-Film Electrodes

A rather new application field for laser material processing is the development of 3D structures in LIBs based on nanoscaled materials and thin films [71–82]. For this purpose, LA and LS of thin-film electrodes made of LCO, SnO_2 and LMO were recently investigated. LCO, or lithium cobalt oxide, is an appropriate model system because it is well established and still the most commonly used cathode material in LIBs [83]. LMO and SnO_2 were studied in detail in order to investigate chemical and mechanical degradation effects in electrode materials. Furthermore, the passivation of laser-generated 3D LMO surfaces with thin films such as indium tin oxide (film thicknesses of 10–50 nm) has also been investigated with respect to a reduction of chemical degradation during electrochemical cycling.

14.5.1.1 Laser annealing

LA was successfully applied to structured and unstructured LCO and LMO thin films [76, 79] in order to adjust the crystalline phase and grain size. In the case of LCO it was shown that suitable annealing temperatures are in the range of $T = 400°C–700°C$. Temperatures lower than or equal to $400°C$ led to an insufficient phase conversion, while temperatures equal to or above $700°C$ led to the formation of a contamination phase (Co_3O_4). LA processes were also developed for rf magnetron sputtered LMO (Li-Mn-O) thin films with the aim to form a spinel-like phase [80]. The Raman spectrum for an as-deposited Li-Mn-O thin film is depicted in Fig. 14.10a-ii. Applying an annealing temperature of $T = 600°C$ for $t = 100$ s on the film, significant changes within the Raman spectra could be observed (Fig. 14.10a-iii). The characteristic peaks for the electrochemical inactive Li_2MnO_3 phase could be assigned [84, 85]. With an increase in LA temperature of up to $T = 680°C$ and a fixed annealing time of $t = 100$ s (Fig. 14.10a-iv), Raman spectroscopy indicates a spinel-like Li-Mn-O phase showing the typical bands that indicate stretching vibrations of manganese and oxygen compounds at 629 cm^{-1} (A_{1g} species), the shoulder around ~590 cm^{-1} (F_{2g}), as well as a weak band at 482 cm^{-1} (F_{2g} species) [84, 86]. The intensity of the shoulder around 583 cm^{-1} can be correlated with the average oxidation state of manganese and, therefore, increases upon deintercalation of lithium [84]. This could be one reason for differences between the spectra for annealed films with the reference powder (Fig. 10a-i) when taking into account the lithium deficiency of the sputtered thin film. Similar Raman spectra could be observed for an annealing temperature of $T = 600°C$ by applying an annealing time of $t = 2000$ s (Fig. 14.10a-v). The thin films showing the spinel phase were cycled and analyzed with respect to the composition of the formed SEI layer [80]. Cyclic voltammetry (CV) scans identified that film annealing at $T = 600°C$ for $t = 2000$ s leads to characteristic redox peaks for spinel thin films (Fig. 14.10b). SEM and XPS analyses showed that the SEI layer was formed on top of a laser-annealed spinel-like cathode surface [80].

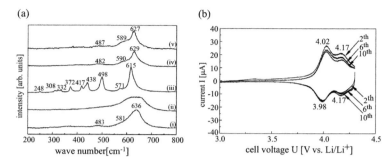

Figure 14.10 Effect of laser annealing on crystallinity and electrochemical performance. (a) Raman spectra of $Li_{0.88}Mn_{1.98}O_4$ thin films (ii–v) and $LiMn_2O_4$ reference powder (i). As-deposited film (ii). Laser annealing was performed for $t = 100$ s at $T = 600°C$ (iii), for $t = 100$ s at $T = 680°C$ (iv), and for $t = 2000$ s at $T = 600°C$ under ambient air (v). (b) Cyclic voltammograms ($Li_{0.88}Mn_{1.98}O_4$, laser-annealed for $t = 2000$ s at $T = 600°C$). Reprinted from Ref. [80], Copyright (2012), with permission from Elsevier.

14.5.1.2 LS of LCO thin films

LS of LCO can be realized via classical laser direct ablation or via self-organized structuring (Fig. 14.11). The formation of laser-induced self-organized conical surface structures on LCO thin films can be explained by selective material ablation and subsequent material redeposition [71, 76]. In recent research it was nearly possible to avoid material loss during patterning, while the height of the created cones could be increased up to 8.4 μm for a thin-film thickness of 3 μm. Appropriate surface structures and orientations of lithium intercalation planes should increase lithium diffusion significantly [87]. The freestanding microstructures (Fig. 14.11) contain little residual stress, and expansion during electrochemical cycling can be easily compensated, leading to reduced crack formation and better cycling stability [88]. In Ref. [89] the transfer of that structuring technology to thick films made of LCO tape-cast electrodes was successfully demonstrated.

Electrochemical cycling was applied to laser-structured and unstructured thin films. The theoretical capacity of 140 mAh/g was used for the calculation of the C-rate. To analyze the high power capability of the samples, the charging current was increased stepwise. After 11 charge/discharge cycles at C/20, the current rate was increased to C/5 for another 11 cycles. Finally, 100 cycles at 1C were measured. The results are shown in Fig. 14.12. The unstructured films showed a higher initial discharge capacity compared to the structured and laser-annealed films. After 11 cycles at the lowest charging rate the capacity dropped by 27%. During the next cycles at C/5 a further decrease in capacity of 31% was measured. After a few cycles at 1C the capacity was reduced to below 5 mAh/g. Although the initial capacity of the laser-structured thin films was lower, the capacity increased to about 140 mAh/g during the first five cycles. This may be due to run-in effects (e.g., formation of a solid electrolyte interphase). During the 11 cycles at the lowest current rate the capacity of the structured thin film increased by about 3%. At C/5 a slight decrease in capacity of 3% could be observed. After 100 cycles at 1C the capacity reached values of 78 mAh/g. The improvement of battery performance by LS can be attributed to different processes. Through the increased surface area of the structured films more lithium diffusion planes are accessible, which in turn leads to a higher lithium diffusion rate at high charging rates. This leads to a maximum lithium diffusion length of 1 μm, which is significantly smaller than the maximum diffusion length of 3.5 μm of the unstructured films. Concomitantly, the freestanding cones contain little residual stress and expansion during electrochemical cycling can be easily compensated, leading to reduced crack formation and better cycling stability.

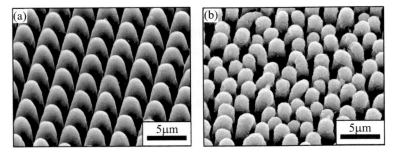

Figure 14.11 SEM images of laser-structured LiCoO$_2$ thin films using mask imaging (a) with a laser fluence of 3 J/cm^2 and self-organized surface structures and (b) with laser fluences of 0.5 J/cm^2 and 2 J/cm^2. Sixty laser pulses at a repetition rate of 100 Hz were applied. Reproduced from Ref. [81] with permission of Springer.

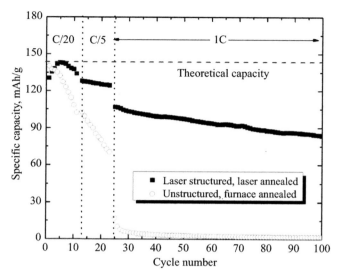

Figure 14.12 Discharge capacity as a function of the cycle number of unstructured and laser-structured LiCoO$_2$ thin films after furnace annealing for 3 h or laser annealing at 600°C for 250 s, respectively. Laser structuring was performed with a wavelength λ = 248 nm in ambient air. Reproduced from Ref. [90] with permission of Springer.

14.5.1.3 LS of SnO$_2$ thin films

LS was also applied for a SnO$_2$ thin-film anode (see Fig. 14.13). The periodic structures had a height of 3 μm and a pitch distance down to 4 μm. The minimum spacing between periodical arranged structures was 400 nm (Fig. 14.13c,d). The ablated grooves (Fig. 14.13a,b) had a width of 2–10 μm and a pitch distance of 5–40 μm. The results of electrochemical cycling between 0.02 V and 1.2 V at a charge/discharge rate of C/2 are depicted in Fig. 14.13e. The unstructured thin film shows poor cycling behavior with the capacity of less than 50 mAh/g after 20 cycles. In comparison, the laser-structured thin films exhibited significantly better performance. Capacities of >700 mAh/g—which are in good agreement with the theoretical capacity—could be obtained for more than 20 cycles. After 50 cycles, capacities between 200 mAh/g and 400 mAh/g could be retained. It becomes clear that even for the line structures with structure sizes of 20 μm, a considerable improvement of battery performance is observed. This result suggests that significant mechanical stress is formed throughout the thin film, which can be reduced by applying freestanding structures (during cycling of SnO$_2$: volume changes of up to 359% are expected). A reduction of the structure dimension leads to further improvement of cycling stability, while the best results were obtained for the freestanding conical structures shown in Fig. 14.13d.

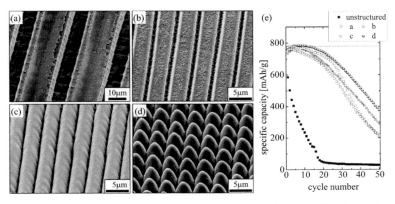

Figure 14.13 (a–d) SEM images of laser-structured SnO$_2$ thin films with different structure sizes. (e) Corresponding specific discharge capacity of laser-structured SnO$_x$ thin-film electrodes as a function of the cycle number at a charge/discharge rate of C/2.

14.5.1.4 LS of LMO thin films

Figure 14.14 shows the result of the laser patterning process on LMO cathode thin films using a grating mask. Spherical surface structures were formed with smooth surfaces and without any debris. Laser ablation was performed under ambient air using a laser fluence of $\varepsilon = 1.6\,\text{J/cm}^2$, $N = 60$ laser pulses, and a repetition rate of $\nu_{\text{rep}} = 100$ Hz. Figure 14.15 shows a cross-sectional focused-ion-beam (FIB) image of the laser-structured all-solid-state cell grown by rf sputtering. First, a 3.3 μm thick Li-Mn-O cathode was sputter-deposited on silicon. A silicon substrate was used in order to break the sample for cross-sectional views. Within the laser patterning process 3D spherical surface structures were formed. In third and fourth steps, LVSO was used as solid electrolyte as well as aluminum (Al) was deposited on top of the surface structures with a maximum LVSO layer thickness of about 1.4 μm and a maximum Al layer thickness of about 700 nm. The challenge was to reach a complete coverage of these materials to the cathode and besides sporadic cracks this could be achieved in a first experimental approach. For the FIB procedure platinum (Pt) was deposited on top of the 3D structures. The combination of laser patterning processes and rf magnetron sputtering can be realized to achieve new 3D battery designs.

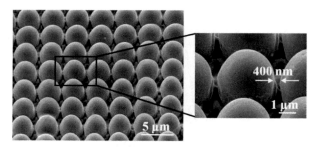

Figure 14.14 Spherical 3D surface structures formed via direct laser structuring of Li-Mn-O thin films (λ = 248 nm, energy density ε = 1.6 J/cm^2, pulse number N = 60, repetition rate v_{rep} = 100 Hz, and ambient air). Reproduced from Ref. [91] with permission from Japan Laser Processing Society.

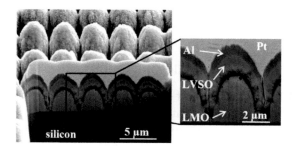

Figure 14.15 Cross-sectional FIB image showing the principle setup of a three-dimensional all-solid-state cell grown by rf magnetron sputtering of lithium vanadium silicon oxide (LVSO) electrolyte and an aluminum (Al) anode material on top of the laser-generated spherical lithium manganese oxide (LMO) structures. Reproduced from Ref. [91] with permission from Japan Laser Processing Society.

14.5.2 LS of Thick-Film Electrodes

The electrodes, anodes, and cathodes in state-of-the-art LIBs are complex multimaterial systems that are provided with defined material components, grain sizes, porosities, and pore size distributions in the micrometer and submicrometer ranges. Thick-film electrodes are formed from slurries of active material powders, binders, solvents, and additives and are fed to coating machines to be deposited on current collector foils. Thick-film electrodes reveal a high porosity in the range of 30–50%, which is necessary to enable suitable wetting of the active material with a liquid electrolyte. The active material and other constituents are mixed by battery manufacturers according to special and generally inaccessible recipes. Currently, secondary Li ion cells utilize cathode materials such as layered compounds, like $LiNi_{1/3}Mn_{1/3}Co_{1/3}O_2$ (NMC), $LiFePO_4$ (LFP), and LCO. Differences can be found in the type of lithium diffusion paths: the olivine-type LFP and some silicon derivatives provide 1D paths whereas the layered compounds 2D paths and the spinel-type 3D paths. Another distinction can be the kinetics of the diffusion. The practical capacities also differ within those materials, counting 148 mAh/g for NMC, 140 mAh/g for LCO, and 130 mAh/g for LFP. As counterelectrodes, graphite anodes are used in state-of-the-art LIBs with a practical capacity of about 330 mAh/g.

Due to the fact that a significant amount of inactive material, such as metallic current collector, separator material, binder, carbon black, liquid electrolyte, or pouch material (for cells with a pouch cell design), is used in a state-of-the-art Li ion cell, the gravimetric energy density is rather low. While on the material level (NMC) a gravimetric energy density of about 650 Wh/kg can be reached, the practical energy density drops down to values in the range of 100–150 Wh/kg. For a successful and economical use of LIBs in a future lightweight actuating device a significant increase of the practical energy density is required. It is assumed that only a systematic and concerted approach on the material level ("material challenge") and cell architecture ("engineering challenge") can achieve such high energy densities, in the range of or above 300 Wh/kg. The development of 3D electrode architectures in thick-film electrodes is a rather new approach for improving battery performance. For this purpose, we established different types of laser processing for increasing the active surface area, that is, laser-assisted self-organizing structuring and direct structuring (Fig. 14.16). The first step of these processes can be applied tothin-film and thick-film electrodes that have small electrode footprint areas (coin cells). The second process is suitable for small and large electrode footprint areas (pouch cells). We used excimer laser ablation at a wavelength of 248 nm to produce self-organized surface structures (Fig. 14.16a) on lithium cobalt oxide and NMC thick-film electrodes. We also applied direct LS—with either a 200 ns fiber laser or an ultrafast fiber laser (380 fs)—to form 3D microstructures (Fig. 14.16b) [92].

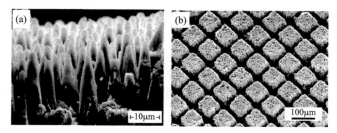

Figure 14.16 SEM of microstructures laser-generated in composited electrode cathode materials: self-organized microstructures by excimer laser ablation (a) and micropillars obtained by direct femtosecond laser structuring (b) [93].

14.5.3 LS Turns Electrodes into Superwicking

A main issue in cell production is the electrolyte wetting of LIBs, which is realized by time- and cost-consuming vacuum and storage processes at elevated temperatures [94]. The liquid electrolyte has to be forced into micro- and nanosized pores of the composite electrode material (Fig. 14.17). In fact, the use of current electrolyte filling processes results in insufficient wetting of the electrode surfaces, which is one drawback leading to a certain rate of production failure, lowered cell capacity, or reduced battery lifetime. At KIT, a cost-efficient laser-based technology for the realization of microcapillary structures in separators [95] and thick-film tape-cast electrodes was developed to achieve a tremendous acceleration of the wetting process and to shorten the time span for cell manufacturing. For thick-film composite electrodes an appropriate structure design delivers the most efficient capillary transport [96]. Electrochemical analysis showed that a steep rise of capacity retention at high charging and discharging currents and an improved cell lifetime can be obtained in comparison to standard cells with unstructured battery materials. This significant increase in battery lifetime is caused by efficient and instantaneous liquid electrolyte transport that is enabled by the laser-generated microcapillary structures. The laser-generated capillary structures in electrode materials increase cell reliability and shorten battery production times. Improved cycle lifetimes and increased capacity retention also mean that high-power batteries for second-life applications become a possibility.

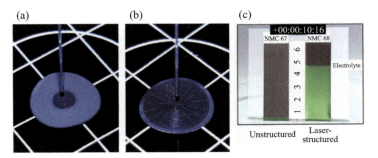

Figure 14.17 Liquid electrolyte wettability of NMC thick-film electrodes: single-drop electrolyte (3 µL) wetting behavior on unstructured (a) and laser-structured (b) thick-film electrodes. Snapshot (c) of video imaging of the capillary rise in dependence of wetting time (~10 s).

Electrodes with laser-structured 3D architectures can provide a homogeneous and fast wetting behavior (Fig. 14.17b) and finally lead to an improved electrochemical performance. The capillary rise of liquid electrolyte in thick-film electrodes (Fig. 14.17c) can be very well described by the classical Washburn, which is described elsewhere [96]. For the formation of capillary structures, ns laser ablation as well as ultrafast laser processing were investigated. For ns laser radiation (λ = 1064 nm and pulse length 200 ns) the laser beam energy is absorbed at the material surface and due to heat conduction the temperature of the surrounding composite material increases. The binder material used for tape-cast electrodes (~5 wt%) is poly(vinylidene fluoride) (PVDF), which has a low decomposition temperature, in the range of 250°C–350°C [97]. Therefore, the PVDF binder matrix spontaneously evaporates and active particles are removed from the laser beam interaction zone. A similar ablation process for metal/polymer composite materials has already been described by Slocombe et al. [98]. Nanosecond laser ablation is not appropriate for each type of electrode material. For example, ns-LS of LFP electrodes leads to melt formation and, therefore, to an undesired modification of the active material. Therefore, the use of ultrafast laser materials processing is necessary for this type of material. Furthermore, the ablation efficiency of LFP increases by a factor of 3 by using fs or ps laser ablation in comparison to ns laser ablation [99]. For battery production costs it is important to reduce the amount of ablated material, which in turn means that small capillary widths and high aspect ratios are preferred. By using ultrafast laser ablation, it is shown that the aspect ratio can be significantly increased and that the loss of active material can be reduced from 20% to values below 5% [100]. At KIT, this structuring process was conducted under ambient air conditions and the ablated material was removed through an exhaust.

Capacity retention and cell lifetime can be illustrated by plotting the discharge capacity as a function of cycle numbers for Li ion cells with both laser-structured and unstructured NMC electrodes (Fig. 14.18). The discharge capacity of the laser-structured NMC cell drops to 80% of the initial capacity after 2290 cycles (Fig. 14.18), while for the Li ion cell with unstructured electrodes, the cell lifetime (80% of its initial capacity) is reached after only 1140 cycles. Furthermore, the discharge capacity of the cell with the laser-structured NMC electrode reaches a value of 108 mAh/g after 2290 cycles, indicating that efficient liquid electrolyte transport improves the electrochemical performance for these cells. These benefits are achieved without cost- and time-consuming storage steps, thanks to the microcapillary structures produced by LS of the electrodes.

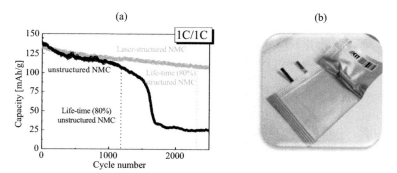

Figure 14.18 Specific discharge capacities (a) of NMC pouch cells (b) with laser-structured and unstructured electrodes as a function of cycle number (1C charge/1C discharge).

14.6 Challenges and Future Directions

Although laser-based processing techniques have been successfully employed for the fabrication of energy storage systems, these techniques present certain challenging issues. Three of the most relevant challenges are presented in the following paragraphs.

First of all, the thickness of the PLD-grown electrodes is limited to less than a few microns due to the increased internal cell resistance characteristic of vapor-deposited electrode films. Thus, the capacity of the thin-film microbatteries is not sufficient to operate many microelectronic devices. One possible way to increase the electrode thickness with relatively low internal resistance loss might be through codeposition of conductive carbon with the cathode and anode materials by alternatively ablating carbon and electrode targets during PLD. These carbon-containing electrodes would enhance the conductivity of the electrodes, reducing the internal cell resistance loss, and allowing improved capacity.

Another challenging issue in designing microscale batteries is to develop appropriate packaging processes that can achieve a reliable seal to protect the electrode materials with minimum weight and volume. Typical commercial coin cell batteries may not be used as micropower sources in many microelectronic devices due to their larger volume. In this matter, microbatteries can be designed to be embedded within the laser-machined pockets on a chip. LIFT processes are practical and easily customizable, since they can print battery materials directly on top of devices and they can also fabricate pockets for embedded microbatteries. Thus, the LIFT techniques will play an important role in the development of the next generation of microelectronic devices.

Finally, integration of the microbatteries into other microelectronic devices on a chip must be achieved for the development of advanced microelectronic devices. One of main challenges to achieving this goal is to develop leak-free electrolytes, such as solid-state composite polymer electrolytes. Although PLD and LIFT techniques were successfully utilized to produce solid-state electrolytes, the ionic conductivity of these solid-state electrolytes is lower than that of liquid electrolytes, which may lower the energy density of the resulting microbatteries. Thus, considerable research on developing solid-state electrolytes with improved properties will be required to produce the micropower sources used for the next generation of microelectronic devices.

14.7 Summary

We have reviewed three laser-based processing techniques for the fabrication of energy storage and power generation devices. Firstly, the use of PLD technique was described to deposit numerous types of cathodes, anodes, and solid-state electrolytes for thin-film microbatteries. By applying sequential PLD processes, all-solid-state thin-film microbatteries were successfully fabricated. Secondly, the use of LIFT techniques was demonstrated to print active electrochemical materials for the fabrication of various micropower systems, including ruthenia-based planar ultracapacitors, rechargeable thick-film Li ion microbatteries, and solid-state electrolyte membranes. The LIFT process was also utilized to print sequential layers of cathode, solid-state electrolyte, and anode layers inside laser-micromachined pockets to build embedded all-solid-state Li ion microbatteries. Thirdly, the LS and LA processes were applied to build 3D architectures into thin-film and thick-film electrodes for Li ion microbatteries, which improved their electrochemical performance, including cycling and high current discharge rate, by increasing their overall active surface area.

As described in the introduction, laser-based processes have been realized as practical techniques to integrate energy storage and micropower sources with microdevices. In principle, these laser-based processes can be used at the beginning of, during, or after the fabrication of a device. For example, with PLD the growth of thin films would most likely take place at the beginning of the device fabrication steps, since the films will have to be patterned after deposition and other steps will be required before a device can be produced. On the other hand, the LIFT process can be applied at any stage of the fabrication sequence, even after the device has been completed, since LIFT is additive and nonlithographic in nature. Thus, laser printing of a microbattery for a particular device could take place after the device is already manufactured, allowing the embedding of micropower sources into existing devices. Finally, laser surface modification, such as LS and LA, can be applied at any stage of the fabrication process and can be applied to only a specific region of a structure or device, which opens a wide range of possibilities. Such flexibility and adaptability make the use of laser-based processing techniques for energy storage applications highly applicable for solving the challenge of developing fully integrated autonomous microelectronics.

Acknowledgments

This work was supported by the Office of Naval Research (ONR) through the Naval Research Laboratory Basic Research Program. Special thanks to Dr. Mike Ollinger and Dr. Tom Sutto for their help in the fabrication of embedded microbatteries and solid-state electrolytes, Ray Auyeung for his help with the LIFT process, and Dr. Johannes Pröll for his great contributions in developing the LS and LA processes of battery materials. This work received funding from the European Union's Horizon 2020 research and innovation program under the Marie Sklodowska-Curie grant agreement no. 644971. Finally, the support for laser materials processing by the Karlsruhe Nano Micro Facility (KNMF, http://www.knmf.kit.edu/), a Helmholtz research infrastructure at the Karlsruhe Institute of Technology (KIT), is gratefully acknowledged.

References

1. Mazor, H., Golodnitsky, D., Bustein, L., Gladkich, A., and Peled, E. (2012). Electrophoretic deposition of lithium iron phosphate cathode for thin-film 3D-microbatteries, *J. Power Sources*, **198**, pp. 264–272.
2. Wang, C., Zheng, W., Yue, Z., Too, C. O., and Wallace, G. G. (2011). Buckled, stretchable polypyrrole electrodes for battery applications, *Adv. Mater.*, **23**, pp. 3580–3584.
3. Sun, K., Wei, T.-S., Ahn, B. Y., Seo, J. Y., Dillon, S. J., and Lewis, J. A. (2013). 3D printing of interdigitated Li-ion microbattery architectures, *Adv. Mater.*, **25**, pp. 4539–4543.
4. Bates, J., Dudney, N. J., Neudecker, B., Ueda, A., and Evans, C. D. (2000). Preferred orientation of polycrystalline $LiCoO_2$ films, *J. Electrochem. Soc.*, **147**, pp. 59–70.

5. Wang, Z. L. (2010). Toward self-powered sensor networks, *Nano Today*, **5**, pp. 512–514.
6. Beidaghi, M., and Gogotsi, Y. (2014). Capacitive energy storage in micro-scale devise: recent advances in design and fabrication of micro-supercapacitors, *Energy Environ. Sci.*, **7**, pp. 867–884.
7. Koeneman, P. B., Busch-Vishniac, I. J., and Wood, K. L. (1997). Feasibility of micro power supplies for MEMS, *J. Microelectromech. Syst.*, **6**, pp. 355–362.
8. Ferrari, S., Loveridge, M. Beatti, S. D. Jahn, M., Dashwood, R. J., and Bhagat, R. (2015). Latest advances in the manufacturing of 3D rechargeable lithium microbatteries, *J. Power Sources*, **286**, pp. 25–46.
9. Oudenhoven, J. F. M., Baggetto, L., and Notten, P. H. L. (2011). All-solid-state lithium-ion microbatteries: a review of various three-dimensional concepts, *Adv. Energy Mater.*, **1**, pp. 10–33.
10. Jeyaseelan, A. V., and Rohan, J. F. (2009). Fabrication of three-dimensional substrates for Li microbatteries on Si, *Appl. Surf. Sci.*, **256**, pp. S61–S64.
11. Sutto, T. E., Ollinger, M., Kim, H., Arnold, C. B., and Piqué, A. (2006). Laser transferable polymer-ionic liquid separator/electrolytes for solid-state rechargeable lithium-ionmicrobatteries, *Electrochem. Solid-State Lett.*, **9**, pp. A69–A71.
12. Simon, P., and Gogotsi, Y. (2008). Materials for electrochemical capacitors, *Nat. Mater.*, **7**, pp. 845–854.
13. Nitta, N., Wu, F., Lee, J. T., and Yushin, G. (2015). Li-ion battery materials: present and future, *Mater. Today*, **18**, pp. 252–264.
14. Goriparti, S., Miele, E., De Angelis, F., Fabrizio, E. D., Zaccaria, R. P., and Capiglia, C. (2014). Review on recent progress of nanostructured anode materials for Li-ion batteries, *J. Power Sources*, **257**, pp. 421–443.
15. Eason, R. (2006). *Pulsed Laser Deposition of Thin Films* (Wiley, New York).
16. Chrisey, D. B., and Hubler, G. K. (1994). *Pulsed Laser Deposition of Thin Films* (Wiley, New York).
17. Phipps, C. R. (2006). *Laser Ablation and its Applications* (Springer, New York).
18. Julien, C., Haro-Poniatowski, E., Camacho-Lopez, M. A., Escobar-Alarcon, L., and Jimenez-Jarquin, J. (2000). Growth of $LiMn_2O_4$ thin films by pulsed laser deposition and their electrochemical properties in lithium microbatteries, *Mater. Sci. Eng. B*, **72**, pp. 36–46.

19. Iriyama, Y., Inaba, M. Abe, T., and Ogumi, Z. (2001). Preparation of c-axis oriented thin films of LiCoO$_2$ by pulsed laser deposition and their electrochemical properties, *J. Power Sources*, **94**, pp. 175–182.

20. Xia, H., Lu, L., and Ceder, G. (2006). Substrate effect on the microstructure and electrochemical properties of LiCoO$_2$ thin films grown by PLD, *J. Alloys Compd.*, **417**, pp. 304–310.

21. Kim, W. S. (2004). Characteristics of LiCoO$_2$ thin film cathodes according to the annealing ambient for the post-annealing process, *J. Power Sources*, **134**, pp. 103–109.

22. Kuwata, N., Kumar, R., Toribami, K. Suzuki, T., Hattori, T., and Kawamura, J. (2006). Thin film lithium ion batteries prepared only by pulsed laser deposition, *Solid State Ionics*, **177**, pp. 2827–2832.

23. Xia, H., and Lu, L. (2007). Texture effect on the electrochemical properties of LiCoO$_2$ thin films prepared by PLD, *Electrochim. Acta*, **52**, pp. 7014–7021.

24. Xia, H., Lu, L., and Meng, Y. S. (2008). Growth of layered LiNi$_{0.5}$Mn$_{0.5}$O$_2$ thin films by pulsed laser deposition for application in microbatteries, *Appl. Phys. Lett.*, **92**, p. 011912.

25. Simmen, F., Lippert, T., Novák, P., Neuenschwander, B., Döbeli, M., Mallepell, M., and Wokaun, A. (2008). The influence of lithium excess in the target on the properties and compositions of Li$_{1+x}$Mn$_2$O$_{4-\delta}$, *Appl. Phys. A*, **93**, pp. 711–716.

26. Perkins, J. D., Bahn, C. S., Parilla, P. A., McGraw, J. M., Fu, M. L., Duncan, M., Yu, H., and Ginley, D. S. (1999). LiCoO$_2$ and LiCo$_{1-x}$Al$_x$O$_2$ thin film cathodes grown by pulsed laser ablation, *J. Power Sources*, **81–82**, pp. 675–679.

27. Wang, G. X., Lindsay, M. J., Ionescu, M., Bradhust, D. H., Dou, S. X., and Liu, H. K. (2001). Physical and electrochemical characterization of LiNi$_{0.8}$Co$_{0.2}$O$_2$ thin-film electrodes deposited by laser ablation, *J. Power Sources*, **97–98**, pp. 298–302.

28. Ramana, C. V., Zaghib, K., and Julien, C. M. (2007). Pulsed-laser deposited LiNi$_{0.8}$Co$_{0.15}$Al$_{0.05}$O$_2$ thin films for application in microbatteries, *Appl. Phys. Lett.*, **90**, p. 021916.

29. Lu, Z. G., Lo, M. F., and Chung, C. Y. (2008). Pulsed laser deposition and electrochemical characterization of LiFePO$_4$-C composite thin films, *J. Phys. Chem.*, **112**, pp. 7069–7078.

30. Yan, B., Liu, J., Song, B., Xiao, P., and Lu, L. (2013). Li-rich thin film cathode prepared by pulsed laser deposition, *Sci. Rep.*, **3**, p. 3332 (1–5).

31. Biserni, E., Xie, M., Brescia, R., Scarpellini, A., Hashempour, M., Movahed, P., George, S. M., Bestetti, M., Li Bassi, A., and Bruno, P. (2015). Silicon algae with carbon topping as thin-film anodes for lithium-ion microbatteries by two-step facile method, *J. Power Sources*, **274**, pp. 252–259.

32. Radhakrishnan, G., Adams, P. M., Foran, B., Quinzio, M. V., and Brodie, M. J. (2013). Pulsed laser deposited Si on multilayer graphene as anode material for lithium ion batteries, *APL Mater.*, **1**, p. 062103 (1–6).

33. Kuwata, N., Kawamura, J., Toribami, K., Hattori, T., and Sata, N. (2004). Thin-film lithium-ion battery with amorphous solid electrolyte fabricated by pulsed laser deposition, *Electrochem. Commum.*, **6**, pp. 417–421.

34. Goriparti, S., Miele, E., De Angelis, F., Di Fabrizio, E., Zaccaria, R. P., and Capiglia, C. (2014). Review on recent progress of nanostructured anode materials for Li-ion batteries, *J. Power Sources*, **257**, pp. 421–443.

35. Ryu, J. H., Kim, J. W., Sung, Y. E., and Oh, S. M. (2004). Failure modes of silicon powder negative electrode in lithium secondary batteries, *Electrochem. Solid-State Lett.*, **7**, pp. A306–A309.

36. Zhao, S., Fu, Z., and Qin, Q. (2002). A solid-state electrolyte lithium phosphorus oxynitride film prepared by pulsed laser deposition, *Thin Solid Films*, **415**, pp. 108–113.

37. Zhao, S., and Qin, Q. (2003). Li-V-Si-O thin film electrolyte for all-solid-state Li-ion battery, *J. Power Sources*, **122**, pp. 174–180.

38. Kawamura, J., Kuwata, N., Toribami, K., Sata, N., Kamishma, O., and Hattori, T. (2004). Preparation of amorphous lithium ion conductor thin films by pulsed laser deposition, *Solid State Ionics*, **175**, pp. 273–276.

39. Ohta, N., Takada, K., Osada, M., Zhang, L., Sasaki, T., and Watanabe, M. (2005). Solid electrolyte, thio-LISICON thin film prepared by pulsed laser deposition, *J. Power Sources*, **146**, pp. 707–710.

40. Sakurai, Y., Sakuda, A., Hayashi, A., and Tatsumisago, M. (2011). Preparation of amorphous Li_4SO_4-Li_3PO_4 thin films by pulsed laser deposition for all-solid-state lithium secondary batteries, *Solid State Ionics*, **182**, pp. 59–63.

41. Sakuda, A., Hayashi, A., Hama, S., and Tatsumisago, M. (2010). Preparation of highly lithium-ion conductive 80Li_2S-20P_2O_5 thin-film electrolytes using pulsed laser deposition, *J. Am. Ceram. Soc.*, **93**, pp. 765–768.

42. Kim, H., Auyeung, R. C. Y., and Piqué, A. (2007). Laser-printed thick-film electrodes for solid-state rechargeable Li-ion microbatteries, *J. Power Sources*, **165**, pp. 413–419.
43. Ollinger, M., Kim, H., Sutto, T. E., and Piqué, A. (2006). Laser printing of nanocomposite solid-state electrolyte membranes for Li microbatteries, *Appl. Surf. Sci.*, **252**, pp. 8212–8216.
44. Ollinger, M., Kim, H., Sutto, T. E., Martin, F., and Piqué, A. (2006). Laser direct-write of polymer nanocomposites, *J. Laser Micro/Nanoeng.*, **1**, pp. 102–105.
45. Kim, H., Pröll, J., Kohler, R., Pfleging, W., and Piqué, A. (2012). Laser-printed and processed $LiCoO_2$ cathode thick films for Li-ion microbatteries, *J. Laser Micro/Nanoeng.*, **7**, pp. 320–325.
46. Arnold, C. B., Wartena, R. C., Pratap, B., Swider-Lyons, K. E., and Piqué, A. (2002). Direct writing of planar ultracapacitor by laser forward transfer processing, *Proc. SPIE*, **4637**, pp. 353–360.
47. Wartena, R. C., Curtright, A. E., Arnold, C. B., Piqué, A., and Swider-Lyons, K. E. (2004). Li-ion microbatteries generated by a laser direct-write method, *J. Power Sources*, **126**, pp. 193–202.
48. Kim, H., Kushto, G. P., Arnold, C. B., Kafafi, Z. H., and Piqué, A. (2004). Laser processing of nanocrystalline TiO_2 for dye-sensitized solar cells, *Appl. Phys. Lett.*, **85**, pp. 464–466.
49. Kim, H., Auyeung, R. C. Y., Ollinger, M., Kushto, G. P., Arnold, C. B., Kafafi, Z. H., and Piqué, A. (2006). Laser-sintered mesoporous TiO_2 electrodes for dye-sensitized solar cells, *Appl. Phys. A*, **83**, pp. 73–76.
50. Piqué, A., Auyeung, R. C. Y., Kim, H., Metkus, K. M., and Mathews, S. A. (2008). Digital microfabrication by laser decal transfer, *J. Laser Micro/Nanoeng*, **3**, pp. 163–169.
51. Duocastella, M., Kim, H., Serra, P., and Piqué, A. (2012). Optimization of laser printing of nanoparticles suspensions for microelectronic applications, *Appl. Phys. A*, **106**, pp. 471–478.
52. Arnold, C. B., Serra, P., and Piqué, A. (2007). Laser direct-write techniques for printing of complex materials, *MRS Bull.*, **32**, pp. 23–31.
53. Calvert, P. (2001). Inkjet printing for materials and devices, *Chem. Mater.*, **13**, pp. 3299–3305.
54. Kang, H., Soltman, D., and Subramanian, V. (2010). Hydrostatic optimization of inkjet-printed films, *Langmuir*, **26**, pp. 11568–11573.
55. Soltman, D., Smith, B., Kang, H., Morris, S. J. S., and Subramanian, V. (2010). Methodology for inkjet printing of partially wetting films, *Langmuir*, **26**, pp. 15686–15693.

56. Piqué, A., Auyeung, R. C. Y., Smith, A. T., Kim, H., Mathews, S. A., Charipar, N. A., and Kirleis, M. A. (2013). Laser transfer of reconfigurable patterns with a spatial light modulator, *Proc. SPIE*, **8608**, p. 86080K.

57. Piqué, A., Kim, H., Auyeung, R. C. Y., and Smith, A. (2013). Laser forward transfer of functional materials for digital fabrication of microelectronics, *J. Imaging Sci. Technol.*, **57**, p. 040101.

58. Wang, J., Auyeung, R. C. Y., Kim, H., Charipar, N. A., and Piqué, A. (2010). Three-dimensional printing of interconnects by laser direct-write of silver nanopastes, *Adv. Mater.*, **22**, pp. 4462–4466.

59. Kim, H., Piqué, A., Charipar, K. M., Auyeung, R. C. Y., and Duocastella, M. (2013). Laser printing of conformal and multi-level 3D interconnects, *Appl. Phys. A*, **113**, pp. 5–8.

60. Kim, H., Melinger, J. S., Khachatrian, A., Charipar, N. A., Auyeung, R. C. Y., and Piqué, A. (2010). Fabrication of terahertz metamaterials by laser printing, *Opt. Lett.*, **35**, pp. 4039–4041.

61. Auyeung, R. C. Y., Kim, H., Birnbaum, A. J., Zalaludinov, M., Mathews, S. A., and Piqué, A. (2009). Laser decal transfer of freestanding microcantilevers and microbridges, *Appl. Phys. A*, **97**, pp. 513–519.

62. Auyeung, R. C. Y., Kim, H., Charipar, N., Birnbaum, A., Mathews, S., and Piqué, A. (2011). Laser forward transfer based on a spatial light modulator, *Appl. Phys. A*, **102**, pp. 21–26.

63. Mathews, S. A., Auyeung, R. C. Y., Kim, H., Charipar, N. A., and Piqué, A. (2013). High-speed video study of laser-induced forward transfer of silver nano-suspensions, *J. Appl. Phys.*, **114**, p. 064910 (1–9).

64. Arnold, C. B., Wartena, R. C., Swider-Lyons, K. E., and Piqué, A. (2003). Direct-write planar microultracapacitors by laser engineering, *J. Electrochem. Soc.*, **150**, pp. A571–A575.

65. Sarangapani, S., Tilak, B., and Chen, C. (1996). Materials for electrochemical capacitors, *J. Electrochem. Soc.*, **143**, pp. 3791–3799.

66. Dmowski, W., Egami, T., Swider-Lyons, K. E., Love, C. T., and Rolison, D. R. (2002). Local atomic structure and conduction mechanism of nanocrystalline hydrous RuO_2 from x-ray scattering, *J. Phys. Chem. B*, **106**, pp. 12677–12683.

67. McKeown, D. A., Hagans, P. L., Carette, L. P. L., Russell, A. E., Swider, K. E., and Rolison, D. R. (1999). Structure of hydrous ruthenium oxides: implications for charge storage, *J. Phys. Chem. B*, **103**, pp. 4825–4832.

68. Long, J. W., Dunn, B., Rolison, D. R., and White, H. S. (2004). Three-dimensional battery architectures, *Chem. Rev.*, **104**(10), pp. 4463–4492.

69. Notten, P. H. L., Roozeboom, F., Niessen, R. A. H., and Baggetto, L. (2007). 3-D integrated all-solid-state rechargeable batteries, *Adv. Mater.*, **19**(24), pp. 4564–4567.

70. Oudenhoven, J. F. M., Baggetto, L., and Notten, P. H. L. (2011). All-solid-state lithium-ion microbatteries: a review of various three-dimensional concepts, *Adv. Energy Mater.*, **1**(1), pp. 10–33.

71. Ketterer, B., Vasilchina, H., Seemann, K., Ulrich, S., Besser, H., Pfleging, W., and Kaiser, T. (2008). Development of high power density cathode materials for Li-ion batteries, *Int. J. Mater. Res.*, **99**(10), pp. 1171–1176.

72. Kohler, R., Besser, H., Hagen, M., Ye, J., Ziebert, C., Ulrich, S., Pröll, J., and Pfleging, W. (2011). Laser micro-structuring of magnetron-sputtered SnO_x thin films as anode material for lithium ion batteries, *Microsyst. Technol.*, **17**(2), pp. 225–232.

73. Kohler, R., Bruns, M., Smyrek, P., Ulrich, S., Przybylski, M., and Pfleging, W. (2010). Laser annealing of textured thin film cathode material for lithium ion batteries, *Proc. SPIE*, **7585**, p. 75850O (1–11).

74. Kohler, R., Pröll, J., Ulrich, S., Przybylski, M., and Pfleging, W. (2011). Laser processing of SnO_2 electrode materials for manufacturing of 3D micro-batteries, *Proc. SPIE*, **7921**, p. 79210P (1–11).

75. Kohler, R., Pröll, J., Ulrich, S., Trouillet, V., Indris, S., Przybylski, M., and Pfleging, W. (2009). Laser-assisted structuring and modification of $LiCoO_2$ thin films, *Proc. SPIE*, **7202**, p. 720207 (1–11).

76. Kohler, R., Smyrek, P., Ulrich, S., Bruns, M., Trouillet, V., and Pfleging, W. (2010). Patterning and annealing of nanocrystalline $LiCoO_2$ thin films, *J. Optoelectron. Adv. Mater.*, **12**(3), pp. 547–552.

77. Pröll, J., Kohler, R., Adelhelm, C., Bruns, M., Torge, M., Heißler, S., Przybylski, M., Ziebert, C., and Pfleging, W. (2011). Laser modification and characterization of Li-Mn-O thin film cathodes for lithium-ion batteries, *Proc. SPIE*, **7921**, p. 79210Q (1–14).

78. Pröll, J., Kohler, R., Mangang, A., Ulrich, S., Ziebert, C., and Pfleging, W. (2012). 3D structures in battery materials, *J. Laser Micro/Nanoeng.*, **7**(1), pp. 97–104.

79. Pröll, J., Kohler, R., Torge, M., Ulrich, S., Ziebert, C., Bruns, M., Seifert, H. J., and Pfleging, W. (2011). Laser microstructuring and annealing processes for lithium manganese oxide cathodes, *Appl. Surf. Sci.*, **257**, pp. 9968–9976.

80. Pröll, J., Kohler, R., Mangang, A., Ulrich, S., Bruns, M., Seifert, H. J., and Pfleging, W. (2012). Diode laser heat treatment of lithium manganese oxide films, *Appl. Surf. Sci.*, **258**(12), pp. 5146–5152.

81. Kohler, R., Pröll, J., Bruns, M., Ulrich, S., Seifert, H. J., and Pfleging, W. (2013). Conical surface structures on model thin-film electrodes and tape-cast electrode materials for lithium-ion batteries, *Appl. Phys. A*, **112**(1), pp. 77–85.

82. Pröll, J., Weidler, P. G., Kohler, R., Mangang, A., Heissler, S., Seifert, H. J., and Pfleging, W. (2013). Comparative studies of laser annealing technique and furnace annealing by X-ray diffraction and Raman analysis of lithium manganese oxide thin films for lithium-ion batteries, *Thin Solid Films*, **531**, pp. 160–171.

83. Zhang, Y., Chung, C. Y., and Min, Z. (2008). Growth of HT-LiCoO$_2$ thin films on Pt-metatized silicon substrates, *Rare Met.*, **27**(3), pp. 266–272.

84. Julien, C. M., and Massot, M. (2003). Lattice vibrations of materials for lithium rechargeable batteries III. Lithium manganese oxides, *Mater. Sci. Eng.*, **100**, pp. 69–78.

85. Park, S. H., Sato, Y., Kim, J.-K., and Lee, Y.-S. (2007). Powder property and electrochemical characterization of Li$_2$MnO$_3$ material, *Mater. Chem. Phys.*, **102**(2–3), pp. 225–230.

86. Julien, C. M., and Massot, M. (2003). Lattice vibrations of materials for lithium rechargeable batteries I. Lithium manganese oxide spinel, *Mater. Sci. Eng.*, **97**, pp. 217–230.

87. Winter, R., and Heitjans, P. (2001). Li+ diffusion and its structural basis in the nanocrystalline and amorphous forms of two-dimensionally ion-conducting Li$_x$TiS$_2$, *J. Phys. Chem. B*, **105**(26), pp. 6108–6115.

88. Hudaya, C., Halim, M., Pröll, J., Besser, H., Choi, W., Pfleging, W., Seifert, H. J., and Lee, J. K. (2015). A polymerized C-60 coating enhancing interfacial stability at three-dimensional LiCoO$_2$ in high-potential regime, *J. Power Sources*, **298**, pp. 1–7.

89. Kohler, R., Pröll, J., Bruns, M., Ulrich, S., Seifert, H. J., and Pfleging, W. (2013). Conical surface structures on model thin-film electrodes and tape-cast electrode materials for lithium-ion batteries, *Appl. Phys. A*, **112**, 77–85.

90. Pfleging, W., Kohler, R., Südmeyer, I., and Rohde, M. (2013). *Laser Micro and Nano Processing of Metals, Ceramics, and Polymers* (Springer, Berlin, Heidelberg).

91. Pröll, J., Kohler, R., Mangang, A., Ulrich, S., Ziebert, C., and Pfleging, W. (2012). 3D structures in battery materials, *J. Laser Micro/Nanoeng.*, **7**(1), pp. 97–104.

92. Kin, J. S., Pfleging, W., Kohler, R., Seifert, H. J., Kim, T. Y., Byun, D., Jung, H. G., Choi, W., and Lee, J. K. (2015). Three-dimensional silicon/carbon core-shell electrode as an anode material for lithium-ion batteries, *J. Power Sources*, **279**, pp. 13–20.

93. Pfleging, W., Mangang, M., Zheng, Y., and Smyrek, P. (2013). Laser structuring for improved battery performance, *SPIE Newsroom*, pp. 1–3.

94. Wood, D. L., Li, J. L., and Daniel, C. (2015). Prospects for reducing the processing cost of lithium ion batteries, *J. Power Sources*, **275**, pp. 234–242.

95. Pröll, J., Schmitz, B., Niemöeller, A., Robertz, B., Schäfer, M., Torge, M., Smyrek, P., Seifert, H. J., and Pfleging, W. (2015). Femtosecond laser patterning of lithium-ion battery separator materials: impact on liquid electrolyte wetting and cell performance, *Proc. SPIE*, **9351**, p. 93511F-7.

96. Pfleging, W., and Pröll, J. (2014). A new approach for rapid electrolyte wetting in tape cast electrodes for lithium-ion batteries, *J. Mater. Chem. A*, **2**(36), pp. 14918–14926.

97. Choi, J., Morikawa, E., Ducharme, S., and Dowben, P. A. (2005). Comparison of crystalline thin poly(vinylidene (70%)-trifluoroethylene (30%)) copolymer films with short chain poly(vinylidene fluoride) films, *Mater. Lett.*, **59**(28), pp. 3599–3603.

98. Slocombe, A., and Li, L. (2000). Laser ablation machining of metal/polymer composite materials, *Appl. Surf. Sci.*, **154**, pp. 617–621.

99. Mangang, M., Pröll, J., Tarde, C., Seifert, H. J., and Pfleging, W. (2014). Ultrafast laser microstructuring of $LiFePO_4$ cathode material, *Proc. SPIE*, **8968**, p. 89680M (1–9).

100. Smyrek, P., Pröll, J., Seifert, H. J., and Pfleging, W. (2016). Laser-induced breakdown spectroscopy of laser-structured $Li(NiMnCo)O_2$ electrodes for lithium-ion batteries, *J. Electrochem. Soc.*, **163**(2), pp. A19–A26.

Index

ablated material, 52, 135–36, 157–58, 160–61, 163, 167, 173, 181–82, 317, 324–26, 331, 442, 446, 532
ablated species, 26, 52, 134–36, 139–40, 158, 178, 288, 415, 442
ablating, 68, 96, 99, 161, 163, 166, 533
ablation, 49–52, 96–100, 106–9, 158–59, 163–64, 317–18, 320–21, 323–27, 331–32, 341–43, 427–28, 443–44, 447–48, 453–54, 481
  direct, 524
  local, 194
  minimal, 407
  nanoscale, 382
  photophysical, 339
  single-pulse, 333, 342
  solvent, 219
  thin Ag film, 434, 453
  thin-film, 453
ablation depth, 89–90, 98, 161–62, 164, 322, 331–43, 335, 342–44, 435
ablation plume, 49, 158, 182, 323, 331, 338, 426–29, 433, 439–46, 454–55
  ejected, 426
  expanding, 143
  luminous, 158
  nascent, 165–66, 171, 342
  structured fs, 158
ablation threshold, 98, 106, 109, 117, 144, 161, 170, 330, 474–75, 488, 505

absorbance, 137, 204, 226–27, 297, 382, 481
absorption, 87, 92, 133–34, 199, 203, 225–26, 232–33, 279, 296–97, 341–42, 365, 372–73, 377, 382, 481
  free carrier, 362
  high, 209, 216, 227, 230, 236
  hot carrier, 134
  inhomogeneous radiation, 408
  laser-sustained, 134
  laser target, 91
  linear, 360
  local, 235
  low, 223
  second pulse energy, 342
  significant, 479
  simultaneous, 361
  single-photon, 361, 364–65, 373
  three-photon, 365, 373
  time-resolved X-ray, 162, 187, 456
  two-photon, 361
absorption coefficient, 88, 200, 231, 360
  effective, 365
  high, 475
  linear optical, 223
  low linear, 479
adatom, 27–45, 53–60, 63–64, 67, 101, 111
adatom density, 35, 39, 63
adatom diffusion, 31–32, 34–35, 37–41, 51, 54
adatom diffusivity, 32, 37, 40, 45
adherence, 209, 246, 248, 254–57, 268

adhesion, 48, 101, 112–13, 255–56, 284, 384, 432, 474, 484–85, 489
adsorption, 3, 27, 29, 31–35, 51, 433
AFM, *see* atomic force microscopy
aggregation, 13–14, 162, 216, 282, 298, 364
alloys, 47, 61–62, 73, 77, 83, 234, 310, 335, 463
amorphization, 221–22, 224, 475
angular distribution, 90, 171–73, 179, 442
annealing, 45, 115, 226, 257, 373, 523
anodes, 267, 499, 502–3, 507–10, 512, 520, 526, 528, 533–34
Arrhenius law, 18, 31, 36, 54–55
atomic force microscopy (AFM), 170, 203–4, 222, 254–55, 259–61, 264–66, 325, 331, 404–7, 409–10, 480, 485
atomic plume, 159–60, 168–70, 172–73, 175–79
atomistic simulations, 425, 435–36, 452–53
atom transfer radical polymerization (ATRP), 282, 285, 292–93, 296–97, 303
ATR-IR, *see* attenuated total reflectance infrared
ATRP, *see* atom transfer radical polymerization
attenuated total reflectance infrared (ATR-IR), 255

background gas, 156–58, 168, 172, 175–79, 182, 319, 425–30, 43, 442–45, 454, 506
backscattering, 95, 122, 368
bandgap, 133, 140, 213, 226, 230, 232, 259, 360–61, 379–80, 416
batteries, 259, 500, 506, 508, 510–14, 518, 520–22, 525–26, 528–33, 535

high-power, 531
microscale, 533
thick-film, 499
thin-film, 500, 505, 510, 512, 522
beams, 198, 264, 321–24, 364–65, 378, 402, 409, 473–74
charged particle, 3
energetic ion deposition, 68
energetic particle, 4
linearly polarized, 483
projected, 514
BG, *see* bioglass
binding energy, 27, 29, 35, 39, 228
biochips, 357–58, 375, 384, 386
biocompatibility, 264, 278, 375
bioglass (BG), 257–58
Boltzmann constant, 26, 91, 340
bonding, 7, 22, 24, 28–29, 136, 256–58, 432
chemical, 6, 162
molecular, 254–55
preferential, 292
bonds, 6–7, 23, 27–34, 36, 87, 137, 226, 228–29, 235, 247, 296, 364, 373
breakdown, 284–85, 299, 303, 336, 359–61, 366, 378, 448
bubble, 203, 331, 338, 447, 454, 469
bulk material, 2, 12, 24, 115–16, 251

capacitors, 280, 284, 300, 302–3, 502, 513–14
carbon, 110, 112, 136, 213, 266, 502, 529
amorphous, 101, 137
conductive, 533
diamondlike, 137
glassy, 137
graphitic, 224
graphitized, 508
carbon nanotubes, 224

multiwalled, 211, 221
single-walled, 211
cathodes, 499, 502–3, 505–6, 512, 517–20, 522–23, 527–30, 533–34
cavitation bubble, 322, 337–38, 446–48
CCD, *see* charge-coupled device
ceramics, 47, 113–14, 126, 252, 258, 265, 280, 421
charge-coupled device (CCD), 91, 94, 158, 260, 324, 513
chemical bond, 22, 24, 105, 133, 371
chemical etching, 358, 372
chemical vapor deposition (CVD), 46, 247, 287, 508–9
click chemistry, 283
clusters, 4, 13, 15–19, 23, 27, 29, 37, 41, 101, 110, 134, 160, 166, 182, 288, 326, 331, 426–27, 429, 435, 440–41, 443–44, 453–54
coalescence, 14, 22, 39, 41–43, 49, 58–59, 110–11, 115, 137, 144, 437, 444, 446, 452, 454
coatings, 42, 137, 214, 224, 246–48, 252, 254, 256–58, 264–65, 338, 489, 509, 512
codeposition, 227, 230, 533
coefficient, 17, 31, 197, 254, 431–32
  accommodation, 31
  characteristic kinetic, 31
  dynamical viscosity, 236
  friction, 137
  heat diffusion, 89
  metal diffusion, 114
  sticking, 31
  thermal expansion, 9–10, 48, 257
coefficient of variation (CV), 197, 280
cohesive energy, 98, 105, 108–9

collisions, 55, 60, 92, 139, 433, 442, 444
condensation, 4–5, 8–9, 13–15, 26, 31–32, 35, 40, 46, 52, 58, 319, 326, 443, 449, 453
conduction band, 340, 361, 479
conductivity, 299, 301–2, 330, 503, 510, 512, 533
confinement, 53, 136, 156, 177–78, 227, 427, 445, 449, 452, 454–55
contaminations, 46–47, 197, 266, 426
core-shell QDs, 219–20, 222
coupling
  electron–phonon, 84, 184, 320, 412, 431–32
  electron-to-lattice, 90
  laser energy, 336
  laser plasma, 86
  laser radiation, 245
  mechanical, 254–55
  plasmon, 344
CP, *see* critical point
craters, 118–19, 146, 194–95, 325, 331–34, 336, 399–400, 403–407, 409–10, 416–18, 420–21
critical point (CP), 165–66, 340
crystal defects, 29, 428
crystalline, 4, 10–11, 28–29, 42–44, 50, 55, 110–14, 136, 216, 218, 220, 247, 264, 286, 290
crystallinity, 44, 113, 179, 288, 506, 524
crystals, 3, 6–11, 13–14, 20, 22, 24, 29–30, 45, 60, 288, 369
cubic
  body-centered, 61
  face-centered, 61, 112
CV, *see* coefficient of variation
CV, *see* cyclic voltammetry
CVD, *see* chemical vapor deposition
cyclic voltammetry (CV), 523

decomposition, 46, 167, 209, 223, 234–35, 325, 340, 437–38, 444, 448, 452–54, 531
defects, 4, 9, 12, 27, 36, 44, 61, 116, 209, 247, 360, 368
delay time, 94–97, 383, 409–12, 416, 418–19
DELI, *see* differential evanescent light intensity
density, 34, 56–57, 63–64, 88, 90–91, 165–66, 169, 338–42, 377–78, 431, 434, 437, 439, 441, 450–51
  average, 51, 263
  carrier, 406, 421
  critical, 372, 434
  current, 301, 511, 518–19, 521
  decay, 92
  dislocation, 44
  high, 56–57, 66, 86, 110, 161, 262–63, 505
  high energy storage, 280, 285
  information, 488
  laser energy surface, 88
  lateral, 431
  low kink, 30
  packing, 48
  photon, 377
  saturation, 37, 62
deposit, 1, 9–10, 14, 21, 41–49, 54, 60, 66, 170, 173, 180, 216, 234, 324, 508, 512
deposition, 1–5, 23, 25–29, 31–32, 34–37, 39–46, 48, 50–64, 66–68, 86, 100–102, 104, 114, 132–33, 143–44, 146, 207–9, 213–14, 223–25, 505, 512
  chemical vapor, 46, 247, 287, 509
  conventional MAPLE, 234
  physical vapor, 46, 79, 287
  pulsed, 63–64, 81
  sequential, 114
deposition flux, 26, 31, 35–37, 43, 45, 47–48, 50–53, 55–57, 62–64, 66, 68, 111, 113

deposition pulse, 49, 51–52, 55, 62–64, 66–67
deposition rate, 3, 40, 47, 51, 56–57, 60, 62, 65–66, 86, 111, 171, 177–78, 181–82, 227, 254
desorption, 3, 31–32, 35, 51, 63
dewetting, 114–17, 251, 271
diamond-like carbon (DLC), 137, 152, 252, 271
dielectric constant, 259, 280, 283–86, 299, 301, 304, 406
dielectrics, 85, 108, 283, 285, 334, 336, 344, 401, 474, 481
differential evanescent light intensity (DELI), 260–61
diffraction, 203, 481
diffusion, 3, 27–29, 32–34, 36–42, 43, 45, 52–56, 62–64, 111, 113–14, 235, 326, 431
diffusivity, 43, 45, 53–54, 56, 60, 62, 64–65, 67, 115
dispersion, 168, 234, 236, 285, 289–93, 301
distilled water, 212, 214, 227–32, 236, 374
DLC, *see* diamond-like carbon
double pulse (DP), 317, 319–21, 323–24, 328–29, 332–34, 336–38, 340–41, 343–44, 399, 402, 404, 409, 411, 426
double-pulse laser ablation (DP-LA), 320, 329, 332, 334–38, 344
DP, *see* double pulse
drug delivery, 147, 278, 345
drying
  freeze, 246
  spray, 288
dry laser cleaning, 202, 206

EAM, *see* embedded atom method
EDX, *see* energy-dispersive X-ray
effective surface area (ESA), 267
Ehrlich–Schwoebel barriers, 34, 42, 59, 64

Index | **549**

electric field, 91, 133, 194, 296, 301, 303, 258, 367, 400, 406–7
electrodes, 296, 502, 513–19, 521–22, 528–33, 543
  active, 513
  carbon-containing, 533
  laser-printed, 516, 519
  small biased metallic, 94
  state-of-the-art, 522
  tape-cast, 524, 530–31
  thick-film composite, 530
  unstructured, 532
electrolyte, 267, 502, 509, 511, 513, 515, 520, 528, 530–31
  gel polymer, 516–17
  leak-free, 533
  single-drop, 531
  solid-state composite polymer, 533
  solid-state polymer, 520
  sulfuric acid, 515
electromagnetic fields, 143, 193, 260, 267
electron cooling, 89, 404
electron heat, 151, 157, 330, 339, 363, 412, 430–32, 463
electron heating, 362
electron temperature, 87, 134, 157, 320, 363, 412–13, 421, 430–32
embedded atom method (EAM), 432, 438, 450
emission, 3, 91–95, 137, 158–60, 169–72, 230, 323, 328–31, 333, 343, 440, 465, 479
energy barriers, 16, 18, 31–34, 36, 43
energy density, 131, 133, 137, 280, 286, 299, 301, 304, 326–27, 331, 437–38, 520, 522, 528–29, 533
energy deposition, 156, 359, 403, 406, 408, 435
energy-dispersive X-ray (EDX), 141

energy exchange, 31, 135, 140, 431, 434
energy storage, 210, 266, 280–81, 284, 286, 299–301, 303–4, 500–2, 513–14, 533–34
energy transfer, 57, 87, 156, 327, 437, 479
enthalpy, 5–6, 101, 118, 432
entropy, 5
EOS, *see* equation of state
epitaxial growth, 10, 20, 23, 42, 53, 59–61, 65, 111–12, 114
epitaxy, 4, 10, 25, 65, 111–12
equation of state (EOS), 339–40, 405, 446, 448
equilibrium, 2–9, 13, 18–21, 23, 26, 41, 46, 60, 63, 90, 112, 136, 139, 158, 165
ESA, *see* effective surface area
etching, 2, 204, 360, 370–71, 374–75
evaporation, 27, 31, 47–48, 52–53, 55, 89, 111, 197, 338–39, 384, 405, 505
excimer laser ablation, 530
excitation, 86, 134, 204, 267–68, 335, 373, 381, 399–400, 406, 412, 419, 478
expansion, 95, 136, 140, 159–61, 163, 167, 169, 171–72, 174–75, 182, 335, 338–39, 444, 446, 524–25
  adiabatic, 90, 442
  continuous, 268
  dynamic, 169
  early time, 169
  fast, 167
  finite, 19
  forward-directed, 50
  forward-peaked, 172
  free, 175
  hemispherical, 176
  initial, 93
  initial isothermal, 90

initial material, 162
late time, 169
rapid, 68, 167
explosive boiling, 222, 224, 234, 437

fabrication
    facile, 288
    laser-assisted, 400
    nanoscale, 386
face-centered cubic (fcc), 61–62, 78, 112
fcc, see face-centered cubic 61–62, 78, 112
FCPA, see Fiber Chirped Pulse Amplification
FDTD, see finite-difference time domain
femtosecond laser ablation (fs-LA), 155–82, 440
femtosecond laser-assisted etching (FLAE), 373, 383–85
femtosecond laser, 68, 320, 336, 339, 360, 370, 372, 472
femtosecond pulses, 156, 319, 473, 478–79, 481
Fermi electron distribution, 87
Fermi level, 87
ferroelectric, 47, 482, 488
FESEM, see field emission scanning electron microscopy
FIB, see focused ion beam
Fiber Chirped Pulse Amplification (FCPA), 363
field emission scanning electron microscopy (FESEM), 225, 402, 404–405, 416
films, 12, 25, 41, 43, 45, 61, 64, 77, 114, 132, 138, 144, 213, 215, 225, 227, 234, 303, 487, 512, 522–25
finite-difference time domain (FDTD), 200

FLAE, see femtosecond laser-assisted etching
fluence, 52, 62, 88–92, 98–99, 105–9, 116–19, 161–62, 164–66, 168–74, 217, 219, 221–24, 234, 327, 341–43, 416–17, 419–20, 426, 431, 435–36, 438–41, 443, 449–53, 474–77, 479–80, 482, 488
fluorescence, 139, 381–83
flux, 26–27, 35, 37, 45–46, 54, 58, 62–63, 76, 86, 104, 108, 110–11, 504–505
FM, see Frank–van der Merwe, 20–22
focused ion beam (FIB), 527–28
Fourier transform infrared (FTIR), 225, 295–97
fragmentation, 117, 158, 165, 167, 222, 334, 337, 339–43
Frank–van der Merwe (FM), 20–21
free electrons, 87, 133, 320, 340, 361–62, 371, 373, 479
free energy, 3–9, 12, 14–17, 19–21, 23–24, 27–28, 36, 101, 256, 484
freestanding structures, 514, 527
frequencies
    angular, 302
    pulsing, 62–63
    radio, 257
    second harmonic, 417
    thermal oscillation, 31
    vibration, 31, 33, 54, 214
    vibration lattice, 32
fs-LA, see femtosecond laser ablation
fs laser pulses, 156–58, 165–66, 168, 172, 182, 262, 267, 274, 410, 436, 438–39, 441, 448–51
    double, 404, 409, 416, 421
    ultrashort, 264
fs pulses, 90, 156–57, 161, 164, 173, 180, 263, 379, 435

Index | 551

FTIR, see Fourier transform infrared
fuel cells, 318, 345
full-width at half-maximum (FWHM), 203, 338
functional groups, 228–29, 283
  nitrogen-containing, 229
  oxygen-containing, 228, 231
functionalities, 208, 278, 359, 383, 426, 472
functional materials, 1, 114, 183, 237, 269, 512, 514, 540
FWHM, see full-width at half-maximum

gas environments, 49, 126, 426, 430, 433, 439, 443, 454
gas pressure, 132, 135–39, 144–45, 160, 175–77, 179, 181, 214, 445, 510
gas sensors, 210, 241, 259
Gaussian beam, 198–99, 323, 364, 378, 435–36
gel permeation chromatography (GPC), 298
Gibbs condition, 7
Gibbs criterion, 7
GISAXS, see grazing incidence small-angle X-ray scattering
GPC, see gel permeation chromatography
grafting, 282–83, 297
grafting from, 281, 283, 292
grafting to, 281, 283, 292
gratings, 418, 421
  Bragg, 369–70, 383
  periodic, 406
  superficial, 474
grazing incidence small-angle X-ray scattering (GISAXS), 477
growth modes, 4–5, 10, 12, 14, 20–23, 25–26, 37–38, 41–42, 46, 50, 53, 59, 66, 506

growth rates, 30, 35, 46, 48, 51–52, 57, 65, 100, 139
growth temperature, 26, 35, 45, 53–55, 57, 65–66

HA, see hydroxyapatite
HAZ, see heat-affected zone
hcp, see hexagonal close-packed
heat, 47, 87–88, 114, 116, 233, 235, 300, 363–64, 389, 446, 504
  latent, 88–89
heat-affected zone (HAZ), 68, 327, 358–59, 362, 385
heat conduction, 90, 403, 430, 531
heat conductivity, 344, 363, 412, 432
heating-cooling processes, fast, 359
heat transfer, 224, 233, 403
Helmholtz equations, 340
Hertz–Knudsen equation, 339
heteroepitaxy, 10, 13, 20–22, 67
heterogeneous nucleation, 17, 19, 36, 251, 289–90
heterostructures, 9, 23, 48–49, 278
hexagonal close-packed (hcp), 61–62, 202
highly oriented pyrolytic graphite (HOPG), 104
high-performance liquid chromatography (HPLC), 147
high-resolution scanning electron microscopy, 216
high-resolution transmission electron microscopy (HRTEM), 220–21, 224
HM, see hydrodynamic modeling
homoepitaxy, 10, 22
homogeneous nucleation, 17, 19, 36, 290, 340, 343
HOPG, see highly oriented pyrolytic graphite
hopping, 3, 28, 31–33, 42, 54

HPLC, *see* high-performance liquid chromatography
HRTEM, high-resolution transmission electron microscopy
hydrodynamic modeling (HM), 163–65, 171, 448
hydrodynamics, 339, 400, 403, 405, 478
hydrophobic, 214, 248–53
hydroxyapatite (HA), 256, 273
hyperthermal, 43, 48, 51–54, 56–57, 64–65, 67

ICCD, *see* intensified charge-coupled device
imaging, 94, 146, 170–71, 194, 259–60, 318
　fast, 158
　mask, 525
　time-resolved, 161, 328
　video, 513, 531
immobilization, 207–9, 225, 228, 292, 297
implantation, 52, 102, 106–10, 119
implants, 254, 256–58, 264, 273
impurities, 19, 29, 32, 36, 39, 44–45, 247
　impingement ratio, 141
　organic, 287
incident laser, 88, 401, 423, 477, 481
incident pulses, 400, 403–4, 407, 414, 416
inert gases, 132, 135, 137–38, 143, 153
inhomogeneous, 43, 139, 161–62, 403, 408, 521
in situ polymerization, 281, 285, 301
instability, 24, 116–17, 246, 454
instantaneous flux, 51, 104, 106
integrated optical density (IOD), 260

intensified charge-coupled device (ICCD), 158–59, 168, 178, 324
intensity, 51, 58–59, 91, 143, 169–72, 228–29, 262, 264, 328–30, 332–35, 357–61, 378, 404–405, 409–10, 523–24
interaction, 7, 54–55, 86, 175, 182, 247–48, 338, 357–60, 400–1, 417–19, 427, 433–35, 444, 452–53, 478–79
　chemical, 115, 255
　dipole, 255
　dynamic, 454
　electron–phonon, 404
　electron-to-phonon, 87
　interatomic, 428, 432–33
　laser–material, 425, 428, 473
　laser–matter, 247
　molecular, 255
　multiphoton, 360
　mutual, 143
　nonstationary, 87
　nonthermodynamic, 133
　pairwise, 341
　phonon-to-phonon, 87
　plume-gas, 136
　short-time, 52
　strain-mediated, 279
　strong, 194, 504
intercalation, 253, 502
interface, 3, 6–8, 12, 14, 17, 19–25, 36, 43, 50, 106, 109–10, 113, 141, 255–56, 370
　channel–polymer, 384
　cluster–substrate, 17
　coherent, 11
　dense, 447
　ideal, 10
　metal-oxide, 101
　metal–substrate, 432
　polar/nonpolar, 48
　polymer/filler, 284
　sharp, 10, 110, 161
　smooth, 49

Index | 553

substrate–film, 432
interfacial energies, 6, 10, 19–20, 60, 101, 248
interference, 161, 367, 401, 406, 414, 417, 421, 473–74, 477–78
  surface plasmon, 403
  wave, 407
interlayer transport, 42, 59, 76
interpulse delay, 318, 320–21, 323–24, 326, 329–38, 342–44, 404, 409–11, 416, 418, 420–21
interpulse delay time, 409–11, 413–15, 417–20
inverse bremsstrahlung, 133–34
inverse-MAPLE, 207, 234–36
IOD, *see* integrated optical density
ionization, 86, 99, 108, 133–34, 331, 361–62, 481, 488
ions, 91, 99–100, 107, 109, 366
IR-MAPLE, 231–33, 236
irradiated target, 3, 49, 426, 431, 435, 440, 447–48
irradiation, 88, 155, 157, 161, 180, 203, 234–35, 318–21, 324, 326–27, 337–38, 360–61, 366, 370–74, 376, 378, 383–84, 399–401, 403–404, 406, 409, 411, 450–52, 473–75, 477–83, 485–89
islands, 4, 14, 20–25, 32, 34, 38–43, 54, 56–60, 62–67, 76, 101, 111, 144, 146, 148, 222
isotropic, 7, 11, 250, 366, 371

jet mill, 286–87
junctions, 141, 148, 381

Keldysh parameter, 362
kinetic energy, 26–27, 31, 33, 48, 52–54, 67, 86, 95–96, 98–99, 106–10, 131, 134, 335, 505
  average, 33, 54, 431
  high, 106

incident average, 51
  low, 54
  peak, 65
  reduced, 150
  residual, 144
kinetic Monte Carlo simulations, 19, 104
kinetics, 1–3, 8, 19, 35, 41, 55, 91, 101, 430, 529
kinks, 28–30, 32, 36, 38
Knudsen layer, 134
Kreibig's evaluation, 371

LA, *see* laser annealing
lab-on-a-chip, potential, 386
LaMer diagram, 291
laser
  classical, 524
  gold-coated glass substrate, 515
  high-repetition-rate, 379
  irradiating, 488
  low-repetition-rate, 379
  nanosecond, 372
  polarized, 484
  short-pulse, 320
  ultrashort, 339
  ultrashort-pulse, 400
  unique, 323
  waveguide, 369
laser ablation, 47, 49, 52, 155–57, 171, 245, 248, 288–89, 425–29, 438–41, 443, 445–46, 448–51, 453–54, 472
  direct, 426
  near-field, 194, 196
  subtractive femtosecond, 383
laser annealing (LA), 499, 522, 524, 526
laser beam, 198–99, 213–14, 247, 322, 324–25, 360–61, 364–65, 384, 401, 414, 418–20, 472–75, 483, 503–504, 512
laser-decal transfer (LDT), 514

laser energy, 49, 65, 92, 133, 135, 234, 289, 323, 336, 340, 364–65, 372, 403, 419, 436, 452
  absorbed, 247, 341, 430, 505
laser excitation, 87, 401, 405–6, 429, 437, 450
laser fluence, 89, 92, 96, 98–99, 106–10, 118–19, 143–44, 169–70, 173, 217–24, 326, 364, 372, 416–17, 426, 435–37, 449, 453, 455, 476–77, 504, 525
laser-generated plasma, 135, 147, 153
laser-induced breakdown spectroscopy (LIBS), 156, 320–23, 335, 344
laser-induced forward transfer (LIFT), 156, 180, 472, 499–501, 512–17, 519–20, 533–35
laser-induced periodic surface structures (LIPSS), 471–85, 487–89
laser-initiated liquid-assisted colloidal (LILAC), 195–96, 200, 202–4
laser interactions, 217, 223, 344, 429–30, 432, 434, 454–55
laser interference lithography (LIL), 472–74
laser irradiation, 89, 115–19, 133, 159, 194, 196, 219, 224, 231–34, 326, 373, 383–84, 452, 454, 475, 477, 479, 481, 487, 489
laser pulses, 86–88, 90, 94–97, 100–106, 113–14, 116–17, 156–60, 179–81, 232–35, 327–29, 360, 401–2, 406, 416–19, 443–44
laser radiation, 50, 133, 209, 217, 227, 230, 232–33, 245, 247, 478

incident, 86, 200, 209, 223, 230, 233, 236, 417
incident UV, 227
laser sources, 155, 162, 170, 198, 211–12, 216–17, 224, 230, 321, 383, 426, 471
laser spot, 94, 146, 231, 435–38, 440
laser structuring (LS), 113, 499–500, 522, 525–26, 532, 534–35
laser wavelength, 99, 131, 134, 137, 210, 215, 318, 322, 401, 471, 474–76, 479–80, 483, 485, 488–89
layer-by-layer growth, 14, 20–21, 24, 42, 57, 60–61, 64–66, 78
LCO, *see* lithium cobalt oxide
LDT, *see* laser-decal transfer
Lennard–Jones potential, 164
LIBS, *see* laser-induced breakdown spectroscopy
LIBs, *see* Li ion batteries
LIFT, *see* laser-induced forward transfer
Li ion batteries (LIBs), 500, 502, 508, 510, 521–22, 528–30
Li ion microbatteries, 501, 505, 510, 514, 516–18, 520, 534
LIL, *see* laser interference lithography
LILAC, *see* laser-initiated liquid-assisted colloidal
LIPSS, *see* laser-induced periodic surface structures
liquid nitrogen temperature (LNT), 213, 216, 225
lithium cobalt oxide (LCO), 502–3, 505–6, 510, 516, 519–20, 522–24, 529
lithium manganese oxide (LMO), 502–3, 505–6, 522–23, 527–28
lithium vanadium silicon oxide (LVSO), 510–11, 527–28

lithography, 2, 195–96, 370, 472
LMO, *see* lithium manganese oxide
LNT, *see* liquid nitrogen temperature
local thermodynamic equilibrium (LTE), 91
LP (Langmuir probe), 94, 96, 144–46, 148–50
LPLD, 49, 111
LS, *see* laser structuring
LTE, *see* local thermodynamic equilibrium
LVSO, *see* lithium vanadium silicon oxide

Mach–Zehnder interferometer (MZI), 369, 382
MAPLE, *see* matrix-assisted pulsed laser evaporation
mass spectrometry (MS), 99, 147
　secondary ion, 255
　time-of-flight, 336
master oscillator power amplifier (MOPA), 198
material ablation, 317, 336–38, 343, 524
material decomposition, 158, 163, 167, 431, 438
material ejection, 163, 426, 436–37, 443, 449
material irradiation, 321, 400
matrix
　copolymer, 520
　dielectric, 102, 104, 106
　volatile, 235
　water ice, 231
matrix-assisted pulsed laser evaporation (MAPLE), 49, 68, 207, 209–11, 213–19, 221, 223–24, 225–27, 229–31, 234–36, 239, 472
Maxwell's curl equations, 200
Maxwell's electromagnetic theory, 200

MBE, *see* molecular beam epitaxy
MCMB, *see* microcarbon microbead
MD, see molecular dynamics
MD simulation, 163–66, 171, 173, 339, 341, 428–29, 432, 434–35, 437, 439, 443, 449, 452, 454–55
melting, 44–45, 47, 88–89, 98, 115–16, 118–19, 133, 158, 219, 222–24, 342, 363, 366, 370, 400, 405, 428, 432, 434, 436, 450–51, 475, 488
MEMS, *see* microelectromechanical systems
metal ablation, 327, 337
metal films, 114–16, 118, 432
metallic NPs, 236, 265
metal NPs, 86, 102, 104, 108–10, 112, 117
metal-oxide semiconductor (MOS), 141, 354
microbatteries, 500–2, 516–18, 520, 522, 533–35
microcarbon microbead (MCMB), 517, 520
microchannels, 374, 381–85
microelectromechanical systems (MEMS), 500, 536
microelectronics, 195, 473, 500–1, 519, 533–35, 540
microfabrication, 360, 372–73, 376
microfluidic, 360, 370–71, 375–76, 379–83, 397
micromachining, 323, 344, 359, 515
micropower, 499–501, 514, 533–34
microscopy, 161, 195, 254–55, 478
　optical, 193, 325, 332
　probing near-field, 194
　scanning electron, 117, 141, 180, 195, 482
　secondary electron, 255
　transmission electron, 102, 139, 219, 259, 296

microstructures, 1, 42, 44, 247, 337, 371, 376, 379, 400, 506, 524, 530
Mie scattering theory, 200
misfit strain, 9–11, 36, 60
MLG, see multilayer graphene
molecular beam epitaxy (MBE), 47–48, 50–53, 56–58, 61–62, 64–65, 111
molecular dynamics (MD), 19, 55, 163–67, 171, 173, 180, 319–20, 339, 341, 428–32, 434–35, 437, 439, 443, 449, 452, 454–55
MOPA, see master oscillator power amplifier
MOS, see metal-oxide semiconductor
MPA, see multiphoton absorption
MS, see mass spectrometry
multilayer graphene (MLG), 508
multiphoton absorption (MPA), 357, 361, 384, 481, 488
multiwalled carbon nanotube (MWCNT), 211, 221–25
MWCNT, see multiwalled carbon nanotube
MZI, see Mach–Zehnder interferometer

nanocomposite films, 214–15, 293, 304
nanocomposites, 100, 207, 210, 213, 215, 229, 239, 281–86, 295, 299, 301, 304, 520
nanoentities, 207, 209–10, 217, 223, 227, 230, 236
nanogratings, 366–67, 421, 489
nanolithography, 251, 473
nanomaterials, 208–9, 211, 245–47, 265, 288–89, 304, 337, 344
nano-objects, 248, 318–19, 344, 473

nanoparticles (NPs), 86, 101–19, 132, 137, 142–45, 147, 150, 160–61, 179–82, 207–9, 213–19, 234–36, 265, 267, 282–94, 296–98, 317–20, 322–23, 325–33, 335–37, 408–11, 426–29, 445–49
 compound, 131
 contamination-free, 426
 dielectric, 194
 dried, 294
 environmentally friendly, 426
 functionalized, 293–94
 high-dielectric-constant, 284
 hollow, 288
 initiator-immobilized, 292
 inorganic, 284
 metal, 85
 metal-oxide, 288
 metastable, 445
 polymer shell, 281
 self-assembled, 132
 semiconductor, 278
 silica, 278
 small, 449
 structured BT@PLA, 285
 ungrafted, 293
nanopatterning, 194–95, 197, 203, 205, 473
nanophotonics, 195, 379
nanosecond laser ablation (ns-LA), 86, 92, 156–57, 160, 171, 531–32
nanostructures, 48, 104, 115, 117, 132, 138, 140, 144–46, 197, 219, 252, 259, 262, 266–68, 274, 277–78, 288–89, 337, 379, 383–84, 400–1, 407, 414, 421, 484–85, 489
nanotechnology, 114, 246, 318, 344
nanotubes, 132, 221–22, 224
nascent plume, 163, 165, 167, 341
Navier–Stokes equation, 405

near-field laser ablation (NFLA), 194–96, 202
near-field optics, 193–94, 204
NFLA, see near-field laser ablation
nonequilibrium, 3, 35, 50, 157, 339, 427–28, 479, 505
nonlinear absorption, 358, 360, 363, 370, 376
NP, see nanoparticles
NP plume, 159–60, 163, 167, 169–70, 173–79, 328, 330
ns-LA, see nanosecond laser ablation
nucleation, 2, 4–5, 8–9, 13–19, 21–22, 24, 34–40, 42–43, 45, 55–60, 63–67, 106, 110–11, 290–91, 446–47
nucleation density, 34–36, 43, 54, 57–59, 63–64
nucleation theory, 5
  classical, 14, 18–20
  mean-field, 40

ODC, see oxygen-deficient center
OES, see optical emission spectroscopy
optical constant, 196, 206, 242, 255
optical density, 145, 162, 260
optical emission, 158, 164, 440, 466
optical emission spectroscopy (OES), 91, 94–96, 121, 158–59, 163, 185, 191, 328, 331, 335–36
optical microscopy, 193, 325, 332
optical penetration depth, 88, 90, 116, 157, 223, 231, 336
optical properties, 115, 132–33, 135, 137, 143–46, 148, 258, 266, 279, 318, 320, 336, 369, 381, 401
optical waveguides, 260, 370, 381–82

optics, 210, 358, 363, 378
Ostwald ripening, 39–40, 76, 292
overlayer, 10, 12, 22, 427, 448
oxygen-deficient center (ODC), 372–73
oxygen hole centers, nonbridging, 360
oxygen pressure, 181, 230, 267

PCE, see power conversion efficiency
periodicities, 52, 366, 383, 407–8, 416, 419, 421, 478, 481, 485
permittivity, 280, 285, 362
PET, see poly(ethylene terephthalate)
PGMA, see poly(glycidyl methacrylate)
phase
  gas/vapor, 4, 35
  solid, 14, 287
  vapor/gas, 26, 28
phase explosion, 134, 158, 165, 167, 233, 325, 339, 437–40, 444, 452–53, 455
photoionization (PI), 90, 99–100, 361–62
photonics, 85, 142, 358, 473
photons, 88, 162, 199, 234, 361, 365, 372, 377–78
photoresists, 357, 359, 376–77, 384–85
  negative, 379
  negative-tone, 376
  positive-tone, 376
photovoltaics, 85, 208, 210, 215
physical vapor deposition (PVD), 46–48, 50, 52, 68, 132, 287, 505, 512
PI, see photoionization
plasma, 53, 86, 90–96, 98–100, 119, 134–36, 139, 146, 150, 197, 246, 320, 322, 329, 334, 336–37

anisotropic, 134
dense, 134
laser-induced, 49, 98, 337
laser-produced, 142, 155, 182, 326, 458
subcritical, 320
ultrashort, 267
plasma expansion, 92–93, 131, 135, 324, 328
plasma plume, 49–50, 52–53, 65, 90, 93, 131–32, 134–37, 139, 150, 170, 342, 504–5
plasmonics, 142–43, 153–54
plasmons, 133, 268, 413, 421
PLD, see pulsed laser deposition, 4–5, 47–62, 64–68, 100–111, 114–15, 132–34, 139–40, 143–44, 209–10, 247, 256–57, 265–68, 499–501, 503–512, 533–34
plume, 93, 135–36, 139–40, 143–45, 157–60, 163, 165, 167–79, 320, 324, 326–31, 334–35, 341–45, 427, 440, 442–46
plume emission, 136, 158–59, 168, 171–72, 332, 336
plume expansion, 68, 132, 135, 144, 168, 171, 175, 179, 319, 325, 427, 433, 444–45, 447
PMMA, see poly(methyl methacrylate)
polarization, 200, 247, 280, 321, 324, 341, 400, 406–8, 421, 471, 475, 481, 483, 485
polycrystalline, 43, 61, 101, 299
poly(ethylene terephthalate) (PET), 280, 475–77, 479–80, 483–84
poly(glycidyl methacrylate) (PGMA), 292–94, 297–304
polymer chains, 282–84, 298, 475
polymer films, 298, 471, 474, 481, 485–86, 488–89

polymerization, 282, 284–85, 292, 298, 377–78
polymer nanocomposites, 284–85, 295
polymers, 48, 209, 223, 252, 280–81, 283–85, 293–94, 298, 355, 358–59, 377, 379, 401, 421, 472–79, 481–82, 484, 488
polymer substrates, 472–73, 485–86, 489
poly(methyl methacrylate) (PMMA), 211, 282–85, 301, 481
polystyrene (PS), 280, 282–83, 285, 475, 485
poly(vinylidene-fluorideco-trifluoroethylene) (PVDF-TrFE), 482, 486–87, 497
pores, 380, 439, 530
porosity, 44, 48, 141, 256, 264, 284, 516, 529
porous structures, 267–68, 499, 513, 516, 518–19
potential energies, 7, 33–34, 55, 254, 436, 439, 450–51
power conversion efficiency (PCE), 215
power law, 58, 104, 116, 302–3
universal, 302, 304
Poynting vector, 200–201
pressure, 5, 65–66, 134–35, 137, 149–50, 175–78, 180, 182, 196–97, 253–54, 267, 295–96, 326, 340, 351, 437–39, 446, 452, 459
ambient, 232
atmospheric, 177, 222–23, 268
base, 402
high, 92, 132, 135, 138, 150, 178–79, 182, 288, 443, 504
inner plume, 136
intermediate, 442
larger, 177–78
laser-induced, 427

low, 442
moderate, 182
negative, 438
partial, 8–9
positive, 438, 453
residual, 324, 402
pressure waves, 320, 359–60, 388, 432, 444
propagation, 131, 135, 160, 168, 171, 175, 178, 180, 321–22, 360, 377, 431, 439, 442, 444
backward, 136
braked, 175
shock wave, 339
PS, see polystyrene
pulsed laser ablation, 131, 288, 452
pulsed laser deposition, see PLD
pulse energy, 65, 90, 321, 324, 336, 358, 363, 367–68
pulses, 68, 86, 89–90, 117, 119, 131–32, 134, 143, 191, 195, 198, 210, 253, 261, 263, 317, 319–20, 328, 330–31, 335, 341, 344, 358, 403, 412, 426, 475, 477, 479, 481, 514
pump-probe experiments, 323, 336, 440
PVD, see physical vapor deposition
PVDF-TrFE, see poly(vinylidene-fluorideco-tri-fluoroethylene)

QD, see quantum dot
quantum dot (QD), 25, 210–11, 214, 220, 222, 258, 279

radiation, 134, 136, 146, 172, 210, 474
radical polymerization
free, 377
radical initiating, 296
radio frequency identification (RFID), 500

RAFT, see reversible addition-fragmentation chain transfer
Raman scattering, 267, 484
Raman spectra, 489, 523–204
Raman spectroscopy, 132–33, 138, 265, 523
Raman spectrum, 150, 523
rarefaction wave, 325, 334, 341, 343
reduction, 18, 39, 99, 104–5, 178, 180, 227–28, 233, 249, 253, 263, 288, 323, 362, 522, 527
reflection high-energy electron diffraction (RHEED), 48
refractive index, 88, 196–97, 199–201, 204, 358, 360, 362–63, 366–67, 369–70, 378, 474, 479
relaxation, 23, 54, 66, 158, 161, 165, 182, 247, 330, 399, 404, 427, 437, 439, 453
resistance, 94, 104, 214, 247, 253–54, 264, 450, 518, 533
reversible addition-fragmentation chain transfer (RAFT), 282, 285, 303
RFID, see radio frequency identification
RHEED, see reflection high-energy electron diffraction
ring-opening polymerization (ROP), 285
ripple formation, 320, 401, 421, 475
RIR-MAPLE, 207, 214–15, 230–31, 236, 240
RMS, see root mean square, 264–65
room temperature (RT), 33, 61, 111, 113–14, 136, 144, 231, 253, 265–66, 292–93, 296, 299, 301, 303, 508–9
root mean square (RMS), 264–65

Index

ROP, *see* ring-opening polymerization
RT, *see* room temperature
rutile phase, 218, 267

SAXS, *see* small-angle X-ray scattering
scanning electron microscopy (SEM), 117, 141, 180, 195, 482
scanning near-field optical microscopy (SNOM), 193–94
scanning polarization force microscopy (SPFM), 254
scanning tunneling optical microscope (STOM), 260
scattering, 87, 135, 139, 144, 194, 199–200, 268, 384, 442, 481
   electron–phonon, 412
self-sputtering, 95, 108, 122, 124
SEM, *see* scanning electron microcopy
semiconductors, 22, 140–41, 156, 160, 162, 181, 247, 258, 399, 401–2, 406, 416, 421, 474, 478
sensitivity, 259, 265, 322, 385
sensors, 143, 149, 213, 266
   analytical, 318, 345
   biochemical, 210
   wireless, 501
   wireless network, 519
SERS, *see* surface-enhanced Raman spectroscopy
shock waves, 143, 342, 428–29, 433, 442–44
SI-ATRP, surface-initiated atom transfer radical polymerization
simulations, 58, 143, 163–66, 171, 173, 200, 319, 339, 341, 403, 425–26, 428–37, 439–41, 443–45, 448–55
   numerical, 222–24, 231–33, 318, 479
   theoretical, 407–8, 410

single-walled carbon nanotube (SWCNT), 211
size exclusion chromatography, 295
SK, *see* Stranski–Krastanow
SLM, *see* spatial light modulator
small-angle X-ray scattering (SAXS), 447–48
SNOM, *see* scanning near-field optical microscopy 193–94
solid-state electrolytes, 499, 505, 510–12, 520, 527, 533–35
solvent matrix, 209, 211–12, 220–22, 225, 227–32
solvothermal method, 288–92
spallation, 339, 341, 427–28, 437, 440, 445, 448, 450, 452–54
spatial confinement, 135, 139, 246, 426–27, 445, 448, 454
spatial distribution, 49, 85, 95, 116, 138, 148, 321, 364–65, 377, 408, 444
spatial light modulator (SLM), 402, 540
spatial resolution, 115, 162, 254, 364, 377–78, 386
spectroscopy, 162, 255, 265, 295, 299
SPFM, *see* scanning polarization force microscopy
spin coating, 197–98, 214
SPP, *see* surface plasmon polariton
SPR, *see* surface plasmon resonance
sputtering, 4, 26, 47–48, 52–53, 68, 86, 100, 105–6, 108–10, 134, 427, 505, 512, 527
step-edge barrier, 28, 33, 37, 42, 61, 64
step-edges, 28–30, 33–34, 36–38, 60, 64
stoichiometry, 2, 47, 49, 505–6
STOM, *see* scanning tunneling optical microscope

strain, 3–5, 9–14, 17, 21–25, 43–44, 50, 112, 253, 340
strain energy, 3, 9–10, 12, 14, 22–25, 41
Stranski–Krastanow (SK), 20–25, 42
stress, 9–10, 179, 251, 259, 334, 343, 437, 439–40, 450, 453, 524–25
stress wave, 431, 437, 439, 462
stretching vibrations, 296, 523
sublimation, 46, 222–24, 232–33
substrate, 23, 29, 36, 54, 61, 102, 113, 117, 150, 179–80, 194, 208–9, 236, 260, 286, 486, 489, 508, 512–13
substrate surface, 21–23, 25–27, 29, 31, 45–46, 54–57, 118–19, 195, 198, 203, 209, 221–24, 227, 230, 233–34
substrate temperature, 26, 31, 37, 43, 45, 47, 51, 53–54, 57, 86, 101, 113–14, 261, 266, 402
subsurface voids, 426, 436–37, 452–54
subtractive, 357–59, 381, 383
superlattices, 23, 62, 77, 80
supersaturation, 4, 9, 14–15, 18–19, 22, 26–27, 35–36, 51–52, 54, 56–57, 59, 66–67, 505
surface diffusion, 3, 22, 27–28, 32–33, 35–38, 42, 44–45, 55–56, 67, 150
surface energy, 5–8, 12, 15, 21–23, 27, 39, 67, 246, 248, 250, 252, 255, 284, 434
surface-enhanced Raman spectroscopy (SERS), 132–33, 143, 146–51, 265, 267–68, 484–86, 489
surface-initiated atom transfer radical polymerization (SI-ATRP), 292–93, 304

surface mobility, 26, 43, 45–46, 57, 60, 106
surface nanostructures, 132–33, 142–43, 150
surface nanostructuring, 137, 146, 453, 455
surface plasmon excitation, 403, 406
surface plasmon polariton (SPP), 143, 349, 478
surface plasmon resonance (SPR), 145–47, 267, 318
surface tension, 250, 271, 340, 379, 403
SWCNT, *see* single-walled carbon nanotube

target material, 47, 49–50, 52, 90–91, 156, 163–64, 210, 247, 254, 339, 400, 430, 432
target surface, 86, 90, 92–95, 134–35, 157–60, 162–63, 168–70, 172, 175–76, 178–79, 323, 325, 328–29, 338, 341
TAS, *see* tensile adhesion strength
TEM, *see* transmission electron microcopy
temporal profiles, 386, 410, 412, 419
tensile adhesion strength (TAS), 256
terrace step kink (TSK), 28
TGA, *see* thermogravimetric analysis
theoretical models, 250, 254, 319, 403, 406, 409, 411, 413, 420, 442
theory, 24, 235, 415
  classical, 1
  classical Mie, 200
  first-principle, 478
  kinetic, 19
  thin-film instability, 440
thermal annealing, 114–15, 225

thermal conductivity, 88, 219, 320
thermal deposition, 51, 56–60, 66
thermal diffusion, 232, 363, 365
thermal evaporation, 26, 46, 48, 51, 60, 68, 215
thermalization, 55, 87, 135, 247, 403–4, 430
thermodynamic equilibrium, 2, 4–5, 14, 91
thermodynamic nonequilibrium, 5
thermodynamics, 1–3, 5–6, 8, 10, 13, 15, 20–21, 24, 28, 35, 49–50, 59, 69
thermogravimetric analysis (TGA), 296–98
thick-film electrodes, 516, 519, 522, 529–31, 534, 539
thick-film microbatteries, 501, 518–19
thin-film cathodes, 505–7
thin-film electrodes, 501, 522, 527
thin-film microbatteries, 499–501, 505, 510, 512, 533–34
thin films, 111, 114–19, 179–80, 213–15, 217, 227–29, 231–32, 335, 338, 476, 479, 503–506, 508, 510–12, 522–29
  composite, 227–28
  continuous, 208
  doped, 338
  laser-structured, 525–26
  sputtered, 523
threshold fluence, 99, 116, 218, 452–53
time-gated imaging, 168, 172
time-of-flight secondary ion mass spectrometry (ToF-SIMS), 255
timescales, 31, 49, 52, 54–56, 62, 67–68, 87, 133–34, 157–61, 169, 330, 445, 453, 455
ToF-SIMS, *see* time-of-flight secondary ion mass spectrometry
toxicity, 318, 345

TPA, *see* two-photon absorption
TPP, *see* two-photon polymerization
transmission electron microscopy (TEM), 102, 139, 219, 259, 296
transparent materials, 357–59, 361, 363, 366, 374, 386
trapping, 33, 205
traps, 33–34, 38, 195
TSK, *see* terrace step kink
TTM, see two-temperature model
TTM-MD method, 341, 426, 429, 431, 435
two-photon absorption (TPA), 361, 365, 377
two-photon polymerization (TPP), 358, 373, 376–81, 373–86
two-temperature model (TTM), 157, 163–64, 319, 330, 339, 341, 400, 426–28, 475

UHV, *see* ultrahigh vacuum
ULP, *see* ultrafast laser pulse
  higher-repetition-rate, 363
  identical, 362
  polarized, 366
  single-shot, 362
ultracapacitors, 495–98, 510, 512, 530
ultrafast laser pulse (ULP), 357–77, 379, 381, 383, 385–86
ultrafast lasers, 357, 359, 361, 375, 385–86
ultrahigh vacuum (UHV), 402, 500
ultrashort laser pulses, 182, 288
ultraviolet (UV), 108, 137, 174, 204, 207, 210, 216, 225–27, 229, 233, 267, 338, 371, 470, 478
UV, *see* ultraviolet
UV laser, 137–38, 140, 230, 232–33
UV light, 213, 234, 372
UV-MAPLE, 213, 216, 219–23, 225, 227–31, 233, 236

vacuum, 6–8, 134–35, 137, 159–60, 162, 171–72, 175, 293–95, 318–21, 334–35, 421–25, 435, 439–42, 446, 448–51
  cost-consuming, 526
  high, 157–58, 175, 181
  ultrahigh, 402, 504
vacuum chamber, 49, 197, 216, 225, 324, 402, 503
valence band, 361
vaporization, 88, 203, 222, 334, 432
vapor phase, 3, 8, 162, 218, 223, 449, 453
vapor pressure, 8, 47, 204, 225, 227, 340, 504–505
variation
  angular, 173–74
  temporal, 163, 175, 178
velocity, 26, 90, 92, 95–97, 117, 139, 163, 169–71, 176, 197, 267, 328, 428, 431, 451
vicinal surfaces, 29, 37
viscosity, 204, 434, 516
voids, 41, 44, 342, 358, 366–67, 439, 450–53
volatility, 98, 236, 520
Volmer–Weber (VW), 20–21, 23, 101
voltage, 3, 502, 516, 518, 521
  breakdown, 280, 296
volume expansion, 503, 508–9
voxels, 377–79, 514
VW, see Volmer–Weber, 20–21, 23

water, deionized, 216, 234, 294
water-assisted ablation, 374, 392
water environment, 434, 448–51, 453–54
water ice, 231–32, 242
water matrix, 230–35
water sublimation, 232–33

waveguides, 260, 358, 369–70, 382–83
  curved, 382
  sorting, 383
  symmetric, 363
wavelength, 149–50, 158, 160, 170, 172, 223, 230–31, 321, 364, 372, 382, 401–2, 478–79, 482–83, 487–88
  appropriate, 475
  central, 164
  incident, 200
  long, 137, 481
  nonresonant, 366
  ripple, 401, 409, 411, 414–15
wavenumber, 149
waves, 205, 260, 341–42, 417, 437, 473
  evanescent, 260
  excited surface, 478
  scattered, 401, 421, 477
  second harmonic, 417
wave vector, 408
Wenzel approaches, 252
Wenzel model, 250
Wenzel roughness factor, 250
Wenzel state, 250
wet etching, 358, 376
wet grinding, 286
wettability, 248, 250–51, 254, 268
wetting, 21–23, 27, 214, 246, 248, 250–53, 271, 529–31
  induced, 14, 20–21, 251, 374, 514
  insufficient, 530
  partial, 113
wetting layer, 23–24, 429
Wulff construction, 7
Wulff–Kaichew theorem, 113

XAFS, see X-ray absorption fine structure
XPS, see X-ray photoelectron spectroscopy

X-ray absorption fine structure (XAFS), 162–64
X-ray absorption spectrum, 163
X-ray diffraction (XRD), 216
X-ray photoelectron energy, 162
X-ray photoelectron spectroscopy (XPS), 227, 229, 232, 255, 523
X-ray probe, 162
X-ray radiography, 446
X-ray scattering, 447, 477, 488
XRD, *see* X-ray diffraction

Young's equation, 248

ZBL, *see* Ziegler–Biersack–Littmark
Zeldovich expression, 18
Ziegler–Biersack–Littmark (ZBL), 433
zones
 boundary, 44
 heated, 88
 high-pressure, 334
 laser beam interaction, 531
 low-mobility, 44
 unstable, 166–67

PGSTL 01/02/2018